Tributes
Volume 37

Argumentation-based Proofs of Endearment
Essays in Honor of Guillermo R. Simari
on the Occasion of his 70th Birthday

Volume 27
Why is this a Proof? Festschrift for Luiz Carlos Pereira
Edward Hermann Haeusler, Wagner de Campos Sanz and Bruno Lopes, eds.

Volume 28
Conceptual Clarifications. Tributes to Patrick Suppes (1922-2014)
Jean-Yves Béziau, Décio Krause and Jonas R. Becker Arenhart, eds.

Volume 29
Computational Models of Rationality. Essays Dedicated to Gabriele Kern-Isberner on the Occasion of her 60th Birthday
Christoph Beierle, Gerhard Brewka and Matthias Thimm, eds.

Volume 30
Liber Amicorum Alberti. A Tribute to Albert Visser
Jan van Eijck, Rosalie Iemhoff and Joost J. Joosten, eds.

Volume 31
"Shut up," he explained. Essays in Honour of Peter K. Schotch
Gillman Payette, ed.

Volume 32
From Semantics to Dialectometry. Festschrift in Honour of John Nerbonne.
Martijn Wieling, Martin Kroon, Gertjan van Noord, and Gosse Bouma eds.

Volume 33
Logic and Computation. Essays in Honour of Amílcar Sernadas
Carlos Caleiro, Fransciso Dionísio, Paula Gouveia, Paulo Mateus and João Rasga, eds.

Volume 34
Models: Concepts, Theory, Logic, Reasoning, and Semantics. Essays Dedicated to Klaus-Dieter Schewe on the Occasion of his 60th Birthday
Atif Mashkoor, Qing Wang and Bernhrd Thalheim, eds.

Volume 35
Language, Evolution and Mind. Essays in Honour of Anne Reboul
Pierre Saint-Germier, ed.

Volume 36
Logic, Philosophy of Mathematics and their History.
Essays in Honor of W. W. Tait
Erich H. Reck, ed.

Volume 37
Argumentation-based Proofs of Endearment. Essays in Honor of Guillermo R. Sim on the Occasion of his 70th Birthday
Carlos I. Chesñevar, Marcelo A. Falappa, Eduardo Fermé, Alejandro J. García, Ar G. Maguitman, Diego C. Martínez, Maria Vanina Martinez, Ricardo O. Rodríguez, Gerardo I. Simari, eds.

Tributes Series Editor
Dov Gabbay   dov.gabbay@kcl.ac.uk

Argumentation-based Proofs of Endearment

Essays in Honor of Guillermo R. Simari
on the Occasion of his 70th Birthday

edited by

Carlos I. Chesñevar

Marcelo A. Falappa

Eduardo Fermé

Alejandro J. García

Ana G. Maguitman

Diego C. Martínez

Maria Vanina Martinez

Ricardo O. Rodríguez

Gerardo I. Simari

© Individual authors and College Publications 2018. All rights reserved.

ISBN 978-1-84890-292-3

College Publications
Scientific Director: Dov Gabbay
Managing Director: Jane Spurr

http://www.collegepublications.co.uk

Cover design by Laraine Welch
Illustration by Diego C. Martinez

Printed by Lightning Source, Milton Keynes, UK

All rights reserved. No part of this publication may be reproduced, stored in a retrieval system or transmitted in any form, or by any means, electronic, mechanical, photocopying, recording or otherwise without prior permission, in writing, from the publisher.

Preface

About Guillermo Simari

This volume is dedicated to Guillermo Simari as a *festschrift* in honor of his extensive academic careeer and contributions in Argumentation. In fact, Guillermo has been one of the "founding fathers" of this fascinating subdiscipline of Artificial Intelligence, which emerged at the end of the 80s in the last century based on the work of many researchers and philosophers (such as Rescher, Kyburg, Pollock, Doyle, Loui, and many others). Although originally proposed within the realms of logic, philosophy, and law, argumentation came to capture wide interest in computing to understand and meet the challenges of a number of applications characterized by the lack of certain, consistent, and complete information, and when numerical (e.g. statistical) information is not available or is only partially available.

Guillermo finished his PhD at Washington University (St. Louis, USA) in 1989, and returned to Universidad Nacional del Sur in Bahía Blanca, Argentina, to work as a full-time professor and researcher. He was the only person in the country with a PhD in Computer Science in those times, and from the very beginning he was willing to disseminate his knowledge on the subject among a group of motivated undergraduate students (including myself!). Looking 30 years back in time, the mid 90s was a very hectic period to be at Universidad Nacional del Sur if you wanted to learn something about AI. In those years, the Department of Computer Science was created, and Guillermo was one of those who was leading the process, as well as the creation of postgraduate programs in Computer Science at the university. At the same time, many of us learned about Guillermo's vegetarianism, and it became clear that his mentoring exceeded AI, since many changed their food tendencies in this direction and, little by little, we turned out to have the largest "vegetarian AI community" in Argentina! Guillermo created the GIIA (Grupo de Investigación en Inteligencia Artificial) in the mid 90s, where my now colleagues and friends Marcelo Falappa, Alejandro García, Pablo Fillotrani, and I could meet regularly and present papers, exchange ideas, and have fruitful discussions. Fernando Tohmé, Claudio Delrieux, and many other people (now colleagues and friends) were also part of that team in those years. It might sound strange, but... yes, there was no Internet at that time!

Together with Guillermo we went through the experience of learning how to publish our research results in the academic arena, starting with local conferences first, and moving ahead towards more sophisticated international challenges (international conferences such as AAAI, ECAI, IJCAI, KR and many others) and different and prestigious academic journals. In 2011 Guillermo and I met in Barcelona, Catalunya (Spain) for the 22nd International Joint Conference on Artificial Intelligence (IJCAI 2011). Ricardo Rodríguez, a collegue and friend from Universidad de Buenos Aires, was also there; together with Guillermo they were striving to get IJCAI to be held in Buenos Aires in 2015.

There were many strong competitors (Brazil, Australia, and the United States), and Guillermo was in charge of presenting Argentina as a candidate. It was July 20th when the decision was made, in something that I remember as an Oscar Award ceremony—when we heard that the winner was Argentina, we could not believe it. It was one of those moments in life in which a mix of feelings rushes through your body; we jumped and shouted in joy, as if we had won the final match of the FIFA World Cup (in the context of AI, of course). Thus, 2015 turned out to be particularly significant for all of us, as Guillermo and Ricardo managed to coordinate and organize, along with the help of many other people, the first IJCAI that took place in South America, and particularly in Buenos Aires, Argentina. The event was a great success, and it made us very proud to be there sharing the magic of being at the "top of the mountain" in Artificial Intelligence. It was indeed a well-deserved hallmark in Guillermo's career.

To a large extent, Guillermo incarnates the American metaphor of the self-made man in the academic world. A man who came back to his home country to start a new scientific endeavor out of nothing but himself, planting seeds in form of ideas, dreams, projects, and visions in a bunch of motivated graduate students in a University that provided the fertile ground needed for such an enterprise. After 30 years of continuous hard work and leadership, Guillermo's harvest resulted in one of the largest research groups in Artificial Intelligence in Latinamerica, with well-deserved international prestige, involving more than 50 people (including current PhD students, researchers, and postdocs) with research connections worldwide. Most importantly, his work was also oriented to strengthening Computer Science across the country, since he was the advisor of many young students from other provinces in Argentina who came to Bahía Blanca to obtain their PhD's in Computer Science. As a consequence, nowadays there are new seeds of Guillermo's work flourishing in the universities of San Luis, Santiago del Estero, Entre Ríos, and Neuquén, among other places, leaving a long-lasting legacy for the AI community in Argentina.

For all those working in argumentation (or thinking about getting into it), this collection of papers provides an enjoyable and interesting insight into this vibrant area of knowledge to which Guillermo has contributed so much in his prolific career. Congratulations Guillermo on your 70th anniversary!

Carlos Iván Chesñevar
Institute of Computer Science and Engineering
(ICIC CONICET UNS)
Department of Computer Science and Engineering
(DCIC UNS)
Bahía Blanca, Argentina

About this book

This book comprises several chapters that consider the broad area of argumentation from different perspectives. It is well known that strength is indeed a central issue in argumentation frameworks. The first chapter of the book by Baroni and Giacomin, entitled *"On the Notion of Argument Strength: A Discussion Paper"*, analyzes various facets of argument strength, suggesting that it can be better understood as a family of related notions rather than a single concept, and discussing the possibility to identify some common traits among them. Aggregation is particularly relevant for information fusion and plays also a significant role when characterizing attacks. In *"From Individual to Collective Argumentation: Aggregating Acyclic Argumentation Frameworks"*, Tohmé and Bodanza prove the existence of aggregation processes that have as input a finite number of acyclic attack relations over the same set of arguments, providing as an output an acyclic relation over them, satisfying all the conditions of the well-known Arrows Theorem.

Attacks are also a major issue in *"Structure-based Semantics of Argumentation Frameworks with Higher-order Attacks and Supports"*, the chapter by Cayrol, Fandino, Farinas del Cerro and Lagasquie-Schiex who propose a generalization of Dungs abstract argumentation framework that allows to represent higher-order attacks and supports (ie., attacks or supports whose targets are other attacks or supports). A semantics is proposed that accounts for the acceptability of the resulting arguments and the validity of their interactions, with some additional properties connecting the proposal with Argumentation Frameworks With Recursive Attack and Suport (ASAF).

Defeasible Logic Programming (DeLP), developed by García and Simari, is well-known as a multipurpose logic programming framework for argumentative reasoning. In *"Extending Defeasible Logic Programming with Informant-based Argumentation"*, Cohen, Gottifredi, Tamargo, and García extend DeLP to introduce an informant-based argumentative approach for deliberative agents that can receive domain knowledge from other agents in a multi-agent setting. Agents are thus able to argue for and against the consideration of informant agents, and resolving conflicts based on the credibility attributed to each information source, in order to determine the warranted beliefs the agents have.

Most argumentation frameworks rely on dialectical proof procedures for determining the ultimate acceptance of warranted beliefs. In *"Infinite Arguments and Semantics of Dialectical Proof Procedures"*, Thang and Dung study the semantics of dialectical proof procedures. As these proof procedures are in general sound but not complete with respect to admissibility semantics, a natural question here is how to provide a more precise semantical characterization of what they compute. Based on a new notion of infinite arguments, they introduce a stronger notion of admissibility, referred to as strong admissibility, and show that dialectical proof procedures are in general sound and complete with respect to strong admissibility.

Possibilistic Defeasible Logic Programming (P-DeLP) is a formalism that combines features from argumentation theory, logic programming, and a unified treatment of possibilistic uncertainty and fuzziness. In *"A P-DeLP Instantiation of a Dynamic Argumentation Framework for Decision Making"*, Ferreti and Errecalde show how to use this formalism to instantiate an existing Abstract Dynamic Argumentation Framework (DAF) for decision making, which has been shown to have a choice behavior consistent with Classical Decision Theory. On the one hand, by instantiating this decision framework the authors prove its adequacy as an abstract framework. On the other hand, a new concrete decision framework is defined for decision making based on P-DeLP, whose choice behavior is also consistent with Classical Decision Theory. The authors also model an application example from the domain of cognitive robotics as a case study.

DeLP is also the source of inspiration for Cecchi and Fillotrani in the chapter *"Complexity of DeLP: Current Status and Moving Forward"* The $\mathcal{GS}$ semantics is a declarative tri-valued game-based semantics for DeLP that is sound and complete for DeLP's proof theory. In this paper the authors review the latest complexity results of important decision problems in DeLP under $\mathcal{GS}$, and also provide extensions for some of them. Furthermore, they introduce DeLP as a query language and study data and combined complexity of query answering in this context for important problems.

Dov and Michael Gabbay offer a novel insight based on networks in their chapter *"Argumentation as Informal Input"*. Given a network (S, R), with $R \subseteq S^2$, the nodes of S are viewed as containing information and the relationship xRy as x transmitting information to y. The authors argue that such networks provide a more general account of attack and defense, and general argument exchange on issues between participants in social media (such as Facebook or Twitter), as well as being able to simulate the traditional Dung approach to argumentation. A general semantics for such networks is also defined.

Abstract argumentation frameworks (AFs) are one of the central formalisms in argumentation theory. In *"On the Functional Completeness of Argumentation Semantics"*, Giacomin, Linsbichler, and Woltran provide a systematic analysis of semantics for AFs by connecting two recent lines of research: the work on input/output frameworks and the study of the expressiveness of semantics. To do so they consider the following question: given a function describing an input/output behavior by mapping extensions (resp. labelings) to sets of extensions (resp., labelings), is there an AF with designated input and output arguments realizing this function under a given semantics? For the major semantics the authors give exact characterizations of the functions that are realizable in this manner.

Belief revision and argumentation are also interconnected, as shown by Hansson in *"Argumentation, Reasoning and Belief Revision"*. In this article, the author analyzes how to include argumentation and reasoning in models of belief change. This implies that these models have to be adjusted so as to avoid the common assumption that the epistemic agent has already drawn all the valid conclusions from the available beliefs. A simple model is proposed in which this adjustment has been performed to make room for deductive inference, reasoning, and argumentation. Indications are given of how the model can be extended to cover non-deductive and non-monotonic inferences as well.

The connection between non-monotonic reasoning and argumentation in its deductive form is analzed by Hunter in *"Non-monotonic Reasoning in Deductive Argumentation"*. Argumentation is a non-monotonic process and consequently new information can cause a change in the conclusions drawn. However, the base logic does not need to be non-monotonic; indeed, most proposals for structured argumentation use a monotonic base logic (e.g., some form of modus ponens with a rule-based language, or classical logic). In his article, Hunter analyzes several relevant issues in capturing defeasible reasoning in argumentation (including choice of base logic and modeling of defeasible knowledge), also presenting insights and tools to be harnessed for research in non-monotonic logics.

Liao and Van der Torre explore how the defense relation among abstract arguments can be used to encode the reasons for accepting arguments. In their article *"Defense Semantics of Argumentation: Encoding Reasons for Accepting Arguments"*, they introduce a novel notion of defense and defense graphs, as well as a defense semantics along with a new notion of defense equivalence for argument graphs. Defense equivalence is contrasted with standard equivalence and strong equivalence; based on the defense semantics, two kinds of reasons for accepting arguments are identified (direct reasons and root reasons, as well as a notion of root equivalence of argument graphs).

In *"Two Short Notes on Argument"*, Loui first fulfills a promise made in Simari and Loui's well-known 1992 paper *"A mathematical treatment of defeasible reasoning and its implementation"*—the central, revised definition of specificity from the unpublished *"Corrigendum to Poole's Rules and a Lemma of Simari-Loui"* is now published in this book. Second, he makes a new start on formalizing the logic of dialectical dialogue—again, this is work from quite some years ago that started when Loui visited UNS in the early 1990s.

Contrasting DeLP with other frameworks has also been the subject of study in the research community. In *"On the Relationship between DeLP and ASPIC+"*, Cohen and Parsons analyze the relationship between DeLP and ASPIC+. The fact that these systems are different is well known, but what is less known is exactly how these systems differ, and, perhaps more interestingly, the ways in which they are the similar. The paper sets the foundations for discussing these issues for future explorations.

In *"Characterizing Abduction with Implicit Background Theory"*, León and Pino Pérez study explanatory relations called abductive relations. They characterize two families in terms of their axiomatic behavior: the weakly reflexive explanatory relations and the ordered explanatory relations. Both families present tight relationships with the framework of Credibility limited revision. These relationships allow to establish semantical representations for each family. A very interesting consequence of their representation results is that the proposed axiomatizations allow to overcome the background theory present in most axiomatizations of abduction.

Prakken adds an interesting perspective to handling support relationships in his paper *"Modelling Support Relations Between Arguments in Debates"*. His paper develops a formal approach within the ASPIC+ framework and compares it to approaches using bipolar abstract argumentation frameworks, arguing that it is crucial to look at the structure of arguments in order to arrive at a proper model of debate evaluation.

In *"Towards Conditional Logic Semantics for Abstract Dialectical Frameworks"*, Kern-Isberner and Thimm introduce an approach for integrating conditional logic semantics with abstract dialectical frameworks. More precisely, an abstract dialectical framework is interpreted as a conditional logic knowledge base; by applying the Z-inference relation a new semantics for abstract dialectical frameworks can be obtained. Sample translations are discussed, as well as theoretical results pertaining to a characterization of different notions of consistency.

Last but not least, Mancarella and Toni present a proposal for modelling argument graphs in DeLP, based on the notion of argument graphs used in Assumption-based Argumentation (ABA). In their article *"Argument Graphs for Defeasible Logic Programming"* the authors argue that both Defeasible Logic Programming and standard Assumption-based Argumentation obtain arguments from rules (albeit of different kinds), and define these arguments in terms of deductions (albeit of different kinds). Relying on a variant of ABA that uses a notion of argument graphs (representing "non-bloated" sets of "rule-minimal" arguments only), the authors consider a similar variant of DeLP in which arguments are graphs, and discuss the potential benefits of using this variant.

November 2018, Bahía Blanca,
Buenos Aires, Funchal.

Carlos I. Chesñevar
Marcelo A. Falappa
Eduardo Fermé
Alejandro J. García
Ana G. Maguitman
Diego C. Martínez
Maria Vanina Martínez
Ricardo O. Rodríguez
Gerardo I. Simari

Table of Contents

On the Notion of Argument Strength: A Discussion Paper............. 1
    Pietro Baroni and Massimiliano Giacomin

From Individual to Collective Argumentation: Aggregating Acyclic
Argumentation Frameworks 19
    Fernando A. Tohmé and Gustavo A. Bodanza

Structure-based Semantics of Argumentation Frameworks with
Higher-order Attacks and Supports................................ 43
    Claudette Cayrol, Jorge Fandinno, Luis Farinas del Cerro and M-Christine Lagasquie-Schiex

Extending Defeasible Logic Programming with Informant-based
Argumentation 73
    Andrea Cohen, Sebastian Gottifredi, Luciano H. Tamargo and Alejandro J. García

Infinite Arguments and Semantics of Dialectical Proof Procedures...... 91
    Phan Minh Thang and Phan Minh Dung

A P-DeLP Instantiation of a Dynamic Argumentation Framework for
Decision Making 103
    Edgardo Ferretti and Marcelo Errecalde

Complexity of DeLP: Current Status and Moving Forward 123
    Laura Cecchi and Pablo Fillottrani

Argumentation as Information Input 145
    Dov Gabbay and Michael Gabbay

On the Functional Completeness of Argumentation Semantics 197
    Massimiliano Giacomin, Thomas Linsbichler and Stefan Woltran

Argumentation, Reasoning, and Belief Revision 227
    Sven Ove Hansson

Non-monotonic Reasoning in Deductive Argumentation 239
    Anthony Hunter

Defense Semantics of Argumentation: Encoding Reasons for Accepting
Arguments 263
    Beishui Liao and Leendert van der Torre

Two Short Notes on Argument: (1) Corrected Specificity (2)
Dialectical Refinement 281
    Ronald P. Loui

On the Relationship Between DeLP and ASPIC$^+$ 295
  Simon Parsons and Andrea Cohen

Characterizing Abduction with Implicit Background Theory 327
  María Victoria León and Ramón Pino Pérez

Modelling Support Relations between Arguments in Debates 351
  Henry Prakken

Towards Conditional Logic Semantics for Abstract Dialectical
Frameworks 369
  Gabriele Kern-Isberner and Matthias Thimm

Argument Graphs for Defeasible Logic Programming 383
  Paolo Mancarella and Francesca Toni

On the Notion of Argument Strength: A Discussion Paper

Pietro Baroni and Massimiliano Giacomin

Dipartimento di Ingegneria dell'Informazione
Università degli Studi di Brescia
Via Branze 38, Brescia, Italy
{pietro.baroni|massimiliano.giacomin}@unibs.it

Abstract. The notion of argument strength is pervasive but somehow elusive at the same time in formal argumentation literature. In this paper we analyze various facets of argument strength, suggesting that it can be better understood as a family of related notions rather than a single concept, and discussing the possibility to identify some common traits among them.

Keywords: Formal Argumentation · Argument Strength

1 Introduction

Arguments are not all equally strong and this quite intuitive fact has been recognized early in formal argumentation literature. For instance in [31] it is observed that "being more particular about the evidence makes an argument stronger", while in [30] it is admitted that treating reasons as if they were all of equal strength is a simplifying assumption. Strength appears therefore to be unquestionably an essential feature of arguments both at the informal and the formal level.

In spite of this rather obvious consideration, it has to be acknowledged that, somewhat surprisingly, there is still not an established general theory of argument strength, and the study and formalization of this notion have witnessed a comparatively slower progress with respect to other key concepts, like, for instance, the one of argument acceptability in the context of abstract argumentation frameworks and their semantics [19, 6].

It might even be argued that the investigation of argument strength has been somehow neglected for some years, with more attention being paid to other aspects of the argumentation process and a (typically simple) relation of preference among arguments being assumed to be able to capture and synthesise their different strengths. An inversion of this trend has been observed in recent years with an increasing number of works devoted to subjects implicitly or explicitly related to the notion of strength and a series of workshops on Argument Strength having been initiated, with the first edition held in Bochum in 2016 (homepages.ruhr-uni-bochum.de/defeasible-reasoning/

Argument-Strength-2016.html) and the second one in Toulouse in 2018 (www.math-info.univ-paris5.fr/AMANDE-ArgStrength-2018/). The variety of the approaches presented in the literature, and specifically in these workshops, provides a further confirmation of the richness and diversity inherent to this notion, which as is typical for basic essential concepts, lends itself to different interpretations and tends to elude any univocal and universally accepted characterization.

In this light, one may wonder whether it is appropriate to conceive argument strength as a single entity, though rich of nuances, or it should be rather be acknowledged that the term strength is being used in a rather liberal manner to refer to a multiplicity of different, though somehow related, features of arguments. To put it in other words, and ascribing some personification to the concept, it may be suggested that there is no some very eclectic person named *Argument Strength* whose character is so changeable to escape any simple description, but rather that there is a whole large family of Strengths, whose members, though connected by some degree of kinship, have distinct identities and may show complementary, if not opposite, individual features.

Following this interpretation, this paper discusses, without any claim of exhaustiveness, some distinctions that may lead to identify different members of the argument strength family and then tries to point out some general features that may be considered to be part of the genetic inheritance of the whole family. Whenever possible, we complement our analysis with the identification of potential connections with other research fields.

The paper is organized as follows. Section 2 discusses the two main types of strength representation, namely scale-based and ranking-based, while Section 3 examines two prominent interpretations of the notions of strength, namely the structural and the relational ones. Section 4 turns to the analysis of contextual aspects affecting the evaluation of strength, while Section 5 concerns the identification of possible commonalities among different notions of strength. Section 6 concludes.

2 Scale-based vs. ranking-based strength

A first basic distinction concerns the representation adopted to capture the fact that arguments have different strength. Letting $\mathcal{A}$ be the set of arguments considered in a given context, two basic choices are available:

- adopting a scale $\mathbb{S}$ (either numerical or qualitative) of possible strength values and representing strength as a function $s : \mathcal{A} \to \mathbb{S}$;
- capturing strength in terms of a ranking between arguments, i.e. of a relation $\preceq \subseteq \mathcal{A} \times \mathcal{A}$, where, given $\alpha, \beta \in \mathcal{A}$, $\alpha \preceq \beta$ means that β is at least as strong as α.

Some simple considerations and further distinctions about these two basic alternatives are worth commenting.

First, any numerical scale is, by nature, totally ordered, and qualitative scales are typically totally ordered too. Thus, unless differently specified, for any scale $\mathbb{S}$ we will assume that it is equipped with a reflexive, transitive and anti-symmetric relation $\leq$, such that for any $x, y \in \mathbb{S}$, with $x \neq y$, exactly one and only one of the relations $x \leq y$, $y \leq x$ holds, a fact which is represented by the usual shortcut $x < y$ or $y < x$, respectively. Further a scale can be finite or infinite, and in case it is infinite, it may admit or not the presence of a minimum and/or of a maximum value, i.e. of a value v such that for any $x \in \mathbb{S}$, $x \neq v$, $v < x$ ($x < v$, respectively). For instance if $\mathbb{S}$ is the set of non-negative integer numbers, it includes a minimum value, namely 0, but no maximum value, while the whole set of integer numbers has no minimum nor maximum (unless the special values $+\infty$ and $-\infty$ are added to the set). Similarly, if $\mathbb{S}$ is a real interval, it includes a minimum and/or maximum value depending on whether the interval is open or closed on its bounds, e.g. the closed real interval $[0, 1]$ has 0 (respectively 1) as minimum (maximum) value, while the open interval $]0, 1[$ includes no minimum nor maximum.

Turning to rankings, two minimal requirements for $\preceq$ can be identified, namely that it is reflexive ($\alpha \preceq \alpha$ for every $\alpha \in \mathcal{A}$) and transitive ($\alpha \preceq \beta$ and $\beta \preceq \gamma$ imply $\alpha \preceq \gamma$ for every $\alpha, \beta, \gamma \in \mathcal{A}$), i.e. that $\preceq$ is a preorder. Optionally, these properties can be strengthened in various way, in particular by assuming that $\preceq$ is total (for every $\alpha, \beta \in \mathcal{A}$ either $\alpha \preceq \beta$ or $\beta \preceq \alpha$).

Having recalled these basic notions, it can be observed that, under our assumptions, for any scale-based strength representation s there is an induced (total) ranking $\preceq_s$ defined for every $\alpha, \beta \in \mathcal{A}$ as $\alpha \preceq \beta$ iff $s(\alpha) \leq s(\beta)$. On the other hand, given an argument ranking $\preceq$ one can think of devising a scale-based assignment $s_\preceq$ *realizing* the ranking i.e. such that $\alpha \preceq \beta$ implies $s_\preceq(\alpha) \leq s_\preceq(\beta)$.

On the basis of these simple considerations one might be led to think that the choice between scale-based and ranking-based representation is more a matter of taste than of substance, and indeed typically in the literature proposals of both kinds are proposed just as available options in the large design space of approaches to strength representation, without much discussion at a substantial level about the distinctive features, and possible merits and limits, of the various choices.

In the following we argue however that in many cases different representational choices actually imply also some (possibly implicit) different ontological choices about the notion of strength, which in turn should be connnected to and motivated by the argumentation context addressed in a theoretical investigation or practical application.

2.1 Incomparability

A ranking admits in general that two arguments α and β are incomparable, i.e. that neither $\alpha \leq \beta$ nor $\beta \leq \alpha$ holds, while with a scale-based representation comparability between every pair of arguments is ensured (and, in a sense, enforced).

Whether incomparable arguments (with respect to strength) may exist is a basic question, whose answer does not appear easy. On the one hand, one might say that comparing very different arguments, possibly produced in very different contexts with different purposes by different subjects is meaningless and impossible. Consider for instance comparing an argument raised by a candidate in a political campaign with an argument which is part of a research debate among some scientists. Intuitively, the criteria for the evaluation of the strength of these arguments can be very different and it appears that comparing them is not sensible and void of practical use. On the other hand, one might interpret strength as a universal notion, applicable to every argument in a context-independent manner and hence ensuring also universal comparability, even between arguments one actually needs not to compare. In this sense, one could advocate some similarity with other supposedly universal notions, like in particular probability, which when assessed numerically, provides a sort of universal measure of uncertainty and then supports the possibility to compare the likelihood of any pair of events, as different and unrelated as they may be. It could be observed however, that also the interpretation of the notion of probability has been, and still is, a subject of substantial debates, and that also in the field of probability theory the requirement of providing a numerical assessment involving a total ordering has been considered too strong in some contexts, leading to the investigation of qualitative (i.e. ranking-based) probabilities.

The foundational debate about this issue being potentially open-ended, we limit ourselves to remark that the issue of numerically representing partial orders has a very general nature across many application domains. This is due to the fact that partial orders, being simple and not demanding, are a particularly convenient knowledge elicitation and representation tool, suitable to capture the most basic attitudes of an agent with respect to a variety of entities, like for instance preferences and utilities, in addition to the already discussed strength and uncertainty.

In this respect we mention that some basic general results on the relationships between numerical and ranking-based representations are provided in [26]. Here, the remark on incomparability previously stated is expressed in formal terms by observing that, given a partial order P on a set X, "it is, of course, impossible to find a real-valued function $f(\cdot)$ such that xPy *if and only if* $f(x) > f(y)$, since P may be incomplete, and the relation of non-comparability N (defined as xNy if and only if not xPy and not yPx) need not be transitive." Further, two basic properties for numerical representations of partial orderings are introduced, namely completeness and transparency.

A real-valued function $f(\cdot)$ defined over X is a *complete representation* of a partial ordering P if and only if there is a function λ defined on $\mathbb{R}^2$ such that for all $x, y \in X$: $\lambda(f(x), f(y)) > 0 \leftrightarrow xPy$.

A complete representation is therefore faithful, in the sense that it reflects exactly the relationships encoded in the partial ordering but it is not "readable" since it achieves this goal indirectly through a further function λ.

Sacrificing exactness to readability leads to the notion of transparency: a real-valued function $f(\cdot)$ defined over X is a *transparent quasi-representation* of a partial ordering P if and only if for all $x, y \in X$: $xPy \to f(x) > f(y)$.

In general the same function $f(\cdot)$ cannot be both complete and transparent and this impossibility result confirms the gap between numerical and ranking-based representations in general. As this gap and related issues have been already considered in other areas like economics, decision theory, and probability theory we conclude this subsection by suggesting that studies on different forms of argument strength may benefit from existing results in these more consolidated fields.

2.2 Information grain and meaning

A ranking-based representation is inherently course grained. Consider three arguments α, β and γ and assume that $\alpha < \beta < \gamma$. This ordering tells that β has an intermediate strength between α and γ but cannot tell, for instance, whether the difference in strength is small or big or whether the strength of β is closer to the one of α or to the one of γ. A scale-based and in particular a numerical representation is necessary if such kind of finer-grained information needs to be captured and one wants to assess differences, proportions, and in general manipulate strength values with operations like sum, multiplication and do on.

Numbers however are more demanding in terms of precision, and the well-known question "Where do the numbers come from?", typically raised to point out the difficulty of eliciting probability values in some contexts and to question the meaningfulness of their precision, may well apply to argument strength values too.

As a matter of fact the study of qualitative (also called comparative) probability, conceptually similar to ranking-based strength, is an active field with a long tradition [33, 16]. Moreover hybrid approaches mixing qualitative and quantitative aspects according to the application needs and to the information actually available have been considered in probabilistic contexts [17] and could be investigated for argument strength too.

At a more substantial level, however, it is worth recalling that numbers do not carry a meaning per se and more specifically, that when the same numbers are used to assess different quantities they may have different meanings and rely on different assumptions, which in turn determine which operations and transformations are legally applicable to the number themselves.

In particular, measurement theory [24] concerns the problem of assigning numbers to things in a meaningful way and traditionally distinguishes several types of scales, differing in the properties of the measured things which are reflected in the assigned numbers. In particular the following types of scales are traditionally considered in measurement theory [32]:

- Nominal (e.g. football players' numbers). This is the simplest and poorest scale. Essentially numbers are used as plain symbols whose unique relevant property is their identity. Two things are assigned the same number (symbol) if they have the same value of the considered attribute.
- Ordinal (e.g. academic grades). Numbers are used as an ordered set with no quantitative meaning. Things are assigned numbers such that the order of the numbers reflects an order relation defined on the attribute.
- Interval (e.g. temperatures in degrees Celsius or Fahreneit). Things are assigned numbers such that differences between the numbers reflect differences of the attribute.
- Log-interval (e.g. density=mass/volume). Things are assigned numbers such that ratios between the numbers reflect ratios of the attribute.
- Ratio (e.g. temperature in degrees Kelvin). Things are assigned numbers such that differences and ratios between the numbers reflect differences and ratios of the attribute.
- Absolute (e.g. probability). Things are assigned numbers such that all properties of the numbers reflect analogous properties of the attribute.

In the four latter scales the quantitative nature of numbers plays some role, but only in the absolute scale it does so in an unconstrained way, while the use of other scales requires some awareness to avoid gross errors. To give an almost trivial example, it is meaningless to claim that it was twice as warm today as yesterday because it was 40 degrees Fahrenheit today but only 20 degrees yesterday. Fahrenheit is not a ratio scale, and there is no meaningful sense in which 40 degrees is twice as warm as 20 degrees. More subtle examples concerns the use of statistics, for instance if one assumes an interval scale only statistics that are invariant or equivariant under change of origin or unit of measurement are meaningful. The mean and the standard deviation satisfy these requirements and therefore make sense for temperatures expressed in degrees Celsius or Fahrenheit, but the coefficient of variation is meaningless because it lacks such invariance or equivariance.

As to our knowledge, in the literature on numerical argument strength no discussion about the issues mentioned above has been carried out and it has been taken for granted that strength values correspond to an absolute scale, i.e. the strongest assumption on the correspondence between the properties of numbers and the properties of strength has commonly (and implicitly) been made. While it has not to be excluded that this assumption is well justified, it might be useful to discuss and motivate it more explicitly and it might also be interesting to consider approaches to the representation of strength based on weaker kinds of scales.

2.3 Extreme values and saturation

Rankings are by nature open-ended, for every argument α in principle it is always possible to state that there is some other argument β such that $\alpha < \beta$ (or $\beta < \alpha$). On the other hand, as already mentioned, a scale may include a top (say $\top$) and/or a bottom (say $\bot$) value which give possibly rise to a situation of saturation: if $s(\alpha) = \top$, for every other argument β it will hold that $s(\beta) \leq s(\alpha)$ (and analogously for $s(\alpha) = \bot$).

While, technically speaking, saturation could be regarded as a small detail, we suggest that conceptually it makes a crucial difference and that the possibility or not of saturation should be considered an important design choice rather than a sort of side effect of the adopted scale.

If present, numerical strength saturation should be seen as the counterpart of a cognitive phenomenon occurring in the context of the argumentation process, and, at a conceptual level, reasonable motivations both to support and to exclude the existence of this phenomenon can be identified.

On the one hand, one can appeal to analogous saturation phenomena to advocate saturation in argument strength. For the bottom level one may observe that you cannot kill a dead or, on a very different matter, that a temperature below the absolute zero cannot be reached. For the top level, you cannot go beyond an efficiency of 100%, cannot make some acknowledged truth more true, or an event cannot have a probability greater than 1. In this perspective one may consider saturation justified by the fact that there are some extreme positive or negative conditions pushing the assessment of the strength of an argument to such a limit that no further improvement or worsening is possible. Cognitively, this may correspond to the fact that, in such extreme conditions, it is impossible (or useless) to appreciate further differences and one enters a sort of flat insensitivity area. Of course these rather generic considerations would require some empirical confirmation from psychological studies, but it is a matter of fact that saturation is present in many literature approaches to the representation of strength.

On the other hand, one might counterargue that when dealing with a computational model of the argumentation process, saturation can be regarded as undesirable since it inherently introduces some arbitrariness and in fact corresponds, as previously mentioned, to a loss of sensitivity which can be regarded as unnecessary. Consider for instance a situation where the strength of an argument is somehow related to its positive or negative interactions with other arguments (say its supporters and attackers). Whatever the situation of the argument at a given moment, one can imagine a modification where its strength is potentially increased (by the addition of a supporter) or decreased (by the addition of an attacker). Similarly, in a situation where strength is related to opinions, say pro or con votes, expressed by the general public about the arguments, one may always imagine a situation where a further pro or con vote is added. Encompassing a saturation effect requires to identify a point where these additions start having no further influence on the strength value and the choice of this point appears to involve inherently some arbitrariness. In the same line, one might argue that saturation is a simplification strategy adopted by people in view of their limited cognitive resources, but that this simplification is not required for computer-based reasoners which should be allowed to evaluate even small differences when they exist.

The discussion above suggests, among other things, that encompassing saturation in a strength model might also be related to the distinction between descriptive or normative strength models, a further important issue which we will not enter in this paper. In the next section we will move from representational issues and their connections with some essential properties of strength, to interpretation issues, namely to (some of) the different meanings that argument strength can assume.

3 Structural vs. relational strength

In engineering (and in other fields as well) strength concerns building something solid, by designing and constructing it using appropriate materials and/or components. In this perspective the strength of an artifact can be evaluated looking at its structure and at its constituent parts and assessing them. Strength is then an intrinsic property of an object, which can be evaluated by analysing the object itself *in isolation*.

In sport (and other fields too) strength is instead a relational property. For instance a tennis player is considered strong if s/he is able to beat other players, i.e. strength is evaluated on the basis of the outcomes of the matches one plays and could not be assessed in absence of other players. In this perspective strength in isolation simply does not exist since the conflicts with peers are its main ingredient. Another difference is that, in this case, there is no absolute reference and strength becomes a more dynamic notion: your strength depends on the set of the opponents you are confronted with and changes with them. In the tennis example the assessment of your strength (say your position in the world ranking) may rise or fall, even if you play exactly in the same way, just because, for instance, some older better player retires or some new player able to beat you comes on stage.

These two general perspectives on the notion of strength have counterparts in alternative interpretations of argument strength too.

On the one hand, the strength of an argument can be assessed in isolation on the basis of its structure and of its components, typically the facts and rules used to build it. For instance in rule-based argumentation, under the assumption that the rules included in an agent's knowledge base have different strength, the well-known weakest link principle states that the strength of an argument is determined by the weakest rule or premise used in its construction, while, according to the last link principle, only the last inference step counts. Examples of articulated approaches to structural evaluation of argument strength in specific application contexts are given for instance by [5], where the strength of arguments included in the preamble of EU (European Union) directives is discussed, and by [13], concerning arguments used for system assurance in engineering.

On the other hand, the strength of an argument can be assessed on the basis of its relationships (typically of attack and support) with other arguments. In the literature this corresponds to strength evaluations based on some abstract argumentation formalism [19, 4, 15] where the internal structure of the arguments is ignored/abstracted away and only the links between arguments play a role. An example of a simple criterion that has been advocated in this context, in particular for the case where only attacks between arguments are considered, is that arguments which do not receive attacks should be assigned the highest strength level (called *void precedence* in [2]). Examples of relational evaluation of argument strength are given by [2, 14].

At a formal level, the two above interpretations of the notion of argument strength can be cleanly distinguished, as they can be associated with separate formalization levels: intrinsic strength evaluation occurs at the level of structured argumentation formalisms, like ASPIC+ [29] or DeLP [22], while relational strength evaluation occurs at the level of abstract argumentation fornalisms, as mentioned previously.

At a conceptual level, however, this clean separation may sound questionable. As to the intrinsic evaluation based on structural properties, one might argue that even if attackers or supporters do not play an explicit role in the strength assessment (indeed no attackers and supporters might have been identified), actually at a deeper level the evaluation implicitly refers to the potential attackers or supporters one may imagine. To give an example, consider two unrelated arguments both dealing with some medical issues coming respectively from a Nobel prize and from a scientifically unqualified web blogger and assume no other arguments are taken into account. Both arguments are then unattacked but one is intuitively led to ascribe a higher strength to the one coming from the Nobel prize because of a more authoritative source. This authoritativeness might be given however a dialectical interpretation as it may correspond to the fact that one can imagine that the argument coming from the Nobel prize has much more potential to collect support and/or to withstand attacks with respect to the one coming from the blogger (see for instance [21] for a notion of potential rebuttal).

The above example suggests that one could consider an attempt to reduce intrinsic evaluation to some form or relational evaluation, but this leads to a vicious circle. In fact, in its pure form, abstract relational evaluation assumes that all arguments are equal (as all the concrete properties that may differentiate them are ignored) before they enter the relational arena. This appears to be an extremely strong assumption, which is hardly tenable in most contexts, and in particular when natural arguments are considered, and clashes also with some formalization practices commonly adopted in the literature. Take two arguments with contradictory conclusions, which therefore have to be considered in conflict. If they are regarded as equal, this conflict can be formalized at an abstract level as a mutual attack between them, which, if no other arguments are involved, corresponds to a situation where no definite winner can be identified. However, sometimes one would like to capture at the abstract level that arguments at the structured level are not all equal and, in the example, there could be reasons two identity one of the two arguments as a clear winner. Then a unidirectional relation of attack is represented at the abstract level, with the direction of the attack being determined by an evaluation of which argument is stronger at the structured level.

Thus, while one may argue that intrinsic strength could (or should) find a deeper interpretation in terms of relational strength, one is also led to observe that the definition of the relationships holding between argument themselves depends on some underlying strength assessment. Leaving the issue of further analyzing and possibly breaking this conceptual knot to future developments, we conclude this section by the more plain observation that the two interpretations of argument strength can be simply regarded as pertaining to two subsequent steps of the argumentation process (argument construction and argument relational evaluation) and, as such, can be easily combined together.

This happens in formalisms like Quantitative Argumentation Debate frameworks [10] and Weighted Argumentation Graphs [3] where arguments are assumed to have an initial strength, depending on their structural properties and assigned before their relations with other arguments are considered, and this initial value is taken into account in the determination of the final strength which also depends on the connections with other arguments. In this context the final strength of an argument is typically defined as a function of the initial strength of the argument itself and of the final strength of its attackers and supporters and it is normally assumed that the final strength coincides with the initial strength for arguments with no attackers nor supporters. The formal study of this family of combined strength evaluation is rapidly progressing and it might contribute to the clarification of the conceptual issues mentioned above too. In this context, a point which may require further attention is the possibility to consider that the relations of attack and support too might have different intensity and so some notion of "strength" could be defined for them too (see for instance [27, 20]).

4 Strength in context

Our discussion up to now has not mentioned any context-dependent factors influencing argument strength. Argumentation however often, if not always, occurs in a context, and therefore whatever interpretation and representation of argument strength one may adopt, it is hard to assume that it can be treated as an acontextual entity.

Consider for instance the notion of social evaluation of arguments [25] where a first evaluation of arguments is based on the votes expressed by the members of some social system (e.g. the users of a social network). Here strength is not meant to be an absolute property of the arguments but rather depends on the subjective (and possibly fluctuating) opinions of the community of voters. As a variation of a previous example, suppose now that two conflicting arguments about the same subject (e.g. vaccines) are raised by a Nobel prize and a web blogger. Assume also that the first argument is based on a long-term collection of experimental evidence, while the latter refers to some specific, recent, and suspected to be fake, news which anyway has got a big mediatic impact. While it is very likely that the Nobel prize's argument is considered to be stronger in the context of a scientific debate, it is possibly the case that the blogger's argument is considered instead to be much stronger not only among his/her followers, but possibly even by the general public or by any set of not scientifically trained people. As a matter of fact, the infamous *Stamina* [1] case in Italy in 2013 has shown how arguments void of any scientific base may prevail over the opinions of experts in political decisions at the highest level.

Providing models of this kind of phenomena is obviously of great interest for the formal argumentation community and calls for some context dependent notion of strength which takes into account other factors in addition to (and possibly in combination with) argument structure and/or inter-argument relationships.

The approach of social abstract argumentation [25] can be regarded as a step in this direction, with many opportunities of furter development being apparent. In particular, the notion of audience is used in value-based argumentation [11] to represent the fact that different people may have different value orderings and as a consequence have different views on which argument prevails over another in a conflict. Formally, an audience determines the actual direction of the attack relation between two conflicting arguments according to the audience-specific ordering of the values promoted by the arguments themselves. The study of a generalized definition of audience, applicable to a large spectrum of forms of argument strength, looks a potentially very fruitful direction of future investigation.

The context where argumentation occurs does not consist of the audience only, however, but of course also of the agent(s) producing the arguments to be evaluated. While monological argumentation [12] refers to the case where a single agent produces and assesses arguments inside his/her mind (typically for reasoning purposes), dialogical argumentation concerns the case where different, typically self-interested, agents produce and exchange arguments while pursuing some dialectical goal (e.g. persuading another agent and/or an audience of a given claim, reaching an advantageous agreement, and so on). In such a scenario, each agent may/should choose opportunistically which arguments to put forward and the effectiveness of a given argument with respect to the achievement of the desired goals depends also on the arguments chosen by the other agents. These considerations lie at the basis of the notion of game-theoretical argument strength, first introduced in [28]. The basic idea is that the argument exchange between two agents is considered as a game and the notion of strength of an argument can be related to the expected reward in this game when an agent commits to put forward that specific argument (possibly together with others). In more detail, in the scenario considered in [28] each agent has to choose a strategy, namely a set of arguments, and makes a move, by putting forward the chosen sets of arguments. On the basis of its move and of the other agent's move, each agent receives a payoff, determined by a reward function. The strenght of an argument α is then defined as the expected optimal payoff when an agent is committed to include α in its move. Providing more technical details about game-theoretical argument strength is beyond the scope of the present paper, it suffices to say that it provides a furter alternative context-aware perspective on the notion of strength which can be combined with other ones. In particular, given the arguments played by the agents in the game, the reward they receive can be related to some basic structural or relational evaluation of these sets of arguments themselves. In particular the latter approach is adopted in [28] where the reward function depends on the attacks between arguments.

It is also useful to note that recently [7] it has been evidenced that game-theoretical strength itself admits several variants and can therefore be conceived as a family of related notions. In particular, it has been observed that in [28] an asymmetrical structure of the game has been considered, where an agent is committed to include in its move a specific argument α (the one whose strength has to be evaluated) while there is no analogous constraint on the move of the other agent, which can freely play any set of arguments. Thus the strength of α is assessed in "absolute terms" with respect to all possible game configurations. In some contexts however both agents may have an agenda, i.e. one (or possibly more) arguments they are committed to play. Think for instance of a political debate between two candidates who have to defend their already announced political programs. To model such a context a symmetric configuration of the game, where each player is committed to play a specific argument, is more appropriate. This gives rise to a game-theorical notion of argument strength which is relative to another argument (the one chosen by the other agent) rather than absolute, i.e. an argument has not a strength value on its own but rather a potentially different strength value for every other argument it can be confronted with.

We conclude this section by observing that game-theoretical argumentation not only adds another branch to the family of argument strengths but also provides hints suggesting possible extensions of the branches considered before.

First, one may wonder whether a notion of relative strength can be considered also outside the game-theoretical perspective. Indeed, combining the relational perspective with some kind of head-to-head comparison looks reasonable and feasible in principle. The structural perspective, encompassing argument evaluation "in isolation" may look less suitable to a relative notion of strength, but one cannot exclude that a structural comparison between distinct arguments leading to a relative strength assessment is, at least in some cases, possible.

Second, in the game-theoretical approach the notion of strength can be straightforwardly extended to sets of arguments, by considering for an agent the commitment to include in its move several arguments, rather than just one. The strength of an argument is then the special case where the set an agent is committed to include in its move is a singleton. One can then wonder whether a collective rather than individual notion of strength can be considered also in other perspectives. Again, this looks feasible and rather natural in the relational perspective, where evaluation based on attacks and supports can be directly and easily extended to set of arguments (as a matter of fact, in traditional abstract argumentation many fundamental properties based on the attack relation, like conflict-freeness and admissibility, are collective), while extending the structural perspective to sets of arguments appears less immediate and possibly to require a more significant creativity leap.

5 Looking for strength DNA

In the previous sections we have discussed (some of) the various ramifications of the notion of argument strength. Altogether it is reasonable to conclude that, a complete census of its members being hard to accomplish, one can anyway safely claim that the family of strengths is a quite numerous one. Having said enough about differences, in this section we focus on commonalities. First, we assume that some commonalities must exist based on the informal consideration that the members of the same family share at least part of their genetic inheritance. Denying any commonality would mean denying that the word strength carries any meaning in this context and assuming that it is used somehow erratically to refer to completely disconnected things. While it cannot be excluded that this is the case, we try to provide some hints in the opposite direction.

Starting from the structural perspective, we are not aware of any general analysis in the literature about some formalism-independent characterization of argument strength. Generic strength properties have been however been introduced in the context of specific structured argumentation approaches. They typically refer to aspects related to argument construction, like the subargument relation between argument or, in rule-based approaches, the distinction between strict and defeasible rules. For instance, in [34] argument strength is defined in terms of an *order of conclusive force* $\leq$ which, in addition to reflexivity and transitivity, satisfies three conditions:

- there are no infinite chains of arguments such that $\alpha_1 < \alpha_2 \ldots < \alpha_n < \ldots$;
- if α is a subargument of β then $\beta \leq \alpha$;
- if the last rule used in the construction of an argument α is strict then there is a direct subargument β of α such that $\beta \leq \alpha$.

While the first condition is a finiteness requirement, the others represent two rather intuitive properties: an argument cannot be stronger than its subarguments and inference steps which are not affected by any uncertainty do not decrease the strength of an argument with respect to its subarguments.

A notion of *reasonable* argument ordering is also introduced in ASPIC+ [29] where some differences and commonalities with Vreeswjik's ordering can be identified[1]. In particular, there is no ordering constraint between an argument and its subarguments, while the role of elements (facts or rules) not affected by uncertainty is more emphasised: not only a strict inference step does not affect strength like in [34], but also if an argument is built using infallible (i.e. noy affected by uncertainty) elements only then it is strictly stronger than any other argument using some uncertain fact and rule. Moreover infallible arguments are assumed to be all equally strong.

In the context of the DeLP formalism [22] two criteria for comparing arguments are considered.

The first one, called *specificity*, states that an argument is preferred to another one if it has a greater information content (intuitively, it is more *precise*) or makes less use of rules (intuitively, it is more *concise*). More formally, specificity refers to the sets of facts that, together with a set of rules, are used to derive a conclusion, called *activating sets*. An argument α is more specific than an argument β if the sets of facts activating α are also activating sets of β but there is some activating set of β which is not an activating set of α.

The second criterion is based on rule priorities: it is assumed that a preference relation over rules is given, and then an argument α is preferred to another argument β if there is no rule used for β which is strictly preferred to a rule used in α and there is a rule used in α which is strictly preferred to a rule used in β.

Differently from the ones previously recalled, these criteria do not directly mention the subargument relation nor the use of strict rules. Implications on the ordering preference between an argument and its subargument can however be derived from these criteria, while, as to strict rules, it can be observed that in DeLP they have essentially the same role as in other formalisms, but formally are not part of the definition of an argument. An interesting remark is that the second criterion, like the orderings considered in [29, 34], allows in principle the comparison of any pair of arguments, while the specificity criterion is meant to compare only arguments which are somehow related (as they must share some activation sets).

Even this very limited sample of three approaches to strength comparison in structured argumentation reveals an inherent difficulty, as the technical differences in the underlying formalism give rise to differences in the proposed characterizations of strength and therefore also to difficulties in their comparison. Nonetheless, some general intuitive commonalities can be identified as some very general notions, like the subargument relation and the distinction between strict and non-strict inference steps, appear to be part of a common background, though their formalizations differ in different proposals. This

---

[1] We do not recall directly the formulation of the properties of reasonable ordering in ASPIC+ as it involves some specific technical aspects which are beyond the scope of this paper.

leaves open the hope that a higher-level common characterization of strength for structured argumentation is possible. Semi-structured argumentation formalisms like the one proposed in [8] might be helpful in this respect.

The situation is rather different considering strength in abstract argumentation. Here the simplicity of the underlying formalization (a set of abstract arguments and one or more binary relations like attack or support among them) makes much easier the definition and analysis of general strength properties. In fact, several recent works have been devoted to this topic, with analyses covering a variety of cases like different forms of abstract frameworks (with attack relation only, with support relation only, or with both), and the presence or not of an initial strength assessment. While the properties considered in the literature are quite diverse (and surveying them is far beyond the scope of this paper) most of them adhere to a general pattern, sketched below:

- two arguments α and β are considered, belonging to the same framework or to two distinct but somehow related frameworks;
- a relevant comparison involving the parents (attackers and/or supporters) or more generally the ancestors of α and β is identified;
- from the above comparison a relation is derived (typically an inequality) between the strengths of α and β.

The comparison criteria about the parents can be quite various, ranging from emptiness vs. non-emptyness, to set inclusion, to cardinality comparison, to some aggregation operator on their strength values, and so on.

For instance the property of *void precedence*, whose content has been already recalled, states for frameworks with attacks only that if the set of parents (attackers) of α is empty and the set of parents of β is not empty then $s(\alpha) > s(\beta)$.

The generic pattern introduced above encompasses a very large space of related properties, also because several "small" variants of the same basic idea are possible. In particular, similar but formally different properties can be obtained e.g. considering a strict vs. a non-strict inequality or including/excluding from the set of considered parents those which have a minimum (null) value of strength (when such a value exists). This gives rise to the problem of possibly having too many available properties, among which one may wonder which are the fundamental ones and whether a synthetic view is possible. A contribution towards answering these questions is provided in [9] where two generic principles of balance and monotonicity (both admitting a strict and a non-strict version) are introduced for bipolar argumentation frameworks and it is shown that these principles imply a large set of properties previously introduced in the literature, along with many variants of these properties not explicitly considered before.

Altogether, though for different reasons, the quest for the DNA of argument strength appears to be at a preliminary status, open for long term investigation.

Considering the fact that, in the more general setting, an argument receives both positive and negative influences from its supporters and attackers, we suggest that this investigation might benefit from looking for possible connections with the study of bipolar decision making, where the evaluation of the decision options shows some significant structural similarities with the evaluation of arguments in a bipolar framework (see for instance [18, 23]).

Ideally one may hope that at some point a principled and articulated classification of the family of strengths is achieved. A possibly inspiring example to this purpose is the organization of the family of uncertainty measures where, starting from the very general notion of *capacity*, progressively more specific measures are identified by adding further requirements leading to define (in reverse order of inclusion) coherent imprecise probabilities, 2-monotone capacities, belief functions, possibility/necessity measures and (as another subclass of belief functions) traditional precise probabilities.

6 Conclusions

With this paper we have tried to provide an unavoidably partial discussion of some conceptual issues, basic distinctions and open questions concerning the prolific family of argument strengths. Strength is a very general notion, not easy to characterize as it is often used to synthesise effectively in a single word a multiplicity of factors: it can be referred to entities either material or immaterial and formal or informal, like in particular human achievements, relationships, and feelings. This book is dedicated to our friend Guillermo Simari, a man whose strong professional achievements are accompanied by an even stronger unique touch of humanity pervading every aspect of his activity. We are grateful to Guillermo for the many occasions of cooperation we have had with him, for the many things we have learned from him, and most importantly for the precious friendship he has offered to us, which we look forward to make stronger and stronger.

References

1. Alison Abbott. Italian stem-cell trial based on flawed data. *Nature News*, July, 2013.
2. Leila Amgoud and Jonathan Ben-Naim. Ranking-based semantics for argumentation frameworks. In *Proc. of the 7th Int. Conf. on Scalable Uncertainty Management (SUM 2013)*, pages 134–147, 2013.
3. Leila Amgoud, Jonathan Ben-Naim, Dragan Doder, and Srdjan Vesic. Acceptability Semantics for Weighted Argumentation Frameworks. In *Proc. of the 26th Int. Joint Conf. on Artificial Intelligence (IJCAI 2017)*, pages 56–62, 2017.
4. Leila Amgoud, Claudette Cayrol, Marie-Christine Lagasquie-Schiex, and P. Livet. On bipolarity in argumentation frameworks. *International Journal of Intelligent Systems*, 23(10):1062–1093, 2008.
5. Corina Andone and Florin Coman-Kund. The importance of argument strength in policy-making: The case of european union directives. In *First Workshop on Argument Strength*, Institute of Philosophy II, Ruhr-University Bochum, 2016.

6. Pietro Baroni, Martin W. A. Caminada, and Massimiliano Giacomin. An introduction to argumentation semantics. *Knowledge Engineering Review*, 26(4):365–410, 2011.
7. Pietro Baroni, Giulia Comini, Antonio Rago, and Francesca Toni. Abstract Games of Argumentation Strategy and Game-Theoretical Argument Strength. In *Prof. of the 20th Int. Conf. on Principles and Practice of Multi-Agent Systems (PRIMA 2017)*, pages 403–419, 2017.
8. Pietro Baroni, Massimiliano Giacomin, and Beishui Liao. A general semi-structured formalism for computational argumentation: Definition, properties, and examples of application. *Artificial Intelligence*, 257:158–207, 2018.
9. Pietro Baroni, Antonio Rago, and Francesca Toni. How many properties do we need for gradual argumentation? In *Proc. of the 32nd AAAI Conf. on Artificial Intelligence (AAAI 2018)*, pages 1736–1743, 2018.
10. Pietro Baroni, Marco Romano, Francesca Toni, Marco Aurisicchio, and Giorgio Bertanza. Automatic evaluation of design alternatives with quantitative argumentation. *Argument & Computation*, 6(1):24–49, 2015.
11. Trevor J. M. Bench-Capon. Persuasion in practical argument using value-based argumentation frameworks. *J. of Logic and Computation*, 13(3):429–448, 2003.
12. Philippe Besnard and Anthony Hunter. *Elements of Argumentation*. The MIT Press, 2008.
13. Robin Bloomfield and Kate Netkachova. Argument strength: an engineering perspective. In *First Workshop on Argument Strength*, Institute of Philosophy II, Ruhr-University Bochum, 2016.
14. Gerhard Brewka. Weighted abstract dialectical frameworks. In *First Workshop on Argument Strength*, Institute of Philosophy II, Ruhr-University Bochum, 2016.
15. Gerhard Brewka and Stefan Woltran. Abstract dialectical frameworks. In *Proc. of the 12th Int. Conf. on Principles of Knowledge Representation and Reasoning (KR 2010)*, pages 102–111, 2010.
16. James P. Delgrande and Bryan Renne. The logic of qualitative probability. In *Proc. of the 24th Int. Joint Conference on Artificial Intelligence (IJCAI 2015)*, pages 2904–2910, 2015.
17. Marek J. Druzdzel and Linda C. van der Gaag. Elicitation of probabilities for belief networks: Combining qualitative and quantitative information. In *Proc. of the 11th Annual Conference on Uncertainty in Artificial Intelligence (UAI '95)*, pages 141–148, 1995.
18. Didier Dubois, Hélène Fargier, and Jean-François Bonnefon. On the Qualitative Comparison of Decisions Having Positive and Negative Features. *J. of Artificial Intelligence Research (JAIR)*, 32:385–417, 2008.
19. Phan Minh Dung. On the acceptability of arguments and its fundamental role in nonmonotonic reasoning, logic programming, and n-person games. *Artificial Intelligence*, 77(2):321–357, 1995.
20. Paul E. Dunne, Diego C. Martínez, Alejandro Javier García, and Guillermo Ricardo Simari. Computation with varied-strength attacks in abstract argumentation frameworks. In *Proc. of the 3rd Int. Conf. on Computational Models of Argument (COMMA 2010)*, pages 207–218, 2010.
21. James Freeman. Comparative argument strength: A formal inquiry. In *First Workshop on Argument Strength*, Institute of Philosophy II, Ruhr-University Bochum, 2016.

22. Alejandro Javier García and Guillermo Ricardo Simari. Defeasible logic programming: An argumentative approach. *Theory and Practice of Logic Programming*, 4(1-2):95–138, 2004.
23. Michel Grabisch and Christophe Labreuche. Bi-capacities - I: definition, Möbius transform and interaction. *Fuzzy Sets and Systems*, 151(2):211–236, 2005.
24. David J. Hand. Statistics and the theory of measurement, with discussion. *J. of the Royal Statistical Society, Series A*, 159:445–492, 1996.
25. João Leite and João Martins. Social Abstract Argumentation. In *Proc. of the 22nd Int. Joint Conf. on Artificial Intelligence (IJCAI 2011)*, pages 2287–2292, 2011.
26. Mukul Majumdar and Amartya Sen. A note on representing partial orderings. *The Review of Economic Studies*, 43(3):543–545, 1976.
27. Diego C. Martínez, Alejandro Javier García, and Guillermo Ricardo Simari. An abstract argumentation framework with varied-strength attacks. In *Proc. of the 11th Int. Conf. on Principles of Knowledge Representation and Reasoning (KR 2008)*, pages 135–144, 2008.
28. Paul-Amaury Matt and Francesca Toni. A Game-Theoretic Measure of Argument Strength for Abstract Argumentation. In *Poic. of the 11th European Conf. on Logics in Artificial Intelligence (JELIA 2008)*, pages 285–297, 2008.
29. Sanjay Modgil and Henry Prakken. A general account of argumentation with preferences. *Artificial Intelligence*, 195:361 – 397, 2013.
30. John Pollock. How to reason defeasibly. *Artificial Intelligence*, 57(1):1–42, 1992.
31. Guillermo R. Simari and Ronald P. Loui. A mathematical treatment of defeasible reasoning and its implementation. *Artificial Intelligence*, 53(2–3):125–157, 1992.
32. Stanley S. Stevens. On the theory of scales of measurement. *Science*, 103:677–680, 1946.
33. Cesareo Villegas. On qualitative probability. *The American Mathematical Monthly*, 74(6):661–669, 1967.
34. Gerard A. W. Vreeswijk. Abstract argumentation systems. *Artificial Intelligence*, 90(1-2):225–279, 1997.

From Individual to Collective Argumentation: Aggregating Acyclic Argumentation Frameworks

Fernando A. Tohmé[1] and Gustavo A. Bodanza[2]

[1] Departamento de Economía, Universidad Nacional del Sur and INMABB, UNS-CONICET, Bahía Blanca
[2] Departamento de Humanidades, Universidad Nacional del Sur and IIESS, UNS-CONICET, Bahía Blanca

Abstract. An argumentation framework consists of a class of arguments and an abstract attack relation among them. The same arguments, but under different attack relationships, yield different frameworks. We analyze here how to merge frameworks with a common set of arguments; we proceed by *aggregating* the attack relations. A positive result ensues for certain classes of attack relations. In particular, in the case of *acyclic* attack relations, where the aggregation is obtained under the same conditions as Arrow's Theorem. These conditions involve certain properties of *fairness* of the aggregation process. The acyclicity of the attack relations implies that, if the class of winning coalitions in the voting-based aggregation process is a *proper prefilter*, a single attack relation can be obtained. While many properties of the individual frameworks are preserved in the consequent merged framework, this procedure does not yield the same results when the aggregation is applied on the *extensions* of argumentation frameworks.

1 Introduction

Defeasible reasoning embodies the intuition that it is possible to reach conclusions even in the presence of competing lines of argumentation. This is achieved by comparing the alternative conclusions in terms of their support. In abstract terms, the support of conclusions is given by a set of arguments. The arguments that remain undefeated in a series of comparisons are declared warranted. The literature presents several formalisms capturing this intuition ([35], [15], [30], [4], [5], [32]) but we will restrict our analysis to Abstract Argumentation Frameworks of [19], which abstract away all the features of argumentation, except the attack relations among entities called "arguments". Alternative semantics have been introduced for these argumentation systems, based on the idea that they are essentially defined by the arguments that survive all possible attacks of other arguments in the system. The classes of surviving arguments are called *extensions* of the system.

While many different aspects of abstract argumentation have been thoroughly examined in the literature, one that only recently has been explored is how to merge different attack criteria among the same arguments [16] [33], [3]. When a group or society has to make a collective decision, its members will engage in some form of argumentation. But the assessment of which argument defeats another depends on the individual points of view: "each agent may have her own view on what an attack is and as a consequence, an agent may believe that one argument attacks another argument, while another agent may believe that this is not the case. Merging argumentation systems can be used to define (sets of) arguments acceptable by the group." ([17], p. 614). In other words, since the attack among arguments cannot be objectively stated, the next best idea is to aggregate the individual subjective criteria into a single one.

This problem is not just of academic importance. As pointed out by Guillermo Simari, in the increasingly interconnected network of sources of information, knowledge can be explicitly represented in the form of argumentation systems. But then, the multiplicity of alternative systems poses the problem of their interaction and the assessment of the ensuing conclusions. An earlier interest in this topic lead to a predecessor of the current paper [36]. More recently, the first survey of all the relevant literature on the topic was published [9].

For a simple example of this problem, consider the following fictional story in which two arguments are presented to a political party's central committee in order to choose a presidential candidate:

A: John Smith is a decent human being who suffered torture while in jail under a dictatorial regime just because he fought for a better society. Now he has the chance to carry out his ideals, and we must support him in as a candidate to presidency.

B: John Smith is a bad guy. He was a guerrilla fighter who killed innocent people seeking power. He can only ruin the country and hence we must reject him as a candidate.

This situation could be modeled as an argumentation framework in which arguments A and B conflict each other. This is a natural view for an impartial observer who knows that Mr. Smith indeed was a guerrilla fighter who ended up being imprisoned and tortured by a dictatorial regime. Nevertheless, it is also natural to conceive that some committee members will only accept that B defeats A, others think that A wins over B, and finally others recognize the mutual defeat between these two arguments. Each of these claims amounts to a different attack relation. A warrant for either one, none or both arguments can only obtain by first coalescing these attack relations into a single one.

The connections between this problem and phenomena arising in the application of e-democracy or multi-agent systems are straightforward. The process of making collective decisions involves arguing, bargaining or voting in different mixtures of the three procedures, of which none can be reduced to another

([20], pp. 5-8.). Our proposal is based on the idea that the aggregation of votes on attack criteria over arguments for and against some alternatives is more reasonable than just voting over the alternatives. The latter case involves an aggregation of preferences on the alternatives while the former, the topic of this paper, involves the aggregation of the *structure* of attacks among arguments about the alternatives. Nevertheless, we will apply some *formal* aspects of the aggregation of preferences to analyze the aggregation of attack relations.

Note that our proposal does not see the problem of aggregating attack relations in the light of *judgment aggregation* [26], which involves the acceptance or rejection of a claim in terms of the number of votes cast on the pieces of judgments that lead either to that conclusion or to its negation. In that framework the opposition among judgments arises from the fact that their conjunction leads to a logical contradiction. This approach has already been taken in the literature: by representing attack relations as *labelings* [13] the problem of aggregating them, with their three "truth values", is closer to 3-value judgement aggregation [3]. We, instead, will analyze aggregation focusing on the *decisive sets*. The results we obtain in this way complement those obtained by the aggregation of labelings, with which it shares significant commonalities [33].

In our view, an important aspect of the aggregation of attack criteria (which may differ in other ways than only by giving way to contradictions) is that it may lead to the rejection, or even the reinstatement, of some arguments as a consequence of the acceptance of other arguments. For instance, consider the acceptance of the following argument in the setting of the example presented above:

C: John Smith was judged in an international court and acquitted of the charges of criminal behavior.

Accepting this argument and its attack on argument B should lead to reject B and reinstate A, even upon a previous acceptance of the attack of B on A.

The processes by which a single preference ordering may be obtained from a class of individual preferences over the same alternatives have been studied as the main problem in social decision-making. The outcome is said the *aggregation* of the individual preferences [28]. In the same token, if each of the individual attack relations among arguments represents an individual *criterion* of warrant, because it determines the extensions that will obtain, aggregation procedures can be applied over them. Any such aggregation process will weigh up the different criteria and determine which extensions will actually arise. While many ways of aggregating criteria can be postulated, a natural form is by means of pairwise voting [6]. That is, each alternative attack relation "votes" over pairs of arguments, and the winning attack is incorporated in the aggregated relation.

Nevertheless, this procedure may fail to satisfy some generally accepted constraints over the aggregation process [12, 7]. These constraints are properties that are assumed to be verified by "fair"[3] aggregation processes. Arrow's Impossibility Theorem [1] shows that four conditions that capture abstractly the properties of a fair aggregation process, cannot be simultaneously satisfied. While these conditions can be imposed on the merging of argumentation frameworks and a version of Arrow's theorem may ensue, attack relations and preference relations are different in many respects. On one hand, Arrow's proof is valid for the case of reflexive and transitive relations over the alternatives. Attack relations usually are not seen as reflexive nor transitive. This point is worth to be emphasized, since it involves the reason why an Arrow-like result may not arise in argumentation systems. This fact makes our purpose non trivial.

Both attack and preference relations are superficially similar. The difference between aggregating individual preferences and attack criteria arises from their corresponding order-theoretic characterizations. While preferences are usually assumed to be *weak orders* (i.e., reflexive, transitive and complete relations),[4] attack relations are free to adopt any configuration. Besides, preference relations are expected to have maximal elements, while this is not the case for attack relations. If A attacks B and B attacks C, it is customarily accepted that not only A does not (necessarily) attack C, but also that A "defends" C, which implies that A and C can be jointly warranted. So, while a preference relation may lead to the choice of its maximal elements, an attack relation can lead to the choice of a maximal (w.r.t. $\subseteq$) set of "defensible" arguments, i.e. an extension.

The worst case scenario is realized when a collectively established attack relation yields several extensions, while the best one arises when only one is obtained. In the first case, the choice of justified arguments could be focused on the intersection of all the extensions leaving the remaining arguments under dispute. In the second case, all the arguments within the extension are indisputable. An interesting question is to determine the conditions that lead to this ideal situation. First, it is known that *well-founded* argumentation frameworks have a unique extension which, moreover, belongs to all the classes of extensions defined by Dung ([19]). Well-founded argumentation frameworks have no infinite sequences of attacks, in particular *cycles*. Hence, our question can be rephrased as: how can a "fair" aggregation mechanism lead to a collective well-founded attack relation? We want to explore whether aggregating individual well-founded frameworks may lead to a collective well-founded framework. In our view, asking for the individual argumentation frameworks to be well-founded seems sensible and not too demanding. It amounts to ask the agents' criteria to be unambiguous: each argument must be sanctioned either *in* or *out* univocally under the same criterion. This condition is also crucial for the computational implementation of reasoning systems that have to yield

---

[3] Here 'fair' has to be understood in an informal sense, different from its technical meaning in Social Choice Theory.

[4] It is worth to note that other relations can be aggregated as well. For instance, graphs (seen as connections among nodes) [21].

yes/no answers to queries to knowledge bases. Acyclicity in the implicit attack relation is a feature of many working computational reasoning systems. For instance DeLP ([22],[23], [37]), which reaches its conclusion by building a tree without repeated arguments.

In this paper we prove the existence of aggregation processes that have as input a *finite* number of acyclic attack relations over the same set of arguments and as output an acyclic relation over them, satisfying all the conditions of Arrow's Theorem. In fact, like in [12] we show that the class of winning coalitions of attack criteria constitutes an algebraic structure called a *proper prefilter*. As it has been discussed in the literature on Arrow's Theorem, a proper prefilter implies the existence of a *collegium*[5] of attack relations. Each member of the collegium belongs to a winning coalition, while the collegium itself does not need to be one. Each collegium member, by itself, cannot determine the outcome of the aggregation process, but can instead veto the behaviors that run contrary to her prescription. The outcome of the aggregation can be seen as the agreement in not vetoing attacks by the representatives of the different winning coalitions. In this sense it indicates a slight consensus among the attack relations.

This result can be interpreted as meaning that, even in the case of "equal opportunity" aggregation procedures, there will exist some fragment of the individual attack relations that will become imposed on the aggregate one. But while in social contexts this is rather undesirable (in the literature the members of the collegium are called *hidden dictators*), in the case of argument systems it is far more reassuring, since it indicates that when the attack relations are minimally rational, a consensual outcome may arise.

While voting may seem a too simple way of aggregating attack relations, they are only instances of our main result, that states that there exist aggregation procedures satisfying very restrictive conditions. These conditions, except for one, involve the comparison between different profiles (with the same number of entries) of attack relations on a common set of arguments. A natural instance of alternative profiles arises when each one is an assessment of an expert. While this may impose an interpretation of how attack relations may arise, the important thing to note is that if a positive result obtains under these conditions, they will apply to less restrictive settings.

A less brighter perspective arises from the fact that pairwise voting does not yield the same results for the aggregation of attack relations as for the aggregation of extensions. More precisely, if the "voters" are attack relations, and they constitute a proper prefilter, the aggregated attack relation may not yield the same extensions as if the individual attack relations cast their votes for the extensions supported by them. While somehow negative, this claim makes clear that voting over attack relations is quite different than voting on

---

[5] This term is drawn from [12], and is now customary in the analysis of voting in Political Science [34].

individual arguments (whether they belong or not to the extension of an attack relation).

The structure of this paper is as follows. Section 2 presents all the relevant concepts of argumentation à la Dung, particularly of the problem of aggregating attack relations. Section 3 restricts the ways in which attack relations can be aggregated. Section 4 introduces the concept of *decision sets* and shows how they can induce an aggregation procedure satisfying Arrow's conditions. Section 5 examines the differences between the aggregation of attack relations and that of extensions. Section 6 and 7 presents the discussion of the results found in the previous section and states some problems still open for further research. These ideas were, as said, already presented in [36], but the current version improves its predecessor in several ways:

– The literature review is far more thorough now, relating our research to recent developments in the discipline.
– More examples are included, in order to provide better motivations for the notions introduced here.
– The analysis of the aggregation of extensions has been enlarged, illuminating the differences with the aggregation of attack relations.
– a key step in the proof of Proposition 3 had a mistake in the previous version. Now it has been corrected and the equivalence claimed in that proposition has been solidly established.

That is, this new version sets the discussion on a larger body of recent developments and amends infelicities of the previous one.

2 Aggregating Attack Relations

Dung defines an argumentation framework as a pair $AF = \langle AR; \to \rangle$, where AR is a set of abstract entities called 'arguments' and $\to \subseteq AR \times AR$ denotes an attack relation among arguments. This relation is used to determine which arguments are "better" than others. Different intuitions yield alternative sets of "best" arguments called *extensions* of AF. These extensions are seen as the semantics of the argumentation framework, i.e. the classes of arguments that can be deemed as the outcomes of the whole process of argumentation. Dung introduces the notions of *preferred*, *stable*, *complete*, and *grounded* extensions, each corresponding to different requirements on conflict-free sets of arguments.

Definition 1. (Dung ([19])). *In any argumentation framework AF an argument σ is said* acceptable *w.r.t. a subset S of arguments of AR, in case that for every argument τ such that $\tau \to \sigma$, there exists some argument $\rho \in S$ such that $\rho \to \tau$. A set of arguments S is said* admissible *if each $\sigma \in S$ is acceptable w.r.t. S, and if S is conflict-free, i.e., the attack relation does not hold for any pair of arguments belonging to S. A preferred extension is any maximally admissible set of arguments of AF. A complete extension of AF is any conflict-free subset of arguments which is a fixed point of $\Phi(\cdot)$, where*

$\Phi(S) = \{\sigma : \sigma \text{ is acceptable w.r.t. } S\}$, *while the* grounded extension *is the least (w.r.t. $\subseteq$) complete extension. Moreover, a* stable extension *is a conflict-free set S of arguments which attacks every argument not belonging to S.*

Interestingly, if the number of arguments is finite and the attack relation is acyclic, the framework is said to be *well-founded* and has only one extension that is grounded, preferred stable and complete (*cf.* [19], theorem 30, pp. 331). The main application of argumentation frameworks is the field of *defeasible reasoning*. Roughly, arguments are structures that support certain conclusions (claims). The extensions include the arguments, and more importantly their conclusions, that become warranted by a reasoning process that considers the attack relation.

We consider, instead, for a given n an *extended* argumentation framework $AF^n = \langle AR; \rightarrow_1, \ldots, \rightarrow_n \rangle$. Each $\rightarrow_i$ is a particular attack relation among the arguments in AR, representing either different criteria or different assessments (made each one by a different "expert") according to which arguments are evaluated one against another. Such extended frameworks may arise naturally in the context of defeasible reasoning, since there might exist more than one criterion of defeat among arguments.

The determination of *preferred, complete* or *grounded* extensions in an argumentation framework is based upon the properties of the single attack relation. There are no equivalent notions for an extended argumentation framework, except for those corresponding to an *aggregate* argumentation framework $AF^* = \langle AR; \mathcal{F}(\rightarrow_1, \ldots, \rightarrow_n)\rangle$, where $\mathcal{F}(\rightarrow_1, \ldots, \rightarrow_n) = \rightarrow$, i.e. $\mathcal{F}(\rightarrow_1, \ldots, \rightarrow_n)$ is the aggregated attack relation of AF^*. That is, AF^* is a an argumentation framework in which its attack relation arises as a function of the attack relations of AF^n. Notice that $\mathcal{F}$ may be applied over any extended argumentation framework with n attack relations. It embodies a method that yields a single attack relation up from n alternatives.

To postulate an aggregate relation addresses the problem of managing the diversity of criteria, by yielding a single approach. This is of course analogous to a social system, in which a unified criterion must by reached. While there exist many alternative ways to aggregate different criteria, most of them are based in some form of voting. In fact, the best known case of $\mathcal{F}$ is *plurality voting*. Unlike political contests in which for each pair $A, B \in AR$ a majority selects either $A \rightarrow B$ or $B \rightarrow A$, we allow for a third alternative in which the majority votes for the absence of attacks between A and B. Formally:

- $A \rightarrow B$ if $|\{i : A \rightarrow_i B\}| > \max(|\{i : B \rightarrow_i A\}|, |\{i : B \not\rightarrow_i A \land A \not\rightarrow_i B\}|)$.

- $B \rightarrow A$ if $|\{i : B \rightarrow_i A\}| > \max(|\{i : A \rightarrow_i B\}|, |\{i : B \not\rightarrow_i A \land A \not\rightarrow_i B\}|)$.

- $(A, B), (B, A) \notin \rightarrow$ (i.e. A does not attack B, nor B does attack A in $\rightarrow$), otherwise.

For instance, if out of 100 individual relations, 34 are such that A attacks B, while 33 verify that B attacks A and the rest that there is no attack relation between A and B, plurality voting would yield that $A \to B$. That is, it only matters which alternative is satisfied by more individual relations than the other two.

Example 1. Consider the following framework in which $AR = \{A, B, C\}$ and the arguments are:

A : "Symptoms x, y and z suggest the presence of disease d_1, so we should apply therapy t_1";

B : "Symptoms x, w and z suggest the presence of disease d_2, so we should apply therapy t_2";

C : "Symptoms x and z suggest the presence of disease d_3, so we should apply therapy t_3".

Assume these are the main arguments discussed in a group of three agents (M.D.s), 1, 2 and 3, having to make a decision on which therapy should be applied to some patient. Suppose that each agent i, $i \in \{1, 2, 3\}$, proposes an attack relation $\to_i$ over the arguments as follows:

$A \to_1 B \to_1 C$ (agent 1 thinks that it is not convenient to make a joint application of therapies t_1 and t_2 or t_2 and t_3; moreover she thinks that B is more specific than C, hence B defeats C, and that, in the case at stake, symptom y is more clearly present than symptom w.[6] Furthermore, this implies that argument A defeats B),

$A \to_2 C$, $B \to_2 C$ (agent 2 thinks that it is not convenient to apply therapies t_1 together with t_3 or t_2 joint with t_3; moreover she thinks that symptoms y and w are equally present in the case at stake. Furthermore, both A and B are more specific than argument C, hence both A and B defeat C),

$A \to_3 C \to_3 B$ (agent 3 thinks that it is not convenient to apply t_1 together with t_3 or t_2 with t_3; moreover she thinks that symptom w is not clearly detectable, hence C defeats B, but A is more specific than C, hence A defeats C).

According to plurality voting we obtain $\to$ over AR:

$A \to C$ since A attacks C under $\to_2$ and $\to_3$.

$B \to C$ since B attacks C under $\to_1$ and $\to_2$.

A and B do not attack each other under $\to$ since they don't do so under $\to_2$ and $\to_3$.

---

[6] Defeat criteria among arguments based on specificity or preferred evidence are well known in the argument systems literature. See, for instance, [27].

In this example, plurality voting picks out one of the individual attack relations, showing that $\to = \to_2$. In any of the extension semantics introduced by [19], arguments A and B become justified under the aggregated attack relation, supporting the decision of applying both therapies t_1 and t_2.

On the other hand, as the following example shows, plurality voting may yield cycles of attacks up from acyclic individual relations.

Example 2. Consider the following three attack relations over the set $AR = \{A, B, C\}$:

$C \to_1 B \to_1 A$,

$A \to_2 C \to_2 B$, and

$B \to_3 A \to_3 C$.

We obtain $\to$ over AR as follows:

$A \to C$ since A attacks C under $\to_2$ and $\to_3$.

$B \to A$ since B attacks C under $\to_1$ and $\to_3$.

$C \to B$ since C attacks B under $\to_1$ and $\to_2$.

Thus, $\to$ yields a cycle $A \to C \to B \to A$. An analogous phenomenon, for preference relations, is known in the literature on voting systems as the **Condorcet's Paradox**.

Another way of aggregating attack relations is by restricting plurality voting to a *qualified voting* aggregation function. It fixes a given class of relations as those that will have more weight in the aggregate. Then, the outcome of plurality voting over a pair of arguments is imposed on the aggregate only if the fixed attack relations belong to the largest coalition. Otherwise, in the attack relation none of the arguments attacks the other. That is, given a set $U \subset \{1, \ldots, n\}$:[7]

$A \to B$ iff $|\{i : A \to_i B\}| > \max(|\{i : B \to_i A\}|, |\{i : B \not\to_i A \wedge A \not\to_i B\}|)$ and $U \subseteq \{i : A \to_i B\}$.

$B \to A$ iff $|\{i : B \to_i A\}| > \max(|\{i : A \to_i B\}|, |\{i : B \not\to_i A \wedge A \not\to_i B\}|)$ and $U \subseteq \{i : B \to_i A\}$.

The situation $(A, B), (B, A) \notin \to$ (i.e. A does not attack B, nor B does attack A in $\to$) can arise as follows:

either $|\{i : B \not\to_i A \wedge A \not\to_i B\}| \geq \max(|\{i : A \to_i B\}|, |\{i : B \to_i A\}|)$ and $U \subseteq \{i : B \not\to_i A \wedge A \not\to_i B\}$,

---

[7] Instead of just caring about *how many* individuals vote for or against an issue, qualified voting takes into account *who* the voters are. Moreover, it leads to an outcome only upon the agreement of the 'qualified voters'.

or if U is not a subset of either $\{i : A \rightarrow_i B\}$, $\{i : B \rightarrow_i A\}$ or $\{i : B \not\rightarrow_i A \wedge A \not\rightarrow_i B\}$.

Example 3. Consider again the individual attack relations in Example 1. If $U = \{2, 3\}$ we have that $A \rightarrow C$, since $A \rightarrow_2 C$ and $A \rightarrow_3 C$. Again $(A, B), (B, A) \not\in \rightarrow$ because $(A, B), (B, A) \not\in \rightarrow_2$ and $(A, B), (B, A) \not\in \rightarrow_3$; but, we also have $(B, C) \not\in \rightarrow$ because although there exists a plurality for B attacking C ($\{1, 2\}$), $C \rightarrow_3 B$, i.e., there is no consensus among the members of U on B and C.

There exist in the literature more involved aggregation procedures, which focus on additional information on the individual attack relations [16, 14]. But our point is to show that even simple procedures may fail to yield the result we seek. While plurality voting fails to output an acyclic attack relation up from a profile of acyclic individual ones, qualified voting fails to be "fair" by giving extra decision power to some individual relations.

3 Conditions on Aggregation Functions

While different schemes of aggregation of attack relations can be postulated, the seminal work of Kenneth Arrow [1] points towards a higher degree of abstraction. Instead of looking for particular functional forms, the goal is to set general constraints over aggregation processes and see if they can be jointly fulfilled. We carry out a similar exercise in the setting of extended argumentation frameworks, in order to investigate the features of aggregation processes that ensure that a few reasonable axioms are satisfied.

Arrovian analyses are carried out in terms of an aggregation process that, up from a family of *weak orders* (complete, transitive and reflexive orderings), yields a weak order over the same set of alternatives. This is because both individual and social *preference* relations are represented as weak orders. But attack relations cannot be assimilated to *preference* orderings, since attacks do not satisfy necessarily any of the conditions that define a weak order. So for instance, reflexivity in an attack relation would mean that each argument attacks itself. While isolated cases of self-attack may arise, this is not a general feature of attack relations. The same is true of transitivity, which means that if, say $A \rightarrow B$ and $B \rightarrow C$ then $A \rightarrow C$. In fact, in many cases of interest, $A \rightarrow B \rightarrow C$ can be interpreted as indicating that A *defends* B. Finally, completeness is by no means a necessary feature of attacks, since there might exist at least two arguments A and B such that neither $A \rightarrow B$ nor $B \rightarrow A$.

Therefore, the difference of our setting with the usual Arrovian context is quite significant.

Let us begin with a few properties that we would like to be satisfied by any aggregation function adapted to our setting [12]. Below, we will use the alternative notation $\rightarrow_{\mathcal{F}}$ instead of $\mathcal{F}(\rightarrow_1, \ldots, \rightarrow_n)$ when no confusion could arise.

Pareto condition. For all $A, B \in AR$ if for every $i = 1, \ldots, n$, $A \to_i B$ then $A \to_{\mathcal{F}} B$.

Positive Responsiveness. For all $A, B \in AR$, and two n-tuples of attack relations, $(\to_1, \ldots, \to_n)$, $(\to'_1, \ldots, \to'_n)$, if $\{i : A \to_i B\} \subseteq \{i : A \to'_i B\}$ and $A \to_{\mathcal{F}} B$, then $A \to_{\mathcal{F}} pB$, where $\to'_{\mathcal{F}} = \mathcal{F}(\to'_1, \ldots, \to'_n)$.

Independence of Irrelevant Alternatives. For all $A, B \in AR$, and given two n-tuples of attack relations, $(\to_1, \ldots, \to_n)$, $(\to'_1, \ldots, \to'_n)$, if $\to_i = \to'_i$ for each i, over (A, B), then $\to_{\mathcal{F}} = \to'_{\mathcal{F}}$ over (A, B).

Non-dictatorship. There does not exist i_0 such that for all $A, B \in AR$ and every $(\to_1, \ldots, \to_n)$, if $A \to_{i_0} B$ then $A \to_{\mathcal{F}} B$.

All these requirements were intended to represent the abstract features of a fair collective decision-making system. To understand them it is interesting to resort to the illustration given in Examples 1 and 2 of [33] and [3]. Namely, consider that each $i \in \{1, \ldots, n\}$ is a detective, evaluating the same pieces of evidence on a case. Each one holds a different view on how this pieces can be put together, in the form of a individual attack relation. The goal is to find a single attack relation (a "case" to be made) up from that information. In this setting it is easier to understand that these conditions (except for the Pareto condition) involve different profiles: each detective may postulate a different attack relation, giving rise to another profile. While this may run against interpreting the *is* as criteria, if we find a positive result, it will also apply to that case.

In any case, let us see why these conditions imply the fairness of the aggregation process.

The *Pareto condition* indicates that if all the attacks relations coincide over a pair of arguments, the aggregated attack should also agree with them. That is, if all the individual attack relations agree on some arguments, this agreement should translate into the aggregate attack relation.

The *positive responsiveness condition* just means that the aggregation function should yield the same outcome over a pair of arguments if some attack relation previously dissident over them now changes towards an agreement with the others. It can be better understood in terms of political elections: if a candidate won an election, she should keep winning in an alternative context in which somebody who voted against her now turns to vote for her.

The axiom of *independence of irrelevant alternatives* states that if there is an agreement over a pair of arguments among alternative n-tuples of attacks, this should be also be true for the aggregation function over both n-tuples. Again, some intuition from political elections may be useful. If the individual preferences over two candidates a and b remain the same when a third candidate c arises, the rank of a and b should be the same in elections with and without c. That is, the third party should be irrelevant to the other two.

Finally, the *non-dictatorship condition* just stipulates that no fixed entry in the n-tuples of attacks should become the outcome in every possible instance.

That is, there is no 'dictator' among the individual attack relations. We have the following proposition:

Proposition 1.
Both the plurality and the qualified voting (with $|U| \geq 2$) aggregation functions satisfy the four axioms.

Proof. Plurality voting:

- (Pareto): if for all $A, B \in AR$ if for every $i = 1, \ldots, n$, $A \to_i B$ then trivially $|\{i : A \to_i B\}| > \max(|\{i : B \to_i A\}|, |\{i : B \not\to_i A \wedge A \not\to_i B\}|)$ which in turn implies that $A \to_\mathcal{F} B$.
- (Positive Responsiveness): if for all $A, B \in AR$, and two n-tuples of attack relations, $(\to_1, \ldots, \to_n)$, $(\to'_1, \ldots, \to'_n)$, if $A \to_\mathcal{F} B$, this means that $|\{i : A \to_i B\}| > \max(|\{i : B \to_i A\}|, |\{i : B \not\to_i A \wedge A \not\to_i B\}|)$ and therefore, if $\{i : A \to_i B\} \subseteq \{i : A \to'_i B\}$ it follows[8] that $|\{i : A \to'_i B\}| > \max(|\{i : B \to'_i A\}|, |\{i : B \not\to'_i A \wedge A \not\to'_i B\}|)$ which in turn implies that $A \to'_\mathcal{F} B$.
- (Independence of Irrelevant Alternatives): suppose that for any given $A, B \in AR$, and two n-tuples of attack relations, $(\to_1, \ldots, \to_n)$, $(\to'_1, \ldots, \to'_n)$, $\to_i = \to'_i$ for each i, over (A, B). Without loss of generality assume that $|\{i : A \to_i B\}| > \max(|\{i : B \to_i A\}|, |\{i : B \not\to_i A \wedge A \not\to_i B\}|)$ then, $A \to_\mathcal{F} B$. But then $|\{i : A \to'_i B\}| > \max(|\{i : B \to'_i A\}|, |\{i : B \not\to'_i A \wedge A \not\to'_i B\}|)$, which implies that $A \to'_\mathcal{F} B$. That is, $\to_\mathcal{F} = \to'_\mathcal{F}$ over (A, B).
- (Non-dictatorship): suppose there were a i_0 such that for all $A, B \in AR$ and every $(\to_1, \ldots, \to_n)$, if $A \to_{i_0} B$ then $A \to_\mathcal{F} B$. Consider in particular that $A \to_{i_0} B$ while $|\{i : B \to_i A\}| = n - 1$, i.e. except i_0 all other attack relations have B attacking A. But then $B \to_\mathcal{F} A$. Contradiction.

The proof for qualified voting, when $|U| \geq 2$, is quite similar:

- (Pareto): if for all $A, B \in AR$ if for every $i = 1, \ldots, n$, $A \to_i B$ then trivially $|\{i : A \to_i B\}| > \max(|\{i : B \to_i A\}|, |\{i : B \not\to_i A \wedge A \not\to_i B\}|)$ and $U \subseteq \{i : A \to_i B\}$ which implies that $A \to_\mathcal{F} B$.
- (Positive responsiveness): if for all $A, B \in AR$, and two n-tuples of attack relations, $(\to_1, \ldots, \to_n)$, $(\to'_1, \ldots, \to'_n)$, if $A \to_\mathcal{F} B$, this means that $|\{i : A \to_i B\}| > \max(|\{i : B \to_i A\}|, |\{i : B \not\to_i A \wedge A \not\to_i B\}|)$ and $U \subseteq \{i : A \to_i B\}$. Therefore, if $\{i : A \to_i B\} \subseteq \{i : A \to'_i B\}$ it follows that $|\{i : A \to'_i B\}| > \max(|\{i : B \to'_i A\}|, |\{i : B \not\to'_i A \wedge A \not\to'_i B\}|)$ and $U \subseteq \{i : A \to'_i B\}$ which in turn implies that $A \to'_\mathcal{F} B$.

---

[8] Notice that if we denote $i^+_{AB} = |\{i : A \to_i B\}|$, $i^-_{AB} = |\{i : B \to_i A\}|$ and $i^0_{AB} = |\{i : B \not\to_i A \wedge A \not\to_i B\}|$), we have that $i^+_{AB} + i^-_{AB} + i^0_{AB} = n = i^{+\prime}_{AB} + i^{-\prime}_{AB} + i^{0\prime}_{AB}$, where $i^{+\prime}_{AB}$, $i^{-\prime}_{AB}$ and $i^{0\prime}_{AB}$ are the corresponding cardinalities for the family of relations $\{\to'_i\}$.

- (Independence of Irrelevant Alternatives): suppose that for any given $A, B \in AR$, and two n-tuples of attack relations, $(\rightarrow_1, \ldots, \rightarrow_n)$, $(\rightarrow'_1, \ldots, \rightarrow'_n)$, $\rightarrow_i = \rightarrow'_i$ for each i, over (A, B). Without loss of generality assume that $|\{i : A \rightarrow_i B\}| > \max(|\{i : B \rightarrow_i A\}|, |\{i : B \not\rightarrow_i A \land A \not\rightarrow_i B\}|)$ and $U \subseteq \{i : A \rightarrow_i B\}$ then, $A \rightarrow_{\mathcal{F}} B$. But then $|\{i : A \rightarrow'_i B\}| > \max(|\{i : B \rightarrow'_i A\}|, |\{i : B \not\rightarrow'_i A \land A \not\rightarrow'_i B\}|)$ and also $U \subseteq \{i : A \rightarrow'_i B\}$ which implies that $A \rightarrow'_{\mathcal{F}} B$. That is, $\rightarrow_{\mathcal{F}} = \rightarrow'_{\mathcal{F}}$ over (A, B).
- (Non-dictatorship): suppose there were a i_0 such that for all $A, B \in AR$ and every $(\rightarrow_1, \ldots, \rightarrow_n)$, if $A \rightarrow_{i_0} B$ then $A \rightarrow_{\mathcal{F}} B$. Consider in particular that $A \rightarrow_{i_0} B$ while $|\{i : B \rightarrow_i A\}| = n - 1$, i.e. except i_0 all other attack relations have B attacking A. If $i_0 \notin U$, $B \rightarrow_{\mathcal{F}} A$, while if $i_0 \in U$, $(A, B) \notin \rightarrow_{\mathcal{F}}$. In either case we have a contradiction.

Compare this result with Arrow's general impossibility theorem, which can be stated in this way: if each $\rightarrow_i$ were complete and transitive, then every aggregation function $\rightarrow_{\mathcal{F}}$ satisfying the Pareto condition, Positive Responsiveness and Independence of Irrelevant Alternatives would be dictatorial. Clearly, the positiveness of our result is due to the fact that the attack relations do not need to be complete nor transitive.

Let us now consider the difference between qualified and plurality voting. While the former seems more lacking in "fairness", it does not generate a variant of Condorcet's Paradox: acyclic individual attack criteria always lead to a collective acyclic attack criterion. As argued in the Introduction, the requirement of acyclicity is not excessively demanding. The acyclicity of the individual attack criteria is intended to represent the commitment of the agents in not being ambiguous about their votes: each individual relation leads to a univocal decision on whether any argument is *in* or *out*. This procedure ensures that the aggregate attack criterion has the same property:

Proposition 2.
If $\mathcal{F}$ is a qualified voting aggregation function and each $\rightarrow_i$ is acyclic, then $\rightarrow_{\mathcal{F}}$ is acyclic.

Proof. Suppose that $\rightarrow_{\mathcal{F}}$ has a cycle of attacks, say $A^0 \rightarrow_{\mathcal{F}} A^1 \rightarrow_{\mathcal{F}} \ldots \rightarrow_{\mathcal{F}} A^k \rightarrow_{\mathcal{F}} A^0$. By definition of qualified voting, for $j = 0 \ldots k - 1$ $A^j \rightarrow_{\mathcal{F}} A^{j+1}$ if and only if $U \subseteq \{i : A^j \rightarrow_i A^{j+1}\}$. By the same token, $A^k \rightarrow_{\mathcal{F}} A^0$ iff

$$U \subseteq \{i : A^k \rightarrow_i A^0\}.$$

That is, for each $i \in U$, $A^0 \rightarrow_i A^1 \rightarrow_i \ldots \rightarrow_i A^k \rightarrow_i A^0$. But this contradicts that each individual attack relation is acyclic.

4 Decisive Sets of Attack Relations

We can now analyze the main claim in our paper, namely the existence of a *positive* result, contrary to Arrow's theorem. For this, we need an auxiliary concept, that of *decisive* sets (*i.e.*, winning coalitions) of attack relations. Interestingly, the structure of this class exhibits (in relevant cases) clear algebraic features that shed light on the behavior of the aggregation function. Formally: $\Omega \subset \{1,\ldots,n\}$ is a *decisive* set if for every possible n-tuple $(\rightarrow_1,\ldots,\rightarrow_n)$ and every $A, B \in AR$, $A \rightarrow_i B$ for every $i \in \Omega$ implies $A \rightarrow_{\mathcal{F}} B$ (i.e. $A \; \mathcal{F}(\rightarrow_1,\ldots,\rightarrow_n) \; B$).

An intuitive meaning of this definition can be found, considering again that the detective story in [33] and [3]. A decisive set of detectives would be such that, no matter what cases they (or the others) present, the aggregate case incorporates all their assessments.

We will focus on classes of decisive sets, $\bar{\Omega} = \{\Omega^j\}_{j \in J}$ that fully determine aggregate attack relations. That is, given any profile $(\rightarrow_1,\ldots,\rightarrow_n)$, $A \rightarrow_{\mathcal{F}} B$ iff $\{i : A \rightarrow_i B\} \in \bar{\Omega}$. As we have already seen in the case of qualified voting aggregation functions, if not every member of a decisive set agrees with the others over a pair of arguments, the aggregate attack relation should not include the pair. But this is so unless any other decisive set can force the pair of arguments into the aggregate attack relation.[9]

Example 4. In Example 1, each of $\{1,2\},\{2,3\}$ *is a decisive set under the plurality rule, since they include more than half of the agents that coincide with pairs of attacks in the aggregate attack relation. On the other hand, for the qualified voting function of Example 3,* $U = \{2,3\}$ *is decisive, but not* $\{1,2\}$ *nor* $\{1,3\}$.

If we recall that the U is a decisive set for qualified voting, we can conjecture that there might exist a close relation between the characterization of an aggregation function and the class of its decision sets. Furthermore, if a function satisfies Arrow's axioms and yields an acyclic attack relation up from acyclic individual attack relations, it can be completely characterized in terms of the class of its decision sets:[10]

Proposition 3.
Consider an aggregate attack relation $\mathcal{F}$ that for every n-tuple $(\rightarrow_1,\ldots,\rightarrow_n)$ of acyclic attack relations yields an acyclic $\rightarrow_{\mathcal{F}}$. It verifies the Pareto condition, Positive Responsiveness, Independence of Irrelevant Alternatives, and Non-Dictatorship if and only if its class of decisive sets $\bar{\Omega} = \{\Omega^j\}_{j \in J}$, where J is a set of indices, verifies the following properties:

- $\{1,\ldots,n\} \in \bar{\Omega}$.
- *If* $O \in \bar{\Omega}$ *and* $O \subseteq O'$ *then* $O' \in \bar{\Omega}$.
- *Given* $\bar{\Omega} = \{\Omega^j\}_{j \in J}$, *where* $J = |\bar{\Omega}|$, $\cap \bar{\Omega} = \bigcap_{j=1}^{J} \Omega^j \neq \emptyset$.
- *No* $O \in \bar{\Omega}$ *is such that* $|O| = 1$.

---
[9] Of course, in a qualified voting function U is always a decisive set.
[10] The proofs are specific for acyclic attack relations and have substantial differences with the proofs for order relations, like those in [12].

Proof.
$\Rightarrow$) We will begin our proof noticing that $1 \leq |\bar{\Omega}| \leq 2^n$. On one hand it includes only a finite number of decisive sets. On the other, by Pareto, it has to include at least one: the grand coalition $\{1,\ldots,n\}$ must be decisive.

By Positive Responsiveness, if a set O is decisive and $O \subseteq O'$, if the attack relations in $O' \setminus O$ agree with those in O, the result will be the same, and therefore O' becomes decisive too.

By Independence of Irrelevant Alternatives (IIA), if over a pair of arguments A, B, the attack relations remain the same then the aggregate attack relation will be the same over A, B. We will prove that this implies that $\cap \bar{\Omega} \neq \emptyset$. To start, let us start assuming that there are at least two decisive sets $O, W \in \bar{\Omega}$ such that $O \cap W = \emptyset$. Take any family of attack relations (they can always be constructed) $\{\rightarrow_i\}_{i=1}^n$ such that for each $i \in O$, $A \rightarrow_i B$, while for each $j \in W$, $B \rightarrow_j A$. By IIA, as long as these properties are preserved, the corresponding $\rightarrow_\mathcal{F}$ will be the same over A and B. But, from the decisiveness of O it follows that $A \rightarrow_\mathcal{F} B$ while by the decisiveness of W it must be that $B \rightarrow_\mathcal{F} A$. Contradiction. Then there must exist at least one common i in O and W. Furthermore, if $\cap \bar{\Omega} = \emptyset$, there is no $\bar{i}$ such that $\rightarrow_\mathcal{F} \subseteq \rightarrow_{\bar{i}}$. But $\rightarrow_\mathcal{F}$ includes other attacks than those in each individual attack relation. Without loss of generality, consider an extended argument framework over n arguments and a profile in which the attack relations over them is such that each one constitutes a linear chain of attacks:

$A^1 \rightarrow_1 A^2 \ldots \rightarrow_1 A^n$,
$A^2 \rightarrow_2 \ldots A^n \rightarrow_2 A^1$,
$\ldots$,
$A^n \rightarrow_n A^1 \ldots \rightarrow_n A^{n-1}$.

Then, over each pair A^j, A^k, $\rightarrow_\mathcal{F}$ has to coincide with some of the individual attack relations. In particular for each pair of arguments A^j, A^{j+1}. But also on A^n, A^1. Therefore, $\rightarrow_\mathcal{F}$ yields a cycle (see Example 2): $A^1 \rightarrow_\mathcal{F} A^2 \ldots \rightarrow_\mathcal{F} A^n \rightarrow_\mathcal{F} A^1$ But this contradicts the assumption that $\rightarrow_\mathcal{F}$ is acyclic. Then, $\cap \bar{\Omega} \neq \emptyset$

Finally, a dictator i_0 is such that $\{i_0\} \in \bar{\Omega}$. Therefore, non-dictatorship implies that there is no $O \in \bar{\Omega}$ such that $|O| = 1$.

$\Leftarrow$) The *Pareto condition* follows from the fact that $\{1,\ldots,n\} \in \bar{\Omega}$. That is, if for a given pair $A, B \in AR$, $A \rightarrow_i B$ for every $i = 1,\ldots,n$, since $\{1,\ldots,n\}$ is decisive, it follows that $A \rightarrow_\mathcal{F} B$.

Positive Responsiveness follows from the fact that if O is decisive and $O \subseteq O'$, O' is also decisive. This is so since, given any $A, B \in AR$, and two n-tuples of attack relations, $(\rightarrow_1,\ldots,\rightarrow_n)$, $(\rightarrow'_1,\ldots,\rightarrow'_n)$, if $\{i : A \rightarrow_i B\} \subseteq \{i : A \rightarrow'_i B\}$ and $A \rightarrow_\mathcal{F} B$, then $\{i : A \rightarrow_i B\}$ is decisive, and therefore $\{i : A \rightarrow'_i B\}$ is also decisive, and then $A \rightarrow'_\mathcal{F} B$.

Independence of Irrelevant Alternatives is obtained from the fact that $\cap \bar{\Omega} \neq \emptyset$. Suppose this were not the case. That is, consider a pair $A, B \in AR$, and two

n-tuples of attack relations, $(\to_1, \ldots, \to_n)$, $(\to'_1, \ldots, \to'_n)$, such that $\to_i = \to'_i$ over (A,B), but $\to_{\mathcal{F}} \neq \to'_{\mathcal{F}}$ over (A,B). Take $\bar{i} \in \cap \bar{\Omega} \neq \emptyset$. If $A \to_{\mathcal{F}} B$, consider a decisive O such that for every $i \in O$, $A \to_i B$, in particular $A \to_{\bar{i}} B$. On the other hand if $A \not\to'_{\mathcal{F}} B$, say $B \to'_{\mathcal{F}} A$, consider W such that for every $i \in W$, $B \to'_i A$, in particular $B \to'_{\bar{i}} A$, which contradicts that $\to_{\bar{i}} = \to'_{\bar{i}}$ (the same is true if we assume that A has no attack relation with B under $\to'_{\mathcal{F}}$).

Non-dictatorship follows from the fact that no set with a single criterion is decisive and therefore, no single attack relation can be imposed over the aggregate for every profile of attack relations.

Finally, notice that since there exists $\bar{i} \in \cap \bar{\Omega}$ over each pair of arguments A, B, $\to_{\mathcal{F}}$ either coincides with $\to_{\bar{i}}$ or $(A,B) \notin \to_{\mathcal{F}}$. Since $\to_{\bar{i}}$ has no cycles of attacks, $\to_{\mathcal{F}}$ will also be acyclic.

When $\bar{\Omega}$ satisfies the properties described in Proposition 3, we say that $\bar{\Omega}$ is a *prefilter* over $\{1, \ldots, n\}$ [12].[11] Moreover, if the class of decision sets for an aggregation function has this structure, it aggregates acyclic attack relations into an acyclic relation, satisfying Arrow's conditions.

Example 5. Over the set $\{\to_1, \to_2, \to_3\}$ of attack relations as given in Example 1 ($\{1,2,3\}$, for short), the only possible proper prefilters are:

$\bar{\Omega}^I = \{\{1,2\}, \{1,2,3\}\}$.

$\bar{\Omega}^{II} = \{\{1,3\}, \{1,2,3\}\}$.

$\bar{\Omega}^{III} = \{\{2,3\}, \{1,2,3\}\}$.

$\bar{\Omega}^{IV} = \{\{1,2\}, \{2,3\}, \{1,2,3\}\}$.

$\bar{\Omega}^V = \{\{1,3\}, \{2,3\}, \{1,2,3\}\}$.

$\bar{\Omega}^{VI} = \{\{1,2\}, \{1,3\}, \{1,2,3\}\}$.

Notice that the corresponding aggregation functions[12] $\mathcal{F}^I$, $\mathcal{F}^{II}$ and $\mathcal{F}^{III}$ are qualified voting functions. As it can be seen, there is a class of attacks that constitutes an "oligarchy": it participates in *all* the decisive sets.

To see how the other three functions act, just consider $\mathcal{F}^{IV}$ over

$$A \to_1 B \to_1 C$$

$$A \to_2 C \,,\, B \to_2 C$$

$$A \to_3 C \to_3 B$$

Then, $\to_{\mathcal{F}} = \mathcal{F}^{IV}(\to_1, \to_2, \to_3)$ is defined as follows:

---

[11] If there exists $O \in \bar{\Omega}$ such that $O = \cap \bar{\Omega}$, $\bar{\Omega}$ is called a *filter*. Otherwise, it is a *proper* prefilter.

[12] Those aggregation functions which are fully described by the decisive sets in the voting process are called *simple* [2].

$A \to_\mathcal{F} C$ since while there is no agreement in $\{1,2\}$, $\{2,3\}$ agree in that A attacks C.

$B \to_\mathcal{F} C$ since B attacks C in $\to_1$ and $\to_2$, although there is no agreement in $\{2,3\}$.

$(A,B),(B,A) \not\subseteq \to_\mathcal{F}$ since $(A,B),(B,A) \not\subseteq \to_2$ and $(A,B),(B,A) \not\subseteq \to_3$, i.e., there is agreement in $\{2,3\}$ although not in $\{1,2\}$.

That means that $\mathcal{F}^{IV}$ behaves like a majority function over the profile $(\to_1, \to_2, \to_3)$. The same conclusion can be drawn for $\mathcal{F}^V$ and $\mathcal{F}^{VI}$.

Notice that, $\mathcal{F}^{IV}$ is actually a plurality function only in the case that $\to_\mathcal{F}$ is acyclic. That is, it behaves like the plurality function in well-behaved cases. Instead, for the individual attack relations in Example 2 it yields an acyclic order $A \to_\mathcal{F} C \to_\mathcal{F} B$, which is *not* the outcome of the plurality function. Therefore, we should actually say that $\mathcal{F}^{IV}$, $\mathcal{F}^V$ and $\mathcal{F}^{VI}$ are *acyclic plurality* functions.

Notice also that any $\bar{i} \in \cap \bar{\Omega}$ acts as a "hidden dictator", in the sense made precise in the following result:

Proposition 4.
If $\bar{i} \in \cap \bar{\Omega}$, and $\mathcal{F}_{\bar{\Omega}}$ is the aggregation function characterized by the prefilter $\bar{\Omega}$ then $\to_{\mathcal{F}_\Omega} \subseteq \to_{\bar{i}}$.

Proof. Suppose that given $A, B \in AR$, we have, without loss of generality, that $A \to_{\bar{i}} B$. Let us consider two cases:

- There exists a decisive set $O \in \bar{\Omega}$ such that for every $i \in O$ (by definition $\bar{i} \in O$), $A \to_i B$. Then $A \to_{\mathcal{F}_\Omega} B$, and therefore $\to_{\mathcal{F}_\Omega}$ coincides with $\to_{\bar{i}}$ over (A,B).

- There does not exist any decisive set O in which for every $i \in O$, $A \to_i B$. Then, neither $A \to_{\mathcal{F}_\Omega} B$ nor $B \to_{\mathcal{F}_\Omega} A$ can be obtained. Therefore $\bar{i}$ *vetoes* $B \to_i A$, although it cannot imposes $A \to B$. In this case $\to_{\mathcal{F}_\Omega} \subset \to_{\bar{i}}$ over (A,B).

5 Aggregation and Extensions of Attack Relations

The analysis of argumentation systems is usually carried out in terms of their *extensions*. The existence and properties of the extensions can be ascertained according to the properties of the attack relation. In the case that several alternative attack relations compete over the same class of arguments, the class of extensions may vary from one to another. The structure of extensions of such an argumentation system should not be seen as just the enumeration of the classes corresponding to each attack relation but should arise from the same aggregation process we have discussed previously. That is, it should follow from the properties of the aggregate attack relation.

In particular, since our main results concern the aggregation of acyclic attack relations into an acyclic aggregate one, we will focus on the case of well-founded argumentation frameworks (*cf.* [19], p. 10). We can say, roughly, that their main feature is the absence of cycles of attack among their arguments. For them, all the types of extensions described by Dung coincide. Furthermore, they all yield a single set of arguments (*cf.* [19], theorem 30, p. 331).

To see how such a single extension of an argument system over a family of individual attack relations may obtain, let us recall that if for each i, $\rightarrow_i$ has no cycles of attack, an aggregate relation $\rightarrow_{\mathcal{F}_\Omega}$, obtained through an aggregation function $\mathcal{F}$ with a prefilter of decisive sets $\bar{\Omega}$, is acyclic as well. The following result is an immediate consequence of this claim.

Proposition 5.
Consider an aggregate argument framework $AF^* = \langle AR; \mathcal{F}(\rightarrow_1, \ldots, \rightarrow_n)\rangle$. If each $\rightarrow_i$ ($i = 1, \ldots, n$) is acyclic and $\mathcal{F}$ is such that its corresponding class of decisive sets $\bar{\Omega}$ is a prefilter, then $\rightarrow_\mathcal{F} = \mathcal{F}(\rightarrow_1, \ldots, \rightarrow_n)$ is acyclic and AF^* has a single extension which is grounded, preferred and stable.

Furthermore (notice the single extension assumption):

Corollary 1.
If $AF^* = \langle AR; \mathcal{F}(\rightarrow_1, \ldots, \rightarrow_n)\rangle$ has a single extension when each $\rightarrow_i$ ($i = 1, \ldots, n$) is acyclic, then if $\mathcal{F}$ is such that its corresponding class of decisive sets $\bar{\Omega}$ is a prefilter, it also verifies the Pareto condition, Positive Responsiveness, Independence of Irrelevant Alternatives, and Non-Dictatorship.

Proof. Immediate. If AF^* has a single extension and $\mathcal{F}$ is such that its corresponding class of decisive sets $\bar{\Omega}$ is a prefilter, then by Proposition 5 the aggregate attack $\rightarrow_\mathcal{F}$ is acyclic. Then, the claim follows from Proposition 3.

Another question that arises from our analysis of the semantics of aggregate frameworks is whether there exists a sensible notion of aggregation of extensions that could correspond to the aggregation of attack criteria. In this paper we have focused on the path that goes from several attack relations to a single aggregate one and from it to its corresponding extension. An alternative would be to go from several attack relations to their corresponding extensions and from there on to a single extension. If, furthermore, these two alternative paths commute (in category-theoretic terms), the aggregation of attack relations would be preferable in applications since it is simpler to aggregate orderings than families of sets.

A natural alternative would be to start with the prefilter that underlies the aggregate attack relation and try to apply the decisive sets on the individual extensions to obtain a single one. That is:

Aggregation of extensions. Consider an aggregate argument framework $AF^* = \langle AR; \mathcal{F}(\rightarrow_1, \ldots, \rightarrow_n)\rangle$, in which each $\rightarrow_i$ ($i = 1, \ldots, n$) is acyclic.

Therefore, for each i, $\to_i$ yields a single extension (which is grounded, preferred, stable and complete, E^i. Given a prefilter $\bar{\Omega}$, an *aggregate extension* $E^{\mathcal{F}}$ is such that $A \in E^{\mathcal{F}}$ iff there exists $O \in \bar{\Omega}$ such that for every $i \in O$, $A \in E^i$.

Notice that this characterization of $E^{\mathcal{F}}$ is analogous to that of $\to_{\mathcal{F}}$. That is, $(A, B) \in \to_{\mathcal{F}}$ iff there exists $O \in \bar{\Omega}$ such that for every $i \in O$, $(A, B) \in \to_i$.

If we call $E^{\to_{\mathcal{F}}}$ the extension of $\to_{\mathcal{F}}$, the commutation of prefilter-aggregation paths would imply that $E^{\to_{\mathcal{F}}} = E^{\mathcal{F}}$. The fact is that this *aggregation-extension* commutation fails even in very simple settings:

Example 6.
Consider four attack relations each yielding an extension (either grounded, preferred, stable or complete) as follows:

$\to_1 = \{(A, B), (B, C)\}$, *with extension* $E^1 = \{A, C\}$,

$\to_2 = \to_3 = \{(B, C)\}$, *with* $E^2 = E^3 = \{A, B\}$,

$\to_4 = \emptyset$, *with* $E^4 = \{A, B, C\}$.

If $\bar{\Omega} = \{\{1, 2\}, \{1, 2, 3\}, \{1, 2, 4\}, \{1, 2, 3, 4\}\}$ *we have that* $\to_{\mathcal{F}} = \{(B, C)\}$ *and its extension is* $E^{\to_{\mathcal{F}}} = \{A, B\}$. *But the aggregate extension is* $E^{\mathcal{F}} = \{A\}$.

The discrepancy between $E^{\to_{\mathcal{F}}}$ and $E^{\mathcal{F}}$ exhibited in this example shows clearly that *aggregation* cannot be expected to commute with *extension*. The reason is that the process of obtaining the extension of an acyclic attack relation involves a loss of information.[13] Furthermore, voting over arguments in extensions adds another layer of information loss, since it totally disregards the relations among the arguments. Then, it is natural to realize that the set of arguments obtained by aggregating individual extensions will differ from the extension of $\to_{\mathcal{F}}$.

Finally, let us note that it is not clear whether the more sensible way to proceed is to aggregate first the attack relation and then to obtain the extension of the aggregate attack relation or the other way around:

Example 7.
Consider three attack relations:

$\to_1 = \{(A, D), (D, G), (A, B), (A, C)\}$, *with extension* $E^1 = \{A, G\}$,

$\to_2 = \{(B, D), (D, G), (B, A), (B, C)\}$, *with* $E^2 = \{B, G\}$,

$\to_3 = \{(C, D), (D, G), (C, B), (C, A)\}$, *with* $E^3 = \{C, G\}$.

If $\bar{\Omega} = \{\{1, 2\}, \{2, 3\}, \{1, 2, 3\}\}$ *we have that* $\to_{\mathcal{F}} = \{(D, G)\}$ *and the corresponding extension is* $E^{\to_{\mathcal{F}}} = \{A, B, C, D\}$. *But the aggregate extension is* $E^{\mathcal{F}} = \{G\}$. *Here we have that* D, *which is deemed defeated according to each individual extension, is warranted by* $E^{\to_{\mathcal{F}}}$, *while* G, *which is warranted in all the individual extensions is excluded from* $E^{\to_{\mathcal{F}}}$.

---

[13] The same extension may arise for different attack relations.

This example reminds the *floating argument* problem of argumentation systems. The floating argument here would be G, since it "floats" over the status of acceptance of A, B and C. Some authors would argue that G should be warranted not because it is defended by an acceptable argument, but because it is defended by a set of arguments $\{A, B, C\}$ which, although somehow uncertain (on which argument in the set will be chosen), eliminates the attacks on G [24]. If so, this claim may suggest that the process of applying *simple* (i.e. decision set-based) aggregation procedures may work better on extensions than on the attack relations themselves, at least for cases like this one. But this is matter of a more philosophical inquiry, namely on which arguments should be deemed warranted in controversial cases like this one.

6 Discussion

As indicated by Brown in [11], the fact that the class of decisive sets constitutes a prefilter is an indication of the existence of a *collegium*, a "shadow" decisive set, being its members interspersed among all the actual decisive sets. Their actual power comes not from being able to enforce outcomes but from their ability to veto alternatives that are not desirable for them. In the current application, the existence of a collegium means that there exists a class of attack relations that by themselves cannot determine the resulting attack relation, but can instead block (veto) alternatives.

This feature still leaves many possibilities open, but as our examples intended to show, there are few aggregation functions that may adopt this form, while at the same time satisfying Arrow's conditions. The main instance is constituted by the acyclic plurality function, but qualified voting functions yield also fair outcomes. The difference is that with the acyclic plurality functions one of the several attack relations in AF^n is selected, while with qualified voting functions, new attack relations may arise. But these new attack relations just combine those of the winning coalitions, and therefore can be seen as resulting from the application of generalized variants of the plurality function. That is, if two or more rules belong to all the decisive sets, their common fragments plus the non-conflicting ones add up to constitute the aggregate attack relation. In a way or another, the attack relations that are always decisive end up acting as hidden dictators in the aggregation process.

Explicit dictators arise in aggregation processes in other branches of non-monotonic reasoning. Doyle and Wellman [18], in particular, suggested to translate Reiter's defaults into total preorders of autoepistemic formulas, representing preferences over worlds. Reasoning with different defaults implies to find, first, the aggregation of the different preorders. These authors show that it is an immediate consequence of Arrow's theorem that no aggregation function can fulfill all the properties that characterize fairness (i.e. the equivalents of the Pareto condition, Positive Responsiveness, Independence of Irrelevant Alternatives, and Non-Dictatorship). In terms of decisive sets, it means that $\hat{\Omega}$ constitutes a *principal ultrafilter*.[14] Since the number of defaults is assumed to be finite, it follows that there exists one of these default rules, say R^* that belongs to each $U \in \hat{\Omega}$. Of course, the existence of R^* violates the Non-Dictatorship condition, and consequently the actual class of formulas that arise in the aggregation are determined by R^*.

It can be said that Doyle and Wellman's analysis is concerned with the *generation* of a class of arguments arising from different default rules while we, instead, concentrate on the *comparison* among arguments in an abstract argumentation framework. But Dung [19] has shown that Reiter's system can be rewritten as an argumentation framework, and therefore both approaches can be made compatible. In this sense, Doyle and Wellman's result can be now interpreted as indicating that $\bar{\Omega}$ (over attack relations) is **not** a *proper prefilter* over $\{1,\ldots,n\}$ (where these indexes range over the attack relations determined each by a corresponding default rule). Instead, as said, it constitutes a *principal ultrafilter* and therefore it implies the existence of a "dictatorial" attack relation, that is imposed over the framework.

The approach most related to ours is Coste-Marquis et al. [16], which already presented some early ideas on how to merge Dung's argumentation frameworks. The aim of these authors is to find a set of arguments collectively warranted, addressing two main problems. One is the individual problem faced by each agent while considering a set of arguments different to that of other agents. The other is the aggregate problem of getting the collectively supported extension. The second one is the most clearly related to our approach, but differs in that the authors postulate a specific way of merging the individual frameworks.

Other relevant contributions are [3] and [33], in which the aggregation is carried out on labellings of attack relations. Another contribution, based on the use of labellings is [31], in which strategic behavior is taken into account with the goal of obtaining an aggregate framework that cannot be manipulated by the users. Other contributions are surveyed in [9].

7 Further Work

A relevant question arises from our analysis of the semantics of aggregate frameworks when the attack relations are acyclic. Namely, whether there exists a sensible notion of aggregation of extensions that could correspond to the aggregation of attack criteria. In this paper we have focused on the path that goes from several attack relations to a single aggregate one and from it to its corresponding extension.

An interesting question is whether it is possible to find other aggregation procedures, under further constraints than Arrow's conditions, that may lead to the commutation of aggregation paths. That is, ensuring the equivalence of results that obtain by aggregating first the attack relations with those that

---

[14] Notice that $\hat{\Omega}$ denotes the decisive set over *default rules* and therefore should not be confused with $\bar{\Omega}$, the class of decisive sets over *attack relations*.

get from the aggregation of extensions. In this respect, the problem resembles the "discursive dilemma" in which several judgments (i.e. pairs of the form ⟨*premises, conclusion*⟩), are aggregated componentwise, that is, a pair formed by an aggregate premises set and an aggregate conclusion is obtained, but it does not constitute an acceptable judgment. Nevertheless, there are reasons to be pessimistic. It has been proven ([8]) that in argumentation, both aggregation paths, i.e. through the aggregation of attack relations and by the aggregation of extensions, commute only in very restricted settings. The search for other sensible constraints under which both aggregation procedures might coincide is matter of further work.

Pigozzi [29] postulates a solution to the discursive dilemma based on the use of operators for merging belief bases in AI [25]. To pose a *merge* operation as an *aggregation* one involves to incorporate a series of trade-offs among the several alternatives that hardly will respect Arrow's conditions, as it is well known in the literature on political systems (see [2]). In relation to this issue, a wider point that we plan to address is to systematize the *non*-fair aggregation procedures that could be applied to the aggregation of attack relations in argument frameworks. The idea would be to lesser the demands on the aggregation function and to see which features arise in the aggregate. It seems sensible to think that depending on the goals of the aggregation process, one or another function should be chosen.

Another question to investigate is the connections between the correspondence of the aggregate attack relation among arguments and the relation of *dominance* among alternatives ([10]). Alternative A *dominates* alternative B iff the number of individuals for which A is preferred to B is larger than the number of individuals for which B is preferred to A. This implies that the dominance relation is asymmetric. Although it is not commonly assumed in the literature that attack relations are (strictly) asymmetric, it follows from our definition of majority voting over pairs of arguments that the resulting attack relations will have this property (even when cycles of order > 2 may occur). Dominance relations lead to the choice of stable sets; notice that aggregate attacks obtained by means of prefilters differ from *tournaments*, in which the dominance relations are also complete. A stable set is such that none of its elements dominates another, and every alternative outside the set is dominated by some of its elements. The correspondence between stable semantics in argumentation frameworks and stable sets was previously studied by Dung ([19]). It is natural, so, to inquire about the relationship between our majoritarian voting aggregation mechanism on attack relations and stable sets in argumentation frameworks.

References

1. Kenneth J. Arrow. *Social Choice and Individual Values*. Yale University Press, 1970.

2. David Austen-Smith and Jeffrey S. Banks. *Positive Political Theory I: Collective Preference*. University of Michigan Press, 2000.
3. Edmond Awad, Richard Booth, Fernando A. Tohmé, and Iyad Rahwan. Judgment aggregation in multi-agent argumentation. *Journal of Logic and Computation*, 27:227 – 259, 2017.
4. Trevor J. M. Bench-Capon and Paul E. Dunne. Argumentation in artificial intelligence. *Artificial Intelligence*, 171(10-15):619–641, 2007.
5. Philippe Besnard and Anthony Hunter. *Elements of Argumentation*. MIT Press, Cambridge, MA, 2008.
6. Duncan Black. *The Theory of Committees and Elections*. Cambridge University Press, 1958.
7. Douglas Blair and Robert Pollack. Acyclic collective choice rules. *Econometrica*, 50:931–943, 1982.
8. Gustavo Bodanza and Marcelo Auday. Social argument justification: Some mechanisms and conditions for their coincidence. In Claudio Sossai and Gaetano Chemello, editors, *Symbolic and Quantitative Approaches to Reasoning with Uncertainty, 10th European Conference, ECSQARU*, pages 95–106. Springer Verlag (LNCS vol. 5590), 2009.
9. Gustavo Bodanza, Fernando Tohmé, and Marcelo Auday. Collective argumentation: A survey of aggregation issues around argumentation frameworks. *Argument and Computation*, 8:1–34, 2017.
10. Felix Brandt, Felix Fischer, and Paul Harrenstein. The computational complexity of choice sets. In Dov Samet, editor, *Proceedings of the 11th Conference on Theoretical Aspects of Rationality and Knowledge (TARK)*, pages 82–91. Presses Universitaires de Louvain, 2007.
11. Donald J. Brown. An approximate solution to arrow's problem. *Journal of Economic Theory*, 9:375–383, 1974.
12. Donald J. Brown. Aggregation of preferences. *Quarterly Journal of Economics*, 89:456–469, 1975.
13. Martin Caminada. Semi-stable semantics. In Paul Dunne and Trevor Bench-Capon, editors, *Proceedings of the 1st International Conference on Computational Models of Argument (COMMA)*, pages 121–130. IOS Press, 2006.
14. Martin Caminada and Gabriella Pigozzi. On judgment aggregation in abstract argumentation. *Autonomous Agents and Multi-Agent Systems*, 1(22):64–102, 2011.
15. Carlos Chesñevar, Ana Maguitman, and Ronald P. Loui. Logical models of argument. *ACM Computing Surveys*, 32:337–383, 2000.
16. Sylvie Coste-Marquis, Caroline Devred, Sbastien Konieczny, Marie-Christine Lagasquie-Schiex, and Pierre Marquis. On the merging of dung's argumentation systems. 171:730–753, 07 2007.
17. Sylvie Coste-Marquis, Caroline Devred, Sébastien Konieczny, Marie-Christine Lagasquie-Schiex, and Pierre Marquis. Merging argumentation systems. In Manuela Veloso and Subbarao Kambhampati, editors, *Proceedings of the 20th National Conference on Artificial Intelligence and the 17th Innovative Applications of Artificial Intelligence Conference*, pages 614–619. AAAI Press, 2005.
18. John Doyle and Michael Wellman. Impediments to universal preference-based default theories. *Artificial Intelligence*, 49:97–128, 1991.
19. Phan Minh Dung. On the acceptability of arguments and its fundamental role in nonmonotonic reasoning, logic programming and n-person games. *Artificial Intelligence*, 77:321–358, 1995.

20. Jon (ed.) Elster. *Deliberative democracy*. Cambridge University Press, Cambridge, 1998.
21. Ulle Endriss and Umberto Grandi. Graph aggregation. *Artificial Intelligence*, 245:86–114, 2017.
22. Alejandro J. García and Guillermo R. Simari. Defeasible logic programming: An argumentative approach. *Theory and Practice of Logic Programming (TPLP)*, 4:95–138, 2004.
23. Alejandro J García and Guillermo Ricardo Simari. Defeasible logic programming: Delp-servers, contextual queries, and explanations for answers. *Argument & Computation*, 2014.
24. Hadassa Jakobovits and Dirk Vermeir. Robust semantics for argumentation frameworks. *Journal of Logic and Computation*, 9:215–261, 1999.
25. Sebastien Konieczny and Pino-Pérez. Propositional belief base merging or how to merge beliefs/goals coming from several sources and some links with social choice theory. *European Journal of Operational Research*, 160:785–802, 2005.
26. Christian List and Philip Pettit. Aggregating sets of judgments. two impossibility results compared. *Synthese*, 140:207–235, 2004.
27. Ronald P. Loui. Defeat among arguments: a system of defeasible inference. *Computational Intelligence*, 3:100–106, 1987.
28. Hervé Moulin. Social choice. In Robert Aumann and Sergiu Hart, editors, *Handbook of Game Theory, vol. 2*. North-Holland, 1994.
29. Gabriella Pigozzi. Belief merging and the discursive dilemma: An argument-based account to paradoxes of judgment aggregation. *Synthese*, 152:285–298, 2006.
30. Henry Prakken and Gerard Vreeswijk. Logical systems for defeasible argumentation. In Dov Gabbay, editor, *Handbook of Philosophical Logic, 2nd ed.* Kluwer Academic Publishers, 2002.
31. Iyad Rahwan, Kate Larson, and Fernando Tohmé. A characterisation of strategy-proofness for grounded argumentation semantics. In *Proceedings IJCAI 2009*, pages 251–256. IJCAI Organization, 2009.
32. Iyad Rahwan and Guillermo Ricardo Simari. *Argumentation in Artificial Intelligence*. Springer, Heidelberg, Germany, 2009.
33. Iyad Rahwan and Fernando Tohmé. Collective argument evaluation as judgement aggregation. In Wiebe van der Hoek; Gal Kaminka; Yves Lespérance; Michael Luck; Sandip Sen, editor, *Proceedings of the Ninth International Conference on Autonomous Agents and Multi-Agent Systems (AAMAS 2010)*, pages 417–424. IFAAMAS, 2010.
34. Norman Schofield. *The Spatial Model of Politics*. Routledge, 2008.
35. Guillermo R. Simari and Ronald P. Loui. A mathematical treatment of defeasible reasoning and its implementation. *Artificial Intelligence*, 53:125–157, 1992.
36. Fernando Tohmé, Gustavo Bodanza, and Guillermo Simari. Aggregation of attack relations: A social-choice theoretical analysis of defeasibility criteria. In Sven Hartmann and Gabriele Kern-Isberner, editors, *Foundations of Information and Knowledge Systems, FoIKS 2008*, pages 8–23. Springer Verlag (LNCS vol. 4932), 2008.
37. Ignacio Darío Viglizzo, Fernando A. Tohmé, and Guillermo Ricardo Simari. The foundations of deLP: defeating relations, games and truth values. *Annal of Mathematics and Artificial Intelligence*, 57(2):181–204, 2009.

Structure-based Semantics of Argumentation Frameworks with Higher-order Attacks and Supports*

Claudette Cayrol, Jorge Fandinno,
Luis Farinas del Cerro, and M-Christine Lagasquie-Schiex

IRIT, Université de Toulouse, CNRS, France
{ccayrol,jorge.fandinno,luis.farinas,lagasq}@irit.fr

Abstract. In this paper, we propose a generalisation of Dung's abstract argumentation framework that allows representing higher-order attacks and supports, that is attacks or supports whose targets are other attacks or supports. We follow the necessary interpretation of the support, based on the intuition that the acceptance of an argument requires the acceptance of each supporter. We propose semantics accounting for acceptability of arguments and validity of interactions, where the standard notion of extension is replaced by a triple of a set of arguments, a set of attacks and a set of supports. Our framework is a conservative generalisation of Argumentation Framework with Necessity (AFN). When supports are ignored, Argumentation Framework with Recursive Attacks (RAF) is recovered. Moreover, correspondences with Argumentation Framework with Recursive Attack and Support (ASAF) proposed by Simari and his colleagues are outlined.

Keywords: Abstract argumentation · bipolar argumentation · higher-order interactions.

1 Introduction

Abstract argumentation frameworks have greatly eased the modelling and study of argumentation. Whereas Dung's framework [12] only accounts for an attack relation between arguments, two natural generalisations have been developed in order to allow positive interactions (usually expressed by a support relation) and higher-order interactions (attacks or supports that target other attacks or supports). Here is an example in the legal field, borrowed from [1] that illustrates both generalisations (this example corresponds to a dynamic process of exchange of pieces of information, each one being considered as an "argument").

---

* The second author is funded by the Centre International de Mathmatiques et dInformatique de Toulouse (CIMI) through contract ANR-11-LABEX-0040-CIMI within the program ANR-11-IDEX-0002-02.

Ex. 1 *The prosecutor says that the defendant has intention to kill the victim (argument b). A witness says that she saw the defendant throwing a sharp knife towards the victim (argument a). Argument a can be considered as a support for argument b. The lawyer argues back that the defendant was in a habit of throwing the knife at his wife's foot once drunk. This latter argument (argument c) is better considered* attacking the support *from a to b, than arguments a or b themselves. Now the prosecutor's argumentation seems no longer sufficient for proving the intention to kill.*

Different interpretations for the notion of support were proposed: deductive support [3], evidential support [17], necessary support [15], that are compared in [7, 9]. Recent works have focused on the necessary interpretation, for instance in Argumentation Frameworks with Necessities (AFN) [16], and in [10, 11, 4]. In [18], correspondences are provided between a framework with evidential support and an AFN. In evidential argumentation standard arguments need to be supported by special (called prima-facie) arguments in order to be considered as acceptable. So arguments need to be able to trace back to prima-facie arguments. With the necessary interpretation of support as in AFN, arguments need to be able to trace back to arguments that require no support in order to be considered as acceptable.

It is worth to note that [7, 16, 18] do not allow the representation of higher-order interactions. In contrast, higher-order attacks and higher-order necessary supports have been considered in [10, 11, 4], with different ways for defining acceptability semantics: a translation into a standard Dung's AF [10], meta-argumentation techniques [4], a direct characterization of extension-based acceptability semantics [11].

Very recently, a new framework has been proposed that allows representing higher-order attacks and higher-order evidential supports [6]. In this framework, called Recursive Evidence Based Argumentation Framework (REBAF), the semantics handle both acceptability of arguments and validity of interactions (attacks or supports), and account for the fact that acceptability of arguments may depend on the validity of interactions and vice-versa. As a consequence, the standard notion of extension is replaced by a triple of a set of arguments, a set of attacks and a set of supports, called a structure.

In this contribution,[1] our purpose is to propose a Recursive Argumentation Framework with Necessity (RAFN) where its semantics accounts for acceptability of arguments and validity of interactions, in the case of higher-order attacks and higher-order necessary support. Moreover, we are interested in a conservative generalisation of AFN. Taking advantage of the correspondences that have been established between evidential and necessary support in [18], our methodology and definitions draw on the REBAF of [6].

---

[1] This contribution is an extended version of a paper published in COMMA'2018 conference (same title). The additional material consists of examples, proofs and a comparison with the Recursive Evidence-Based Argumentation Framework (REBAF).

The paper is organized as follows: Section 2 gives some background about necessary support and about the REBAF; the definition and semantics for the RAFN are proposed in Section 3; in Section 4 we prove a one-to-one correspondence with AFN in the case of first-order interactions, and we give a comparison with recent work about recursive attacks and supports [11]; and we conclude in Section 5. Proofs are given in Appendix A and a synthesis of RAFN examples can be found in Appendix B.

2 Background

We next review some basic background about the works the paper is based on: first, an abstract argumentation framework handling first-order necessary supports (AFN), and another one handling first-order evidential supports (EBAF); then two approaches dealing with higher-order attacks and higher-order supports, one for necessary supports (ASAF) and a recent approach for evidential supports (REBAF). We also recall the existing links between AFN and EBAF.

2.1 First-order necessary support (AFN)

Binary necessary support was initially introduced in [15], then discussed in [10, 11, 4] in a more general context (particularly with higher-order interactions).

Let a and b be two arguments, "a necessarily supports b" means that the acceptance of a is necessary to get the acceptance of b, or equivalently that the acceptance of b implies the acceptance of a.

Necessary support has been extended to express the fact that a given argument requires at least one element among a set of arguments. In [16], an Argumentation Framework with Necessities (AFN) is defined as follows:

Def. 1 (AFN [16]) *An* Argumentation Framework with Necessities (AFN) *is a tuple* $\langle \mathbf{A}, \mathbf{R}, \mathbf{N} \rangle$, *where* $\mathbf{A}$ *is a finite and non-empty set of arguments,* $\mathbf{R} \subseteq \mathbf{A} \times \mathbf{A}$ *represents the* attack *relation and* $\mathbf{N} \subseteq (2^{\mathbf{A}} \setminus \varnothing) \times \mathbf{A}$ *represents the* necessity *relation.*

For $E \subseteq \mathbf{A}$, $E\mathbf{N}b$ reads "E is a necessary support for b", which means that if no argument of E is accepted then b cannot be accepted, or equivalently that the acceptance of b requires the acceptance of at least one element of E. Moreover, in AFN semantics, acyclicity of the support relation is required among accepted arguments. Intuitively, in a given extension, support for each argument is provided by at least one of its necessary arguments and there is no risk of a deadlock due to necessity cycles. These requirements have been formalized in [16] and can be reformulated as follows:

Def. 2 (Semantics in AFN) *Given* $AFN = \langle \mathbf{A}, \mathbf{R}, \mathbf{N} \rangle$ *and* $T \subseteq \mathbf{A}$.
- T *is* support-closed *iff for each* $a \in T$, *if* $E\mathbf{N}a$, *then* $E \cap T \neq \varnothing$.

- *Assume that T is support-closed. $a \in T$ is* support-cycle-free *in T iff $\forall E \subseteq \mathbf{A}$ such that $E\mathbf{N}a$, there is $b \in E \cap T$ such that b is support-cycle-free in $T \setminus \{a\}$.*
- *T is* coherent *iff T is support-closed and every $a \in T$ is support-cycle-free in T.*
- *$a \in \mathbf{A}$ is* deactivated *by T iff $\forall C \subseteq \mathbf{A}$ coherent subset containing a, $T\mathbf{R}C$ (i.e. there is $x \in T$ and $c \in C$ such that $x\mathbf{R}c$).*
- *$a \in \mathbf{A}$ is* acceptable *w.r.t. T iff (i) $T \cup \{a\}$ is coherent and (ii) $\forall b \in \mathbf{A}$ such that $b\mathbf{R}a$, b is deactivated by T.*
- *T is* admissible *iff T is conflict-free, coherent, and every a in T is acceptable w.r.t. T.*
- *T is a* complete *extension iff T is admissible and $\forall a \in \mathbf{A}$, if a is acceptable w.r.t. T, then $a \in T$.*
- *T is a* preferred *extension iff T is a $\subseteq$-maximal complete extension.*
- *T is a* stable *extension iff T is complete and $\forall a \in \mathbf{A}$, $a \in \mathbf{A} \setminus T$ iff a is deactivated by T.*
- *T is a* grounded *extension iff T is a $\subseteq$-minimal complete extension.*

The following examples illustrate these semantics.

Ex. 2 *Consider the framework representing an attack from a to b and no necessary support.*

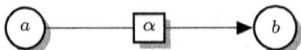

(arguments are in a simple circle, attack names in a square, and attacks are represented with simple arrows)

The unique extension (preferred, stable and grounded) is $\{a\}$. Indeed, the AFN framework is a conservative generalisation of Dung's framework.

Ex. 3 *Consider the framework representing a necessary support from $\{a\}$ to b and no attack.*

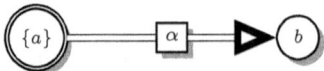

(arguments are in a simple circle and sets of arguments in a double circle, support names are in a square, and supports are represented with double arrows)

$\{\}$ and $\{a\}$ are two admissible extensions. However, due to the necessary support, an admissible set containing b must also contain a. So, $\{b\}$ is not admissible, and the unique complete extension is $\{a,b\}$.

Ex. 4 *Consider the framework representing a cycle of necessary supports between a and b, and no attack. This cycle is represented by $\{a\}\mathbf{N}b$ and $\{b\}\mathbf{N}a$.*

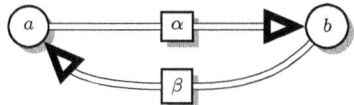

(here, by abuse of notation, since the source of the support is a singleton, it can be represented by a argument – so, by a simple circle –)

There is no non-empty admissible set. Indeed, there is no way to trace back with a chain of supports from a (resp. b) to arguments that require no support.

Ex. 5 Consider the framework representing k necessary supports to a: $a \in \mathbf{A}$, $X_1, \ldots, X_k$ are non-empty subsets of $\mathbf{A}$ such that $X_i \mathbf{N} a, i = 1k$.

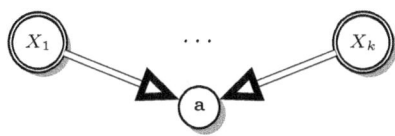

Let E be an admissible set containing a. Then $\forall i = 1k$, at least one argument of X_i must belong to E.

2.2 First-order evidential framework (EBAF)

The evidential understanding of the support relation [18] allows to distinguish between two different kinds of arguments: *prima-facie* and *standard arguments*. *Prima-facie* arguments are justified whenever they are not defeated. On the other hand, *standard arguments* are not directly assumed to be justified and must inherit support from prima-facie arguments through a chain of supports.

Let us recall the formal definition of Evidence-Based Argumentation frameworks [18].[2]

Def. 3 (EBAF) *An* Evidence-Based Argumentation Framework (EBAF) *is a tuple $\langle \mathbf{A}, \mathbf{R}, \mathbf{E} \rangle$, where $\mathbf{A}$ is a finite and non-empty set of arguments, $\mathbf{R} \subseteq (2^{\mathbf{A}} \setminus \varnothing) \times \mathbf{A}$ represents the* attack relation *and $\mathbf{E} \subseteq (2^{\mathbf{A}} \setminus \varnothing) \times \mathbf{A}$ represents the* evidential support relation. *A special argument $\eta \in \mathbf{A}$ is distinguished satisfying that there is no $(B, \eta) \in \mathbf{R} \cup \mathbf{E}$ for any set B nor there is $(B, a) \in \mathbf{R}$ with $\eta \in B$.*

The special argument η serves as a representation of the prima-facie arguments.

Note that the attack relation is not a binary relation. For $X \subseteq \mathbf{A}$, $X\mathbf{R}b$ reads as follows: If all the arguments of X are accepted, then b cannot be accepted.

---

[2] The reader is referred to [18] for the definition of EBAF semantics.

For $X \subseteq \mathbf{A}$, $X\mathbf{E}b$ reads as follows: X "is an evidential support" for b, which means that the acceptance of b requires the acceptance of *all the elements of* X. Moreover, accepted arguments need to trace back to the special argument η.

Ex. 4 (cont'd) *With the evidential understanding of the support, in the associated EBAF, the cycle of support is represented by $\{a\}\mathbf{E}b$ and $\{b\}\mathbf{E}a$.*
- *If $\{\eta\}\mathbf{E}a$ and not $\{\eta\}\mathbf{E}b$, then the admissible sets are $\varnothing$, $\{\eta\}$, $\{\eta,a\}$ and $\{\eta,a,b\}$. Note that $\{\eta,b\}$ is not admissible since b requires support from a. The opposite case ($\{\eta\}\mathbf{E}b$ and not $\{\eta\}\mathbf{E}a$) is similar.*
- *If $\{\eta\}\mathbf{E}a$ and $\{\eta\}\mathbf{E}b$, then the admissible sets are $\varnothing$, $\{\eta\}$, $\{\eta,a\}$, $\{\eta,b\}$, and $\{\eta,a,b\}$. Indeed the support is useless.*
- *Otherwise (there is no prime-facie argument), $\{\eta\}$ is the unique non-empty admissible set.*

Ex. 5 (cont'd) *Let E be an admissible set containing a. With the evidential understanding of the support, in the associated EBAF, then $\exists i = 1k$ such that X_i is included in E.*

2.3 Turning an AFN into an EBAF

In [18], correspondences are provided between a framework with evidential support and an AFN, that preserve the admissible, preferred, grounded, complete and stable semantics. We recall here the main ideas on which these correspondences are based.

First, the difference between necessary support and evidential support is highlighted by Ex.5. If $X_1, \ldots, X_k$ are subsets of $\mathbf{A}$ supporting a in an AFN, checking that an admissible subset S contains a requires to check whether $\forall i = 1k, \exists b_i \in X_i, b_i \in S$. In contrast, if $X_1, \ldots, X_k$ are subsets of $\mathbf{A}$ supporting a in an EBAF, checking that an admissible subset S contains a requires to check whether $\exists i = 1k$ such that $\forall b_i \in X_i, b_i \in S$.

Secondly, in EBAFs, arguments need to be able to trace back to primafacie arguments (represented as η), while in AFNs arguments need to be able to trace back to arguments that require no support (acyclicity principle). So unsupported arguments in AFNs must correspond to arguments supported by η in EBAFs.

The following translation from a given AFN into an EBAF is given in [18]:

Def. 4 (From AFN to EBAF) *Let $\langle \mathbf{A}, \mathbf{R}, \mathbf{N} \rangle$ be an AFN. The corresponding EBAF $\langle \mathbf{A}', \mathbf{R}', \mathbf{E} \rangle$ is defined by:*
- $\mathbf{A}' = \mathbf{A} \cup \{\eta\}$
- *For $(a,b) \in \mathbf{R}$, put $(\{a\},b)$ in $\mathbf{R}'$*
- *Let $a \in \mathbf{A}$ and $X = \{X_1, \ldots, X_k\}$ be the collection of all sets X_i such that $X_i\mathbf{N}a$. If X is empty, add $(\{\eta\},a)$ to $\mathbf{E}$. Otherwise, for all $X' \in (X_1 \times \ldots \times X_k)$ add (X'_s, a) to $\mathbf{E}$, where X'_s denotes the set of all elements in X'.*

It has been proved in [18] that $T \subseteq \mathbf{A}$ is a σ-extension of an AFN iff $T \cup \{\eta\}$ is a σ-extension of the associated EBAF, for $\sigma \in \{admissible, preferred, complete, stable\}$. The following examples illustrate this translation.

Ex. 6 Let $AFN = \langle\{a,b,c\}, \varnothing, \{(\{b\},a),(\{c\},a)\}\rangle$. The EBAF version of AFN is $\langle\{\eta,a,b,c\}, \varnothing, \{(\{\eta\},b), (\{\eta\},c), (\{b,c\},a)\}\rangle$ and represented in Figure 1.

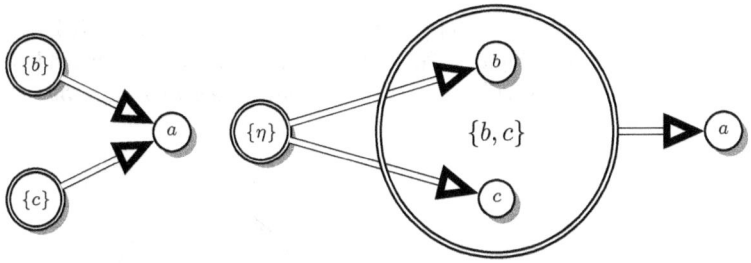

Fig. 1: An AFN and its EBAF version (Ex. 6)

Ex. 7 Let $AFN = \langle\{a,b,c,d\}, \varnothing, \{(\{b,c\},a),(\{d\},a)\}\rangle$. The EBAF version of AFN is $\langle\{\eta,a,b,c,d\}, \varnothing, \{(\{\eta\},b), (\{\eta\},c), (\{\eta\},d), (\{b,d\},a)(\{c,d\},a)\}\rangle$ and represented in Figure 2.

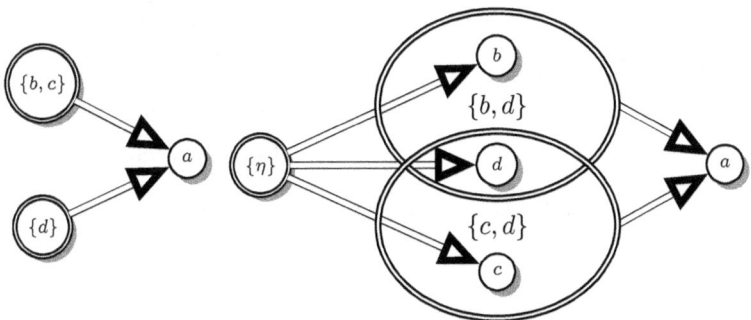

Fig. 2: An AFN and its EBAF version (Ex. 7)

For the translation from a given EBAF to an AFN, attacks must be assumed binary.

Def. 5 (From EBAF to AFN) *Let $\langle \mathbf{A}, \mathbf{R}, \mathbf{E} \rangle$ be an EBAF with binary attacks. The corresponding AFN $\langle \mathbf{A}, \mathbf{R}', \mathbf{N} \rangle$ is defined by:*
- *The set of arguments remains the same*
- *For $(\{a\}, b) \in \mathbf{R}$, put (a, b) in $\mathbf{R}'$*
- *Let $a \neq \eta \in \mathbf{A}$ and $X = \{X_1, \ldots, X_k\}$ be the collection of all sets X_i such that $X_i \mathbf{E} a$. If X is empty, add $(\{a\}, a)$ to $\mathbf{N}$. Otherwise, for every subset X' of $(X_1 \cup \ldots \cup X_k)$ such that $\forall i = 1k, X' \cap X_i \neq \varnothing$, add (X', a) to $\mathbf{N}$.*

The following example illustrates this translation.

Ex. 8 *Let $EBAF = \langle \{\eta, a, b, c\}, \varnothing, \{(\{\eta\}, b), (\{b\}, a), (\{c\}, a)\} \rangle$. $EBAF$ is represented in Figure 3a. The AFN version of $EBAF$ is $\langle \{\eta, a, b, c\}, \varnothing, \{(\{\eta\}, b), (\{c\}, c), (\{b, c\}, a)\} \rangle$ and represented in Figure 3b.*

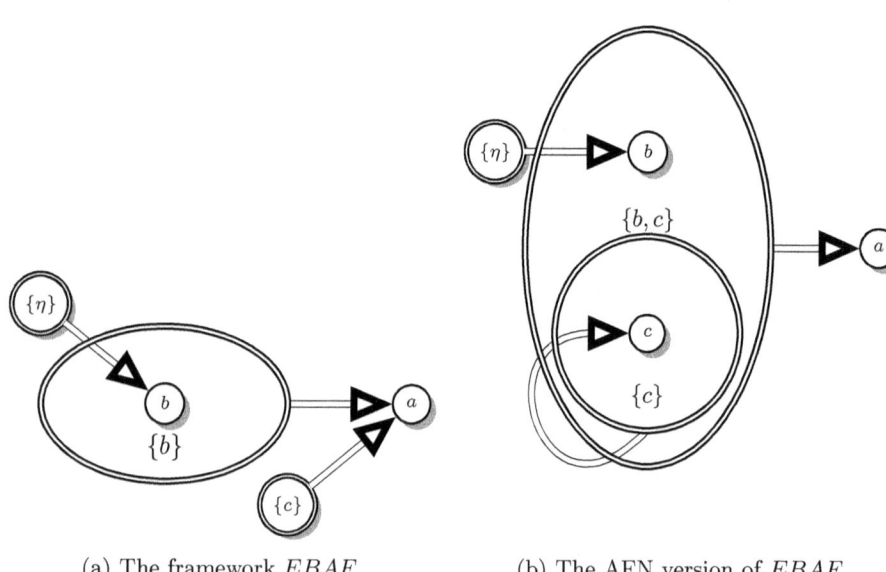

(a) The framework $EBAF$ (b) The AFN version of $EBAF$

Fig. 3: An EBAF and its AFN version (Ex. 8)

2.4 Argumentation Frameworks with Recursive Attack and Support (ASAF)

Argumentation Frameworks with Recursive Attack and Support have been introduced in [10] with a flattening approach for computing semantics and redefined in [11] with a direct definition of semantics.

Def. 6 (ASAF) *An* Argumentation Framework with Recursive Attack and Support (ASAF) *is a triple* $\langle \mathbf{A}, \mathbf{R}, \mathbf{N} \rangle$, *where* $\mathbf{A}$ *is a set of arguments,* $\mathbf{R} \subseteq \mathbf{A} \times (\mathbf{A} \cup \mathbf{R} \cup \mathbf{N})$ *is an attack relation and* $\mathbf{R} \subseteq \mathbf{A} \times (\mathbf{A} \cup \mathbf{R} \cup \mathbf{N})$ *is a support relation. We assume that* $\mathbf{N}$ *is acyclic and* $\mathbf{R} \cap \mathbf{N} = \varnothing$.

In addition to the conflicts already expressed by the attack relation, new conflicts, called defeats, are identified. Two main types of defeat are defined: those that can be inferred directly by looking at the attack relation of the ASAF (unconditional defeats, with 2 sub-types), and those that are conditioned by the existence of supports (conditional defeats, with also 2 sub-types).

Def. 7 (Defeat in ASAF) *Let* $ASAF = \langle \mathbf{A}, \mathbf{R}, \mathbf{N} \rangle$. *Let* $\alpha \in \mathbf{R}$, $X \in \mathbf{A} \cup \mathbf{R} \cup \mathbf{N}$.

Unconditional defeat (u − def): α **u − def** X *iff*
- *either the target of α is X (denoted by α* **d − def** X*),*
- *or α* **d − def** *a with X being an attack and a being the source of X (denoted by α* **i − def** X*).*

Conditional defeat (c − def): *let* $S \subseteq \mathbf{N}$, α **c − def** X *given* S *iff*
- *either there exists a sequence of supports $\Sigma = (\beta_1, \ldots, \beta_n)$ in $ASAF^3$ such that $\beta_n = X$, the target of α is β_1 and S is the support set of Σ^4 (denoted by α* **e − def** X *given* S*),*
- *or α* **e − def** *a with X being an attack and a being the source of X (denoted by α* **ei − def** X *given* S*).*

Note that the source of a defeat is always an attack (and never an argument) and its target is either an argument or another interaction.

The conflict-freeness is defined using all these defeats.

Def. 8 (Conflict-freeness in ASAF) *Let* $ASAF = \langle \mathbf{A}, \mathbf{R}, \mathbf{N} \rangle$. *Let* $S \subseteq \mathbf{A} \cup \mathbf{R} \cup \mathbf{N}$. *$S$ is conflict-free in ASAF iff* $\nexists \alpha, X \in S$, *s.t.* α **u − def** X *and* $\nexists \alpha, X \in S$, $\nexists S' \subseteq S \cap \mathbf{N}$, *s.t.* α **c − def** X *given* S'.

Acceptability of an element X w.r.t. a set of elements S is defined similarly as in Dungs theory: if there exists an interaction α defeating X, then it must exist another interaction β in S that in turn defeats α. Formally, we have:

Def. 9 (Acceptability in ASAF) *Let* $ASAF = \langle \mathbf{A}, \mathbf{R}, \mathbf{N} \rangle$. *Let* $S \subseteq \mathbf{A} \cup \mathbf{R} \cup \mathbf{N}$. *Let* $X \in \mathbf{A} \cup \mathbf{R} \cup \mathbf{N}$. *$X$ is acceptable w.r.t. S iff:*
- *for all $\alpha \subseteq \mathbf{R}$, s.t. α* **u − def** X:
  - *either there exists $\beta \in S$ s.t. β* **u − def** α,
  - *or there exist $\beta \in S$ and $S' \subseteq S$ s.t. β* **c − def** α *given* S'.
- *for all $\alpha \subseteq \mathbf{R}$, for all $T \subseteq \mathbf{N}$, s.t. α* **c − def** X *given* T:
  - *either there exists $\beta \in S$ s.t. β* **u − def** α,

---
[3] *i.e.,* $\forall i = 1, \ldots, n-1$, the target of β_i is the source of β_{i+1}.
[4] *i.e.,* $\{\beta_1, \ldots, \beta_n\}$.

- or there exist $\beta \in S$ and $S' \subseteq S$ s.t. β **c** $-$ **def** α given S',
- or there exist $\beta \in S$ and $\gamma \in T$ s.t. β **u** $-$ **def** γ,
- or there exist $\beta \in S$, $\gamma \in T$ and $S' \subseteq S$ s.t. β **c** $-$ **def** γ given S'.

Finally, these new notions of conflict-freeness and acceptability are used for defining the semantics as usual.

Def. 10 (Semantics in ASAF) *Let $ASAF = \langle \mathbf{A}, \mathbf{R}, \mathbf{N} \rangle$. Let $T \subseteq \mathbf{A} \cup \mathbf{R} \cup \mathbf{N}$.*
- *T is admissible iff T is conflict-free, and every X in T is acceptable w.r.t. T.*
- *T is a complete extension iff T is admissible and $\forall X \in \mathbf{A} \cup \mathbf{R} \cup \mathbf{N}$, if X is acceptable w.r.t. T, then $X \in T$.*
- *T is a preferred extension iff T is a $\subseteq$-maximal admissible extension.*
- *T is a stable extension iff T is conflict-free and $\forall X \in (\mathbf{A} \cup \mathbf{R} \cup \mathbf{N}) \setminus T$, either there exists $\alpha \in T$ s.t. α **u** $-$ **def** X, or there exist $\alpha \in T$ and $T' \subseteq T$ s.t. α **c** $-$ **def** X given T'.*
- *T is a grounded extension iff T is a $\subseteq$-minimal complete extension.*

2.5 Recursive Evidence-Based Argumentation Frameworks (REBAF)

Recently introduced in [6], the Recursive Evidence-Based Argumentation Framework (REBAF) allows representing higher-order attacks and higher-order supports, *i.e.* attacks or supports from arguments to either other arguments or other attacks or supports. It is a generalisation of the Evidence-Based Argumentation Framework (EBAF) [18].

In the REBAF, the semantics handle both acceptability of arguments and validity of interactions (attacks or supports), and account for the fact that acceptability of arguments may depend on the validity of interactions and vice-versa. As a consequence, the standard notion of extension is replaced by a triple of a set of arguments, a set of attacks and a set of supports. We briefly recall the main definitions.

Def. 11 (Recursive EBAF) *A Recursive Evidence-Based Argumentation Framework (REBAF) is a sextuple $\langle \mathbf{A}, \mathbf{R}, \mathbf{S}, s, t, \mathbf{P} \rangle$, where $\mathbf{A}$, $\mathbf{R}$ and $\mathbf{S}$ are three pairwise disjunct sets respectively representing arguments, attacks and supports names, and where $\mathbf{P} \subseteq \mathbf{A} \cup \mathbf{R} \cup \mathbf{S}$ is a set representing the* prima-facie *elements that do not need to be supported. Functions $s : (\mathbf{R} \cup \mathbf{S}) \longrightarrow 2^{\mathbf{A}} \setminus \varnothing$ and $t : (\mathbf{R} \cup \mathbf{S}) \longrightarrow (\mathbf{A} \cup \mathbf{R} \cup \mathbf{S})$ respectively map each attack and support to its source and its target.*

Note that, in contrast with EBAFs, the set **P** may contain several prima-facie elements (arguments, attacks and supports) without any constraint (they can be attacked or supported).

A REBAF can be graphically represented. For instance, a support named α (with $s(\alpha) = \{a\}$ and $t(\alpha) = c$) being the target of an attack β with $s(\beta) = \{b\}$ is represented by:

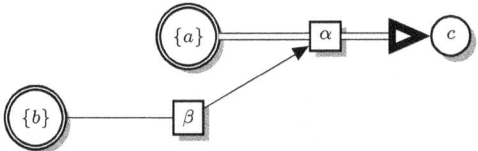

The notion of structure allows characterizing which arguments are regarded as "acceptable" and which attacks and supports are regarded as "valid" with respect to a given argumentation framework. It is the basis of defining the semantics for recursive frameworks.

Def. 12 (Structure [6]) *A structure of the REBAF $\langle \mathbf{A}, \mathbf{R}, \mathbf{S}, s, t, \mathbf{P} \rangle$ is a triple $U = (T, \Gamma, \Delta)$ with $T \subseteq \mathbf{A}$, $\Gamma \subseteq \mathbf{R}$ and $\Delta \subseteq \mathbf{S}$.*

Intuitively, the set T represents the set of "acceptable" arguments w.r.t. the structure U, while Γ and Δ respectively represent the set of "valid attacks" and "valid supports" w.r.t. U.

For the rest of this section, we assume that all definitions are relative to a given REBAF $\langle \mathbf{A}, \mathbf{R}, \mathbf{S}, s, t, \mathbf{P} \rangle$ and to a given structure U of this REBAF.

An element x (argument, attack or support) can be defeated w.r.t. U iff there is a "valid attack" w.r.t. U that targets x and whose source is "acceptable" w.r.t. U.

Def. 13 ([6])
- $Def_X(U) = \{x \in X / \exists \alpha \in \Gamma, s(\alpha) \subseteq T \text{ and } t(\alpha) = x\}$ with $X \in \{\mathbf{A}, \mathbf{R}, \mathbf{S}\}$.
- $\overline{Def_X(U)} = X \setminus Def_X(U)$ for $X \in \{\mathbf{A}, \mathbf{R}, \mathbf{S}\}$.
- $Def(U) = Def_\mathbf{A}(U) \cup Def_\mathbf{R}(U) \cup Def_\mathbf{S}(U)$, *the set of all defeated elements w.r.t. U.*
- $\overline{Def(U)} = (\mathbf{A} \cup \mathbf{R} \cup \mathbf{S}) \setminus Def(U)$.

Concerning the notion of *supported elements* w.r.t. a structure, the prima-facie elements (arguments, attacks, supports) of a given framework are supported w.r.t. any structure. Then, a standard element is supported if there exists a chain of supported supports, leading to it, which is rooted in prima-facie arguments. Formally, the set of supported elements $Supp(U)$ is recursively defined as follows:

Def. 14 ([6]) *Let U_{-x} denote $(T \setminus \{x\}, \Gamma \setminus \{x\}, \Delta \setminus \{x\})$. The set of supported elements w.r.t. U is defined as follows:*
$Supp(U) = \mathbf{P} \cup \{t(\alpha) / \exists \alpha \in (\Delta \cap Supp(U_{-t(\alpha)})) \text{ with } s(\alpha) \subseteq (T \cap Supp(U_{-t(\alpha)}))\}$.

Drawing on the notions of defeated elements and supported elements, the *supportable* elements can be defined. An element is considered as supportable if there exists some non-defeated support with all its source elements non-defeated and regarded as supportable. Formally, an element x is supportable w.r.t. U iff x is supported w.r.t. $U' = (\overline{Def_\mathbf{A}(U)}, \mathbf{R}, \overline{Def_\mathbf{S}(U)})$.

Elements that are defeated w.r.t. U or that are unsupportable w.r.t. U cannot be accepted w.r.t. U. $UnAcc(U) = Def(U) \cup \overline{Supp(U')}$ denotes the set of unacceptable elements w.r.t. U.

Moreover, an attack $\alpha \in \mathbf{R}$ is unactivable[5] w.r.t. U iff either it is unacceptable w.r.t. U or some element in its source is unacceptable w.r.t. U. $UnAct(U) = \{\alpha \in \mathbf{R} / \alpha \in UnAcc(U) \text{ or } s(\alpha) \cap UnAcc(U) \neq \varnothing\}$.

Ex. 9 Consider the framework $\langle \mathbf{A}, \mathbf{R}, \mathbf{S}, s, t, \mathbf{P} \rangle$ where $\mathbf{A} = \{a, b, c, d\}$, $\mathbf{R} = \{\beta, \delta\}$, $\mathbf{S} = \{\alpha\}$ and $\mathbf{P} = \{a, c, \alpha, \beta, \delta\}$ corresponding to the graph depicted in the following figure (prima-facie elements are represented with dashed lines; as the source of each interaction is a singleton, it can be represented by an argument):

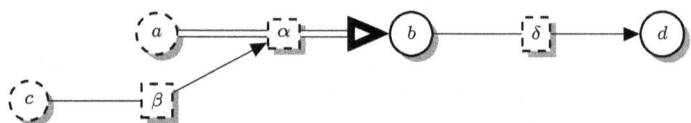

Let U be the structure $(T = \{a, c\}, \Gamma = \{\beta\}, \Delta = \{\alpha\})$. $Supp(U) = \{a, b, c, \beta, \delta\}$. Note that a, c, β and δ are supported because they are prima-facie elements. Let us prove that $b \in Supp(U)$: $b = t(\alpha)$ with $\alpha \in \Delta$ and $s(\alpha) = \{a\} \subseteq T$. As α and a both belong to $\mathbf{P}$, $s(\alpha)$ and α both belong to $Supp(U \setminus \{b\})$. However, b is unsupportable w.r.t. U since α is defeated by β. As a consequence, although the attack δ is supported w.r.t. U, it is also unactivable w.r.t. U.

Finally, an element is acceptable w.r.t. a given structure U iff it is supported w.r.t. U and, in addition, every attack against it can be considered as unactivable w.r.t. U because either some argument in its source or itself has been regarded as unacceptable w.r.t. U.

Def. 15 (Acceptability [6]) Let $x \in \mathbf{A} \cup \mathbf{R} \cup \mathbf{S}$. x is acceptable w.r.t. U iff (i) $x \in Supp(U)$ and (ii) for each attack $\alpha \in \mathbf{R}$ with $t(\alpha) = x$, $\alpha \in UnAct(U)$. $Acc(U)$ denotes the set of all elements that are acceptable w.r.t. U.

Then semantics are defined as follows:

Def. 16 (Semantics in REBAF [6]) A structure $U = (T, \Gamma, \Delta)$ is said:
1. self-supporting iff $(T \cup \Gamma \cup \Delta) \subseteq Supp(U)$,
2. conflict-free iff $T \cap Def_\mathbf{A}(U) = \varnothing$, $\Gamma \cap Def_\mathbf{R}(U) = \varnothing$ and $\Delta \cap Def_\mathbf{S}(U) = \varnothing$,
3. admissible iff it is conflict-free and $(T \cup \Gamma \cup \Delta) \subseteq Acc(U)$,
4. complete iff it is conflict-free and $(T \cup \Gamma \cup \Delta) = Acc(U)$,

---

[5] Intuitively, such an attack cannot be "activated" in order to defeat the element that it is targeting.

5. preferred *iff* it is a $\subseteq$-maximal[6] admissible structure,
6. stable *iff* $(T \cup \Gamma \cup \Delta) = \overline{UnAcc(U)}$.

3 Handling higher-order necessary supports

Our purpose is to propose a framework that allows representing higher-order attacks and higher-order necessary supports, using similar definitions as those at work in the REBAF. First, we provide a definition of a "recursive AFN". Then we show that, in presence of higher-order interactions, the translation from an AFN to an EBAF proposed in [18] and recalled in Section 2.3 cannot be extended. That leads us to provide direct definitions for the semantics of recursive AFNs.

Def. 17 (Recursive AFN) *A Recursive Argumentation Framework with Necessity (RAFN) is a tuple $\langle \mathbf{A}, \mathbf{R}, \mathbf{N}, s, t \rangle$, where $\mathbf{A}$, $\mathbf{R}$ and $\mathbf{N}$ are three pairwise disjunct sets respectively representing arguments, attacks and supports names, s is a function from $\mathbf{R} \cup \mathbf{N}$ to $(2^{\mathbf{A}} \setminus \varnothing)$ mapping each interaction to its source and t is a function from $\mathbf{R} \cup \mathbf{N}$ to $(\mathbf{A} \cup \mathbf{R} \cup \mathbf{N})$ mapping each interaction to its target. It is assumed that $\forall \alpha \in \mathbf{R}, s(\alpha)$ is a singleton.*

Note that, in contrast with ASAF, the source of a support in a RAFN is a *set* of arguments. And, in constrast with REBAF, the source of an attack is a singleton.

3.1 Turning a recursive AFN into a recursive EBAF ?

In the particular case of first-order interactions, a one-to-one correspondence between a REBAF and a finite EBAF has been proved, by considering only one prima-facie argument denoted by η [6]. Besides, correspondences have been provided between an AFN and an EBAF, that preserve the semantics [18] (see Section 2.3): the basic idea is that unsupported arguments in an AFN correspond to arguments supported by the special argument η in an EBAF, or equivalently to prima-facie arguments in a REBAF.

A natural idea would be to generalize the construction proposed in Def. 4. However, in that construction, it is worth to notice that if an argument a receives several supports in an AFN (let α_i denote the support $X_i \mathbf{N} a$), new supports β_j to a are created in the corresponding EBAF (see Ex. 6). Assume now that one of the supports α_i is attacked, it is impossible to know which one of the new supports should be attacked. This point is illustrated on Ex. 10.

Ex. 10 *Let $RAFN = \langle \{a, b, c, d\}, \{\alpha_3\}, \{\alpha_1, \alpha_2\}, s, t \rangle$ with $s(\alpha_1) = \{b\}, s(\alpha_2) = \{c\}, s(\alpha_3) = \{d\}, t(\alpha_1) = t(\alpha_2) = a, t(\alpha_3) = \alpha_1$ (it is obtained*

---
[6] For any pair of structures $U = (T, \Gamma, \Delta)$ and $U = (T', \Gamma', \Delta')$, $U \subseteq U'$ means that $(T \cup \Gamma \cup \Delta) \subseteq (T' \cup \Gamma' \cup \Delta')$.

from the AFN given in Ex. 6 by naming the supports and adding an attack to one of the supports).

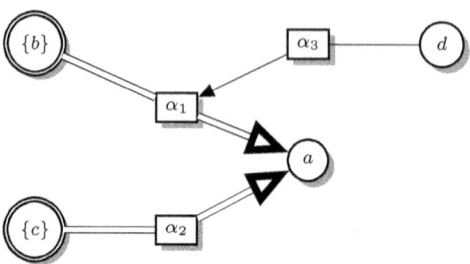

Creating a support β from the set $\{b, c\}$ to a would not enable to take into account the fact that α_1 is attacked. In particular it would not be sound to create an attack from d to β.

Moreover, in the higher-order framework and even in the particular case when each element (argument, attack or support) receives at most one support, whose source is reduced to one argument (case of binary supports), the different understanding of evidential and necessary supports implies that the construction proposed in Def. 4 cannot be extended. This point is illustrated on the two following examples.

Ex. 11 *Consider the RAFN framework $\langle \mathbf{A}, \mathbf{R}, \mathbf{N}, s, t, \rangle$ where $\mathbf{A} = \{a, b, c\}$, $\mathbf{R} = \{\beta\}$ and $\mathbf{N} = \{\alpha\}$, with $s(\alpha) = \{a\}$, $s(\beta) = \{c\}$, $t(\alpha) = b$ and $t(\beta) = \alpha$.*

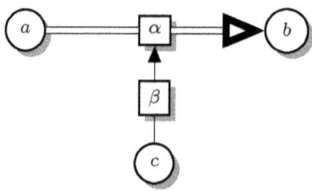

Generalizing Def. 4 would produce the corresponding REBAF $\langle \mathbf{A}, \mathbf{R}, \mathbf{N}, s, t, \mathbf{P} \rangle$ with $\mathbf{P} = \{a, c, \alpha, \beta\}$ (indeed, following the necessary understanding of the support, each element that is not supported can be considered as prima-facie). Let $U = (\{a, c\}, \{\beta\}, \varnothing)$. As shown in [6], with the REBAF semantics, b has no support w.r.t. U as $\alpha \notin U$, so b cannot be accepted. In contrast, considering necessary supports, we should be able to say that b is supported w.r.t. U: As $\alpha \notin U$, there is no necessary support to be considered for ensuring the acceptance of b.

Ex. 12 *Consider now $RAFN'$ obtained from $RAFN$ given in Ex. 11 by replacing the attack β from c to α with a support δ from c to α.*

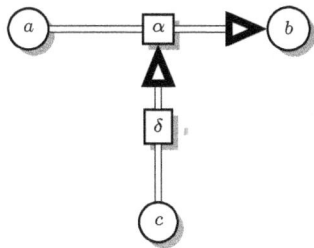

Let $U = (\{a\}, \varnothing, \{\alpha, \delta\})$. With the REBAF semantics, b has no support w.r.t. U, since α has no support from U (as $c \notin U$). In contrast, considering necessary supports, as $c \notin U$, α cannot be valid, so there is no necessary support to be considered for ensuring the acceptance of b, so we should be able to say that b is supported w.r.t. U.

3.2 Semantics of a recursive AFN

Even if a direct translation from RAFN to a REBAF is not possible, the analogy between prima-facie arguments in REBAF and non-supported arguments in RAFN suggests to draw on the REBAF approach. So we next provide direct definitions for the semantics of recursive AFNs, based on the notion of structure, in a similar way as for a REBAF. We keep the definition (and the intuition) for a structure, and the definition for an element being defeated recalled in Section 2.5.

Consider a RAFN $\langle \mathbf{A}, \mathbf{R}, \mathbf{N}, s, t, \rangle$. We recall that $\forall \alpha \in \mathbf{R}$, $s(\alpha)$ is a singleton. A structure of the RAFN is a triple $U = (T, \Gamma, \Delta)$ such that $T \subseteq \mathbf{A}$, $\Gamma \subseteq \mathbf{R}$ and $\Delta \subseteq \mathbf{N}$. Intuitively, the set T represents the set of "acceptable" arguments w.r.t. the structure U, while Γ and Δ respectively represent the set of "valid attacks" and "valid necessary supports" w.r.t. U. An element x (argument, attack or support) can be defeated w.r.t. U iff there is a "valid attack" w.r.t. U that targets x and whose source is "acceptable" w.r.t. U.

Def. 18 *Given a structure $U = (T, \Gamma, \Delta)$.*
- $Def_X(U) = \{x \in X / \exists \alpha \in \Gamma, s(\alpha) \in T \text{ and } t(\alpha) = x\}$ with $X \in \{\mathbf{A}, \mathbf{R}, \mathbf{N}\}$.
- $\overline{Def_X(U)} = X \setminus Def_X(U)$ for $X \in \{\mathbf{A}, \mathbf{R}, \mathbf{N}\}$.
- $Def(U) = Def_{\mathbf{A}}(U) \cup Def_{\mathbf{R}}(U) \cup Def_{\mathbf{N}}(U)$, the set of all defeated elements w.r.t. U.
- $\overline{Def(U)} = (\mathbf{A} \cup \mathbf{R} \cup \mathbf{N}) \setminus Def(U)$.

Concerning the notion of *supported elements* w.r.t. a structure, elements (arguments, attacks, supports) which receive no necessary support do not require any support, so they are supported w.r.t. any structure; this is the equivalent of the set of prima-facie elements used in REBAF: $\mathbf{P} = \{x \in \mathbf{A} \cup \mathbf{R} \cup \mathbf{N} / \text{there is no } \alpha \in \mathbf{N} \text{ with } t(\alpha) = x\}$. Moreover, in an AFN, for $E \subseteq \mathbf{A}$, ENx means that the acceptance of x requires the acceptance of *at least one* element of E. Then, an element x is supported w.r.t. a given structure U if *for each* support α (which can be regarded as supported), the source of α contains *at least one* argument of U that can be regarded as supported. Formally, the set of supported elements $Supp(U)$ is recursively defined as follows:[7]

Def. 19 *Given a structure* $U = (T, \Gamma, \Delta)$.
- $Supp(U) = \{x | \forall \alpha \in \Delta$ *such that* $t(\alpha) = x$, *if* $\alpha \in Supp(U_{-x})$
  then $s(\alpha) \cap (T \cap Supp(U_{-x})) \neq \emptyset\}$.
- U *is* self-supporting *iff* $(T \cup \Gamma \cup \Delta) \subseteq Supp(U)$.

Pursuing the analogy with REBAF, an element of a RAFN is considered as being still supportable as long as *for each* non-defeated support, *there exists at least one* argument in its source, which is non-defeated and regarded as supportable. Formally, an element x is supportable w.r.t. U iff x is supported w.r.t. $U' = (\overline{Def_{\mathbf{A}}(U)}, \mathbf{R}, \overline{Def_{\mathbf{N}}(U)})$.

Now, drawing on these new notions of supported (resp. unsupportable) element, we can keep the definitions used in an REBAF for defining unacceptable elements and unactivable attacks. Namely, elements that are defeated or that are unsupportable are said *unacceptable* (they cannot be accepted). Then an attack $\alpha \in \mathbf{R}$ is *unactivable* (such an attack cannot be "activated" in order to defeat the element that it is targeting) iff either it is unacceptable or its source is unacceptable.

Def. 20 *Given a structure* $U = (T, \Gamma, \Delta)$, *let* $U' = (\overline{Def_{\mathbf{A}}(U)}, \mathbf{R}, \overline{Def_{\mathbf{N}}(U)})$.
- $UnSupp(U) = \overline{Supp(U')}$.
- $UnAcc(U) = Def(U) \cup UnSupp(U)$ *denotes the set of* unacceptable *elements w.r.t.* U.
- $UnAct(U) = \{\alpha \in \mathbf{R}/\alpha \in UnAcc(U) \text{ or } s(\alpha) \subseteq UnAcc(U)\}$ *denotes the set of* unactivable *attacks w.r.t.* U.

Finally, an element is acceptable w.r.t. a given structure U iff it is supported and, in addition, every attack against it can be considered as unactivable because either some argument in its source or itself has been regarded as unacceptable.

Def. 21 (Acceptability) *Given a structure* $U = (T, \Gamma, \Delta)$, $x \in \mathbf{A} \cup \mathbf{R} \cup \mathbf{N}$ *is acceptable w.r.t.* U *iff (i)* $x \in Supp(U)$ *and (ii) for each attack* $\alpha \in \mathbf{R}$ *with* $t(\alpha) = x$, $\alpha \in UnAct(U)$.
$Acc(U)$ *denotes the set of all elements that are acceptable w.r.t.* U.

The two following examples illustrate the previous definitions.

Ex. 13 *Consider the framework* RAFN $= \langle \{a, x, y, z, t\}, \{\beta\}, \{\alpha_1, \alpha_2, \alpha_3\}, s, t \rangle$ *with* $s(\alpha_1) = \{a\}$, $s(\alpha_2) = \{z\}$, $s(\alpha_3) = \{t\}$, $s(\beta) = \{y\}$,

---
[7] Note that, unlike in the REBAF case, the union with the set $\mathbf{P}$ is useless because the universal condition holds vacuous and so, by definition, $\mathbf{P}$ is included in $Supp(U)$.

$t(\alpha_1) = x, t(\alpha_2) = y, t(\alpha_3) = y, t(\beta) = x$.

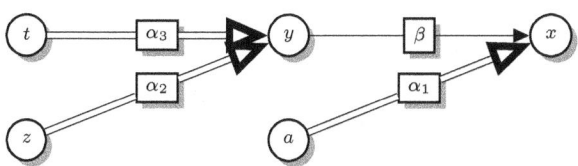

Let $U = (\{a\}, \{\beta\}, \{\alpha_1, \alpha_2, \alpha_3\})$. We have $x \in Supp(U)$. However, $x \notin Acc(U)$ as it is the target of the attack β, $s(\beta) = \{y\}$ and $y \notin UnAcc(U)$. Indeed y is not attacked and $y \in Supp(U')$ since α_2 and α_3 belong to **P** and z and t do not belong to $Def_\mathbf{A}(U)$.

Ex. 14 *Now consider the RAFN obtained by adding an attack γ from a to z in the RAFN of Ex. 13.*

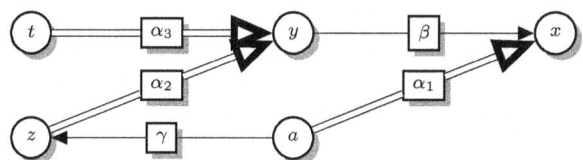

Consider the new structure $U = (\{a\}, \{\beta, \gamma\}, \{\alpha_1, \alpha_2, \alpha_3\})$. With this new structure, we have $z \in Def_\mathbf{A}(U)$. So $y \notin Supp(U')$ and therefore x becomes acceptable w.r.t. U.

Note that the Fundamental Lemma cannot be generalized. Indeed, the function $Supp$ is not monotonic as shown by the following examples.
- Let us first consider $RAFN$ obtained from AFN of Ex. 6 just by naming supports: $s(\alpha_1) = \{b\}$, $s(\alpha_2) = \{c\}$, $t(\alpha_1) = t(\alpha_2) = a$. Let $U = (\{b\}, \emptyset, \{\alpha_1\})$. $a \in Supp(U)$ and $U \cup \{a\}$ is self-supporting. Moreover U is admissible and a and α_2 both belong to $Acc(U)$. However, $a \notin Supp(U \cup \{\alpha_2\})$. So $a \notin Acc(U \cup \{\alpha_2\})$.
- As a second example, let us consider Ex. 12: (a support α from a to b supported by δ from c). Let $U = (\{b\}, \emptyset, \{\alpha, \delta\})$. $b \in Supp(U)$ since $c \notin T$ and so α is not supported. However, $b \notin Supp(U \cup \{c\})$ since $a \notin T$.

Finally, semantics are defined as follows:[8]

Def. 22 (Semantics in RAFN) *A structure $U = (T, \Gamma, \Delta)$ is said:*
1. *conflict-free iff $T \cap Def_\mathbf{A}(U) = \emptyset$, $\Gamma \cap Def_\mathbf{R}(U) = \emptyset$ and $\Delta \cap Def_\mathbf{N}(U) = \emptyset$,*
2. *admissible iff it is conflict-free and $(T \cup \Gamma \cup \Delta) \subseteq Acc(U)$,*
3. *complete iff it is conflict-free and $(T \cup \Gamma \cup \Delta) = Acc(U)$,*

---
[8] As there is no Fundamental Lemma, preferred and stable extensions are assumed to be complete sets.

4. preferred *iff it is a $\subseteq$-maximal complete structure*,
5. stable *iff it is complete and* $\overline{(T \cup \Gamma \cup \Delta)} = UnAcc(U)$,
6. grounded *iff it is a $\subseteq$-minimal complete structure*.

Note that every admissible structure is also self-supporting.

Ex. 10 (cont'd) *Consider the case of two supports α_1 (from b to a) and α_2 (from c to a), α_1 being the target of the attack α_3 from d. We have* $\mathbf{P} = \{b, c, d, \alpha_1, \alpha_2, \alpha_3\}$. *Let us study different structures:*

- *Let $U_1 = (\{a, b, c, d\}, \varnothing, \{\alpha_1, \alpha_2\})$. U_1 is conflict-free (as $\alpha_3 \notin \Gamma_1$) and self-supporting. $b, c, d, \alpha_1, \alpha_2$ belong to $Supp(U_1)$ (as $\mathbf{P} \subseteq Supp(U_1)$. And $a \in Supp(U_1)$ since $\alpha_1, \alpha_2, s(\alpha_1), s(\alpha_2)$ belong to $\mathbf{P}$ hence to $Supp(U_{1-a})$. However, $\alpha_1 \notin Acc(U_1)$. Indeed $\alpha_3 \notin UnAct(U_1)$ as α_3 and $s(\alpha_3)$ both belong to $\mathbf{P}$ and to $\overline{Def(U_1)}$. So U_1 is not admissible.*
- *Let $U_2 = (\{a, c, d\}, \{\alpha_3\}, \{\alpha_2\})$. U_2 is conflict-free (as $\alpha_1 \notin \Gamma_2$) and self-supporting. As c, d, α_2, α_3 belong to $\mathbf{P}$, we just have to prove that $a \in Supp(U_2)$. As α_2 is the unique support in Δ_2 targeting a, due to Def. 19, we just have to prove that if $\alpha_2 \in Supp(U_{2-a})$, then $c \in T_2 \cap Supp(U_{2-a})$. That is true since $c \in (T_2 \cap \mathbf{P})$. U_2 is admissible since U_2 is self-supporting and no element of U_2 is attacked. It is worth to note that in the structure U_2, a is accepted without b being accepted. This is due to the fact that the necessary support α_1 is defeated by U_2 hence unacceptable w.r.t. U_2. So, α_1 does not have to considered as a necessary support for a.*
- $U_3 = (\{a, b, c, d\}, \{\alpha_3\}, \{\alpha_2\})$ *is the unique preferred structure.*

Ex. 15 *Consider the framework* $RAFN = \langle \{a, b, c, d, e\}, \{\alpha_3\}, \{\alpha_1, \alpha_2\}, s, t \rangle$ *with* $s(\alpha_1) = \{b, c\}$, $s(\alpha_2) = \{d\}$, $s(\alpha_3) = \{e\}$, $t(\alpha_1) = t(\alpha_2) = a$, $t(\alpha_3) = b$. *We have* $\mathbf{P} = \{b, c, d, e, \alpha_1, \alpha_2, \alpha_3\}$.

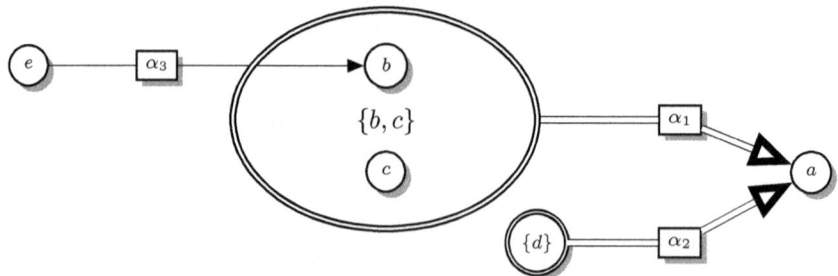

Let us study different structures:
- $U_1 = (\{a, b, d, e\}, \varnothing, \{\alpha_1, \alpha_2\})$. U_1 *is conflict-free (as $\alpha_3 \notin \Gamma_1$) and self-supporting. As $b, d, e, \alpha_1, \alpha_2$ belong to $\mathbf{P}$, we just have to prove that $a \in Supp(U_1)$. Due to Def. 19, we have to consider α_1 and α_2, the supports in Δ_1 that target a. As both of them belong to $\mathbf{P}$, we have to consider their source. $s(\alpha_2) = \{d\} \subseteq \mathbf{P} \cap T_1$, and $s(\alpha_1)$ contains b that is an element of $T_1 \cap \mathbf{P}$. So $a \in Supp(U_1)$. However, $b \notin Acc(U_1)$. Indeed $\alpha_3 \notin UnAct(U_1)$ as α_3 and $s(\alpha_3)$ both belong to $\mathbf{P}$ and to $\overline{Def(U_1)}$. So U_1 is not admissible.*

- $U_2 = (\{a,c,d,e\}, \varnothing, \{\alpha_1, \alpha_2\})$. U_2 is conflict-free. It is also self-supporting (it can be proved as for U_1 replacing b by c) and no element of U_2 is attacked. So U_2 is admissible.
- $U_3 = (\{a,c,d,e\}, \{\alpha_3\}, \{\alpha_1, \alpha_2\})$ is the unique preferred structure. Note that U_3 follows the intuition behind Def. 19, that at least one element in the source of the support α_1 has to be accepted (here c) in order to accept the target (here a).

Ex. 4 (cont'd) Consider the RAFN corresponding to the AFN (α_1 and α_2 being the names of the supports). $U = (\varnothing, \varnothing, \{\alpha_1, \alpha_2\})$ is the unique stable structure. So differently from Dung's approach, it can be the case that an element is not in the stable structure even if it is not defeated by it (it is left out because it is unsupported by the structure).

4 Related works

First, we consider the particular case of RAFN without support, then we compare our framework with AFN and ASAF.

4.1 RAFN without support

In that case, we get exactly the Recursive Argumentation Framework (RAF) described in [5].
Besides, [5] provided correspondences between RAF-structures and AFRA-extensions of [2]. The RAFN without support also corresponds to the REBAF without support (in the particular case of binary attacks).

Moreover, for a RAFN with only first-order attacks and without support, we get a RAF with only first-order attacks. That case has been studied in [5] and proved to be a conservative generalisation of Dung's framework.

4.2 Relation with AFN

We will show that the RAFN framework is a conservative generalisation of the AFN framework. For that purpose, given an AFN, we have to give a translation into a corresponding RAFN, and prove a one-to-one correspondence between complete (resp. preferred, stable, grounded) extensions of the AFN and complete (resp. preferred, stable, grounded) structures of the corresponding RAFN. Let us start by giving the RAFN corresponding to a given AFN. We just have to name the interactions.

Def. 23 Given $AFN = \langle \mathbf{A}, \mathbf{R}, \mathbf{N} \rangle$, its corresponding RAFN is $\langle \mathbf{A}, \mathbf{R}', \mathbf{N}', s', t' \rangle$, where $\mathbf{R}'$ and $\mathbf{N}'$ are two disjunct sets with the same cardinality as $\mathbf{R}$ and $\mathbf{N}$ respectively, and s' and t' map each interaction to their corresponding source and target, that is:

- for $(a,b) \in \mathbf{R}$, and α the associated name in $\mathbf{R}'$, we have $s'(\alpha) = \{a\}$ and $t'(\alpha) = b$.
- for $(X,b) \in \mathbf{N}$, and β the associated name in $\mathbf{N}'$, we have $s'(\beta) = X$ and $t'(\beta) = b$.

Moreover $\mathbf{P}' = \{x \in \mathbf{A} /$ there is no $\alpha \in \mathbf{N}'$ with $t'(\alpha) = x\} \cup \mathbf{R}' \cup \mathbf{N}'$.

Note that in an AFN, each attack (resp. support) can be considered as "valid", as it is neither attacked nor supported. Hence, in the corresponding RAFN, such an interaction must be acceptable w.r.t. any structure. Accordingly, given a set $T \subseteq \mathbf{A}$, by $U_T = (T, \mathbf{R}', \mathbf{N}')$ we denote its corresponding structure.

Consider $AFN = \langle \mathbf{A}, \mathbf{R}, \mathbf{N} \rangle$, and its corresponding $RAFN = \langle \mathbf{A}, \mathbf{R}', \mathbf{N}', s', t' \rangle$. Given $T \subseteq \mathbf{A}$, $U = (T, \mathbf{R}', \mathbf{N}')$, and $x \in \mathbf{A}$, the structure U_{-x} is equal to $(T \setminus \{x\}, \mathbf{R}', \mathbf{N}')$ and exactly corresponds to the structure $U_{T \setminus \{x\}}$. Then $Supp(U_T) = \mathbf{P}' \cup \{x \in \mathbf{A} / \forall \alpha \in \mathbf{N}'$ such that $t'(\alpha) = x$, $s'(\alpha) \cap (T \cap Supp(U_{T \setminus \{x\}})) \neq \varnothing\}$ (as each support belongs to $\mathbf{P}'$). Then the two following propositions hold (all proofs are given in Appendix A).

Prop. 1 *Let $T \subseteq \mathbf{A}$. Let $a \in T$. If T is support-closed in AFN, then $a \in Supp(U_T)$ iff a is support-cycle-free in T. Moreover U_T is self-supporting in RAFN iff T is coherent in AFN.*

Prop. 2 *Let $T \subseteq \mathbf{A}$ and $a \in \mathbf{A}$. If a is acceptable w.r.t. T in AFN, then a is acceptable w.r.t. U_T in RAFN. If T is self-supporting and a is acceptable w.r.t. U_T in RAFN, then a is acceptable w.r.t. T in AFN.*

The semantics of RAFN are given by Def. 22 and rely upon Def. 19. So, using the previous results, it can be proved that T is a conflict-free (resp. admissible, complete, preferred, stable, grounded) extension of an AFN iff the structure U_T is a conflict-free (resp. admissible, complete, preferred, stable, grounded) structure in the corresponding RAFN.

Prop. 3 *Let $T \subseteq \mathbf{A}$.*
1. *T is a conflict-free extension in AFN iff U_T is a conflict-free structure in RAFN.*
2. *T is an admissible extension of AFN iff U_T is an admissible structure in RAFN.*
3. *T is a complete extension of AFN iff U_T is a complete structure in RAFN.*
4. *T is a preferred extension of AFN iff U_T is a preferred structure in RAFN.*
5. *T is a stable extension of AFN iff U_T is a stable structure in RAFN.*
6. *T is a grounded extension of AFN iff U_T is a grounded structure in RAFN*

4.3 Relation with ASAF

We next compare our RAFN semantics with ASAF semantics [11]. This comparaison needs to consider particular cases of RAFN, as ASAF excludes cycles of necessary supports, and assumes that interactions are binary ones (the source of an attack or a support is a unique argument). The common idea is that the extensions may not only include arguments but also attacks and supports. However, several differences can be outlined. First, in ASAF, attacks and supports are combined to obtain extended (direct or indirect) defeats. Conflict-freeness for a set of elements (arguments, attacks, supports) is defined w.r.t. these extended defeats. So the conflict-freeness requirement takes support into account. In contrast, in RAFN, the notions of support and attack are dealt with separately (see Def. 18 and Def. 19). As for acceptability, in ASAF, an element is acceptable w.r.t. a set of elements whenever it can be defended against each defeat. So, in the particular case when there is no attack, each argument would be acceptable w.r.t. any set. In contrast, Def. 21 explicitly requires a support.

Ex. 6 (cont'd) *Consider the corresponding RAFN of AFN:* $\langle \{a,b,c\}, \varnothing, \{\alpha_1, \alpha_2\}, s, t\rangle$ *with* $s(\alpha_1) = \{b\}, s(\alpha_2) = \{c\}, t(\alpha_1) = t(\alpha_2) = a$. *With ASAF semantics, the sets* $\{a, \alpha_1, \alpha_2\}$, $\{a, b, \alpha_1, \alpha_2\}$, $\{a, c, \alpha_1, \alpha_2\}$ *are admissible. In constrast, the structures* $(\{a\}, \varnothing, \{\alpha_1, \alpha_2\})$, $(\{a,b\}, \varnothing, \{\alpha_1, \alpha_2\})$ *and* $(\{a,c\}, \varnothing, \{\alpha_1, \alpha_2\})$ *are not admissible in RAFN.*

Another difference was already pointed in [5], where correspondences have been provided between a RAF and an ASAF without support. Indeed, in an ASAF, an attack is not acceptable whenever its source is not acceptable (Prop.2 in [11]).

Ex. 14 (cont'd) *With RAFN semantics, β is not attacked and not supported so β must belong to each complete structure. With ASAF semantics, if β is acceptable w.r.t. a set E, then y must be also acceptable w.r.t. E. If E is a complete extension, E contains a, γ and α_2. As y is defeated by γ given α_2 it cannot be the case that y is acceptable w.r.t. E. So β cannot belong to any complete extension.*

So, following the work of [5], we propose to define the following mappings.
- Let $\langle \mathbf{A}, \mathbf{R}, \mathbf{N}, s, t\rangle$ be a RAFN. Given a structure $U = (T, \Gamma, \Delta)$, by $E_U = T \cup \{\alpha \in \Gamma$ such that $s(\alpha) \subseteq T\} \cup \Delta$, we denote its corresponding ASAF extension.
- Let $\langle \mathbf{A}, \mathbf{R}, \mathbf{N}, s, t\rangle$ be a RAFN. Given $E \subseteq (\mathbf{A} \cup \Gamma \cup \mathbf{N})$ an ASAF extension, by $U_E = (T_E, \Gamma_E, \Delta_E)$, we denote its corresponding RAFN structure, where $T_E = \mathbf{A} \cap E$, $\Gamma_E = (\mathbf{R} \cap E) \cup \{\alpha \in (\mathbf{R} \cap Acc(U'_E))$ such that $s(\alpha) \notin E\}$ with U'_E denoting the structure $(T_E, \mathbf{R} \cap E, \mathbf{N} \cap E)$ and $\Delta_E = (\mathbf{N} \cap E) \cup (\mathbf{N} \cap Acc(U'_E))$.

Our intuition is that, despite the differences between conflict-free and acceptability requirements, the above mappings will enable to achieve correspondences between ASAF and RAFN for the complete (and also grounded and preferred) semantics.

Ex. 14 (cont'd) *Consider the unique complete structure* $U = (\{a, x, t\}, \{\beta, \gamma\}, \{\alpha_1, \alpha_2, \alpha_3\})$. *The corresponding ASAF extension is* $E_U = \{a, x, t, \gamma, \alpha_1, \alpha_2, \alpha_3\}$. *It can be checked that it is an ASAF complete extension. Conversely, let* $E = \{a, x, t, \gamma, \alpha_1, \alpha_2, \alpha_3\}$. *We have*

$U'_E = (\{a,x,t\},\{\gamma\},\{\alpha_1,\alpha_2,\alpha_3\})$. Obviously, $\beta \in Acc(U'_E)$ as β is neither attacked nor supported. So β can be added to Γ_E and $U_E = U$.

5 Conclusion

In this paper, we have proposed an abstract framework that deals with higher-order interactions, using two types of interaction: attacks and necessary supports. That framework generalises both abstract frameworks with necessities (AFN, see [15]) and recursive abstract frameworks (RAF, see [5]), and so is called RAFN. We have defined semantics accounting for acceptability of arguments and also validity of interactions. As a source of inspiration, we have used the approach presented in [6] that does a similar work for REBAF, another framework dealing with higher-order interactions using evidential supports in place of necessary ones. A notable feature of our approach is that valid interactions remain explicit and distinct from arguments. In other words, semantics are defined directly w.r.t. the original framework without translating it into other frameworks.

In the literature, there exist few works handling higher-order attacks and necessary supports, in a direct way, except the ASAF framework of Cohen et al [10, 11]. However, in ASAF, cycles of support are excluded, and interactions are assumed binary. Our framework is a conservative generalisation of AFN and RAF, and we are also able to outline the differences with ASAF semantics proposed in [11].

In this work, we have defined structure-based semantics in a similar way as done in [6] for evidential support. That paves the way for studying a more general framework capable of taking into account both necessary supports and evidential supports. We aim to address that issue as future work. We also plan to connect RAFN to Logical formalisms, following existing works relating Dung's framework to logic programs and ASP (for instance [13]) or following some other works about logical translation of higher-order frameworks (for instance [14] or [8]).

References

1. Arisaka, R., Satoh, K.: Voluntary manslaughter? a case study with meta-argumentation with supports. In: Proc. of JSAI-isAI 2016. LNCS, vol 10247. pp. 241–252 (2017)
2. Baroni, P., Cerutti, F., Giacomin, M., Guida, G.: AFRA: Argumentation framework with recursive attacks. Intl. Journal of Approximate Reasoning **52**, 19–37 (2011)
3. Boella, G., Gabbay, D.M., van der Torre, L., Villata, S.: Support in abstract argumentation. In: Proc. of COMMA. pp. 111–122 (2010)
4. Cayrol, C., Cohen, A., Lagasquie-Schiex, M.C.: Towards a new framework for recursive interactions in abstract bipolar argumentation. In: Proc. of COMMA. pp. 191–198 (2016)

5. Cayrol, C., Fandinno, J., Fariñas del Cerro, L., Lagasquie-Schiex, M.C.: Valid attacks in argumentation frameworks with recursive attacks. In: Proc. of Commonsense Reasoning. vol. 2052. CEUR Workshop Proceedings (2017)
6. Cayrol, C., Fandinno, J., Fariñas del Cerro, L., Lagasquie-Schiex, M.C.: Argumentation frameworks with recursive attacks and evidence-based support. In: Proc. of FoIKS. vol. LNCS 10833, pp. 150–169. Springer-Verlag (2018)
7. Cayrol, C., Lagasquie-Schiex, M.C.: Bipolarity in argumentation graphs: towards a better understanding. Intl. J. of Approximate Reasoning **54**(7), 876–899 (2013)
8. Cayrol, C., Lagasquie-Schiex, M.C.: Logical encoding of argumentation frameworks with higher-order attacks. In: Proc. of ICTAI. IEEE (2018)
9. Cohen, A., Gottifredi, S., García, A.J., Simari, G.R.: A survey of different approaches to support in argumentation systems. The Knowledge Engineering Review **29**, 513–550 (2014)
10. Cohen, A., Gottifredi, S., García, A.J., Simari, G.R.: An approach to abstract argumentation with recursive attack and support. J. Applied Logic **13**(4), 509–533 (2015)
11. Cohen, A., Gottifredi, S., García, A.J., Simari, G.R.: On the acceptability semantics of argumentation frameworks with recursive attack and support. In: Proc. of COMMA. pp. 231–242 (2016)
12. Dung, P.M.: On the acceptability of arguments and its fundamental role in nonmonotonic reasoning, logic programming and n-person games. Artificial Intelligence **77**, 321–357 (1995)
13. Egly, U., Gaggl, S.A., Woltran, S.: Answer-set programming encodings for argumentation frameworks. Argument & Computation **1**(2), 147–177 (2010)
14. Gabbay, D.M., Gabbay, M.: The attack as strong negation, part I. Logic Journal of the IGPL **23**(6), 881–941 (2015)
15. Nouioua, F., Risch, V.: Argumentation frameworks with necessities. In: Proc. of SUM. pp. 163–176 (2011)
16. Nouioua, F.: AFs with necessities: further semantics and labelling characterization. In: Proc. of the International Conference on Scalable Uncertainty Management (SUM). pp. 120–133 (2013)
17. Oren, N., Norman, T.J.: Semantics for evidence-based argumentation. In: Proc. of COMMA. pp. 276–284 (2008)
18. Polberg, S., Oren, N.: Revisiting support in abstract argumentation systems. In: Proc. of COMMA. pp. 369–376 (2014)

A Proofs

Proof of Prop. 1: Let $T \subseteq \mathbf{A}$.
1. Let $a \in T$. Let us assume that T is support-closed in AFN. By definition, a is support-cycle-free in T iff $\forall E \subseteq \mathbf{A}$ such that ENa, there is $b \in E \cap T$ such that b is support-cycle-free in $T \setminus \{a\}$.
$a \in Supp(U_T)$ means $\forall \alpha \in \mathbf{N'}$ such that $t'(\alpha) = x$, $s'(\alpha) \cap (T \cap Supp(U_{T \setminus \{x\}})) \neq \emptyset$
or equivalently, $\forall \alpha \in \mathbf{N'}$ such that $t'(\alpha) = x$, there is $b \in s'(\alpha) \cap T$ such that $b \in Supp(U_{T \setminus \{x\}})$.
So the proof just follows by induction.
2. Assume that U_T is self-supporting. Then $T \subseteq Supp(U_T)$. It follows that T is support-closed in AFN. Then, due to the first part of the proof, every $a \in T$ is support-cycle-free in T. As T is support-closed, it follows that T is coherent.
Conversely, assume that T is coherent, so that every $a \in T$ is support-cycle-free in T. From the first part of the proof, it follows that $T \subseteq Supp(U_T)$. As $\mathbf{R'}$ and $\mathbf{N'}$ are included in $\mathbf{P'}$, it follows that U_T is self-supporting. □

Lemma 1 *Let $T_1 \subseteq T_2 \subseteq \mathbf{A}$. $Supp(U_{T_1}) \subseteq Supp(U_{T_2})$*

> **Proof:** Let $T \subseteq \mathbf{A}$. $Supp(U_T) = \mathbf{P'} \cup \{x \in \mathbf{A}/\forall \alpha \in \mathbf{N'}$ such that $t'(\alpha) = x, s'(\alpha) \cap (T \cap Supp(U_{T \setminus \{x\}})) \neq \emptyset\}$. If $T_1 \subseteq T_2 \subseteq \mathbf{A}$, we have $T_1 \setminus \{x\} \subseteq T_2 \setminus \{x\} \subseteq \mathbf{A}$. So the proof just follows by induction. □

Lemma 2 *Given AFN and its corresponding $RAFN$, let $T \subseteq \mathbf{A}$ and $x \in \mathbf{A}$. If $x \in Supp(U_T)$, $\forall \alpha \in \mathbf{N'}$ such that $t'(\alpha) = x$, $\exists y \in (s'(\alpha) \cap T \setminus \{x\} \cap Supp(U_{T \setminus \{x\}}))\}$.*

> **Proof:** Let $x \in Supp(U_T)$. $\forall \alpha \in \mathbf{N'}$ such that $t'(\alpha) = x$, $\exists y \in (s'(\alpha) \cap T \cap Supp(U_{T \setminus \{x\}}))\}$. Assume that for some α_0 with $t'(\alpha_0) = x$ we have $(s'(\alpha_0) \cap T \cap Supp(U_{T \setminus \{x\}})) = \{x\}$ (1). Then $x \in T$ and $x \in Supp(U_{T \setminus \{x\}})$. So, as $t'(\alpha_0) = x$, $\exists z \in (s'(\alpha_0) \cap T \setminus \{x\} \cap Supp(U_{T \setminus \{x\}}))$. We have $z \neq x$ and $z \in (s'(\alpha_0) \cap T \cap Supp(U_{T \setminus \{x\}}))$, which is in contradiction with the assumption (1). □

Lemma 3 *Given AFN and its corresponding $RAFN$, let $T \subseteq \mathbf{A}$ and $x \in \mathbf{A}$. If $x \in T$ and $x \in Supp(U_T)$, then $\exists C \subseteq T$ such that $x \in C$ and C is coherent in AFN.*

> **Proof:** Let $x \in Supp(U_T)$. Either $x \in \mathbf{P'}$ or x is supported. In the first case, let $C = \{x\}$. Obviously, C is coherent and $C \subseteq T$, as $x \in T$.
> Let us consider the case when x is supported. Let $\alpha_1, \ldots, \alpha_k$ be the supports of x. For each i, there is $y_i \in (s'(\alpha_i) \cap T \cap Supp(U_{T \setminus \{x\}}))$, and from Lemma 2, it can be assumed that $y_i \in T \setminus \{x\}$. Let $S_1(x) = \{y_1, \ldots, y_k\}$. We have $S_1(x) \subseteq (T \setminus \{x\} \cap Supp(U_{T \setminus \{x\}}))$. We consider $C_1 = \{x\} \cup S_1(x)$.
> - If $S_1(x) \subseteq \mathbf{P'}$, it is easy to see that C_1 is support-closed and every $a \in C_1$ is support-cycle-free in C_1. So C_1 is coherent.

- In the other case, for each $y \in S_1(x)$ such that $y \notin \mathbf{P}'$, as $y \in Supp(U_{T \setminus \{x\}})$, as done for x, we can build a set of arguments $S(y) \subseteq (T \setminus \{x, y\} \cap Supp(U_{T \setminus \{x,y\}}))$. We add all these sets $S(y)$ to C_1.
- This construction is iterated and will end as T is reduced at each step (T then $T \setminus \{x\}$ then $T \setminus \{x, y\}$, ...). □

Lemma 4 *Given AFN and its corresponding RAFN, let $T \subseteq \mathbf{A}$ and $x \in \mathbf{A}$.*
1. *If $x \in UnSupp(U_T)$ then x is deactivated by T in AFN.*
2. *If x is deactivated by T in AFN and $x \in \overline{Def_\mathbf{A}(U_T)}$ then $x \in UnSupp(U_T)$.*

Proof: By definition, $UnSupp(U_T) = \overline{Supp(U'_T)}$ where $U'_T = (\overline{Def_\mathbf{A}(U_T)}, \mathbf{R}', \overline{Def'_\mathbf{N}(U_T)})$. As noted before, $Def'_\mathbf{N}(U_T) = \varnothing$. So, $U'_T = (\overline{Def_\mathbf{A}(U_T)}, \mathbf{R}', \mathbf{N}') = U'_{T'}$ where T' denotes $\overline{Def_\mathbf{A}(U_T)}$. Note that T' contains the arguments that are not attacked by T.
  1. We assume that $x \in UnSupp(U_T)$. By definition, $x \notin Supp(U'_T)$ (1). Assume that x is not deactivated by T in AFN. There exists $C \subseteq \mathbf{A}$ coherent subset containing x, such that T attacks no argument of C. It follows that $C \subseteq T'$. As C is coherent in AFN, due to Prop 1, we have that U_C is self-supporting. So, $C \subseteq Supp(U_C)$. Moreover, due to Lemma 1, we have $Supp(U_C) \subseteq Supp(U'_T)$. By transitivity, we obtain $C \subseteq Supp(U'_T)$. As C contains x, we conclude that $x \in Supp(U'_T)$ which is in contradiction with the assumption (1).
  2. We assume that x is deactivated by T in AFN and $x \in \overline{Def_\mathbf{A}(U_T)} = T'$. Assume that $x \notin UnSupp(U_T)$. Then $x \in Supp(U'_T)$. From Lemma 3, $\exists C \subseteq T'$ such that $x \in C$ and C is coherent in AFN. So C is a coherent set containing x such that T attacks no argument of C. That is in contradiction with x being deactivated by T. □

Proof of Prop. 2: Let $T \subseteq \mathbf{A}$ and $a \in \mathbf{A}$.
1. a is acceptable w.r.t. T in AFN means that $T \cup \{a\}$ is coherent and $\forall b \in \mathbf{A}$ such that $b\mathbf{R}a$, b is deactivated by T.
If $T \cup \{a\}$ is coherent, from Prop 1 and Lemma 2, it is easy to prove that $a \in Supp(U_T)$ (i).
Let $\alpha \in \mathbf{R}'$ with $t'(\alpha) = a$. By definition of RAFN, there is $b \in \mathbf{A}$ such that $s'(\alpha) = \{b\}$ and $b\mathbf{R}a$. As a is acceptable w.r.t. T in AFN, b is deactivated by T. Due to Lemma 4, it follows that either $b \in Def_\mathbf{A}(U_T)$ or $b \in UnSupp(U_T)$, or equivalently either $s'(\alpha) \subseteq Def_\mathbf{A}(U_T)$ or $s'(\alpha) \subseteq UnSupp(U_T)$. In both cases, $\alpha \in UnAct(U_T)$ (ii).
So we have proved that a is acceptable w.r.t. U_T in RAFN.
2. Let us assume that T is self-supporting. a is acceptable w.r.t. U_T in RAFN means that $a \in Supp(U_T)$ and for each attack $\alpha \in \mathbf{R}'$ with $t'(\alpha) = a$, $\alpha \in UnAct(U)$. As attacks are neither attacked nor supported in RAFN, $\alpha \in UnAct(U)$ means $s'(\alpha) \subseteq UnAcc(U_T) = (Def_\mathbf{A}(U_T) \cup UnSupp(U_T))$.
As T is self-supporting, and $a \in Supp(U_T)$, due to Lemma 1, it is easy to prove that $T \cup \{a\}$ is also self-supporting and from Prop 1 it follows that $T \cup \{a\}$ is coherent (i).
Let $b \in \mathbf{A}$ such that $b\mathbf{R}a$. By definition of RAFN, there is $\alpha \in \mathbf{R}'$ such that

$s'(\alpha) = \{b\}$ and $t'(\alpha) = a$. As a is acceptable w.r.t. U_T in $RAFN$, we have that $b \in UnAcc(U_T) = (Def_\mathbf{A}(U_T) \cup UnSupp(U_T))$. We have two cases:
- If $b \in UnSupp(U_T)$, from Lemma 4, we conclude that b is deactivated by T in AFN.
- If $b \in Def_\mathbf{A}(U_T)$, there is $c \in T$ such that $c\mathbf{R}b$ in AFN. This is a particular case when b is deactivated by T. Indeed, if T attacks b, T attacks any coherent set containing b.

In both cases, b is deactivated by T (ii).
So we have proved that a is acceptable w.r.t. T in AFN.

□

Proof of Prop. 3: Let $T \subseteq \mathbf{A}$.

1. *Conflict-free semantics*

Note that $Def_{\mathbf{R}'}(U_T) = Def_{\mathbf{N}'}(U_T) = \varnothing$, as interactions target only arguments in $RAFN$. It follows that U_T is a conflict-free structure in $RAFN$ iff $T \cap Def_\mathbf{A}(U_T) = \varnothing$, which is equivalent to T being conflict-free in AFN, due to the definition of $RAFN$.

2. *Admissible semantics*
- Assume that T is an admissible extension of AFN. By definition, T is conflict-free, coherent, and each argument of T is acceptable w.r.t. T. From the previous item of this proposition, U_T is a conflict-free structure of $RAFN$. From Prop 2, each argument of T is acceptable w.r.t. U_T in $RAFN$. It is also the case for the other elements of the structure, as they are neither attacked nor supported. So, U_T is an admissible structure in $RAFN$.
- Assume that U_T is an admissible structure in $RAFN$. By definition, U_T is conflict-free and each argument of U_T is acceptable w.r.t. U_T in $RAFN$. From the previous item of this proposition, T is a conflict-free in AFN. As U_T is an admissible structure, it is self-supporting, so from Prop 1, we have that T is coherent. Lastly, from Prop 2, we have that each argument of T is acceptable w.r.t. T. So T is an admissible extension of AFN.

3. *Complete semantics*

The result follows from the definitions and Prop 2.

4. *Preferred semantics*

In $RAFN$, U_T is a preferred structure iff it is a $\subseteq$-maximal complete structure. In AFN, T is a preferred extension iff T is a $\subseteq$-maximal complete extension.

Assume that U_T is a preferred structure. Then, from the second item of Prop. 3, T is complete. If T is not $\subseteq$-maximal complete, there is T' $\subseteq$-maximal complete extension strictly containing T. Then, we know that $U_{T'}$ is complete and obviously $U_{T'}$ strictly contains U_T. That is in contradiction with U_T being preferred. So T is a preferred extension.

Conversely, assume that T is a preferred structure in AFN. Then, from the second item of Prop. 3, U_T is complete. If U_T is not $\subseteq$-maximal complete, there is a complete structure U' that strictly contains U_T. As U_T has the form $(T, \mathbf{R}', \mathbf{N}')$, it follows that U' has the form $(T', \mathbf{R}', \mathbf{N}')$ with T' strictly containing T. As U' is complete, we have that T' is complete, which is in contradiction with T being preferred. So U_T is a preferred structure.

5. *Stable semantics*

In AFN, T is a stable extension iff T is complete and $\forall a \in \mathbf{A}$, $a \in \mathbf{A} \setminus T$ iff a is deactivated by T. In $RAFN$, U_T is a stable structure iff U_T is complete and $\overline{(T \cup \mathbf{R}' \cup \mathbf{N}')} = UnAcc(U_T)$.

From the second item of Prop. 3, we know that T is complete in AFN iff U_T is a complete structure of $RAFN$. Moreover, $UnAcc(U_T) = (Def_\mathbf{A}(U_T) \cup UnSupp(U_T))$ so $UnAcc(U_T) \subseteq \mathbf{A}$ (indeed, as each interaction belongs to $\mathbf{P'}$, $UnSupp(U_T) \cap (\mathbf{R'} \cup \mathbf{N'}) = \varnothing$). And as said above in the proof of Prop 2, each argument in $Def_\mathbf{A}(U_T)$ is deactivated by T, so from Lemma 4, we have that $x \in UnAcc(U_T)$ iff x is deactivated by T. It follows that $\forall a \in \mathbf{A}$, ($a \in \mathbf{A} \setminus T$ iff a is deactivated by T) is equivalent to ($a \in \overline{(T \cup \mathbf{R'} \cup \mathbf{N'})}$ iff $a \in UnAcc(U_T)$), in other words, T is a stable extension of AFN iff U_T is a stable structure in $RAFN$.

6. Grounded semantics

In AFN, T is a grounded extension iff T is a $\subseteq$-minimal complete extension. In $RAFN$, U_T is a grounded structure iff U_T is a $\subseteq$-minimal complete structure.

Assume that T is a grounded extension. From the second item of Prop. 3, we know that U_T is complete. If U_T is not a $\subseteq$-minimal complete structure, there is a complete structure $U' = (T', \Gamma', \Delta')$ such that $T' \cup \Gamma' \cup \Delta'$ is strictly included in $T \cup \mathbf{R'} \cup \mathbf{N'}$. As U' is complete, we have $U' = Acc(U')$. As $(\mathbf{R'} \cup \mathbf{N'}) \subseteq \mathbf{P'}$, and as no interaction is attacked, we have $(\mathbf{R'} \cup \mathbf{N'}) \subseteq Acc(U')$, so $(\mathbf{R'} \cup \mathbf{N'}) \subseteq U'$. Hence $U' = U_{T'}$ where $T' \subset T$. As U' is complete, we know that T' is complete in AFN. So we have T' a complete extension of AFN strictly included in T $\subseteq$-minimal complete extension. There is a contradiction. So, we have proved that U_T is a grounded structure.

Conversely, assume that U_T is a grounded structure. From the second item of Prop. 3, we know that T is a complete extension. If T is not a $\subseteq$-minimal complete extension, there is a complete extension T' such that $T' \subset T$. As T' is complete we have that $U_{T'}$ is a complete structure. Obviously we have $U_{T'} \subset U_T$, which is in contradiction with U_T being a $\subseteq$-minimal complete structure. So we have proved that T is a grounded extension.

□

B Synthesis of RAFN examples

In this section, for each example used previously,[9] we give the results obtained with the different approaches in terms of semantics.

Note that, in these examples, all semantics (complete, preferred, stable and grounded) give exactly the same result. It is not always the case (see the two last examples given in this appendix).

Note also that for each example, at least two results are given:
- one corresponding to our approach for RAFN using Def. 21,
- one corresponding to the approach for the ASAF given in [11].

Then, according to the structure of the framework, we also give Dung's extensions, AFN extensions and RAF extensions (NA means "not applicable").

The following examples illustrate some differences between our approach and that of [11]:
- either because of cycles in the argumentation framework: Example 4,
- or because some supports have a set as a source: Examples 7 and 15,
- or because of the definition of defeats in the ASAF: Example 14.

Type of approach	Example number			
	2	3	4	6
RAFN	$(\{a\}, \{\alpha\}, \{\})$	$(\{a, b\}, \{\}, \{\alpha\})$	$(\{\}, \{\}, \{\beta, \alpha\})$	$(\{a, b, c\}, \{\}, \{\alpha_1, \alpha_2\})$
ASAF	$\{a, \alpha\}$	$\{a, b, \alpha\}$	NA	$\{a, b, c, \alpha_1, \alpha_2\}$
AF	$\{a\}$	NA	NA	NA
AFN	$\{a\}$	$\{a, b\}$	$\{\}$	$\{a, b, c\}$
RAF	$(\{a\}, \{\alpha\})$	NA	NA	NA

Type of approach	Example number		
	7	10	11
RAFN	$(\{a, b, c, d\}, \{\}, \{\alpha_1, \alpha_2\})$	$(\{a, b, c, d\}, \{\alpha_3\}, \{\alpha_2\})$	$(\{a, b, c\}, \{\beta\}, \{\})$
ASAF	NA	$\{a, b, c, d, \alpha_3, \alpha_2\}$	$\{a, b, c, \beta\}$
AF	NA	NA	NA
AFN	$\{a, b, c, d\}$	NA	NA
RAF	NA	NA	NA

Type of approach	Example number	
	12	13
RAFN	$(\{a, b, c\}, \{\}, \{\alpha, \delta\})$	$(\{a, y, z, t\}, \{\beta\}, \{\alpha_1, \alpha_2, \alpha_3\})$
ASAF	$\{a, b, c, \alpha, \delta\}$	$\{a, y, z, t, \beta, \alpha_1, \alpha_2, \alpha_3\}$
AF	NA	NA
AFN	NA	$\{a, y, z, t\}$
RAF	NA	NA

---

[9] Except Examples 1 and 5 that are not totally formalized and Examples 8 to 9 that are EBAF or REBAF examples.

Type of approach	Example number	
	14	15
RAFN	$(\{a,x,t\},\{\beta,\gamma\},\{\alpha_1,\alpha_2,\alpha_3\})$	$(\{a,c,d,e\},\{\alpha_3\},\{\alpha_1,\alpha_2\})$
ASAF	$(\{a,x,t,\gamma,\alpha_1,\alpha_2,\alpha_3\})$	NA
AF	NA	NA
AFN	$\{a,x,t\}$	$\{a,c,d,e\}$
RAF	NA	NA

The next two examples illustrate the fact that results can be different depending on the chosen semantics.

Ex. 16 *Consider the following RAFN framework.*

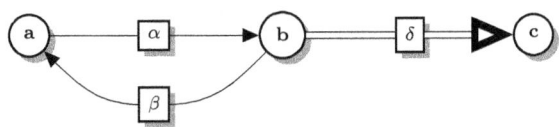

Some interesting structures in our RAFN approach:

complete	$(\{\},\{\alpha,\beta\},\{\delta\}), (\{a\},\{\alpha,\beta\},\{\delta\}), (\{b,c\},\{\alpha,\beta\},\{\delta\})$
preferred	$(\{a\},\{\alpha,\beta\},\{\delta\}), (\{b,c\},\{\alpha,\beta\},\{\delta\})$
stable	$(\{a\},\{\alpha,\beta\},\{\delta\}), (\{b,c\},\{\alpha,\beta\},\{\delta\})$
grounded	$(\{\},\{\alpha,\beta\},\{\delta\})$

Some interesting extensions using ASAF approach [11]:

complete	$\{\delta\}, \{a,\alpha,\delta\}, \{b,c,\beta,\delta\}$
preferred	$\{a,\alpha,\delta\}, \{b,c,\beta,\delta\}$
stable	$\{a,\alpha,\delta\}, \{b,c,\beta,\delta\}$
grounded	$\{\delta\}$

There is no one to one correspondence between our approach and [11]. Indeed, with the approach proposed by [11], α and β cannot be accepted together. This example is also an AFN. In this case, the resulting extensions are:

complete	$\{\}, \{a\}, \{b,c\}$
preferred	$\{a\}, \{b,c\}$
stable	$\{a\}, \{b,c\}$
grounded	$\{\}$

Ex. 17 *Consider the following RAFN framework.*

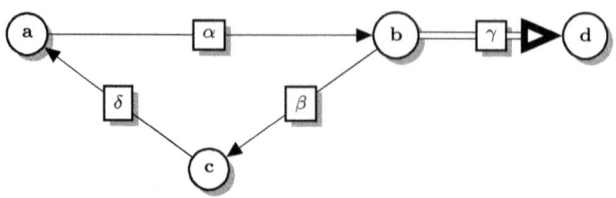

Some interesting structures in our RAFN approach:

complete	$(\{\}, \{\alpha, \beta, \delta\}, \{\gamma\})$
preferred	$(\{\}, \{\alpha, \beta, \delta\}, \{\gamma\})$
stable	no stable
grounded	$(\{\}, \{\alpha, \beta, \delta\}, \{\gamma\})$

Some interesting extensions using ASAF approach [11]:

complete	$\{\gamma\}$
preferred	$\{\gamma\}$
stable	no stable
grounded	$\{\gamma\}$

There is no one to one correspondence between our approach and [11]. Indeed, with the approach proposed by [11], α, β and δ cannot be accepted together.

This example is also an AFN. In this case, the resulting extensions are:

complete	$\{\}$
preferred	$\{\}$
stable	no stable
grounded	$\{\}$

Extending Defeasible Logic Programming with Informant-based Argumentation

Andrea Cohen, Sebastian Gottifredi,
Luciano H. Tamargo, Alejandro J. García

Institute for Computer Science and Engineering (CONICET-UNS)
Department of Computer Science and Engineering, Universidad Nacional del Sur
San Andrés 800 - Campus de Palihue, (8000) Bahía Blanca, Argentina
{ac,sg,lt,ajg}@cs.uns.edu.ar

Abstract. This work introduces an informant-based argumentative approach for deliberative agents that can receive domain knowledge from other agents in a multi-agent setting. Also, agents will be able to model, through the use of backing and detracting rules, reasons for and against the consideration of any piece of information provided by the informant agents. Then, from all this knowledge, agents will be able to build different kinds of arguments, from which different kinds of conflicts may arise. Furthermore, agents will be equipped with a credibility order over informant agents, that may be used within an argument comparison criterion in order to decide between conflicting arguments. As a result, our approach extends Defeasible Logic Programming by enabling to argue for and against the consideration of informant agents, and resolving conflicts based on the credibility attributed to each information source, to finally determine the warranted beliefs the agents have.

1 Introduction

In this chapter we introduce an informant-based argumentative approach for deliberative agents that participate in a multi-agent system. Our motivation is to provide agents in this setting with a knowledge representation and reasoning tool, allowing them to handle the information they send/receive to/from other agents in the group. We assume that agents share tentative information and that each agent obtains information from informant agents that can have different degrees of credibility. In order to reach to conclusions, agents will build arguments based on the information in their knowledge bases, which includes the pieces of information they received from other agents. Then, those arguments could be challenged due to the information they use, but also because of the credibility of their information sources. These two ways of challenging an argument will lead to us to considering two kinds of conflicts or attacks. To address all these issues, we will introduce an extension of Defeasible Logic Programming [3] that combines and extends the approaches of [14] and [2].

Defeasible Logic Programming (DeLP for short) provides a computational reasoning tool that uses an argumentation engine to obtain answers from a knowledge base represented using a logic programming language extended with defeasible rules. DeLP has its roots in the seminal work by Guillermo Simari in his PhD thesis [12] and published in [11], where a mathematical treatment of defeasible reasoning was very clearly presented and explained. The complete formalization of DeLP was later developed in [4], and after that several extensions were developed enhancing both the representational and inferential capabilities of DeLP (see *e.g.*, [5,6]). In particular, as mentioned before, our current proposal will be based on two of those extensions (namely, [14] and [2]).

In [14] an extension of DeLP with the possibility of having defeasible information received from informant agents was introduced. There, the credibility assigned to informants was considered as part of the argument comparison process. As a simple example, consider an agent I_1 that has to decide whether to buy a used smartphone; this agent receives advice from two informant agents I_2 and I_3. Let us also assume that agent I_1 regards I_3 as less credible than I_2. Now, I_2 informs I_1 that "the phone is a good option", whereas I_3 informs I_1 that "the phone is not a good option". With the extension of DeLP developed in [14], agent I_1 will arrive to the conclusion that "the phone is a good option" based on information that was provided by the more credible agent. However, in that approach, reasons for or against the credibility assigned to the informant agents cannot be considered. For instance, suppose we would like to extend the previous example to model that any piece of information provided by I_2 should be taken into account whenever it refers to electronic devices, because I_2 is an expert on that matter; also, that the information provided by I_2 should not be considered when it comes to an item that I_2 owns. With the aim of being able to model and reason with this new kind of information, we will follow the approach of [2].

In [2], the authors proposed an extension of DeLP that allows for the representation of attack and support for defeasible rules. In that work, the extended representational language incorporates two new kind of rules: backing rules and undercutting rules. On the one hand, a backing rule models the notion of *backing* proposed by Toulmin [15] and expresses that *reasons to believe in the body of the rule, provide reasons in favor of using the defeasible rule in its head*. The intuition here is that a backing rule establishes conditions under which the use of the defeasible rule it supports makes sense. On the other hand, an undercutting rule models the notion of *undercut* introduced by Pollock [10] and expresses that *reasons to believe in the body of the rule, provide reasons against using the defeasible rule in its head*. In this case, the intended meaning is that an undercutting rule establishes conditions under which the defeasible rule it is related to should not be used. As will become clear later, these ideas will allow us to define new kinds of rules in our approach: backing and retracting rules. With these new rules, we will be able to define the conditions under which the information received from the different informant agents should (respectively, should not) be considered. Going back to our previous example, we could use a backing rule to express that information provided by I_2 should be taken in account when it refers to electronic devices. Similarly, we could use a retracting rule to express that information given by I_2 should not be considered when it comes to an item that I_2 owns.

The rest of this chapter is organized as follows. In Section 2 we will introduce the elements to represent the agents' knowledge. Section 3 shows how an agent can build different kinds of arguments and identify different kinds of attacks between them. Then, in Section 4, we will introduce the ways in which the conflicts between arguments are resolved, leading to defeats. Moreover, taking those defeats into account, we will discuss how the warranted beliefs of an agent are determined. Finally, Section 5 provides some conclusions and discusses relevant related work.

2 Knowledge Representation

In this section we will introduce our proposal for agents' knowledge representation that will be defined as an *Informant-based DeLP program*. We assume that agents share tentative information in the form of defeasible rules, and that each agent obtains information from informant agents that have different degrees of credibility. Also, backing and retracting rules will be proposed in our formalization, that allow to represent reasons in favor and against defeasible rules provided by some informants.

The assumption of the existence of a total order over informants is not quite realistic in many multi-agent application domains and a similar observation applies to the existence of a global order shared by all agents. With this observation in mind, the approach to be introduced below will consider that every agent has its own partial order defined over its informants, representing the credibility assigned to those informants. Each agent has a knowledge base where every sentence is attached with an agent identifier representing that the corresponding agent is the source of that piece of information. In addition, agents could communicate with their peers for obtaining new information or for sharing their beliefs. Clearly, as agents may disagree with one another, they may store beliefs that are in conflict with another agent's knowledge. Then, when sharing conflicting information, the credibility order among the informant agents can be used to decide which information prevails. Next, we briefly introduce the formalization of credibility orders, which will be used throughout the rest of the paper.

We assume a finite set $\mathbb{I}$ of identifiers for naming informant agents that is shared by all agents. Agents' identifiers will be denoted with an uppercase typewriter letter "I" that can have letters and natural numbers as subscripts (*i.e.*, $\mathbb{I} = \{I_a, I_b, \ldots, I_z, I_1, I_2, \ldots, I_n, \ldots\}$), and each identifier is unequivocally associated with a single agent.

Each agent will have its own *credibility order*, that will be represented by an irreflexive, asymmetric, and transitive binary relation over $\mathbb{I}$ denoted $<_{co}^{I_x}$ (*i.e.*,

$<_{co}^{I_x}$ is a strict partial order over $\mathbb{I}$), where the superscript $I_x \in \mathbb{I}$ stands for the agent identifier this order belongs to. For instance, $<_{co}^{I_a}$ is the credibility order of the agent identified as I_a. In this chapter we will assume that the credibility order relates agents that are sources of information of the same topic; multi-topic or multi-context credibility orders are left as future work. The notation $I_b <_{co}^{I_a} I_c$ means that agent I_a deems agent I_b as less credible than agent I_c (equivalently, under I_a's perspective, I_c is more credible than I_b). Then, the notation $I_b \not<_{co}^{I_a} I_c$ is used to express that, for agent I_a, I_b is not less credible than I_c. Furthermore, if $I_b \not<_{co}^{I_a} I_c$ and $I_c \not<_{co}^{I_a} I_b$, then agent I_a regards I_b and I_c to be incomparable, noted as $I_b \sim_{co}^{I_a} I_c$. It should be noted that agents may change their assessment of one another; as a result, this change would impact their credibility orders, resulting in an update. The dynamic nature of credibility orders is outside the scope of this paper. However, for instance, we could use the formalism proposed in [13], which provides a mechanism for handling the dynamics behind the use of a credibility order, to complement our proposal.

Agents can receive information from different sources, that may not be equally credible. Thus, the information they provide, which may be contradictory, will be considered to be tentative. In this setting, we need a mechanism for dealing with uncertain and conflicting information, able to ultimately determine the beliefs an agent should commit to. In particular, when deciding between contradictory conclusions, the reasoning mechanism may rely on the credibility order the agent has. In our approach, knowledge representation and reasoning will be based on Defeasible Logic Programming (DeLP). Our interest in this particular computational tool is that its language provides the declarative capability of representing weak information in the form of *defeasible rules* and *presumptions*, and its defeasible argumentation inference mechanism allows warranting conclusions in the presence of contradictory information. Below, we will summarily present the elements of DeLP's representational language[1], and introduce the new elements that will be used for representing an agent's knowledge.

The representational language of DeLP is defined in terms of three disjoint sets: a set of *facts*, a set of *strict rules* and a set of *defeasible rules*. These facts and rules are obtained using ground literals, which correspond to ground atoms or negated ground atoms using *strong negation*. As defined in [3] (following the ideas of [1]), the symbol "$\sim$" represents the strong negation, which differs from the *default negation* frequently used in logic programming and the *classical negation* used in classical logic. The aim of having strong negation in DeLP is to allow for the defeasible derivation of potentially contradictory information.

Facts and strict rules in DeLP express non-defeasible or indisputable information, whereas defeasible rules express tentative information that may be used when nothing could be posed against it. As mentioned before, we assume

---

[1] We refer the reader to [3] for a full account of Defeasible Logic Programming.

the knowledge an agent has is tentative since it is provided by informant agents. Therefore, we will only make use of defeasible rules, which are defined next.

Definition 1 (Ground Literal). *Let $\mathcal{L}$ be a set of ground atoms. A ground literal L (or literal, for short) is an atom A or a negated atom $\sim A$, where $A \in \mathcal{L}$.*

Definition 2 (Defeasible Rule). *A defeasible rule is an ordered pair, denoted "Head $\prec$ Body", where the first element (Head) is a ground literal and the second element (Body) is a finite set of ground literals. A defeasible rule with head L_0 and body $\{L_1, \ldots, L_n\}$ ($n \geq 0$) will also be written as $L_0 \prec L_1, \ldots, L_n$.*

A defeasible rule "*Head $\prec$ Body*" expresses that "*reasons to believe in the antecedent Body give reasons to believe in the consequent Head.*" For instance, the defeasible rule "*closed_roads $\prec$ snowing*" represents that "*reasons to believe that is snowing provide reasons to believe that the roads are closed*", whereas "*$\sim$closed_roads $\prec$ snowplows*" represents that "*reasons to believe that snowplows are working provide reasons to believe that the roads are not closed.*" In particular, a defeasible rule with an empty body is noted as "*Head $\prec$* " and is called a *presumption* [8]. Hence, for instance, the presumption "*snowplows $\prec$* " expresses that "*there are defeasible reasons to believe that snowplows are working*".

In order to associate defeasible rules (which represent domain knowledge) to their informant agents, we characterize the notion of *domain object*.

Definition 3 (Domain Object). *Let $\mathbb{I}$ be a finite set of agent identifiers. A Domain object is a tuple $(R, \mathtt{I})$, where $\mathtt{I} \in \mathbb{I}$ and R is a defeasible rule.*

In addition to defeasible rules, we introduce *backing rules* and *detracting rules* to express reasons for and against the consideration of informants, respectively. The addition of these new types of rules will allow to argue about the contexts in which the information provided by the agents should be used or not. Formally:

Definition 4 (Backing Rule). *Let $\mathbb{I}$ be a finite set of agent identifiers. A Backing Rule is an ordered pair, denoted "Head $\leftarrow\!\oplus$ Body", where Head $\in \mathbb{I}$ is the identifier of an informant agent and Body is a finite and non-empty set of literals. A backing rule with head $\mathtt{I}$ and body $\{L_1, \ldots, L_n\}$ ($n \geq 1$) will also be written as $\mathtt{I} \leftarrow\!\oplus L_1, \ldots, L_n$.*

Definition 5 (Detracting Rule). *Let $\mathbb{I}$ be a finite set of agent identifiers. A Detracting Rule is an ordered pair, denoted "Head $\leftarrow\!\otimes$ Body", where Head $\in \mathbb{I}$ is the identifier of an informant agent and Body is a finite and non-empty set of literals. A detracting rule with head $\mathtt{I}$ and body $\{L_1, \ldots, L_n\}$ ($n \geq 1$) will also be written as $\mathtt{I} \leftarrow\!\otimes L_1, \ldots, L_n$.*

Syntactically, the only difference between backing and detracting rules is the use of $\oplus$ and $\otimes$; however, these two types of rules are semantically different. A backing rule "$\text{I} \leftarrow\!\!\oplus Body$" expresses that *"the antecedent Body gives reasons for considering the information provided by the informant I"*. On the contrary, a detracting rule "$\text{I} \leftarrow\!\!\otimes Body$" expresses that *"the antecedent Body gives reasons against considering the information provided by the informant I"*. When convenient, we will refer to backing and detracting rules as *informant rules*.

The entire knowledge of an agent, which will be used to make inferences and construct arguments, is composed of a set of domain objects and a set of informant rules. This knowledge base is called *informant-based DeLP program*.

Definition 6 (Informant-Based DeLP program). *Let $\mathbb{I}$ be a finite set of agent identifiers. An informant-based DeLP program (or IBDP for short) is a pair (Δ, Σ) where Δ is a finite set of domain objects and Σ is a finite set of informant rules.*

Example 1. Let us consider a scenario where an agent named Ana, whose identifier is I_a, has to decide whether to take some days off work or not. She received information from three other agents besides herself: her work partner I_p, her boss I_b, and her assigned client I_c. Ana's knowledge for making such a decision is codified within the IBDP $\text{P}_{\text{I}_a} = (\Delta_{\text{I}_a}, \Sigma_{\text{I}_a})$, where the literals *tdo*, *lwa*, *cip* and *pwp* have the following intended meaning:

$$\begin{aligned} tdo & : \text{ take some days off} \\ lwa & : \text{ low workload ahead} \\ cip & : \text{ client interaction phase} \\ pwp & : \text{ work partner seems to want Ana's position} \end{aligned}$$

$$\Delta_{\text{I}_a} = \begin{Bmatrix} (\sim tdo \prec, \text{I}_a) & (lwa \prec, \text{I}_c) \\ (tdo \prec lwa, \text{I}_b) & (lwa \prec, \text{I}_p) \\ (cip \prec, \text{I}_b) & (pwp \prec, \text{I}_c) \end{Bmatrix} \quad \Sigma_{\text{I}_a} = \begin{Bmatrix} \text{I}_c \leftarrow\!\!\oplus cip \\ \text{I}_p \leftarrow\!\!\otimes pwp \end{Bmatrix}$$

The domain object $(\sim tdo \prec, \text{I}_a)$ represents that "usually, Ana thinks she should not take some days off"; the object $(tdo \prec lwa, \text{I}_b)$ expresses that "according to Ana's boss, if she has low work load ahead, then she can take some days off", whereas the object $(cip \prec, \text{I}_b)$ represents that "Ana's boss has informed her that they are currently in the client interaction phase of the project". Then, the domain objects $(lwa \prec, \text{I}_c)$ and $(lwa \prec, \text{I}_p)$ represent that both Ana's client and work partner assume that "there is low workload ahead". In addition, the domain object $(pwp \prec, \text{I}_c)$ expresses that "Ana's client thinks her partner wants her position". Finally, the set Σ_{I_a} in Ana's IBDP includes a backing rule stating that "the information provided by the client should be considered when the project is in the client interaction phase" and a detracting rule expressing that "she should not consider the information provided by her partner if he is aiming to obtain her position".

Note that in an IBDP there can exist two or more domain objects having the same rule R but different informants; this occurs in Example 1, where the presumption "$lwa \prec$ " has I_c and I_p as associated informants. This does not mean that the agent's knowledge base is redundant; rather, it encodes the fact that the same piece of information was received from different sources. This feature can be considered as an advantage of our approach since, as mentioned before, the credibility order of an agent may change (even dynamically); hence, at any moment, the more credible informant of R could be considered. Furthermore, given the existence of detracting rules, a specific informant of R could be challenged, whereas others are not. Again, a situation like this is evidenced in Example 1, where there exists a detracting rule for the informant I_p of "$lwa \prec$ ", but not for the informant I_c. As a result, the existence of multiple informants may strengthen the position of a piece of domain information in the agent's knowledge base.

It is important to remark that the existence of backing and detracting rules in an IBDP is not mandatory. Thus, the domain objects provided by an agent for which no backing rules exist can be regarded as applicable by default, since there are not explicit requirements for their use. For instance, the informants I_a, I_b and I_p from Example 1 could be considered as applicable since there are no backing rules for them.

In our approach, an agent will be specified in terms of four components: its own *agent identifier*, an *informant-based DeLP program* used to store its knowledge, a *credibility order* among informants, and an *argument comparison criterion* that will be used in the warranting process for deciding the accepted arguments and the justified conclusions. The first three elements have been introduced above. On the other hand, the argument comparison criterion will be discussed further below. An agent is formally defined as follows.

Definition 7 (Agent). *Let $\mathbb{I}$ be a finite set of agent identifiers. An agent is a tuple $(\text{I}, \text{P}_\text{I}, <_{co}^\text{I}, \succ_\text{I})$ where $\text{I} \in \mathbb{I}$, $\text{P}_\text{I} = (\Delta_\text{I}, \Sigma_\text{I})$ is an informant-based DeLP program, $<_{co}^\text{I}$ is a credibility order over $\mathbb{I}$, and $\succ_\text{I}$ is an argument comparison criterion. In addition, for every domain object $(R, \text{I}') \in \Delta_\text{I}$, it holds that $\text{I}' \in \mathbb{I}$. Analogously, for every backing rule $\text{I}'' \leftarrow\oplus Body'' \in \Sigma_\text{I}$ and every detracting rule $\text{I}''' \leftarrow\otimes Body''' \in \Sigma_\text{I}$, it holds that $\text{I}'', \text{I}''' \in \mathbb{I}$.*

Given the characterization of an agent, it can be the case that two different agents $(\text{I}_1, \text{P}_1, <_{co}^{\text{I}_1}, \succ_1)$ and $(\text{I}_2, \text{P}_2, <_{co}^{\text{I}_2}, \succ_2)$ have the same IBDP and the same argument comparison criterion (thus, $\text{P}_1 = \text{P}_2$ and $\succ_1 = \succ_2$) but a different credibility order. In such a case, even though the two agents share the same knowledge, if the comparison criterion is defined in terms of the credibility order, the conflicts arising from the consideration of inconsistent information may be resolved differently for the two agents. Hence, in such a case, the agents' inferences (warranted conclusions) may differ. On the other hand, even though two agents $(\text{I}_3, \text{P}_3, <_{co}^{\text{I}_3}, \succ)$ and $(\text{I}_4, \text{P}_4, <_{co}^{\text{I}_4}, \succ)$ cannot have the same agent identifier, they can share every other component (*i.e.*, $\text{P}_3 = \text{P}_4$, $<_{co}^{\text{I}_3} = <_{co}^{\text{I}_4}$ and $\succ_3 = \succ_4$); clearly, in this case, the two agents will draw the same conclusions.

Example 2. Consider the scenario introduced in Example 1 and the set of informants $\mathbb{I} = \{I_a, I_p, I_c, I_b\}$. Then, agent Ana can be specified as $(I_a, P_{I_a}, <_{co}^{I_a}, \succ_{I_a})$, where P_{I_a} is the IBDP of Example 1 and $<_{co}^{I_a}$ is such that $I_a <_{co}^{I_a} I_c <_{co}^{I_a} I_b$ and $I_p <_{co}^{I_a} I_b$.

3 Arguments and Attacks

A central piece of this formalism that will allow the agent to handle contradictory domain information is the notion of argument. Intuitively, an argument is a structure whose conclusion is obtained from a set of premises, informed by some agents, through the use of a reasoning mechanism. In particular, the claims of the arguments will be the tentative beliefs of the agent. When analyzing an argument for a particular belief, the agent can find other arguments, referred to as counter-arguments, that are in conflict with it. In particular, a conflict may arise because the counter-argument contradicts some domain information (*i.e.*, a premise, an intermediate conclusion or the final claim of the argument) or because it challenges an informant in the original argument. In this situation, it is necessary to have a mechanism for comparing the conflicting arguments to decide which one prevails. This analysis leads to a dialectical process seeking to validate the arguments in conflict. The arguments that survive all possible attacks from their counter-arguments are said to *warrant* their conclusions or claims.

Next, we will show how an agent can build different types of arguments using the domain objects and the informant rules stored in its informant-based DeLP program. As a preliminary notion, before defining the concept of argument, we introduce the *defeasible derivation*.

Definition 8 (Defeasible Derivation). *Let* $\mathsf{P} = (\Delta, \Sigma)$ *be an* IBDP *and* L *a literal. A defeasible derivation of* L *from* $\mathsf{S} \subseteq \Delta \cup \Sigma$, *denoted* $\mathsf{S} \mathrel{\vert\!\sim}_\mathsf{P} L$, *consists of a finite sequence* $L_1, L_2, \ldots, L_n = L$ *of literals, where each literal* L_i ($1 \leq i \leq n$) *is in the sequence because there exists a domain object* $(R, \mathtt{I})$ *in* S *such that:*

1. *either:*
   (a) $R = L_i \prec$, *or*
   (b) $R = L_i \prec B_1, \ldots, B_k$, *and every* B_t ($1 \leq t \leq k$) *is an element* L_j *of the sequence appearing before* L_i ($j < i$); *and*
2. *one of the following conditions holds:*
   (a) *there is no backing rule in* Σ *with head* $\mathtt{I}$, *or*
   (b) *there is a backing rule in* S *with head* $\mathtt{I}$ *and body* $S_1, \ldots, S_m$, *and every* S_p ($1 \leq p \leq m$) *is an element* L_v *of the sequence appearing before* L_i ($v < i$).

A derivation for a literal L is called "defeasible" because as we will show next, there may exist information in contradiction with L, or any of the literals appearing in the sequence, and in certain conditions this could prevent the acceptance of L as a warranted belief. It should be noted that rules from different informants can be combined together to derive a literal. In particular, conditions (a) and (b) in the second clause of Definition 8 state that in order to use a presumption or a defeasible rule R informed by an agent $\mathtt{I}$ in the derivation process, it must be the case that either: there are no explicit requirements for considering $\mathtt{I}$ as a reliable source of information; or the information provided by $\mathtt{I}$ can only be used under the conditions specified by one (any) of the backing rules for $\mathtt{I}$. Also, note that the set S contains all the elements available for obtaining the derivation. However, when looking for the existence of backing rules for $\mathtt{I}$, the entire set of informant rules Σ must be taken into account to ensure that all possible constraints associated with $\mathtt{I}$ are considered. Finally, it is important to observe that from the same IBDP it is possible to obtain several, distinct, defeasible derivations for a given literal. Furthermore, as the following example shows, from a given IBDP it is possible to obtain defeasible derivations for complementary literals (w.r.t. the strong negation "$\sim$").

Example 3. Let us consider the IBDP $\mathsf{P}_{\mathtt{I}_a}$ from Example 1. From that program, it is possible to obtain two defeasible derivations for the literal "*tdo*". On the one hand, the sequence '*lwa, tdo*' is a defeasible derivation for "*tdo*", obtained from the domain objects $(tdo \prec lwa, \mathtt{I}_b)$ and $(lwa \prec, \mathtt{I}_p)$. On the other hand, the sequence '*cip, lwa, tdo*' is another defeasible derivation for "*tdo*", obtained from the domain objects $(tdo \prec lwa, \mathtt{I}_b)$ and $(lwa \prec, \mathtt{I}_c)$, and the backing rule $\mathtt{I}_c \leftarrow \oplus cip$. Note that $\mathsf{P}_{\mathtt{I}_a}$ also allows to derive the literal "$\sim tdo$" using the domain object $(\sim tdo \prec, \mathtt{I}_a)$. That is, from the same IBDP it is possible to obtain defeasible derivations for complementary literals.

Given that, as shown by the preceding example, defeasible derivations for complementary literals can be obtained from an IBDP, we need to identify coherent sets of elements within an IBDP, which will be referred to as *non-contradictory*. The notion of a *contradictory set* of domain objects and informant rules of an IBDP is defined next.

Definition 9 (Contradictory Set). *Let* $\mathsf{P} = (\Delta, \Sigma)$ *be an* IBDP *and* $\mathsf{S} \subseteq \Delta \cup \Sigma$. *We say that the set* S *is* contradictory *if and only if there exist two complementary literals L and $\sim L$ such that* $\mathsf{S}\mid\!\sim_\mathsf{P} L$ *and* $\mathsf{S}\mid\!\sim_\mathsf{P} \sim L$.

Example 4. Let $\mathsf{P}_{\mathtt{I}_a}$ be the IBDP introduced in Example 1. The set $\mathcal{A}_x = \{(tdo \prec lwa, \mathtt{I}_b), (lwa \prec, \mathtt{I}_p), (\sim tdo \prec, \mathtt{I}_a)\}$ is contradictory whereas the set $\mathcal{A}_y = \{(\sim tdo \prec, \mathtt{I}_a), (tdo \prec lwa, \mathtt{I}_b)\}$ is not. As a result, for instance, every superset of $\mathcal{A}_x$ will also be contradictory.

Observe that detracting rules are not used for obtaining defeasible derivations. As will be shown next, they will only be used to build arguments that challenge the consideration of informant agents. The usual definition of argument is then extended to consider backing and detracting rules, when required. As a result, we will distinguish among three different types of arguments: the first type regards arguments for literals, whereas the other two deal with arguments for or against informants, respectively.

Definition 10 (Claim Argument). *Let* $\mathsf{P} = (\Delta, \Sigma)$ *be an* IBDP *and* L *a literal.* $\langle \mathcal{A}, L \rangle$ *is a* claim argument *for the literal* L*, obtained from* P*, if the following conditions hold:*

1. $\mathcal{A} \subseteq (\Delta \cup \Sigma)$,
2. $\mathcal{A} \mathrel{|\!\sim}_\mathsf{P} L$,
3. $\mathcal{A}$ *is non-contradictory, and*
4. $\mathcal{A}$ *is minimal: there is no* $\mathcal{B} \subset \mathcal{A}$ *satisfying conditions 2 and 3.*

Definition 11 (Backing Argument). *Let* $\mathbb{I}$ *be a finite set of agent identifiers,* $\mathsf{P} = (\Delta, \Sigma)$ *an* IBDP *and* $\mathtt{I} \in \mathbb{I}$. $\langle \mathcal{A}, \mathtt{I} \rangle_b$ *is a* backing argument *for the informant* $\mathtt{I}$*, obtained from* P*, if* $\mathcal{A} = \{\mathtt{I} \hookleftarrow\!\oplus L_1, \ldots, L_n\} \cup \mathcal{A}'$ *and the following conditions hold:*

1. $\mathcal{A} \subseteq (\Delta \cup \Sigma)$,
2. $\mathcal{A}' \mathrel{|\!\sim}_\mathsf{P} L_i$ $(1 \leq i \leq n)$,
3. $\mathcal{A}'$ *is non-contradictory, and*
4. $\mathcal{A}'$ *is minimal: there is no* $\mathcal{B} \subset \mathcal{A}'$ *satisfying conditions 2 and 3.*

Definition 12 (Detracting Argument). *Let* $\mathbb{I}$ *be a finite set of agent identifiers,* $\mathsf{P} = (\Delta, \Sigma)$ *an* IBDP *and* $\mathtt{I} \in \mathbb{I}$. $\langle \mathcal{A}, \mathtt{I} \rangle_d$ *is a* detracting argument *for the informant* $\mathtt{I}$*, obtained from* P*, if* $\mathcal{A} = \{\mathtt{I} \hookleftarrow\!\otimes L_1, \ldots, L_n\} \cup \mathcal{A}'$ *and the following conditions hold:*

1. $\mathcal{A} \subseteq (\Delta \cup \Sigma)$,
2. $\mathcal{A}' \mathrel{|\!\sim}_\mathsf{P} L_i$ $(1 \leq i \leq n)$,
3. $\mathcal{A}'$ *is non-contradictory, and*
4. $\mathcal{A}'$ *is minimal: there is no* $\mathcal{B} \subset \mathcal{A}'$ *satisfying conditions 2 and 3.*

Briefly, all arguments share the characteristic of being a minimal and non-contradictory set of rules that allows to defeasibly derive a conclusion or the conditions for or against the consideration of an informant agent. Nevertheless, despite their common features, the three argument types are mutually exclusive. When convenient, we will abstract from an argument's type, referring to it just as an argument (*i.e.*, omitting its type). Then, given an argument built from an IBDP, we can define the notion of *sub-argument* as follows.

Definition 13 (Sub-argument). *Let* $\mathsf{P} = (\Delta, \Sigma)$ *be an* IBDP *and* $\langle \mathcal{A}_1, L_1 \rangle$, $\langle \mathcal{A}_2, L_2 \rangle$ *two arguments built from* P*. We say that* $\langle \mathcal{A}_2, L_2 \rangle$ *is a* sub-argument *of* $\langle \mathcal{A}_1, L_1 \rangle$ *if and only if* $\mathcal{A}_2 \subseteq \mathcal{A}_1$.

Example 5. Given the specification of the agent of Example 2, agent $\mathtt{I}_a$ will be able to build the following arguments from $\mathsf{P}_{\mathtt{I}_a}$:

$\langle \mathcal{A}_1, {\sim}tdo \rangle$, with $\mathcal{A}_1 = \{({\sim}tdo \prec\!\!-\ , \mathtt{I}_a)\}$
$\langle \mathcal{A}_2, tdo \rangle$, with $\mathcal{A}_2 = \{(tdo \prec\!\!- lwa,\ \mathtt{I}_b), (lwa \prec\!\!-\ , \mathtt{I}_c), (cip \prec\!\!-\ , \mathtt{I}_b), \mathtt{I}_c \hookleftarrow\!\oplus cip\}$
$\langle \mathcal{A}_3, tdo \rangle$, with $\mathcal{A}_3 = \{(tdo \prec\!\!- lwa,\ \mathtt{I}_b), (lwa \prec\!\!-\ , \mathtt{I}_p)\}$
$\langle \mathcal{A}_4, lwa \rangle$, with $\mathcal{A}_4 = \{(lwa \prec\!\!-\ , \mathtt{I}_c), (cip \prec\!\!-\ , \mathtt{I}_b), \mathtt{I}_c \hookleftarrow\!\oplus cip\}$
$\langle \mathcal{A}_5, lwa \rangle$, with $\mathcal{A}_5 = \{(lwa \prec\!\!-\ , \mathtt{I}_p)\}$

$\langle \mathcal{A}_6, cip \rangle$, with $\mathcal{A}_6 = \{(cip \prec , \mathtt{I}_b)\}$
$\langle \mathcal{A}_7, pwp \rangle$, with $\mathcal{A}_7 = \{(pwp \prec , \mathtt{I}_c), (cip \prec , \mathtt{I}_b), \mathtt{I}_c \leftarrow\!\!\oplus cip\}$
$\langle \mathcal{A}_8, \mathtt{I}_c \rangle_b$, with $\mathcal{A}_8 = \{(cip \prec , \mathtt{I}_b), \mathtt{I}_c \leftarrow\!\!\oplus cip\}$
$\langle \mathcal{A}_9, \mathtt{I}_p \rangle_d$, with $\mathcal{A}_9 = \{(pwp \prec , \mathtt{I}_c), (cip \prec , \mathtt{I}_b), \mathtt{I}_c \leftarrow\!\!\oplus cip, \mathtt{I}_p \leftarrow\!\!\otimes pwp\}$
Note that the first seven are claim arguments, whereas the last two correspond to a backing and a detracting argument, respectively. In addition, for instance, observe that $\langle \mathcal{A}_4, lwa \rangle$ is a sub-argument of $\langle \mathcal{A}_2, tdo \rangle$ and that the backing argument $\langle \mathcal{A}_8, \mathtt{I}_c \rangle_b$ is a sub-argument of $\langle \mathcal{A}_2, tdo \rangle$, $\langle \mathcal{A}_4, lwa \rangle$ and $\langle \mathcal{A}_7, pwp \rangle$.

The above example illustrates that from a given IBDP it is possible to build arguments that are in conflict with one another, such as $\langle \mathcal{A}_1, \sim tdo \rangle$ and $\langle \mathcal{A}_2, tdo \rangle$, or $\langle \mathcal{A}_1, \sim tdo \rangle$ and $\langle \mathcal{A}_3, tdo \rangle$. These conflicts become evident since the arguments' conclusions correspond to complementary literals. However, the existence of detracting arguments may lead to a new kind of conflict, which challenges the consideration of an informant agent within an argument. The two kinds of conflict that may occur between arguments built from an IBDP, from hereon referred to as *attacks*, are formally defined next.

Definition 14 (Conclusion Attack). *Let $\langle \mathcal{A}_1, L_1 \rangle$ be a claim argument and $\langle \mathcal{A}_2, L_2 \rangle$ any argument. We say that $\langle \mathcal{A}_1, L_1 \rangle$ c-attacks $\langle \mathcal{A}_2, L_2 \rangle$ at the literal L if there exists a claim sub-argument $\langle \mathcal{A}, L \rangle$ of $\langle \mathcal{A}_2, L_2 \rangle$ such that L_1 and L are complementary literals with respect to "$\sim$".*

Definition 15 (Informant Attack). *Let $\langle \mathcal{A}_1, \mathtt{I} \rangle_d$ be a detracting argument and $\langle \mathcal{A}_2, L_2 \rangle$ any argument. We say that $\langle \mathcal{A}_1, \mathtt{I} \rangle_d$ i-attacks $\langle \mathcal{A}_2, L_2 \rangle$ at the informant $\mathtt{I}$ if there exists a domain object $(R, \mathtt{I}) \in \mathcal{A}_2$.*

Whenever an argument $\langle \mathcal{A}_1, L_1 \rangle$ attacks (either c-attacks or i-attacks) another argument $\langle \mathcal{A}_2, L_2 \rangle$, we will say that $\langle \mathcal{A}_1, L_1 \rangle$ is a *counter-argument* of $\langle \mathcal{A}_2, L_2 \rangle$.

Example 6. Let us consider the arguments from Example 5. Argument $\langle \mathcal{A}_1, \sim tdo \rangle$ c-attacks argument $\langle \mathcal{A}_2, tdo \rangle$ at the literal "tdo"; the attacked sub-argument in this case is $\langle \mathcal{A}_2, tdo \rangle$ itself. Observe that, in addition, $\langle \mathcal{A}_2, tdo \rangle$ c-attacks $\langle \mathcal{A}_1, \sim tdo \rangle$ at the literal "$\sim tdo$". Similar c-attacks also exist between $\langle \mathcal{A}_1, \sim tdo \rangle$ and $\langle \mathcal{A}_3, tdo \rangle$. Finally, it holds that $\langle \mathcal{A}_9, \mathtt{I}_p \rangle_d$ i-attacks arguments $\langle \mathcal{A}_3, tdo \rangle$ and $\langle \mathcal{A}_5, lwa \rangle$ at the informant $\mathtt{I}_p$, because $(lwa \prec , \mathtt{I}_p) \in \mathcal{A}_3$ and $(lwa \prec , \mathtt{I}_p) \in \mathcal{A}_5$.

As the preceding example shows, conclusion attacks are symmetric in a sense. That is, if $\langle \mathcal{A}_1, L_1 \rangle$ c-attacks $\langle \mathcal{A}_2, L_2 \rangle$, then the attacked sub-argument $\langle \mathcal{A}, L \rangle$ of $\langle \mathcal{A}_2, L_2 \rangle$ is such that it c-attacks $\langle \mathcal{A}_1, L_1 \rangle$. In particular, if $\langle \mathcal{A}_1, L_1 \rangle$ c-attacks $\langle \mathcal{A}_2, L_2 \rangle$ on its final conclusion L_2 (*i.e.*, L_1 and L_2 are complementary literals with respect to "$\sim$"), then $\langle \mathcal{A}_2, L_2 \rangle$ c-attacks $\langle \mathcal{A}_1, L_1 \rangle$. In contrast, informant attacks need not be symmetric. This is due to the fact that the attacking argument is a detracting argument, and no particular attacked sub-argument is identified. However, this does not prevent detracting arguments from attacking each other. For instance, it could be the case that $\langle \mathcal{A}_1, \mathtt{I}_1 \rangle_d$

i-attacks $\langle \mathcal{A}_2, \mathtt{I}_2 \rangle_d$ at the informant $\mathtt{I}_1$, and $\langle \mathcal{A}_2, \mathtt{I}_2 \rangle_d$ i-attacks $\langle \mathcal{A}_1, \mathtt{I}_1 \rangle_d$ at the informant $\mathtt{I}_2$; what would occur in such a case is that there exist two domain objects $(R, \mathtt{I}_1)$ and $(R', \mathtt{I}_2)$ such that $(R, \mathtt{I}_1) \in \mathcal{A}_2$ and $(R', \mathtt{I}_2) \in \mathcal{A}_1$.

Given that an agent may build multiple arguments, which in turn may have several counter-arguments, in order to determine the agent's beliefs we need to determine what are the non-defeated arguments. To establish whether an argument $\langle \mathcal{A}, L \rangle$ is non-defeated, it is necessary to explicitly account for all its counter-arguments. Let $\langle \mathcal{A}_1, L_1 \rangle, \langle \mathcal{A}_2, L_2 \rangle, \ldots, \langle \mathcal{A}_k, L_k \rangle$ be the counter-arguments of $\langle \mathcal{A}, L \rangle$. If any counter-argument $\langle \mathcal{A}_i, L_i \rangle$ is (according to the argument comparison criterion) "better" than or unrelated to $\langle \mathcal{A}, L \rangle$, then $\langle \mathcal{A}_i, L_i \rangle$ is a candidate for defeating $\langle \mathcal{A}, L \rangle$. However, if argument $\langle \mathcal{A}, L \rangle$ is "better" than $\langle \mathcal{A}_i, L_i \rangle$, then this counter-argument will not be taken in consideration as a defeater for $\langle \mathcal{A}, L \rangle$ Therefore, in order to determine the defeaters of an argument, we will make use of the argument comparison criterion. Then, once all successful attacks are identified, we will be able to determine the acceptance status of the argument and establish whether its conclusion is warranted or not. This issues will be addressed in the following section.

4 Defeats and Warrant

Agents build arguments from their knowledge bases to support their beliefs. However, as discussed in Section 3, the existence of conflicting information within an agent's knowledge base leads to the existence of attacks. Furthermore, since attacks could succeed or fail, it is necessary to have a comparison criterion to determine whether the attacking argument prevails, in which case it becomes a defeater. Informally, an attack will be considered to be effective when the attacked argument is not better than the attacking argument with respect to a comparison criterion. In particular, when comparing arguments built from an agent's IBDP, the comparison criterion "$\succ$" will be the one provided in the agent's specification. In particular, given two arguments $\langle \mathcal{A}_1, L_1 \rangle$ and $\langle \mathcal{A}_2, L_2 \rangle$, $\langle \mathcal{A}_1, L_1 \rangle \succ \langle \mathcal{A}_2, L_2 \rangle$ means that $\langle \mathcal{A}_1, L_1 \rangle$ is better than (or preferred to) $\langle \mathcal{A}_2, L_2 \rangle$; analogously, $\langle \mathcal{A}_1, L_1 \rangle \not\succ \langle \mathcal{A}_2, L_2 \rangle$ means that $\langle \mathcal{A}_1, L_1 \rangle$ is not better than (or not preferred to) $\langle \mathcal{A}_2, L_2 \rangle$. The formal definition, adapted from [3], follows.

Definition 16 (Defeat). *Let $\langle \mathcal{A}_1, L_1 \rangle$ and $\langle \mathcal{A}_2, L_2 \rangle$ be two arguments built from an* IBDP*. We say that $\langle \mathcal{A}_1, L_1 \rangle$ defeats $\langle \mathcal{A}_2, L_2 \rangle$ if one of the following conditions hold:*

1. *$\langle \mathcal{A}_1, L_1 \rangle$ c-attacks $\langle \mathcal{A}_2, L_2 \rangle$ at the literal L, and the attacked sub-argument $\langle \mathcal{A}, L \rangle$ of $\langle \mathcal{A}_2, L_2 \rangle$ is such that $\langle \mathcal{A}, L \rangle \not\succ \langle \mathcal{A}_2, L_2 \rangle$;*
2. *$\langle \mathcal{A}_1, L_1 \rangle$ i-attacks $\langle \mathcal{A}_2, L_2 \rangle$ at the informant L_1, and there is no backing sub-argument of $\langle \mathcal{A}_2, L_2 \rangle$ for the informant L_1; or*
3. *$\langle \mathcal{A}_1, L_1 \rangle$ i-attacks $\langle \mathcal{A}_2, L_2 \rangle$ at the informant L_1, and the backing sub-argument $\langle \mathcal{A}, L_1 \rangle$ of $\langle \mathcal{A}_2, L_2 \rangle$ for the informant L_1 is such that $\langle \mathcal{A}, L_1 \rangle \not\succ \langle \mathcal{A}_1, L_1 \rangle$.*

It is important to observe the difference in the resolution of c-attacks and i-attacks. For c-attacks, it suffices to compare the attacking argument with the attacked sub-argument. In contrast, since i-attacks target an informant and the attacked argument may be such that it has no backing sub-argument for that informant, two cases have to be considered. Then, in the case where there is a backing argument for that informant, the attacking argument (which is a detracting argument) is compared to the corresponding backing argument to decide which one prevails. In that way, backing arguments are intended to defend their associated informants to prevent an i-attack from becoming a defeat.

According to the specification of an agent characterized in Definition 7, the argument comparison criterion is modular. Thus, it can be selected accordingly, depending on the domain of the information represented within the agent's knowledge base. Furthermore, as previously discussed throughout this chapter, the comparison criterion can make use of the agent's credibility order; examples of such criteria are the ones proposed in [14].

Next, for illustration purposes, we introduce the *single rule credibility criterion* from [14]. This criterion will prefer an argument $\langle \mathcal{A}_1, L_1 \rangle$ over an argument $\langle \mathcal{A}_2, L_2 \rangle$ if there is at least one domain object in $\mathcal{A}_1$ whose informant is more credible than the informant of some domain object in $\mathcal{A}_2$, and no informant of the domain objects in $\mathcal{A}_2$ is more credible than one informant of the domain objects in $\mathcal{A}_1$. Formally:

Definition 17 (Single Rule Credibility Criterion). *Let* $(\mathtt{I}, \mathtt{P_I}, <^\mathtt{I}_{co}, \succ_\mathtt{I})$ *be the specification of an agent, and* $\langle \mathcal{A}_1, L_1 \rangle$ *and* $\langle \mathcal{A}_2, L_2 \rangle$ *two arguments built from the* IBDP $\mathtt{P_I}$. *We will say that* $\langle \mathcal{A}_1, L_1 \rangle$ *is preferred to* $\langle \mathcal{A}_2, L_2 \rangle$, *denoted* $\langle \mathcal{A}_1, L_1 \rangle \succ_\mathtt{I} \langle \mathcal{A}_2, L_2 \rangle$, *if and only if:*

1. *There exist* $(R, \mathtt{I}_i) \in \mathcal{A}_1$ *and* $(S, \mathtt{I}_j) \in \mathcal{A}_2$ *such that* $\mathtt{I}_j <^\mathtt{I}_{co} \mathtt{I}_i$; *and*
2. *There are no* $(T, \mathtt{I}_k) \in \mathcal{A}_1$ *and* $(U, \mathtt{I}_l) \in \mathcal{A}_2$ *such that* $\mathtt{I}_k <^\mathtt{I}_{co} \mathtt{I}_l$.

Example 7. Let $(\mathtt{I}_a, \mathtt{P}_{\mathtt{I}_a}, <^{\mathtt{I}_a}_{co}, \succ_{\mathtt{I}_a})$ be the agent specified in Example 2. Let us assume that $\succ_{\mathtt{I}_a}$ corresponds to the *single rule credibility criterion* from Definition 17. Then, using the $\succ_{\mathtt{I}_a}$ comparison criterion, we identify the following defeats:

- $\langle \mathcal{A}_2, tdo \rangle$ defeats $\langle \mathcal{A}_1, \sim tdo \rangle$. In particular, $\langle \mathcal{A}_2, tdo \rangle \succ_{\mathtt{I}_a} \langle \mathcal{A}_1, \sim tdo \rangle$ since $(\sim tdo \prec\!\!\prec\, , \mathtt{I}_a)$, the only domain object in $\mathcal{A}_1$ is such that $\mathtt{I}_a <^{\mathtt{I}_a}_{co} \mathtt{I}_b$ and $\mathcal{A}_2$ contains, among others, the domain object $(tdo \prec\!\!\prec lwa, \mathtt{I}_b)$.

- $\langle \mathcal{A}_3, tdo \rangle$ defeats $\langle \mathcal{A}_1, \sim tdo \rangle$. Here, like in the previous case, it holds that $\langle \mathcal{A}_3, tdo \rangle \succ_{\mathtt{I}_a} \langle \mathcal{A}_1, \sim tdo \rangle$ since $(\sim tdo \prec\!\!\prec\, , \mathtt{I}_a)$, the only domain object in $\mathcal{A}_1$ is such that $\mathtt{I}_a <^{\mathtt{I}_a}_{co} \mathtt{I}_b$ and $\mathcal{A}_3$ contains, among others, the domain object $(tdo \prec\!\!\prec lwa, \mathtt{I}_b)$.

- $\langle \mathcal{A}_9, \mathtt{I}_p \rangle_d$ defeats both $\langle \mathcal{A}_3, tdo \rangle$ and $\langle \mathcal{A}_5, lwa \rangle$. These defeats occur because the attacked arguments have no backing sub-argument for the informant $\mathtt{I}_p$ (*i.e.*, the defeats correspond to the second case of Definition 16).

To determine whether an agent I can accept a belief L, it is necessary to find out if, from the IBDP $\mathtt{P}_\mathtt{I}$, it is possible to build an argument $\langle \mathcal{A}, L \rangle$ that survives all the defeats it receives. Naturally, this will require the consideration of all possible defeaters of $\langle \mathcal{A}, L \rangle$. Then, given a defeater $\langle \mathcal{B}, L' \rangle$, defeaters for it will also have to be considered, as well as the defeaters of those defeaters, and so on. This dialectical analysis will lead to the construction of a tree structure of defeaters, called *dialectical tree*, which gathers all sequences of defeaters starting from a given argument (the root argument, in our example $\langle \mathcal{A}, L \rangle$). Hence, every node in the tree will be a defeater of its parent, and the leaves will correspond to undefeated arguments. As a result, once built, the dialectical tree is marked according to the following criterion: (1) leaf nodes are marked as undefeated and (2) a non-leaf node (*i.e.*, an inner node or the root) is marked as undefeated if all its children are marked as defeated; otherwise it is marked as defeated. In particular, a dialectical tree whose root argument is $\langle \mathcal{A}, L \rangle$ will be noted as $\mathcal{T}_{\langle \mathcal{A}, L \rangle}$. In addition, the mark of an undefeated argument will be noted as **U** and the mark of a defeated argument will be noted as **D**. For full details on the construction of dialectical trees and their marking criterion, we refer the reader to [3].

Example 8. Consider the agent $(\mathtt{I}_a, \mathtt{P}_{\mathtt{I}_a}, <_{co}^{\mathtt{I}_a}, \succ_{\mathtt{I}_a})$ specified in Example 2 and the arguments from Example 5. Figures 1(a), 1(b) and 1(c) respectively depict the marked dialectical trees for arguments $\langle \mathcal{A}_1, \sim tdo \rangle$, $\langle \mathcal{A}_2, tdo \rangle$ and $\langle \mathcal{A}_3, tdo \rangle$. Arguments are depicted with triangles, and defeats are represented as solid arrows between arguments. The circles beside the defeasible rules symbol "$\prec$" in the arguments are used to indicate the informant associated to each domain object in those arguments. Also, each argument is marked as defeated or undefeated with a circle containing a **D** or a **U**, respectively.

It should be noted that a marked dialectical tree embodies a dialectical analysis considering every possible argument an agent can build for and against the root argument in the tree. Hence, if the root argument is marked as **U**, it means that the conclusion of that argument is warranted, and the agent can adopt it as a belief. Moreover, the existence of *one* argument $\langle \mathcal{A}, L \rangle$ marked as **U** in its dialectical tree $\mathcal{T}_{\langle \mathcal{A}, L \rangle}$ is sufficient for the agent to adopt L as a belief. On the contrary, if every argument for L is marked as **D** in its own tree, then the literal L will not be warranted and thus, the agent will not adopt it as a belief. The notion of warranted literal is formalized next:

Definition 18 (Warranted Literal). *Let* $(\mathtt{I}, \mathtt{P}_\mathtt{I}, <_{co}^\mathtt{I}, \succ_\mathtt{I})$ *be the specification of an agent and L a literal. We say that I warrants L as a belief if there exists an argument $\langle \mathcal{A}, L \rangle$ built from $\mathtt{P}_\mathtt{I}$ such that $\langle \mathcal{A}, L \rangle$ is marked as **U** in its dialectical tree* $\mathcal{T}_{\langle \mathcal{A}, L \rangle}$.

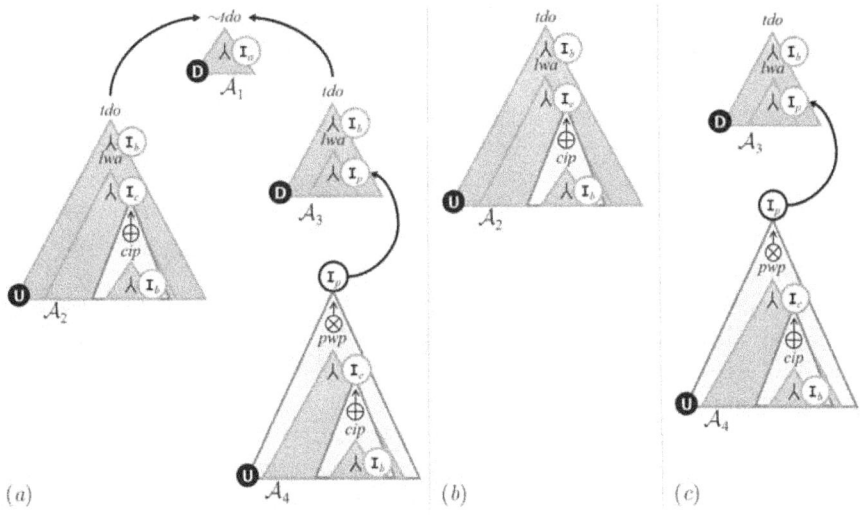

Fig. 1. Dialectical trees (a) $\mathcal{T}_{\langle \mathcal{A}_1, \sim tdo \rangle}$, (b) $\mathcal{T}_{\langle \mathcal{A}_2, tdo \rangle}$ and (c) $\mathcal{T}_{\langle \mathcal{A}_3, tdo \rangle}$.

Example 9. Consider the agent $(\mathtt{I}_a, \mathtt{P}_{\mathtt{I}_a}, <_{co}^{\mathtt{I}_a}, \succ_{\mathtt{I}_a})$ specified in Example 2 and the marked dialectical trees from Example 8, illustrated in Figure 1. Agent $\mathtt{I}_a$ will warrant "*tdo*" as a belief since, as depicted in Figure 1(*b*), argument $\langle \mathcal{A}_2, tdo \rangle$ is marked as **U** in its dialectical tree $\mathcal{T}_{\langle \mathcal{A}_2, tdo \rangle}$; this holds in spite of argument $\langle \mathcal{A}_3, tdo \rangle$ being marked as **D** in its dialectical tree $\mathcal{T}_{\langle \mathcal{A}_3, tdo \rangle}$. Furthermore, agent $\mathtt{I}_a$ does not warrant the literal "$\sim tdo$" as a belief, since the only argument for that literal, $\langle \mathcal{A}_1, \sim tdo \rangle$, is marked as **D** in its dialectical tree $\mathcal{T}_{\langle \mathcal{A}_1, \sim tdo \rangle}$ (see Figure 1(*a*)).

5 Conclusions and Related Work

In this chapter we have presented an argumentative reasoning formalism where the credibility of informants plays a central role. In our approach, defeasible rules (which represent domain knowledge) are associated to their informant agents. Also, we introduced two new kinds of rules, backing and detracting rules, in order to be able to argue about the contexts in which the domain knowledge provided by the informant agents should be used or not. In other words, these rules are used to express reasons for and against the consideration of informants, respectively. From all this knowledge, an agent will be able to construct arguments to support its inferences. In addition, agents are equipped with a credibility order among informant agents and a comparison criterion, which can be used to asses the strength of conflicting arguments.

We have proposed a notion of defeasible derivation that combines the use of defeasible rules and backing rules, and leads to the construction of different types of arguments. The first two types of argument characterize the information they support: domain arguments provide support for domain knowledge, whereas backing arguments provide support for informants. In contrast, arguments of the third type, detracting arguments, are against the consideration of informant agents. As we have shown, these arguments can be in conflict. On the one hand, domain arguments can support complementary information. On the other hand, a detracting argument challenges the information provided by an agent; therefore, it will be in conflict with every domain argument containing information provided by such agent.

We considered that every source of domain information is attributed a certain credibility, leading to a perception of their trustworthiness. Trust in turn may be used to define a preference order on the set of arguments, helping to decide which conclusions are warranted as the agents' beliefs. In particular, in this work we adapted one of the comparison criteria of [14], based on this attributed credibility, which is used to determine what argument prevails in a conflict and leads to a defeat relation between conflicting arguments. Finally, based on the available arguments and the defeat relation on them, we followed the dialectical proof procedure for warranting conclusions proposed in [3], for classical DeLP, to determine the agents' beliefs.

In the literature there exist various approaches that combine results from trust and argumentation (see *e.g.*, [7, 9, 16, 17]); in general, these works are focused on using argumentation to reason about trust. Similarly to them, we allow to argue about the contexts in which the information provided by different informant agents should be considered. Thus, even though we do not change the credibility order over the informant agents, this could be viewed as a form of reasoning about trust. Also, we take the credibility or trustworthiness of information sources into account as part of the decision process (*i.e.*, we perform reasoning with trust). As a result, our approach can be considered to be complementary to theirs. Notwithstanding this, since we know there may exist other similarities and differences between the above mentioned works and ours, we plan to thoroughly analyze and discuss their relationship as part of future work.

Acknowledgment

Guillermo is our academic father and, besides being our teacher, supervisor and mentor, over time has become a great friend. He has been, and continues to be, our guide and example to follow. We will always be grateful to him for his invaluable advice and for sharing his wisdom with us. Today, we are what we are, to a large extent, thanks to Guillermo.

References

1. José Júlio Alferes, Luís Moniz Pereira, and Teodor C. Przymusinski. Strong and explicit negation in non-monotonic reasoning and logic programming. In *JELIA*

1996, pages 143–163, 1996.
2. Andrea Cohen, Alejandro Javier García, and Guillermo Ricardo Simari. A structured argumentation system with backing and undercutting. *Eng. Appl. of AI*, 49:149–166, 2016.
3. A. J. García and G. R. Simari. Defeasible logic programming: An argumentative approach. *Theory and Practice of Logic Programming*, 4(1-2):95–138, 2004.
4. Alejandro J. García. *Defeasible Logic Programming: Definition, Operational Semantics and Parallelism. (Ph.D. Thesis)*. PhD thesis, Computer Science and Engineering Department, Universidad Nacional del Sur, Bahía Blanca, Argentina, December 2000.
5. Alejandro Javier García, Jürgen Dix, and Guillermo Ricardo Simari. Argument-based logic programming. In Guillermo Ricardo Simari and Iyad Rahwan, editors, *Argumentation in Artificial Intelligence*, pages 153–171. Springer, 2009.
6. Alejandro Javier García and Guillermo Ricardo Simari. Defeasible logic programming: Delp-servers, contextual queries, and explanations for answers. *Argument & Computation*, 5(1):63–88, 2014.
7. A. Hunter. Reasoning about the appropriateness of proponents for arguments. In D. Fox and C. P. Gomes, editors, *Proc. of the 23rd AAAI Conference on Artificial Intelligence, AAAI*, pages 89–94. AAAI Press, 2008.
8. Donald Nute. Defeasible reasoning: a philosophical analysis in PROLOG. In J. H. Fetzer, editor, *Aspects of Artificial Intelligence*, pages 251–288. Kluwer Academic Pub., 1988.
9. S. Parsons, P. McBurney, and E. Sklar. Reasoning about trust using argumentation: A position paper. In *ArgMAS*, pages 159–170, 2010.
10. John L. Pollock. Defeasible reasoning. *Cognitive Science*, 11(4):481–518, 1987.
11. G. R. Simari and R. P. Loui. A Mathematical Treatment of Defeasible Reasoning and its Implementation. *Artificial Intelligence*, 53:125–157, 1992.
12. Guillermo R. Simari. *A Mathematical Treatment of Defeasible Reasoning and its Implementation*. PhD thesis, Washington University, Department of Computer Science (Saint Louis, Missouri, EE.UU.), 1989.
13. Luciano H. Tamargo, Alejandro J. García, Marcelo A. Falappa, and Guillermo R. Simari. On the revision of informant credibility orders. *Artificial Intelligence*, 212:36–58, 2014.
14. Luciano Héctor Tamargo, Sebastian Gottifredi, Alejandro Javier García, Marcelo Alejandro Falappa, and Guillermo Ricardo Simari. Deliberative DeLP agents with multiple informants. *Inteligencia Artificial, Revista Iberoamericana de Inteligencia Artificial*, 15(49):13–30, 2012.
15. Stephen E. Toulmin. *The Uses of Argument*. Cambridge University Press, 1958.
16. S. Villata, G. Boella, D. M. Gabbay, and L. Van der Torre. Arguing about trust in multiagent systems. In *Proc. of the 11th Symposium on Artificial Intelligence of the Italian Association for Artificial Intelligence (AIIA'10)*, pages 236–243, 2010.
17. S. Villata, G. Boella, D. M. Gabbay, and L. Van der Torre. Arguing about the trustworthiness of the information sources. In *ECSQARU*, pages 74–85, 2011.

Infinite Arguments and Semantics of Dialectical Proof Procedures

Phan Minh Thang[1] and Phan Minh Dung[2]

[1] International College of Burapha University, Thailand
[2] Department of Computer Science, Asian Institute of Technology, Thailand

Abstract. We study the semantics of dialectical proof procedures. As dialectical proof procedures are in general sound but not complete wrt admissibility semantics, a natural question here is whether we could give a more precise semantical characterization of what they compute? Based on a new notion of infinite arguments, we introduce a stronger notion of admissibility, referred to as strong admissibility, and show that dialectical proof procedures are in general sound and complete wrt strong admissibility.

1 Introduction

Argumentation is a reasoning model in which reasons for conclusions that are drawn as well for resolving conflicts are given explicitly. While abstract argumentation studies the acceptance of arguments, structured argument systems like assumption-based argumentation or defeasible logic programming provide frameworks for structuring arguments based on assumptions and rules ([23, 8, 3, 15, 4, 21, 16]). Argument-based systems are becoming increasingly popular due to their intuitive appeal to the ways humans perform their practical and daily reasoning ([12, 2, 1, 30, 25]).

Dialectical proof procedures for argumentation have been developed either for abstract argumentation ([5, 13, 20, 31, 29, 11]) or for rule-based instances of it like logic programming ([14, 6, 7]) or assumption-based argumentation ([9, 10, 27, 28]). While the procedures for abstract argumentation are sound and under reasonable conditions also complete wrt admissibility semantics, the procedures developed for structured argumentation systems are in general sound but not complete wrt admissible semantics. *A natural question here is whether we could give a more precise semantical characterization of what dialectical proof procedures compute?* The following example illustrates this point.

Example 1. Consider two assumption-based argumentation frameworks below.
$\mathcal{F}_1: \ p \leftarrow \alpha \qquad \overline{\alpha} \leftarrow f \qquad f \leftarrow f \qquad \overline{\beta} \leftarrow$

$\mathcal{F}_2: \ p \leftarrow \alpha \qquad \overline{\alpha} \leftarrow \beta, f \qquad f \leftarrow f \qquad \overline{\beta} \leftarrow$

where α, β are assumptions and $\overline{\alpha}, \overline{\beta}$ are contraries of α, β.

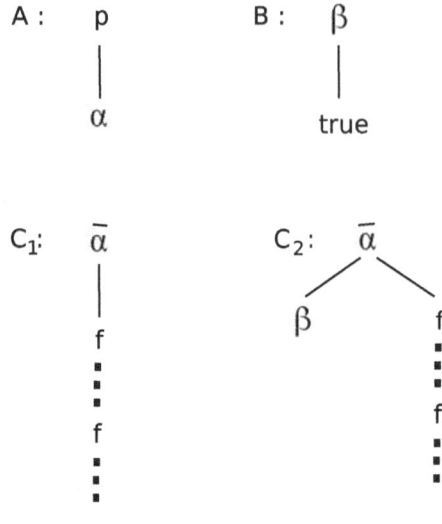

Fig. 1.

The relevant arguments (defined as proof trees) for both programs are given in figure 1. An argument X attacks an argument Y if the conclusion of X is the contrary of an assumption in the support of Y.

The unique argument supporting p in both frameworks is A supported by assumption α. Due to the loop involving f in both frameworks, it is obvious that $\overline{\alpha}$ is not supported as there is no (finite) argument supporting it. Hence there are no attacks against both A and B in both frameworks. $\{A, B\}$ is consequently admissible wrt both frameworks. But dialectical reasoning engines like the proof procedures for logic programming or assumption-based argumentation in [19, 14, 7, 9, 10, 27, 28, 18] fail to deliver A wrt the first framework $\mathcal{F}_1$ as they could not overcome the non-termination of the process to construct an argument supporting $\overline{\alpha}$ due to the loop involving f, though all of them deliver A wrt the second framework $\mathcal{F}_2$ despite the presence of the seemingly similar loop involving f in $\mathcal{F}_2$.

The distinct behavior of the dialectical proof procedures for structured argumentation systems wrt the two frameworks $\mathcal{F}_1, \mathcal{F}_2$ suggests that to understand their semantics, we need to consider the effects of "infinite loops" on their behavior. In this paper, we accomplish this by introducing a new notion of infinite arguments.

For framework $\mathcal{F}_1$, there is an infinite argument for $\overline{\alpha}$ representing by the infinite proof tree C_1. This argument can not be attacked as it is not based on any assumption.

For framework $\mathcal{F}_2$, there is an infinite argument for $\overline{\alpha}$ representing by the infinite proof tree C_2. This argument is based on assumption β and hence is attacked by the argument B.

Therefore, if infinite arguments are taken into account, the set $\{A, B\}$ is not admissible wrt $\mathcal{F}_1$ while it is still admissible wrt $\mathcal{F}_2$.

The example suggests that a stronger notion of admissibility taking into account also the effects of infinite arguments, is needed to characterize the semantics of dialectical proof procedures.

The paper is organized in 4 sections including the introduction. In the following section, we recall the basic notions of abstract and assumption-based argumentation as well as introduce the infinite arguments together with a new notion of strong admissibility. We then present two dialectical proof procedures in section 3 and show that they are sound and complete wrt strong admissibility. We then conclude the paper with a short discussion.

2 Assumption-based Argumentation with Infinite Arguments

2.1 Abstract Argumentation

An argumentation framework ([8]) is a pair $AF = (AR, att)$, where AR is a set of arguments, and att is a binary relation over AR representing the attack relation between the arguments where $(A, B) \in att$ means that A attacks B). A set S of arguments attacks an argument A if some argument in S attacks A;

A set S of arguments is *conflict-free* iff it does not attack itself. S is *self-defensible* iff S attacks each argument attacking S. S is *admissible* iff S is conflict-free and self-defensible. S is a *preferred extension* iff S is maximally (wrt set inclusion) admissible.

An argument A is said to be credulously accepted iff it is contained in at least one preferred extension. It follows that an argument is credulously accepted iff it belongs to an admissible set of arguments.

2.2 Assumption-Based Argumentation

Given a logical language $\mathcal{L}$, an assumption-based argumentation (ABA) framework ([3]) is a triple $\mathcal{F} = (\mathcal{R}, \mathcal{A}, ^-)$ where $\mathcal{R}$ is a set of inference rules of the form $l_0 \leftarrow l_1, \ldots l_n$ (for $n \geq 0$ and $l_n \in \mathcal{L}$), and $\mathcal{A} \subseteq \mathcal{L}$ is a set of assumptions, and $^-$ is a (total) mapping from $\mathcal{A}$ into $\mathcal{L}$, where $\bar{x}$ is referred to as the *contrary* of x.

Assumptions in $\mathcal{A}$ do not appear in the heads of rules in $\mathcal{R}$. Without loss of generality, we assume that the function $\bar{\alpha}$ is one-one, i.e. for two different assumptions $\alpha, \beta, \bar{\alpha}, \bar{\beta}$ are different. For simplification, we also assume that $\bar{\alpha}$ is not an assumption.[3]

From now on until the end of the paper, we assume an arbitrary but fixed assumption-based argumentation framework $\mathcal{F} = (\mathcal{R}, \mathcal{A}, ^-)$.

---

[3] Logic programming is a well-known instance of ABA where the contrary of a negation-as-failure assumption not_p is p.

A *partial proof tree* for a sentence σ is a (possibly infinite) tree with nodes labeled by sentences in $\mathcal{L}$ such that

- the root is labeled by σ
- for every node N labeled by γ if N is not a leaf node then there is an inference rule $\gamma \leftarrow b_1,\ldots,b_m$ ($m \geq 0$) in $\mathcal{R}$ and either $m = 0$ and the child of N is *true* or $m > 0$ and N has m children, labeled by $b_1,\ldots,b_m$ respectively

The *conclusion* of a partial proof tree T, denoted by $Cl(T)$, is the sentence labeling its root. The *support* of a partial proof tree T, denoted by $Sp(T)$, is the set of all sentences labeling the leaves of T and different to *true*.

For a set S of partial proof trees, $Cl(S)$ (resp $Sp(S)$) denotes the set of conclusions (resp. the union of supports) of trees in S.

A *full proof tree* is a partial proof tree whose support consists only of assumptions.

An *argument* for α is a full proof tree for α.

The set of all arguments wrt the ABA framework $\mathcal{F}$ is denoted by $AR_\mathcal{F}$ while the set of all finite arguments is denoted by $AR_{f,\mathcal{F}}$

As infinite arguments do not provide support for their conclusions, it can not be contained in admissible sets. But infinite arguments do represent possible threats to the acceptance other arguments if dialectical proof procedures are deployed to compute their acceptance.[4] This insight motivates the following definition of attack relation between arguments.

Definition 1.
An argument A attacks an argument B iff one of the following conditions holds:

1. *The conclusion of A is the contrary of some assumption in the support of B.*
2. *A and B are identical and infinite.*

The attack relation between arguments in $AR_\mathcal{F}$ is denoted by $att_\mathcal{F}$ while the attack relations between finite arguments is denoted by $att_{f,\mathcal{F}}$.

Define $AF_\mathcal{F} = (AR_\mathcal{F}, att_\mathcal{F})$ and $AF_{f,\mathcal{F}} = (AR_{f,\mathcal{F}}, att_{f,\mathcal{F}})$.

Due to the fact that the infinite arguments always attack them self, the following lemma holds obviously.

Lemma 1. *Let $S \subset AR_\mathcal{F}$ be admissible wrt $AF_\mathcal{F} = (AR_\mathcal{F}, att_\mathcal{F})$. Then S contains only finite arguments.*

Abusing the notation slightly for simplicity, we say that a set of arguments $S \subseteq AR_\mathcal{F}$ is

---

[4] For illustration, consider the arguments in figure 1.

- *strongly admissible* iff it is admissible wrt the argumentation framework $AF_\mathcal{F} = (AR_\mathcal{F}, att_\mathcal{F})$, and
- *admissible* iff it is admissible wrt the argumentation framework $AF_{f,\mathcal{F}} = (AR_{f,\mathcal{F}}, att_{f,\mathcal{F}})$.

It holds obviously

Theorem 1. *If S is strongly admisisble then S is also admissible.*

In figure 1, the set of arguments $\{A, B\}$ is strongly admissible wrt $\mathcal{F}_2$ but not wrt $\mathcal{F}_1$ as B attack C_2 but not C_1.

We define accordingly the notions of *preferred (strongly preferred)* extensions.

For a set of assumptions S, let AR_S be the set of all finite arguments whose supports are subsets of S. We say S is *conflict-free/admissible (strongly admissible) /preferred (strongly preferred)* if AR_S is conflict-free/admissible (strongly admissible)/preferred (strongly preferred) respectively.

For a set S of assumptions and a sentence σ, we write $S \vdash_\mathcal{F} \sigma$ if there is a finite argument for σ whose support is a subset of S.

A sentence $\sigma \in \mathcal{L}$ is said to be *credulously derived* (resp. *strongly credulously derived*) from $\mathcal{F}$ if there is an admissible (resp. strongly admissible) set S of assumption such that $S \vdash_\mathcal{F} \sigma$.

3 Dialectical Proof Procedures

Dialectical proof procedures could be viewed as games between a proponent who is trying to construct an argument for some conclusion and defend it from the attacking arguments constructed by an opponent. Both players construct their arguments by expanding partial proof trees stepwise to the full proof trees. We study shortly two such procedures.

A partial proof tree B is an *immediate expansion* of a partial proof tree A if there is a leaf node N in A labelled by a non-assumption σ and a rule r of the form $\sigma \leftarrow b_1, \ldots, b_m$ such that B is a proof tree obtained from A by adding m children labelled by $b_1, \ldots, b_m$ to N (for m=0, a child node labelled by *true* is added to N). We write $B = \mathbf{exp(A, N, r)}$.

Example 2. Consider the ABA framework below where the associated partial proof trees are given in figures 2, 3.

$h \leftarrow \alpha$ $\quad \overline{\alpha} \leftarrow g, f$ $\quad g \leftarrow \beta$ $\quad \overline{\beta} \leftarrow$
$h \leftarrow q$ $\quad f \leftarrow f$
where α, β are assumptions.

A_1, A_1' are immediate expansions of A_0 (figure 2) and C_{11}, C_2 are immediate expansions of C_1 (figure 3).

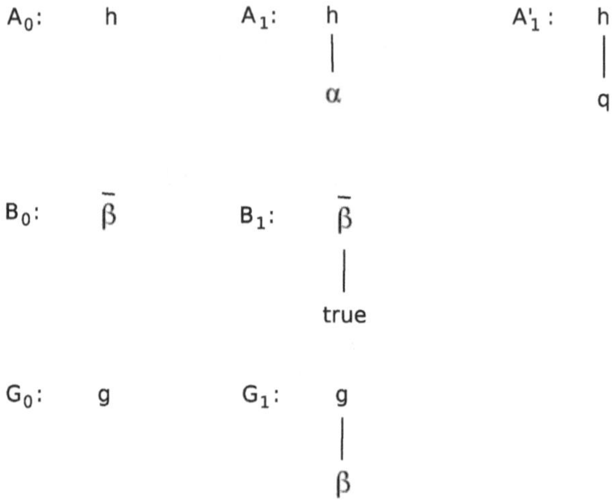

□

A_0:   h A_1:   h A_1':   h
                             | |
                             α   q

B_0: $\bar{\beta}$ B_1: $\bar{\beta}$
                                      |
                                      true

G_0:   g G_1:   g
                             |
                             β

Fig. 2.

Let A be a partial proof tree and N be a leaf node in A labelled by a non-assumption sentence σ. Define $CE(A, N) = \{exp(A, N, r) \,|\, r \text{ is a rule in } \mathcal{F} \text{ with head } \sigma\}$.

We give below a dialectical proof procedure for constructing an admissible set of arguments supporting some sentence σ. The procedure could be viewed as a step-wise construction of the dispute trees in [10].

Definition 2. *A ab-dispute derivation for a sentence σ is a sequence of the form*

$$\langle PA_0, OA_0, SA_0 \rangle, \ldots, \langle PA_n, OA_n, SA_n \rangle$$

where

- *for each i, SA_i is a set of assumptions and PA_i, OA_i are multisets of partial proof trees, and*
- *PA_0 contains exactly one partial proof tree consisting of only the root labeled by σ, and $OA_0 = SA_0 = \emptyset$, and*
- *at step i, one of the dispute parties makes a move satisfying following properties:*

    1. *Suppose the proponent makes a move at step i.*

Infinite Arguments and Semantics of Dialectical Proof Procedures

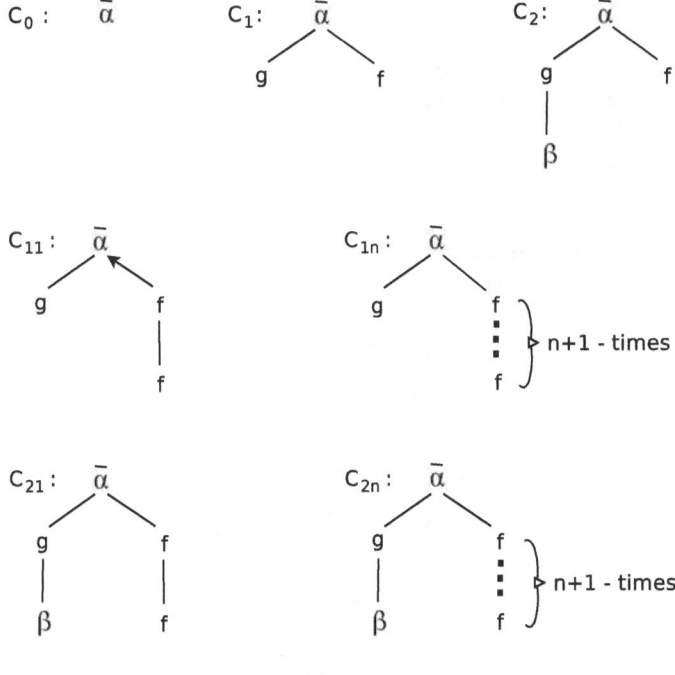

Fig. 3.

(a) The proponent expands a partial proof tree $A \in PA_i$ at a leaf node N in A labeled by non-assumption δ using a rule r with head δ, i.e.,
$PA_{i+1} = (PA_i \setminus \{A\}) \cup \{exp(A, N, r)\}$.
$OA_{i+1} = OA_i$.
$SA_{i+1} = SA_i$

(b) The proponent attacks a partial proof tree $A \in OA_i$ at an assumption $\alpha \in Sp(A) \setminus SA_i$, i.e.,
$PA_{i+1} = PA_i \cup \{\overline{\alpha}\}$.
$OA_{i+1} = OA_i \setminus \{A\}$.
$SA_{i+1} = SA_i$

2. Suppose the opponent makes a move at step i.
   (a) The opponent attacks a proponent partial proof tree $A \in PA_i$ at a leaf node labeled by an assumption $\alpha \in Sp(A) \setminus SA_i$, i.e.,
   $PA_{i+1} = PA_i$
   $OA_{i+1} = OA_i \cup \{\overline{\alpha}\}$.
   $SA_{i+1} = SA_i \cup \{\alpha\}$
   (b) The opponent expands an opponent partial proof tree $A \in OA_i$ at a leaf node N labelled by a non-assumption, i.e.,
   $PA_{i+1} = PA_i$.
   $OA_{i+1} = (OA_i \setminus \{A\}) \cup CE(A, N)$.
   $SA_{i+1} = SA_i$

Definition 3. A ab-dispute derivation $\langle PA_0, OA_0, SA_0\rangle, \ldots, \langle PA_n, OA_n, SA_n\rangle$ is successful if $OA_n = \emptyset$, and PA_n consists only of full proof trees and $SA_n = Sp(PA_n)$.

Example 3. Consider again the argumentation framework in example 2 with the partial arguments in figures 2, 3.

An ab-dispute derivation for h is given in the table below.

PA	OA	SA	Step
A_0	$\emptyset$	$\emptyset$	1a
A_1	$\emptyset$	$\emptyset$	2a
A_1	C_0	α	2b
A_1	C_1	α	2b
A_1	C_{11}	α	2b
A_1	C_{12}	α	2b
A_1	C_{23}	α	1b
A_1, B_0	$\emptyset$	α	1a
A_1, B_1	$\emptyset$	α	success

Theorem 2. 1. Let $\langle PA_0, OA_0, SA_0\rangle, \ldots, \langle PA_n, OA_n, SA_n\rangle$ be a successful ab-dispute derivation for σ. Then PA_n is strongly admissible and σ labels the root of some full proof tree in PA_n.
2. If $\mathcal{L}$ is finite then for each strongly credulously derived sentence σ, there is a successful ab-dispute derivation for σ.

Proof(Sketch)

1. The proof is based on the insight that for each argument attacking an assumption in SA_n there is a step i such that a prefix of it belongs to OA_i that is attacked by the proponent via 2b-step.
2. Let S be a strongly admissible set of argument such that there exists $A \in S$ with $Cl(A) = \sigma$. We show that there exists a successful ab-dispute derivation $\langle PA_0, OA_0, SA_0\rangle, \ldots, \langle PA_n, OA_n, SA_n\rangle$ such that $PA_n \subseteq S$.

In many proof procedures for assumption-based argumentation ([14, 7, 9, 10]), only the assumptions appearing in proponent arguments are of interest, not the arguments them self. In such cases, there is often no need to carry along entire proponent trees. A closer observation of the definition of ab-dispute derivation reveals that to compute $\langle PA_{i+1}, OA_{i+1}, SA_{i+1}\rangle$ from $\langle PA_i, OA_i, SA_i\rangle$, we need only the sentences in the supports of arguments in $PA_i \cup OA_i$.

Hence a simplification of ab-dispute derivation could be obtained by replacing each tree in $PA_i \cup OA_i$ by its support. Further, assumptions in SA_i could be removed from such sets as they are not considered in the following steps. As result, in successful ab-derivations, all elements in PA_n would be empty. We hence do not need to separate the sets in PA_n as they all will be simplified away. This observation leads to a simplification of ab-dispute derivations corresponding to the dispute derivations in the literature

Definition 4. *A standard ab-dispute derivation for a sentence σ is a sequence:*

$$\langle SPA_0, SOA_0, SA_0 \rangle, \ldots, \langle SPA_n, SOA_n, SA_n \rangle$$

where

- *for each i, SA_i is a set of assumptions and SPA_i is a set of sentences and SOA_i is a set of multisets of sentences, and*
- $SPA_0 = \{\sigma\}$, *and* $SOA_0 = SA_0 = \emptyset$, *and*
- *at step i, one of the dispute parties makes a move satisfying the following properties:*
  1. *Suppose the proponent makes a move at step i.*
     (a) *The proponent selects a non-assumption $\delta \in SPA_i$ and a rule r whose head is δ, and*
     $SPA_{i+1} = (SPA_i \setminus \{\delta\}) \cup (bd(r) \setminus SA_i)$,[5]
     $SOA_{i+1} = SOA_i$,
     $SA_{i+1} = SA_i$.
     (b) *The proponent selects a set $S \in SOA_i$ and an assumption $\alpha \in (S \setminus SA_i)$ and*
     $SPA_{i+1} = SPA_i \cup \{\overline{\alpha}\}$,
     $SOA_{i+1} = SOA_i \setminus \{S\}$,
     $SA_{i+1} = SA_i$
  2. *Suppose the opponent makes a move at step i.*
     (a) *The opponent selects an assumption $\alpha \in (SPA_i \setminus SA_i)$, and*
     $SPA_{i+1} = SPA_i \setminus \{\alpha\}$,
     $SOA_{i+1} = SOA_i \cup \{\{\overline{\alpha}\}\}$,
     $SA_{i+1} = SA_i \cup \{\alpha\}$
     (b) *The opponent selects an $S \in SOA_i$ and non-assumption $\delta \in S$, and*
     $SPA_{i+1} = SPA_i$,
     $SOA_{i+1} = (SOA_i \setminus \{S\}) \cup \{S' \mid S' = S \setminus \{\delta\} \cup bd(r) \text{ s.t. } hd(r) = \delta\}$,
     $SA_{i+1} = SA_i$

□

Definition 5. *A standard ab-dispute derivation $\langle SPA_0, SOA_0, SA_0 \rangle, \ldots, \langle SPA_n, SOA_n, SA_n \rangle$ is successful if $SOA_n = SPA_n = \emptyset$.*

□

---

[5] $bd(d)$ is the body of rule r

Example 4. Consider again the argumentation framework in example 2 with the partial arguments in figures 2, 3.

A normal ab-dispute derivation for h is given in the table below.

PA	OA	SA	Step
h	∅	∅	1a
α	∅	∅	2a
∅	$\overline{\alpha}$	α	2b
∅	g,f	α	2b
∅	g,f	α	2b
∅	g,f	α	2b
∅	β,f	α	1b
β	∅	α	1a
∅	∅	α	success

□

It is not difficult to see

Theorem 3. *1. Let $\langle SPA_0, SOA_0, SA_0 \rangle, \ldots, \langle SPA_n, SOA_n, SA_n \rangle$ be a successful standard ab-dispute derivation for σ. Then SA_n is strongly admissible and $SA_n \vdash_\mathcal{F} \sigma$.*

2. If $\mathcal{L}$ is finite then for each strongly credulously derived sentence σ, there is a successful standard ab-dispute derivation for σ.

Proof Follows from theorem 2. □

4 Discussion

We have showed that the inclusion of infinite arguments provides a succinct specification of the semantics of dialectical proof procedures for assumption-based argumentation. We believe that similar results could also hold for dialectical proof procedures wrt other approaches to structured argumentation like ASPIC+ ([21]) or defeasible logic programming ([26, 15, 16]).

[24] has studied infinite arguments within the context of ASPIC+ while [17] has introduced infinite arguments to capture ambiguity blocking and ambiguity propagating proof theories in defeasible logic ([22]). As the frameworks of assumption-based argumentation and defeasible logics are based on distinct concepts, it would be interesting to look at the relationship between our approach and the approach in [17].

5 Acknowledgement

Thank you very much Guillermo for having done so much for the argumentation community and for making me (PMD) working hard all the time just to keep up with you.

References

1. P. Baroni, F. Cerutti, M. Giacomin, and G. Simari, editors. *Proceedings of Conference on Computational Models of Arguments*. IOS Press, 2010.
2. P. Besnard, S. Doutre, and A. Hunter, editors. *Proceedings of Conference on Computational Models of ArgumentsComputational Models of Arguments*. IOS Press, 2008.
3. A. Bondarenko, P.M. Dung, R.A. Kowalski, and F. Toni. An abstract, argumentation-theoretic approach to default reasoning. *Artif. Intell.*, 93:63–101, 1997.
4. M. Caminada and L. Amgoud. On the evaluation of argumentation formalisms. *Artificial Intelligence*, 171:286–310, 2007.
5. C. Cayrol, S. Doutre, and J. Mengin. On decision problems related to the preferred semantics for argumentation frameworks. *J. Log. Comput.*, 13(3):377–403, 2003.
6. Phan Minh Dung. Logic programming as dialogue games. Technical report, Division of Computer Science, Asian Institute of Technol- ogy, Thailand (submitted to LPNMR 1993) http://www.cs.ait.ac.th/ dung/Site/Publications_files/LPasDialog.pdf, 1993.
7. Phan Minh Dung. An argumentation theoretic foundation for logic programming. *Journal of Logic Programming*, 22:151–177, 1995.
8. Phan Minh Dung. On the acceptability of arguments and its fundamental role in nonmonotonic reasoning, logic programming and n-person gamescceptability of arguments and its fundamental role in nonmono- tonic reasoning, logic programming and n-person games. *Artif. Intell.*, 77(2):321–358, 1995.
9. Phan Minh Dung, R.A. Kowalski, and F. Toni. Dialectic proof procedures for assumption-based, admissible argumentation. *Artif. Intell.*, 170(2):114–159, 2006.
10. Phan Minh Dung, R.A. Kowalski, and F. Toni. *Argumentation in AI*, chapter Assumption-based Argumentation. Springer-Verlag, 2009.
11. Phan Minh Dung and Phan Minh Thang. A modular framework for dialectical dispute in argumentation. In *Proc of IJCAI*, 2009.
12. P.E. Dunne and T. Bench-Capon, editors. *Proceedings of Conference on Computational Models of ArgumentsComputational Models of Arguments*. IOS Press, 2006.
13. P.E. Dunne and T.J.M. Bench-Capon. Two party immediate response disputes: Properties and efficiency. *Artif. Intell.*, 149(2):221–250, 2003.
14. K. Eshghi and R.A. Kowalski. Abduction compared with negation by failure. In *In Logic Programming: Proceedings of the Sixth International Conference*, pages 234–254. Cambridge, and Massachusetts, The MIT Press, 1989.
15. A.J. Garcia and G.R. Simari. Defeasible logic programming: An argumentative approach. *TPLP*, 4(1-2):95–138, 2004.
16. A.J. Garcia and G.R. Simari. Defeasible logic programming: Delp servers, contextual queries and explanation for answers. *J. Arguments and Computation*, 2014.
17. G. Governatori, M.J. Maher, G. Antoniou, and D. Billington. Argumentation semantics for defeasible logic. *J. Log. Comput.*, 14(5):675–702, 2004.
18. A.C. Kakas and F. Toni. Computing argumentation in logic programming. *Journal of Logic and Computation*, 1999.
19. J.W. Lloyd. *Foundations of Logic Programming*. Springer Verlag, 1987.

20. S. Modgil and M. Caminada. *Argumentation in AI*. Springer Verlag, 2009.
21. Sanjay Modgil and Henry Prakken. A general account of argumentation with preferences. *Artificial Intelligence*, 195:361–397, 2013.
22. D. Nute. Defeasible logic programming: Delp servers, contextual queries and explanation for answers. In *Proc 20th Hawaii International Conference on System Science*. IEEE Press, 1987.
23. J.L. Pollock. Defeasible reasoning. *Cognitive Science*, 11(4):481–518, 1987.
24. H. Prakken. An abstract framework for argumentation with structured arguments. *J. Arguments and Computation*, 1(2):93–124, 2010.
25. I. Rahwan and G. Simari, editors. *Handbook of Argumentation in AI*. Springer Verlag, 2009.
26. G.R. Simari and R.P. Loui. A mathematical treatment of defeasible reasoning and its implementation. *Artif. Intell.*, 53, 1992.
27. F. Toni. A generalised framework for dispute derivations in assumption-based argumentation. *Artif. Intell.*, 195:1–43, 2013.
28. F. Toni. A tutorial on assumption-based argumentation. *Journal of Arguments and Computation*, 2013.
29. B. Verheij. A labeling approach to the computation of credulous acceptance in argumentation. In M. Veloso, editor, *JCAI 2007*, pages 623–628. Morgan Kaufmann, 2007.
30. B. Verheij, S. Szeider, and S. Woltran, editors. *Proceedings of Conference on Computational Models of Arguments*. IOS Press, 2012.
31. G. Vreeswijk and H. Prakken. Credulous and sceptical argument games for preferred semantics. In Manuel Ojeda-Aciego, Inman P. de Guzmán, Gerhard Brewka, and Luís Moniz Pereira, editors, *JELIA 2000*, volume 1919 of *Lecture Notes in Computer Science*, pages 239–253. Springer, 2000.

A P-DeLP Instantiation of a Dynamic Argumentation Framework for Decision Making[*]

Edgardo Ferretti and Marcelo Errecalde

Departamento de Informática, Universidad Nacional de San Luis, 5700, Argentina
{ferretti,merreca}@unsl.edu.ar

Abstract. Possibilistic Defeasible Logic Programming (P-DeLP) is a formalism which combines features from argumentation theory, logic programming and a unified treatment of possibilistic uncertainty and fuzziness. In this work, we show how we can use this formalism to instantiate an existing Abstract Dynamic Argumentation Framework (DAF) for decision making, which have been proven to have a choice behavior consistent with Classical Decision Theory. On one hand, by instantiating this decision framework we prove its adequacy as an abstract framework. On the other hand, we create a new concrete decision framework for decision making based on P-DeLP, whose choice behavior is also consistent with Classical Decision Theory. The emphasis of the instantiation lies in defining the proper conversion functions which map notions of the DAF to P-DeLP. Also, an application example from the domain of cognitive robotics is modeled.

Keywords: Possibilistic Defeasible Logic Programming · Dynamic Argumentation Frameworks · Decision Making.

1 Introduction

Argumentative reasoning for decision making is an active research trend related to argumentation [16]. Several proposals have been published in the last two decades [3, 6, 7, 9, 13, 15] which roughly differ in two aspects: (a) the underlying argumentation framework used, viz. abstract or concrete, and (b) the way decision making is conceived, that is, whether the decision maker's goals are considered to decide which action to accomplish next, or each alternative posed to decision is conceptualized as a product that the consumer (decision maker) is evaluating to buy (selection); namely, marketing's point of view [17].

In spite of not being recent, the work presented in [13] is an ambitious proposal. A concrete argumentation-based framework to support the decision-making of an agent is defined. This proposal allows priority relation on the sentences of the theory not to be simply a static relation, but a dynamic relation that captures the non-static preferences associated with roles and context.

---
[*] Partially supported by Project PROICO P-31816, UNSL.

Thus, arguments and their strengths depend on the particular context that the agent finds itself. Moreover, motivated by works in Cognitive Psychology, an argumentation-based personality theory for agents has been integrated within the framework.

Other works based on concrete argumentation formalisms, but following the marketing viewpoint to decision making, are those by Ferretti et al. [6, 7]; where Defeasible Logic Programming (DeLP) [11] and Possibilistic Defeasible Logic Programming (P-DeLP) [2, 4], have been used, respectively. Both works share the use of decision rules as device to make decisions, together with the underlying argumentation formalism used to represent knowledge and perform reasoning. The advantage of having an interpreter available like the one built for DeLP [10], is that it makes possible to directly tackle real-world decision-making problems [8]. Conversely, [7] has the advantage over [6] in that explicit multi-criteria aggregation is used within the framework by means of arguments accrual [12].

Regarding abstract argumentation frameworks, notable proposals for decision-making have been introduced in [3] and [15]. The former work remains close to the classical view of decision making in that it leaves aside aspects of practical reasoning, such as goal generation, feasibility and planning, to concentrate on the issue of justifying (based on argumentation) the best decision to make in a given situation; besides, it has a logical view of decision that unifies the treatment of multiple criteria decision and decision under uncertainty. The latter one, proposes to model decisions as sets of literals rather than as single literal, as usual in argumentation-based decision making literature. This conceptualization of decisions which may partially overlap, results in a more finely tuned set of decisions when argumentation systems have to be built to derive arguments about the goals achieved when certain decisions are made.

Another interesting abstract framework for decision making (AADF) which has been proposed recently [9], differ with the above-mentioned works in that the marketing literature approach is followed, and hence, in the way that multi-criteria decision making problems are conceptualized. In particular, [9] introduces a formalism for single-agent decision making that is based on Dynamic Argumentation Frameworks (DAF) [18]. The formalism can be used to justify a choice, which is based on the current situation the agent is involved. Taking advantage of the inference mechanism of the argumentation formalism, it is possible to consider preference relations, and conflicts among the available alternatives for that reasoning. With this formalization, given a particular set of evidence, the justified conclusions supported by warranted arguments will be used by the agents decision rules to determine which alternatives will be selected.

Like in [6, 7], in [9] decision rules are used to make decisions; however, they were reformulated with respect to [6, 7]. In particular, [6] could be considered as an instantiation of [9]. Nonetheless, being rigorous, in order to prove that [6] or any other proposal is an instantiation of the AADF proposed in [9], proper conversions functions must be defined. That is, an instantiation of the AADF consists of defining conversion functions where the original components of the AADF, are mapped to equivalent versions in the concrete formalism. It is worth noting, that depending on the characteristics of the concrete formalism, certain explicit features of the AADF might remain implicit in the concrete formalisms; for instance, the working set of arguments.[1]

Therefore, conversion functions are particular to each concrete formalism. Hence, each concrete formalism intended to instantiate the AADF should define the following elements: (a) The logical representation language; (b) A conversion function for arguments; (c) A conversion function for evidence; (d) A method for generating the strict total order on which is based the arguments comparison criterion; (e) An instantiation function of primitive argumental structures; and (f) A proof that the concept of warranty between both formalisms is equivalent.

In this paper we present an instantiation of the AADF mentioned above, following the aforesaid steps for the P-DeLP formalism. In this way, our contribution is twofold; in the first place, we explicitly show how an instantiation of the AADF proposed in [9] could be accomplished, thus proving its adequacy as an abstract framework. Second, we create a new concrete decision framework for decision making based on P-DeLP, whose choice behavior is also consistent with Classical Decision Theory.

The rest of the article is organized as follows. Section 2 presents the minimal theoretical concepts of the AADF which are involved in the instantiation proposed in our work. Then, Sect. 3 defines the logical representation language of the instantiation. In Sect. 4 the conversion functions mentioned in points (b)-(e) are introduced, while Section 5 states the formalization aimed at demonstrating that the concept of warranty is equivalent. Together with the development of the instantiation, the P-DeLP formalism is introduced conveniently along the different sections. Besides, Sect. 6 models the proposed instantiation on an application domain from the field of cognitive robotics. Finally, Section 7 offers the conclusions.

2 A Dynamic Argumentation Framework for Decision Making: a Brief Overview

As mentioned in the introductory section, in [9], an approach to single-agent decision making based on Dynamic Argumentation Frameworks was proposed. We have denoted this decision framework based on a Dynamic Argumentation Framework, as AADF (Abstract Argumentation Decision Framework).

Dynamic Argumentation Frameworks (or DAF for short) were introduced in [18] and provide a formalization for abstract argumentation systems where the current set of *evidence* dynamically activates arguments that belong to a

---

[1] The working set of arguments contains every argument that is available for use by the reasoning process.

working set of arguments. The main objective of DAFs is to extend Argumentation Frameworks [5] to provide the ability of handling dynamics; for achieving that, at a given moment, the set of available *evidence* determines which arguments are *active* and can be used to make inferences to obtain justified conclusions.

In [9], a DAF is used for representing preference relations and the conflicts among the available alternatives. Four other components complete the formalism: a set X of mutually exclusive *alternatives* which are available to the agent; a set of *distinguished literals* representing different binary preference relations for comparing the alternatives; a *strict total order* over the set of distinguished literals to represent the priority among the preference criteria provided to the agent; and a set of *decision rules* that implement the agent's decision making policy. A key concept that brings together three of the afore-mentioned components is that of epistemic component, stated in Definition 1; but before, other notions on which this definition is built upon, are introduced. In particular, decision rules will not be referred anymore from now on, since they are an independent decision component and the proposed instantiation directly relates to the underlying DAF of the AADF.

As mentioned above, the working set W will contain arguments for reasoning about when an alternative is better than other. Note that in a DAF, an argument $\mathcal{A}$ is a reasoning step for a claim α from a set of premises $\{\beta_1, \ldots, \beta_n\}$ denoted as the pair $\langle\{\beta_1, \ldots, \beta_n\}, \alpha\rangle$. An argument will be *active* if its premises are satisfied based on the current evidence. Given an evidence set E, an argument's premise is satisfied whether it belongs to E, or it is the conclusion of an active argument according to E. In this DAF the set $\bowtie$ will contain the conflicts among arguments in W. Given an argument $\mathcal{A} \in W$, $cl(\mathcal{A})$ denotes the claim of $\mathcal{A}$ and $\overline{cl(\mathcal{A})}$ represents the complement of $cl(\mathcal{A})$ with respect to negation ($\neg$). Finally, the preference function $\mathfrak{pref}$ will consider all the agents' criteria represented by (distinguished) literals in $\mathcal{C}$, and, if it is possible, it will return the argument that is based on a better distinguished literal with respect to the order $>_\mathcal{C}$.

Since the epistemic component is defined in an abstract form, the function $\mathfrak{pref}$ is defined in terms of *argumental structures* (denoted with Σ) which are built with one or more arguments from W. When the set of arguments in an argumental structure is a singleton, that is, $args(\Sigma) = \{\mathcal{A}\}$,[2] the argumental structure is called *primitive*. In order to compare two argumental structures, distinguished literals will be used.

In the AADF, the exchange of arguments resembles a dialogical discussion where different alternatives are compared. As such, it makes sense that the introduction of a new argument by one of the participants should be consistent with the previously posed arguments. Indeed, it is also desirable to require that none of the parties be allowed to introduce an argument already posed by them. The set of all the arguments posed by the proponent is referred as *pro*, while

---

[2] $args(\Sigma)$ denotes the set of arguments belonging to argumental structure Σ.

the set of all the arguments posed by the opponent is referred as *con*. Definition 2 formalizes the intuitions referred above on what is called an *acceptable argumentation line*. Indeed, several argumentation lines starting with the same argumental structure resemble the intuition of a discussion around a topic. This notion known as *dialectical tree* is also formalized next, in Definition 3.

Definition 1 (Epistemic component). *Let X be the set of all the possible candidate alternatives, C be a set of distinguished literals in $\mathcal{L}$ and $>_C$ be a strict total order over C. An epistemic component $\mathcal{K}_\mathcal{A}$, is a DAF $\langle E, W, \bowtie, \mathfrak{pref} \rangle$ where:*

- *The evidence E is a consistent set of sentences of the form same_att(x,y) or $c(x,y)$, such that $x, y \in X$ and $c \in C$.*
- *The working set W will be such that if $c \in C$, $\{x,y\} \subseteq X$ $(x \neq y)$ and better $\notin C$ then:*

$$\langle \{c(x,y)\}, better(x,y) \rangle \in W$$
$$\langle \{c(x,y)\}, \neg better(y,x) \rangle \in W$$
$$\langle \{c(y,x)\}, better(y,x) \rangle \in W$$
$$\langle \{c(y,x)\}, \neg better(x,y) \rangle \in W$$
$$\langle \{same\_att(x,y)\}, \neg better(x,y) \rangle \in W$$
$$\langle \{same\_att(x,y)\}, \neg better(y,x) \rangle \in W$$

- $\bowtie = \{(\mathcal{A}, \mathcal{B}) | \{\mathcal{A}, \mathcal{B}\} \subseteq W, cl(\mathcal{A}) = \overline{cl(\mathcal{B})}\}$.
- *Let Σ_1 and Σ_2 be two argumental structures in W, then*

$$\mathfrak{pref}(\Sigma_1, \Sigma_2) = \begin{cases} \Sigma_1 \text{ if } \forall c \in dlits(\Sigma_2), \exists c' \in dlits(\Sigma_1) \text{ st. } (c',c) \in >_C, \\ \Sigma_2 \text{ if } \forall c \in dlits(\Sigma_1), \exists c' \in dlits(\Sigma_2) \text{ st. } (c',c) \in >_C \\ \epsilon \quad \text{otherwise} \end{cases}$$

where $dlits(\Sigma) \subseteq C$ is the set of distinguished literals that are contained in arguments of an argumental structure Σ.

Definition 2 (Acceptable Argumentation Line). *Given an argumentation line λ in the context of a DAF, $\mathbf{F} = \langle E, W, \bowtie, \mathfrak{pref} \rangle$, λ is acceptable in $\mathbf{F}$ iff it holds that:*

- *There is no repetition of structures in λ (non-circularity), and*
- *Sets* pro *and* con *are consistent with respect to the conflict relation among argumental structures (concordance).*

It is worth mentioning that an acceptable argumentation line is *exhaustive* if it is not possible to insert more argumental structures in the sequence.

Definition 3 (Dialectical Tree). *Given a DAF $\mathbf{F}$ and a set S of exhaustive argumentation lines in $\mathbf{F}$ rooted in Σ_1, such that S is maximal wrt. set inclusion, a dialectical tree for an argumental structure Σ_1 is a tree $\mathcal{T}_\mathbf{F}(\Sigma_1)$ verifying:*

- Σ_1 is the root;
- A structure $\Sigma_{i\neq 1}$ in a line $\lambda_i \in S$ is an inner node, iff has as children all the Σ_j in lines $\lambda_j \in S$ such that $\Sigma_j \Rightarrow \Sigma_i$ and $\lambda^\uparrow[\Sigma_i] = \lambda^\uparrow(\Sigma_j)$;[3]
- The leaves of the tree correspond to the leaves of the lines in S.

Dialectical trees are defined over the working set of arguments, and hence they can contain active and inactive argumental structures. A dialectical tree that contains only active structures is called *active dialectical tree*, and it is denoted $\mathbb{T}_\mathbf{F}(\Sigma)$. Once a dialectical tree has been built for an argumental structure, a marking criterion determines which structures in the tree are defeated and which ones remain undefeated. This criterion is specified by a marking function. Definition 4 introduces the marking function used in the AADF.

Definition 4 (Skeptical Marking Function). *Given an argumental structure Σ_i in a line λ_i in a dialectical tree $\mathcal{T}_\mathbf{F}(\Sigma)$, the skeptical marking function $\mathfrak{m}_e$ is defined as follows: $\mathfrak{m}_e(\Sigma_i, \lambda_i, \mathcal{T}_\mathbf{F}(\Sigma)) = D$ iff $\exists \Sigma_j$ s.t. $\mathfrak{m}_e(\Sigma_j, \lambda_j, \mathcal{T}_\mathbf{F}(\Sigma)) = U$, where Σ_j is a child of Σ_i in $\mathcal{T}_\mathbf{F}(\Sigma)$.*

Once the marking function has been defined, the warranty status of the root of a dialectical tree can be determined, as defined next.

Definition 5 (Warrant). *Given a DAF $\mathbf{F}$ and a marking function $\mathfrak{m}$, an argumental structure Σ from $\mathbf{F}$ is warranted in $\mathbf{F}$, iff $\mathfrak{m}(\Sigma, \lambda, \mathbb{T}_\mathbf{F}(\Sigma)) = U$, where λ is any argumentation line from $\mathbb{T}_\mathbf{F}(\Sigma)$. The conclusion $cl(\Sigma)$ is justified by $\mathbf{F}$.*

To conclude, it is worth mentioning that the notion of warrant is defined on active dialectical trees, since all the reasoning only can be carried out over the set of active arguments.

Considering that this section aims at introducing the minimal theoretical concepts underpinning the instantiation proposed in our work, for more details on the AADF, the interested reader is referred to [9].

3 The Language $\mathcal{L}$

P-DeLP language is based on Possibilistic Gödel Logic (PGL) language, denoted as $\mathcal{L}_G$ [1]. Language $\mathcal{L}_G$ is built from a countable set *Var* of *fuzzy propositional variables* $\{p, q, \ldots\}$, the *connectives* $\wedge$ and $\rightarrow$, and the *truth* constant $\overline{0}$. As usual, a *negation* connective $\neg$ is defined, where $\neg\varphi$ is $\varphi \rightarrow \overline{0}$. In fact, P-DeLP language is a sub-language of $\mathcal{L}_G$, denoted $Horn_{PGL}$, and consists of PGL formulas of the kind $(p_1 \wedge \ldots \wedge p_k \rightarrow q, \alpha)$, with $k \geq 0$, where

---

[3] The defeat relation between argumental structures is denoted as "$\Rightarrow$". Moreover, given an argumentation line $\lambda = [\Sigma_1, \ldots, \Sigma_n]$, the *top segment* of Σ_i $(1 < i \leq n)$ in λ is $[\Sigma_1, \ldots, \Sigma_i]$ and it is denoted as $\lambda^\uparrow(\Sigma_i)$. The *proper top segment* of Σ_i in λ is $[\Sigma_1, \ldots, \Sigma_{i-1}]$ and is denoted as $\lambda^\uparrow[\Sigma_i]$.

$p_1, \ldots, p_k, q \in \mathit{Var}$, and $\alpha \in [0,1]$ is a lower bound on the belief on q in terms of necessity measures.

We shall refer these weighted Horn-rules as PGL *clauses*. Also, we shall refer to the conclusion q and the set of premises $p_1, \ldots, p_k$ as the *head* and the *body*, respectively. We distinguish between two types of formulas in this sub-language: *facts* when $k = 0$ (empty body) which are simply written as q, and *rules* when $k > 0$; denoted as $p_1 \wedge \ldots \wedge p_k \to q$, respectively.

PGL semantics is given by *interpretations* I of the propositional variables into the real unit interval $[0,1]$, which are extended to arbitrary formulas by means of the following rules:

$$I(\overline{0}) = 0,$$
$$I(L_1 \wedge \ldots \wedge L_n) = \min(I(L_1), \ldots, I(L_n)),$$
$$I(\varphi \to Q) = \begin{cases} 1, & \text{if } I(\varphi) \leq I(Q) \\ I(Q) & \text{otherwise.} \end{cases} \quad (1)$$

In order to determine the maximum degree of possibilistic entailment of a PGL clause, there is no need to resort to the whole logical apparatus of general possibilistic logic but only to a single instance of the *generalized modus ponens* rule, that will be denoted as GMP:

$$\frac{(p_1 \wedge \ldots \wedge p_k \to q, \alpha) \quad (p_1, \beta_1), \ldots, (p_k, \beta_k)}{(q, \min(\alpha, \beta_1, \ldots, \beta_k))}$$

In this way, given a set P of PGL clauses, if there exists a finite sequence of clauses $C_1, \ldots, C_m$ such that $C_m = (q, \alpha), q \in \mathit{Var}, \alpha > 0$ and, for each $i \in \{1, \ldots, m\}$ it holds that $C_i \in P$, then C_i is an instance of the axiom or C_i is obtained by applying the above inference rule to previous clauses in the sequence. This will be formally written as $P \vdash_{gmp} (q, \alpha)$. Moreover, the *maximum degree* of deduction is defined as follows.

Definition 6 (Maximum degree of deduction). *The maximum degree of deduction of a goal q from a set of PGL clauses P, is $|q|_P = \max\{\alpha \in [0,1] \mid P \vdash_{gmp} (q, \alpha)\}$.*

In argumentative frameworks, the negation connective allows to identify existing conflicts among different pieces of information, that will be considered in a deliberation process in order to determine which one prevails. In PGL, negation is especial since $\neg \varphi$ always refers to a non-fuzzy proposition. Therefore, it would be ideal to have another *well-formed* negation connective '$\sim$'. By *well-formed*, we mean that $I(q)$ and $I(\sim q)$ do not take simultaneously value 1, for no interpretation I. Hence, q and $\sim q$ would model contradictory information.

In this way, P-DeLP language is Horn_{PGL} language but extended over a set Var^* of propositional variables, where a new variable $\sim p$ is added, for each $p \in \mathit{Var}$. It is worth mentioning that symbol $\sim$, intuitively refers to negation, but is not considered as a proper connective. This is due to the fact that "$\sim p$" is considered like any other propositional variable, but with the particular feature that it would be used to detect contradictions at a syntactic level with respect to p.

Definition 7. *A set of clauses P (in the extended language) is deemed contradictory, denoted $P \vdash \bot$, if for some atom $q \in Var$, $P \vdash_{gmp} (q, \alpha)$ and $P \vdash_{gmp} (\sim q, \beta)$, with $\alpha > 0$ and $\beta > 0$.*

Henceforth, positive and negative propositional variables in Var^*, will be referred as P-DeLP literals. Clauses in P-DeLP, will be PGL clauses defined on the set Var^*, and will be denoted as $(q \leftarrow p_1 \wedge \ldots \wedge p_k, \alpha)$ following the usual style of logic programming. Finally, the notion of *proof* in P-DeLP corresponds to $Horn_{PGL}$ but denoted $\vdash$ instead of $\vdash_{gmp}$.

4 Arguments

P-DeLP distinguishes between *certain* clauses and *uncertain* ones. A clause (φ, α) is referred as certain if $\alpha = 1$ and uncertain otherwise. In fact, a P-DeLP program is a set clauses in which certain information is distinguished from uncertain information, with the additional requirement that certain knowledge is required to be non-contradictory.

Definition 8 (P-DeLP Program). *A P-DeLP program is a pair (Π, Δ), where Π is a non-contradictory finite set of certain clauses, and Δ is a finite set of uncertain clauses.*

The notion of *argument* in P-DeLP refers to a tentative proof (as it relies to some extent on uncertain, possibilistic information) from a consistent set of clauses supporting a given conclusion Q with a necessity degree α.

Definition 9 (Argument). *Given a P-DeLP program (Π, Δ), a set $\mathcal{A} \subseteq \Delta$ of uncertain clauses is an argument for a conclusion Q with necessity degree $\alpha > 0$, denoted $\langle \mathcal{A}, Q, \alpha \rangle$, iff:*

1. *$\Pi \cup \mathcal{A}$ is non contradictory;*
2. *$\alpha = \max\{\beta \in [0, 1] \mid \Pi \cup \mathcal{A} \vdash (Q, \beta)\}$, i.e., α is the greatest degree of deduction of Q from $\Pi \cup \mathcal{A}$;*
3. *$\mathcal{A}$ is minimal with respect to set inclusion, i.e., there is no $\mathcal{A}' \subset \mathcal{A}$ such that $\Pi \cup \mathcal{A}' \vdash (Q, \alpha)$.*

Given a P-DeLP program $\mathcal{P} = (\Pi, \Delta)$, as stated in [4], from a procedural point of view, an argument $\langle \mathcal{A}, Q, \alpha \rangle$ for Q can be built by computing $\mathcal{A}$ and α through the recursive application of the following rules:

1. (INTF) Building arguments from facts:
    (a) **If** $(Q, 1) \in \Pi$ **then** $\mathcal{A} = \emptyset$ and $\alpha = 1$.

(b) **If** $(Q,\beta) \in \Delta$ and $\Pi \cup \{(Q,\beta)\} \not\vdash \bot$ and $\Pi \not\vdash (Q,1)$ **then** $\mathcal{A} = \{(Q,\alpha)\}$ and $\alpha = \beta$.
2. (MPA) Building Arguments by GMP:
   (a) **If** $(Q \leftarrow L_1 \wedge \ldots \wedge L_k, 1) \in \Pi$ and $\langle \mathcal{A}_1, L_1, \beta_1 \rangle, \ldots, \langle \mathcal{A}_k, L_k, \beta_k \rangle$ are arguments and $\Pi \cup \bigcup_{i=1}^{k} \mathcal{A}_i \not\vdash \bot$ and there is no $\mathcal{B} \subset \bigcup_{i=1}^{k} \mathcal{A}_i$ such that $\Pi \cup \mathcal{B} \vdash (Q, \gamma)$ with $\gamma \geq \min(\beta_1, \ldots, \beta_k)$ **then** $\mathcal{A} = \bigcup_{i=1}^{k} \mathcal{A}_i$ and $\alpha = \min(\beta_1, \ldots, \beta_k)$.
   (b) **If** $(Q \leftarrow L_1 \wedge \ldots \wedge L_k, \beta) \in \Delta$ and $\langle \mathcal{A}_1, L_1, \beta_1 \rangle, \ldots, \langle \mathcal{A}_k, L_k, \beta_k \rangle$ are arguments and $\Pi \cup \{(Q \leftarrow L_1 \wedge \ldots \wedge L_k, \beta)\} \cup \bigcup_{i=1}^{k} \mathcal{A}_i \not\vdash \bot$ and there is no $\mathcal{B} \subset \bigcup_{i=1}^{k} \mathcal{A}_i \cup \{(Q \leftarrow L_1 \wedge \ldots \wedge L_k, \beta)\}$ such that $\Pi \cup \mathcal{B} \vdash (Q, \gamma)$ with $\gamma \geq \min(\beta_1, \ldots, \beta_k)$ **then** $\mathcal{A} = \bigcup_{i=1}^{k} \mathcal{A}_i \cup \{(Q \leftarrow L_1 \wedge \ldots \wedge L_k, \beta)\}$ and $\alpha = \min(\beta, \beta_1, \ldots, \beta_k)$.

Given a P-DeLP program $\mathcal{P}$, we will use notation $\mathcal{P} \vdash_\Delta \langle \mathcal{A}, Q, \alpha \rangle$ to indicate that there exists a finite sequence of applications of rules INTF and MPA in order to obtain argument $\langle \mathcal{A}, Q, \alpha \rangle$. Finally, it is worth mentioning that given two arguments $\langle \mathcal{A}, Q, \alpha \rangle$ and $\langle \mathcal{S}, R, \beta \rangle$, $\langle \mathcal{S}, R, \beta \rangle$ is a sub-argument of $\langle \mathcal{A}, Q, \alpha \rangle$, iff $\mathcal{S} \subseteq \mathcal{A}$.

With the theoretical background introduced so far, it is possible to define the conversion function for arguments in W to set of rules in $\Pi \cup \Delta$, such that given a P-DeLP program $\mathcal{P} = (\Pi, \Delta)$, for all argument $\langle \{\beta\}, Q \rangle \in W$, $\mathcal{P} \vdash_\Delta \langle \{\beta\}, Q, \alpha \rangle$ for some $\alpha \in [0,1]$.

Definition 10 (Arguments Conversion). *Given an abstract epistemic component $\mathcal{K}_\mathcal{A} = \langle E, W, \bowtie, \mathfrak{pref} \rangle$, the conversion function of arguments is $\Psi_{Arg} : 2^{2^\mathcal{L} \times \mathcal{L}} \mapsto 2^\mathcal{L}$, such that:*

$$\Psi_{Arg}(W) = \{(\alpha \leftarrow \beta, \gamma) \mid \mathcal{A}_i \in W, int(\mathcal{A}_i) = \langle \{\beta\}, \alpha \rangle \text{ and } \gamma \in [0,1]\}$$

In this way, the set of uncertain clauses obtained from W will be denoted Δ_W and is characterized as follows:[4]

$$\Delta_W = \{(better(X,Y) \leftarrow c_i(X,Y), \alpha_i), (\sim better(Y,X) \leftarrow c_i(X,Y), \alpha_i) \mid c_i \in \mathcal{C}, \alpha_i \in [0,1)\}$$

It could also be the case that certain clauses may arise from the arguments conversion. If so, the set Π_W will contain those rules having literal $same\_att(\cdot, \cdot)$ in their body.

$$\Pi_W \supseteq \left\{ \begin{array}{l} (\sim better(X,Y) \leftarrow same\_att(X,Y), 1), \\ (\sim better(Y,X) \leftarrow same\_att(X,Y), 1) \end{array} \right\}$$

Definition 11 (Evidence Conversion). *Given an abstract epistemic component $\mathcal{K}_\mathcal{A} = \langle E, W, \bowtie, \mathfrak{pref} \rangle$, the conversion function of evidence is $\Psi_E : 2^\mathcal{L} \mapsto 2^\mathcal{L}$, such that:*

---
[4] Following the usual convention [14], whenever is necessary we will use schematic clauses with variables, standing for the set of all ground instances of it.

$$\Psi_E(E) = \{(\beta_i, \alpha_i) \mid \beta_i \in E, \alpha_i \in (0,1]\}$$

From the previous definition it can be noticed that the certainty value assigned to facts must belong to the interval $(0,1]$. This is due to the fact that it makes no sense to have factual information with a null certainty degree. Likewise, P-DeLP allows having factual information not necessarily with certainty degree equal to 1. In this way, depending on the application domain, it might happen that $\Psi_E(E) \subset \Delta_W$ or $\Psi_E(E) \subset \Pi_W$, or $\Psi_E(E) \cap \Delta_W \neq \emptyset \wedge \Psi_E(E) \cap \Pi_W \neq \emptyset$.

Given a program and a particular context, is usual to have arguments supporting contradictory literals. In P-DeLP, this is formalized with the notions of *counter-argument* and *defeat* introduced below.

Definition 12. *Let $\langle \mathcal{A}_1, Q_1, \alpha_1 \rangle$ and $\langle \mathcal{A}_2, Q_2, \alpha_2 \rangle$ be two arguments with respect to a P-DeLP program $\mathcal{P}$. We will say that $\langle \mathcal{A}_1, Q_1, \alpha_1 \rangle$ counterargues $\langle \mathcal{A}_2, Q_2, \alpha_2 \rangle$, iff there exists a subargument (called disagreement subargument) $\langle \mathcal{S}, Q, \beta \rangle$ of $\langle \mathcal{A}_2, Q_2, \alpha_2 \rangle$ such that $Q_1 =\sim Q$.*

Taking into account that arguments are mostly based on uncertain information, conflicts among them can be solved comparing their strengths. Hence, the notion of defeat is equivalent to establishing a *preference criterion* among conflicting arguments. In P-DeLP, this criterion is defined by considering the certainty values that arguments have, as defined next.

Definition 13 (Defeat). *Let $\mathcal{P}$ be a P-DeLP program and let $\langle \mathcal{A}_1, Q_1, \alpha_1 \rangle$ be a counter-argument for $\langle \mathcal{A}_2, Q_2, \alpha_2 \rangle$ with disagreement sub-argument $\langle \mathcal{A}, Q, \beta \rangle$. Argument $\langle \mathcal{A}_1, Q_1, \alpha_1 \rangle$ is a* proper defeater *(respectively,* blocking defeater*) for $\langle \mathcal{A}_2, Q_2, \alpha_2 \rangle$ if $\alpha_1 > \beta$ (respectively, $\alpha_1 = \beta$).*

In [9], the comparison criterion for conflicting arguments is based on priorities over literals representing different binary preference relations for comparing the alternatives; together with a *strict total order* over this set of distinguished literals to represent the priority among the preference criteria provided to the agent. As it could be observed, this is not the case for the proposed instantiation. Nonetheless, an *implicit* strict total order could be generated to restrict the certainty values associated with arguments, so that they respect the priorities of the literals used in the body of the rules composing the argument. In Example 1 we present a possible way of implementing this implicit ordering, but before we state a generic definition to achieve this purpose.

Definition 14 (Strict total Order Generation). *Let $\mathcal{C}$ be the set of distinguished literals and let $>_\mathcal{C}$ be the strict total order defined over $\mathcal{C}$. Given the set Δ_W of uncertain clauses, for all pair of clauses $(\alpha_1 \leftarrow \beta_1, \gamma_1)$, $(\alpha_2 \leftarrow \beta_2, \gamma_2)$, $\{(\alpha_1 \leftarrow \beta_1, \gamma_1), (\alpha_2 \leftarrow \beta_2, \gamma_2)\} \subseteq \Delta_W$, if $\beta_1, \beta_2 \in \mathcal{C} \wedge (\beta_1, \beta_2) \in >_\mathcal{C}$ then $\gamma_1 > \gamma_2$ with $\gamma_1, \gamma_2 \in [0,1)$.*

It is worth mentioning, that function **pref** will be implemented by the comparison criterion of arguments provided by P-DeLP (previously introduced in

Definition 13). Given that this function is defined on argumental structures, below it is presented an instantiation function that obtains the corresponding argument from a primitive argumental structure. Likewise, in Proposition 1 is shown that for those active argumental structures in the DAF, their equivalent arguments can be generated from factual information of the P-DeLP program.

Definition 15 (Instance of a Primitive Argumental Structure). *Given an epistemic component $\mathcal{K}_\mathcal{A}$ and let Σ_i be a primitive argumental structure in $\mathcal{K}_\mathcal{A}$, the instantiation function of primitive argumental structures is $\Psi_{Inst\Sigma}$: $(2^\mathcal{L} \times \mathcal{L}) \mapsto (2^\mathcal{L} \times \mathcal{L})$, such that:*

$$\Psi_{Inst\Sigma}(\Sigma_i) = \langle \{(\alpha \leftarrow \beta, \gamma_1); (\beta, \gamma_2)\}, \alpha, \gamma_3 \rangle \text{ where } \gamma_3 = \min(\gamma_1, \gamma_2),$$
$$args(\Sigma_i) = \{\mathcal{A}\}, \ \Psi_E(pr(\mathcal{A})) = \{(\beta, \gamma_2)\} \text{ and } cl(\Sigma_i) = \alpha.$$

Proposition 1. *Let $\mathcal{K}_\mathcal{A} = \langle E, W, \bowtie, \mathfrak{pref} \rangle$ be an abstract epistemic component. Let $\mathcal{P} = (\Pi_W, \Delta_W)$ be a P-DeLP program such that for all clause $(\alpha \leftarrow \beta, \gamma) \in \Delta_W$, α satisfies the restriction imposed by Definition 14. Let $\Psi_{Inst\Sigma}$ be the instantiation function of primitive argumental structures presented in Definition 15. If the primitive argumental structure Σ_i is active in $\mathcal{K}_\mathcal{A}$ with respect to the evidence E, then given $\Psi_{Inst\Sigma}(\Sigma_i) = \langle \{(\alpha \leftarrow \beta, \gamma_1); (\beta, \gamma_2)\}, \alpha, \gamma_3 \rangle$, it holds that $\{(\beta, \gamma_2)\} \subseteq \Pi_W \cup \Delta_W$.*

Proof. Trivial from Definitions 11 and 15.

5 Proof Theory

Taking into account that both formalisms (viz. P-DeLP and the AADF) have proof theories based on dialectical processes, this makes easier to state the equivalence at the warranty level. In fact, for this proposed instantiation, formalization is almost straight. This is due to the fact that both languages define their proof theories on the notions of *acceptable argumentation line* and *dialectical tree*. In this way, by posing this equivalence we are sure that when a conclusion is justified in one formalism, will also be justified in the other one.

Definition 16 (Argumentation Line). *An argumentation line is a sequence $[\langle \mathcal{A}_0, Q_0, \alpha_0 \rangle, \ldots, \langle \mathcal{A}_n, Q_n, \alpha_n \rangle, \ldots]$ that can be thought of as an exchange of arguments between two parties, a proponent (evenly-indexed arguments) and an opponent (oddly-indexed arguments), where each $\langle \mathcal{A}_i, Q_i, \alpha_i \rangle_{(i>0)}$ is a defeater for the previous argument $\langle \mathcal{A}_{i-1}, Q_{i-1}, \alpha_{i-1} \rangle$ in the sequence.*

As stated in Definition 18, in order to avoid fallacious reasoning, additional constraints are imposed on argumentation lines to be considered rationally acceptable with respect to a P-DeLP program $\mathcal{P}$. Before introducing such definition, the concept of *contradictory arguments set* is posed first.

Definition 17 (Contradictory Arguments Set). *Let $\mathcal{P}$ be a P-DeLP program, a set $S = \bigcup_{i=1}^{n}\{\langle \mathcal{A}_i, Q_i, \alpha_i \rangle\}$ is contradictory with respect to $\mathcal{P}$, iff $\Pi \cup \bigcup_{i=1}^{n}\{\langle \mathcal{A}_i, Q_i, \alpha_i \rangle\}$ is contradictory.*[5]

Definition 18 (Acceptable Argumentation Line). *Given a P-DeLP program $\mathcal{P}$, an argumentation line $\lambda = [\langle \mathcal{A}_0, Q_0, \alpha_0 \rangle, \ldots, \langle \mathcal{A}_n, Q_n, \alpha_n \rangle, \ldots]$ is acceptable iff the following constraints hold:*

- *Non-contradiction: given an argumentation line λ, the set of arguments of the proponent (respectively, the opponent) should be non-contradictory with respect to $\mathcal{P}$.*
- *Non-circularity: no argument $\langle \mathcal{A}_j, Q_j, \alpha_j \rangle$ in λ is a sub-argument of an argument $\langle \mathcal{A}_i, Q_i, \alpha_i \rangle$ in λ, with $i < j$.*
- *Progressive argumentation: if $\langle \mathcal{A}_i, Q_i, \alpha_i \rangle$ is a blocking defeater in λ then $\langle \mathcal{A}_{i+1}, Q_{i+1}, \alpha_{i+1} \rangle$ is a proper defeater in λ.*

The first condition of Definition 18 disallows the use of contradictory information on either side (proponent or opponent). The second condition eliminates the "circular reasoning" fallacy. The last condition enforces the use of a stronger argument to defeat an argument which acts as a blocking defeater. An argumentation line satisfying the above restrictions can be proven to be finite.

In this way, given a P-DeLP program $\mathcal{P}$ and an argument $\langle \mathcal{A}_0, Q_0, \alpha_0 \rangle$, the set of all the acceptable argumentation lines starting with $\langle \mathcal{A}_0, Q_0, \alpha_0 \rangle$, account for a complete dialectical analysis for $\langle \mathcal{A}_0, Q_0, \alpha_0 \rangle$, that is usually represented by a tree structure called *dialectical tree*.

Definition 19 (Dialectical Tree). *Let $\langle \mathcal{A}_0, Q_0, \alpha_0 \rangle$ be an argument w.r.t. a P-DeLP program $\mathcal{P}$. A dialectical tree for $\langle \mathcal{A}_0, Q_0, \alpha_0 \rangle$, denoted $\mathcal{T}_{\langle \mathcal{A}_0, Q_0, \alpha_0 \rangle}$, is a tree structure defined as follows:*

1. *The root node of $\mathcal{T}_{\langle \mathcal{A}_0, Q_0, \alpha_0 \rangle}$ is $\langle \mathcal{A}_0, Q_0, \alpha_0 \rangle$.*
2. *$\langle \mathcal{B}', h', \beta' \rangle$ is an immediate child of $\langle \mathcal{B}, h, \beta \rangle$ iff an acceptable argumentation line $\lambda = [\langle \mathcal{A}_0, Q_0, \alpha_0 \rangle, \ldots, \langle \mathcal{A}_n, Q_n, \alpha_n \rangle, \ldots]$ exists, s.t. $\exists i \in \{0 \ldots n-1\}$ where $\langle \mathcal{A}_i, Q_i, \alpha_i \rangle = \langle \mathcal{B}, h, \beta \rangle$ and $\langle \mathcal{A}_{i+1}, Q_{i+1}, \alpha_{i+1} \rangle = \langle \mathcal{B}', h', \beta' \rangle$.*

Definition 20. *Given a program $\mathcal{P} = (\Pi, \Delta)$ and a goal Q, Q will be justified w.r.t. $\mathcal{P}$ with a maximum certainty degree α, iff there exists an argument $\langle \mathcal{A}, Q, \alpha \rangle$, for some $\mathcal{A} \subseteq \Delta$, such that:*

1. *Every acceptable argumentation line starting with $\langle \mathcal{A}, Q, \alpha \rangle$ has an odd number of arguments; that is, each of these lines end with an argument posed by the proponent that is in favor of Q with at least a certainty degree α; and*
2. *There is no argument $\langle \mathcal{A}_1, Q, \alpha \rangle$, with $\beta > \alpha$, satisfying the previous condition.*

---

[5] Here, the term *contradictory* refers to the one mentioned in Definition 7.

When a goal Q is justified by an argument $\langle \mathcal{A}, Q, \alpha \rangle$, as referred in Definition 20, we will say that $\langle \mathcal{A}, Q, \alpha \rangle$ is *warranted* w.r.t. $\mathcal{P}$, and will be denoted as $\mathcal{P} \mathrel{\mid\!\sim_w} \langle \mathcal{A}, Q, \alpha \rangle$. In this way, given a program $\mathcal{P}$, a P-DeLP interpreter will find an answer for a goal Q, by determining whether Q is supported by some warranted argument $\langle \mathcal{A}, Q, \alpha \rangle$. Thus, the possible answers for a goal could be:

1. YES (with certainty degree α), whenever Q is supported by a warranted argument $\langle \mathcal{A}, Q, \alpha \rangle$;
2. NO (with certainty degree α), whenever $\sim Q$ is supported by a warranted argument $\langle \mathcal{A}, \sim Q, \alpha \rangle$; or
3. UNDECIDED, whenever none of the previous points hold.

The following proposition presents the first step to prove the equivalence between formalisms at the warranty level.

Proposition 2. *If all the argumental structures in an acceptable argumentation line are primitive, then the notion of acceptable argumentation line in a DAF is equivalent to the notion of acceptable argumentation line in P-DeLP.*

Proof. A sequence $[\langle \mathcal{A}_0, Q_0, \alpha_0 \rangle, \ldots, \langle \mathcal{A}_n, Q_n, \alpha_n \rangle, \ldots]$ of P-DeLP arguments, that satisfies the constraints posed in Definition 18 is an acceptable argumentation line, denoted $\lambda_\mathcal{P}$. Given an acceptable argumentation line $[\Sigma_1, \ldots, \Sigma_n]$ in the DAF, where for each argumental structure Σ_i holds that $args(\Sigma_i) = \{\mathcal{A}_i\}$, then its set of pro argumental structures, $\lambda^+ = [\Sigma_1, \Sigma_3, \ldots]$, corresponds to the set of arguments posed by the proponent in $\lambda_\mathcal{P}$. In the same way, its set of cons argumental structures, $\lambda^- = [\Sigma_2, \Sigma_4, \ldots]$, corresponds to the set of arguments posed by the opponent in $\lambda_\mathcal{P}$.

From among the three restrictions that an acceptable argumentation line $\lambda_\mathcal{P}$ must hold in P-DeLP, the first one (non-contradiction) corresponds to the concordance property of Definition 2. Likewise, the second restriction of Definition 18, corresponds to the non-circularity constraint imposed on an argumentation line in Definition 2. In a DAF no more restrictions are posed on argumentation lines to be acceptable, therefore the notion of acceptable argumentation line is equivalent on both formalisms.

The second step to show the existing equivalence at the warranty level, consists of establishing the equivalence on the notion of dialectical tree for both formalisms.

Proposition 3. *Given an abstract epistemic component $\mathcal{K}_\mathcal{A}$ and a maximal set S (w.r.t. set inclusion) of exhaustive argumentation lines of $\mathcal{K}_\mathcal{A}$ rooted in Σ_1, if all the lines in S contain only primitive argumental structures, then the notion of dialectical tree in a DAF is equivalent to the notion of dialectical tree in P-DeLP.*

Proof. Trivial from Definitions 3, 18 and 19, and from Proposition 2.

Finally, it only remains to specify the equivalence when both formalisms dialectically determine that a particular argument is warranted.

Proposition 4. *Let $\mathcal{P}$ a P-DeLP program and let $\mathcal{K}_\mathcal{A}$ be an abstract epistemic component. Let $\mathbb{T}_{\mathcal{K}_\mathcal{A}}(\Sigma)$ be an active dialectical tree in $\mathcal{K}_\mathcal{A}$ and let $\mathcal{T}_{\langle\mathcal{A}_0,Q_0,\alpha_0\rangle}$ be the equivalent dialectical tree $\mathbb{T}_{\mathcal{K}_\mathcal{A}}(\Sigma)$ in $\mathcal{P}$. If $\mathfrak{m}_e(\Sigma,\lambda,\mathbb{T}_{\mathcal{K}_\mathcal{A}}(\Sigma)) = \mathbf{U}$, then the argument $\langle\mathcal{A}_0,Q_0,\alpha_0\rangle$ is warranted in $\mathcal{P}$ and consequently Q_0 is justified.*

Proof. Trivial from Proposition 3 and Definitions 4, 5 and 20.

6 Application Example

In order to clarify the above-mentioned concepts regarding the proposed instantiation, in this section an example is developed. The application domain chosen correspond to a decision making scenario proposed in [6], where a simulated *Khepera* 2 robot has to decide which box to transport next to the storage area.

It is a simple scenario where the robot must collect boxes scattered through the environment. The environment consists of a square arena of 100 units per side which is conceptually divided into square cells of 10 units per side each. A global camera provides to the robot the necessary information to perform its activities. The store is a 30×30 units square on the top-right corner and represents the target area where the boxes should be transported. There are boxes of three different sizes (small, medium and big) spread over the environment. The autonomy of the robot is limited and it cannot measure the state of its battery, thus, the robot cannot perform a globally optimized task. Because of this drawback, a greedy strategy is used to select the next box. Therefore, the robot will prefer its nearer boxes, then the boxes nearer to the store, and finally the smaller ones.

Example 1. Given the situation depicted in Fig. 1, there is a robot ($khep_1$) and five boxes. Three of the boxes (box_1, box_2, box_3) have the same properties: they are small, they are far from the robot and from the store. Box box_4 is medium size, is near to the store and far from $khep_1$; and box_5 is big, is near to the store and near to the robot. Hence, the corresponding evidence set converted to P-DeLP, would be given for the clauses shown in Fig. 2. It is worth mentioning that literals denoted as *same_prop* in [6] and in our running example, match the semantic meaning of literals *same_att* in Definition 1 (cf. [9] for more details).

Given the set $X = \{box_1, box_2, box_3, box_4, box_5\}$ of all the alternatives available for decision, the arguments of the working set W of the abstract epistemic component $\mathcal{K}_\mathcal{A} = \langle E, W, \bowtie, \mathfrak{pref}\rangle$, are shown in Fig. 3.[6] It is worth remembering that black triangles represent inactive arguments, while the white ones represent those which are active according to the set of evidence E.

---

[6] In this figure, literals *same_prop*, *smaller*, *nearer_store*, *nearer_robot*, *better*, ∼*better*, $box_1, box_2, box_3, box_4$ and box_5 are abbreviated as: *sp, sm, ns, nr, bt,* ¬ *b,* 1, 2, 3, 4 and 5, respectively.

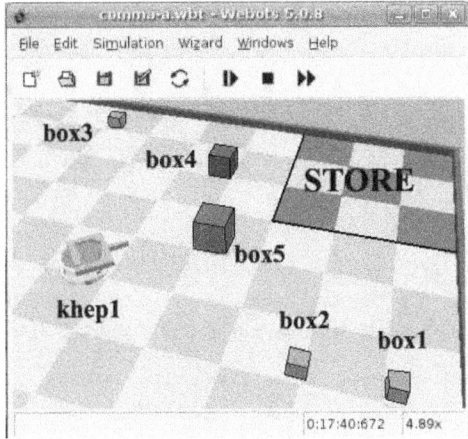

Fig. 1. Simulated experimental environment from [6].

$$\left\{\begin{array}{ll} (same\_prop(box_1, box_2), \alpha_7), & (nearer\_store(box_4, box_3), \alpha_{19}), \\ (same\_prop(box_1, box_3), \alpha_8), & (nearer\_store(box_5, box_1), \alpha_{20}), \\ (same\_prop(box_2, box_3), \alpha_9), & (nearer\_store(box_5, box_2), \alpha_{21}), \\ (smaller(box_1, box_4), \alpha_{10}), & (nearer\_store(box_5, box_3), \alpha_{22}), \\ (smaller(box_1, box_5), \alpha_{11}), & (nearer\_store(box_5, box_4), \alpha_{23}), \\ (smaller(box_2, box_4), \alpha_{12}), & (nearer\_robot(box_4, box_1), \alpha_{24}), \\ (smaller(box_2, box_5), \alpha_{13}), & (nearer\_robot(box_4, box_2), \alpha_{25}), \\ (smaller(box_3, box_4), \alpha_{14}), & (nearer\_robot(box_4, box_3), \alpha_{26}), \\ (smaller(box_3, box_5), \alpha_{15}), & (nearer\_robot(box_5, box_1), \alpha_{27}), \\ (smaller(box_4, box_5), \alpha_{16}), & (nearer\_robot(box_5, box_2), \alpha_{28}), \\ (nearer\_store(box_4, box_1), \alpha_{17}), & (nearer\_robot(box_5, box_3), \alpha_{29}), \\ (nearer\_store(box_4, box_2), \alpha_{18}), & (nearer\_robot(box_5, box_4), \alpha_{30}) \end{array}\right\}$$

Fig. 2. Evidence set $\Psi_E(E)$.

If the arguments conversion function Ψ_{arg} is applied to W, we would obtain the set of rules shown in Fig. 4 (denoted in a schematic way). If Definitions 10 and 11 are analyzed, the conversion functions of arguments and evidence, do not impose further restrictions on the α_i values, than the values belong to intervals $[0, 1]$ and $(0, 1]$, respectively. In fact, it is in Definition 14, where the implicit strict total order is generated, coded in the α_i values. Hence, for the moment, we will assign generic α_i values ($i = 1 \ldots 30$) to the P-DeLP clauses.

Taking into account the preferences of $khep_1$ mentioned above and according to Definition 14, the implicit generation of the strict total order should consider that the strength of an argument using rules (1) or (4) of Fig. 4, be stronger than the strength of another argument based on rule (2) or (5), and (3) or (6). Similarly, an argument that uses rules (2) or (5), should be stronger than a counter-argument based on rules (3) or (6). Therefore, inter-

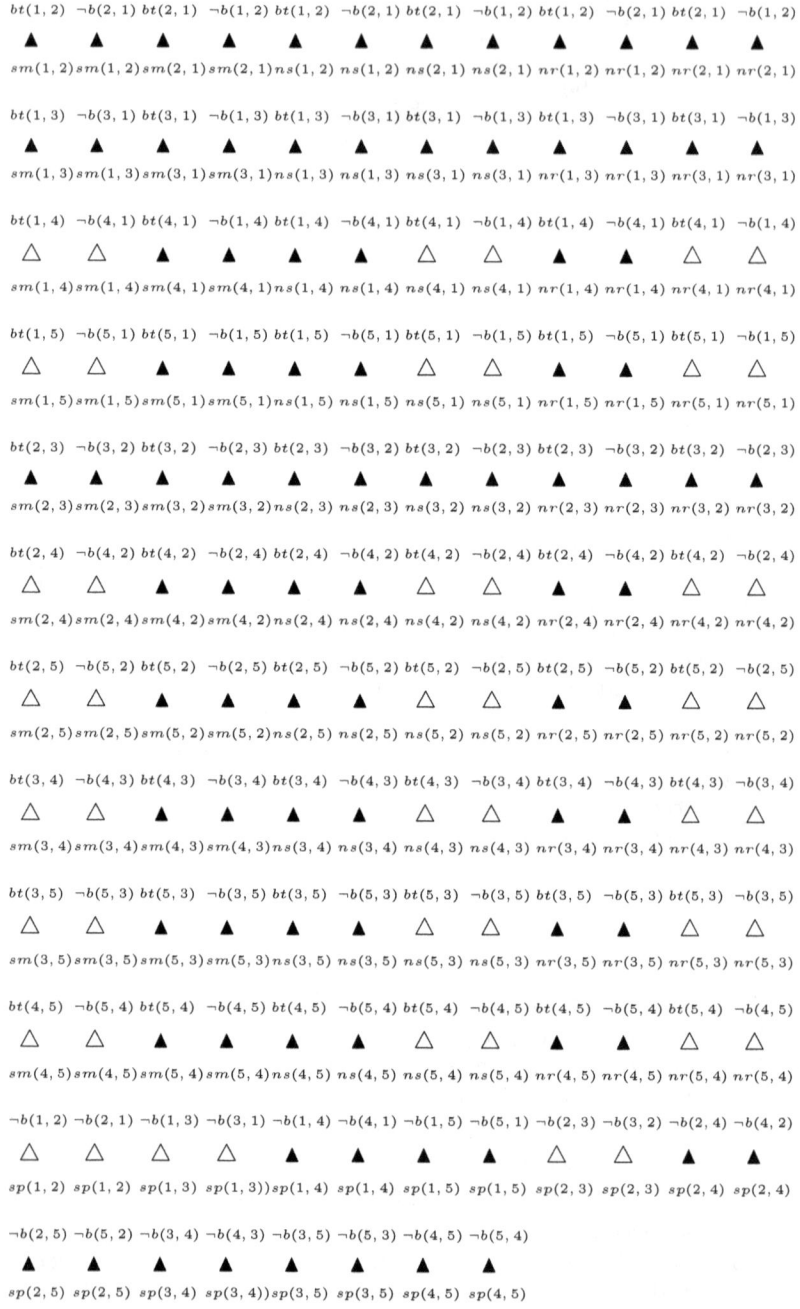

Fig. 3. Active and inactive arguments of the working set W.

$$\Delta_W = \begin{cases} (better(Box, Obox) \leftarrow nearer\_robot(Box, Obox), \alpha_1) & (1) \\ (better(Box, Obox) \leftarrow nearer\_store(Box, Obox), \alpha_2) & (2) \\ (better(Box, Obox) \leftarrow smaller(Box, Obox), \alpha_3) & (3) \\ (\sim better(Box, Obox) \leftarrow nearer\_robot(Obox, Box), \alpha_4) & (4) \\ (\sim better(Box, Obox) \leftarrow nearer\_store(Obox, Box), \alpha_5) & (5) \\ (\sim better(Box, Obox) \leftarrow smaller(Obox, Box), \alpha_6) & (6) \end{cases}$$

$$\Pi_{W_s} = \begin{cases} (\sim better(Box, Obox) \leftarrow same\_prop(Obox, Box), 1) \ (7) \\ (\sim better(Box, Obox) \leftarrow same\_prop(Box, Obox), 1) \ (8) \end{cases}$$

Fig. 4. Rules provided to the robot $khep_1$.

val $[0,1]$ could be divided into three sub-intervals $[0..0.33)$, $[0.33..0.66)$ and $[0.66..1)$ such that $\alpha_1 = \alpha_4 = 0.99$, $\alpha_2 = \alpha_5 = 0.66$ and $\alpha_3 = \alpha_6 = 0.33$. Given that the GMP inference rule uses as support value of the conclusion the minimum certainty degree of all the pieces of information involved, it is convenient to suppose that all factual information in $\Psi_E(E)$ has a certainty degree equal to 1; that is, $\forall i \in \{7, \ldots, 30\}, \alpha_i = 1$. In this way, given the alternatives box_3 and box_4, based on program $\mathcal{P} = (\Pi_W, \Delta_W)$ such that $\Psi_E(E) \subset \Pi_W$, the following conflicting arguments $\langle \mathcal{A}_1, better(box_3, box_4), 0.33 \rangle$ and $\langle \mathcal{A}_2, \sim better(box_3, box_4), 0.66 \rangle$, can be built:

$$\mathcal{A}_1 = \begin{cases} (better(box_3, box_4) \leftarrow smaller(box_3, box_4), 0.33), \\ (smaller(box_3, box_4), 1) \end{cases}$$

$$\mathcal{A}_2 = \begin{cases} (\sim better(box_3, box_4) \leftarrow nearer\_store(box_4, box_3), 0.66), \\ (nearer\_store(box_4, box_3), 1) \end{cases}$$

Argument $\mathcal{A}_2$ is stronger than $\mathcal{A}_1$ and hence the latter is defeated. Besides, box box_4 is also nearer to the robot than box_3; consequently, no more arguments can be generated to support that box_3 is better than box_4. Therefore, argument $\mathcal{A}_2$ is warranted and its conclusion $\sim better(box_3, box_4)$ is justified. Similarly, given alternatives box_3 and box_5, the following conflicting arguments $\langle \mathcal{A}_3, better(box_3, box_5), 0.33 \rangle$ and $\langle \mathcal{A}_4, \sim better(box_3, box_5), 0.66 \rangle$ can be built:

$$\mathcal{A}_3 = \begin{cases} (better(box_3, box_5) \leftarrow smaller(box_3, box_5), 0.33), \\ (smaller(box_3, box_5), 1) \end{cases}$$

$$\mathcal{A}_4 = \begin{cases} (\sim better(box_3, box_5) \leftarrow nearer\_store(box_5, box_3), 0.66), \\ (nearer\_store(box_5, box_3), 1) \end{cases}$$

A dialectical process analogous to the one mentioned above takes place when box_3 is compared against box_5, obtaining a justification for conclusion $\sim better(box_3, box_5)$. Likewise, when alternatives

box_1, box_2 are compared against box_4 and box_5, analogous conclusions are obtained, since these three alternatives have the same features: $\{(same\_prop(box_1, box_2), 1), (same\_prop(box_1, box_3), 1), (same\_prop(box_2, box_3), 1)\} \subset \Psi_E(E)$. Besides, by using rules (7) and (8) the following empty arguments can also be built:

$$\mathcal{A}_7 = \langle \emptyset, \sim better(box_1, box_2), 1\rangle \quad \mathcal{A}_{10} = \langle \emptyset, \sim better(box_3, box_1), 1\rangle$$
$$\mathcal{A}_8 = \langle \emptyset, \sim better(box_2, box_1), 1\rangle \quad \mathcal{A}_{11} = \langle \emptyset, \sim better(box_2, box_3), 1\rangle$$
$$\mathcal{A}_9 = \langle \emptyset, \sim better(box_1, box_3), 1\rangle \quad \mathcal{A}_{12} = \langle \emptyset, \sim better(box_3, box_2), 1\rangle$$

Finally, when alternatives box_4 and box_5 are compared against each other, arguments $\langle \mathcal{A}_5, better(box_4, box_5), 0.33\rangle$ and $\langle \mathcal{A}_6, \sim better(box_4, box_5), 0.99\rangle$ can be built:

$$\mathcal{A}_5 = \left\{ \begin{array}{l} (better(box_4, box_5) \leftarrow smaller(box_4, box_5), 0.33), \\ (smaller(box_4, box_5), 1) \end{array} \right\}$$

$$\mathcal{A}_6 = \left\{ \begin{array}{l} (\sim better(box_4, box_5) \leftarrow nearer\_robot(box_5, box_4), 0.99), \\ (nearer\_robot(box_5, box_4), 1) \end{array} \right\}$$

Like argument $\mathcal{A}_6$ is stronger than $\mathcal{A}_5$ the latter is defeated, and $\mathcal{A}_6$ is warranted thus justifying conclusion $\sim better(box_4, box_5)$.

7 Conclusions

In this work we have presented an instantiation of the AADF proposed in [9] based on P-DeLP, a concrete formalism which combines features from argumentation theory, logic programming and a unified treatment of possibilistic uncertainty and fuzziness.

In order to properly instantiate the above-mentioned AADF, in Sect. 3 it was firstly specified the logical representation language. Then, in Sect. 4 the arguments and evidence conversion functions were defined, together with the instantiation of primitive argumental structures. Also, it was defined how to generate the strict total order needed for the arguments comparison criterion; and in particular, in Sect. 6 –in the context of an application example from the domain of cognitive robotics–, it was proposed a possible implicit codification of this strict total order, that is not necessarily the only one or the best. Finally, in Sect. 5 it was formally proved that the concept of warranty between both formalisms is equivalent.

As mentioned above, by instantiating the AADF we prove its adequacy as an abstract framework for decision making. Moreover, as a consequence of this instantiation we have a secondary contribution in this work; namely, a new concrete decision framework for decision making based on P-DeLP, whose choice behavior is also consistent with Classical Decision Theory. We have not stressed this latter aspect, since the emphasis of the instantiation lay in defining the proper conversion functions which mapped notions of the AADF to P-DeLP, and hence, the choice behavior consistent with Classical Decision Theory is obtained in addition.

If we compare the instantiation obtained, with an existing decision framework based on P-DeLP [7], we can see that the instantiation is a simpler decision framework in that there is no explicit multi-criteria aggregation like in [7], that is achieved by means of arguments accrual [12]. In this respect, the framework proposed in [7] develops a methodology to assign certainty values to uncertain clauses, so that they reflect the existing differences of the alternatives being compared together with the priority that distinguished literals have, in order to respect *maximality*[7] and *non-depreciation*[8] properties, which are relevant in aggregating necessity degrees. In our case, all evidence was modeled as certain clauses, and the rules in charge of generating the arguments were the ones encoding the priority that distinguished literals have.

To conclude, we believe that our P-DeLP instantiation also shortens the gap to analyze how in a future work, the AADF proposed in [9] could be extended so that a proposal with explicit multi-criteria aggregation like [7], could result as an instantiation of it.

References

1. T. Alsinet and L. Godo. A complete calculus for possibilistic logic programming with fuzzy propositional variables. In *Proceedings of the 16th Conference on Uncertainty in Artificial Intelligence (UAI)*, pages 1–10. ACM Press, 2000.
2. Teresa Alsinet, Carlos Iván Chesñevar, Lluís Godo, and Guillermo Ricardo Simari. A logic programming framework for possibilistic argumentation: formalization and logical properties. *Fuzzy Sets and Systems*, 159(10):1208–1228, 2008.
3. L. Amgoud and H. Prade. Using arguments for making and explaining decisions. *Artificial Intelligence*, 173(3-4):413–436, 2009.
4. Carlos I. Chesñevar, Guillermo R. Simari, Teresa Alsinet, and Lluís Godo. A logic programming framework for possibilistic argumentation with vague knowledge. In *AUAI '04: Proceedings of the 20th conference on Uncertainty in Artificial Intelligence*, pages 76–84, Banff, Canada, 2004. AUAI Press.
5. P. M. Dung. On the acceptability of arguments and its fundamental role in nonmonotonic reasoning, logic programming and n-person games. *Artificial Intelligence*, 77(2):321–358, 1995.
6. E. Ferretti, M. Errecalde, A. García, and G. Simari. Decision rules and arguments in defeasible decision making. In Ph. Besnard, S. Doutre, and A. Hunter, editors, *2nd International Conference on Computational Models of Arguments (COMMA)*, pages 171–182. IOS Press, 2008.
7. E. Ferretti, M. Errecalde, A. García, and G. Simari. A possibilistic defeasible logic programming approach to argumentation-based decision-making. *Journal of Experimental & Theoretical Artificial Intelligence*, 26(4):519–550, 2014.

---

[7] Accrual means total certainty only if there is an argument with necessity degree equal to 1.
[8] Accruing arguments result in a necessity degree not lower than any single argument involved in the accrual.

8. E. Ferretti, R. Kiessling, A. Silnik, R. Petrino, and M. Errecalde. *New Trends in Electrical Engineering Automatic Control, Computing and Communication Sciences*, chapter Integrating vision-based motion planning and defeasible decision making for differential-wheeled robots: A case of study with the Khepera 2 robot. LOGOS Verlag, Berlin, Germany, 2010.
9. Edgardo Ferretti, Luciano H. Tamargo, Alejandro J. García, Marcelo L. Errecalde, and Guillermo R. Simari. An approach to decision making based on dynamic argumentation systems. *Artificial Intelligence*, 242:107 – 131, 2017.
10. A. García, N. Rotstein, M. Tucat, and G. Simari. An argumentative reasoning service for deliberative agents. In *Knowledge Science, Engineering and Management: Second International Conference, KSEM 2007*, volume 4798 of *LNCS*, pages 128–139. Springer, 2007.
11. A. García and G. Simari. Defeasible logic programming: An argumentative approach. *TPLP*, 4(1-2):95–138, 2004.
12. Mauro Gómez, Carlos Chesñevar, and Guillermo Simari. Modelling argument accrual in possibilistic defeasible logic programming. In *ECSQARU*, pages 131–143, 2009.
13. Antonis Kakas and Pavlos Moraïtis. Argumentation based decision making for autonomous agents. In *2nd international joint conference on autonomous agents and multiagent systems (AAMAS)*, pages 883–890, Melbourne, Australia, July 14-18 2003. ACM Press.
14. V. Lifschitz. Foundations of logic programming. In Gerhard Brewka, editor, *Principles of Knowledge Representation*, pages 69–127. CSLI Publications, Stanford, California, 1996.
15. J. Müller and A. Hunter. An argumentation-based approach for decision making. In *IEEE 24th International Conference on Tools with Artificial Intelligence, ICTAI 2012*, pages 564–571, 2012.
16. Iyad Rahwan and Guillermo R. Simari, editors. *Argumentation in Artificial Intelligence*. Springer, 2009.
17. John H. Roberts and Gary L. Lilien. Explanatory and predictive models of consumer behavior. In J. Eliashberg and G. L. Lilien, editors, *HandBooks in Operations Research and Management Science*, volume 5. North-Holland, Amsterdam, The Netherlands, 1993.
18. N. Rotstein, M. Moguillansky, A. García, and G. Simari. A dynamic argumentation framework. In *Computational Models of Argument: Proceedings of COMMA 2010*, pages 427–438, 2010.

Complexity of DeLP: Current Status and Moving Forward

Laura Cecchi[1] and Pablo Fillottrani[2]

[1] Faculty of Informatics
Universidad Nacional del Comahue
Buenos Aires 1400, Neuquén Capital, Argentina
lcecchi@fi.uncoma.edu.ar

[2] Department of Computer Science and Engineering
Universidad Nacional del Sur
San Andrés 800, Bahía Blanca, Buenos Aires, Argentina
Comisión de Investigaciones Científicas, Provincia de Buenos Aires
prf@cs.uns.edu.ar

Abstract. Defeasible Logic Programming (DeLP) is a general argumentation based system for knowledge representation and reasoning. Its proof theory is based on a dialectical analysis where arguments for and against a literal interact in order to determine whether this literal is believed by a reasoning agent. The $\mathcal{GS}$ semantics is a declarative trivalued game-based semantics for DeLP that is sound and complete for DeLP proof theory. In this paper we review the latest complexity results of some important decision problems of DeLP under $\mathcal{GS}$, and extend some of them. Furthermore, we introduce DeLP as a query language and we study data and combined complexity of query answering in the context of DeLP for some important problems. As far as we know, data and combined complexity have not been introduced in the context of argumentation systems.

KEYWORDS: Argumentation Systems, Defeasible Logic Programming, Game-based Semantics, Computational Complexity, Data and Combined Complexity

1 Introduction

Defeasible Logic Programming (DeLP) is a general argumentation based system and tool for knowledge representation and reasoning [30][3]. Its proof theory is based on a dialectical analysis where arguments for and against a literal interact

---
[3] The interested reader can find an on-line interpreter for DeLP in http://lidia.cs.uns.edu.ar/DeLP

in order to determine whether this literal is believed by a reasoning agent. The semantics $\mathcal{GS}$ is a declarative trivalued game-based semantics for DeLP that links game-semantics [1] and model-theory. Soundness and completeness of $\mathcal{GS}$ with respect to DeLP proof theory have been proved [15].

The study of the complexity of reasoning in different formalisms has led to a clear understanding of their expressive properties and implementation possibilities. Thus, the applicability of the inference techniques can only be determined, after the tradeoff between expressiveness and tractability of reasoning is thoroughly analysed. DeLP complexity has not been analysed in depth, even thought complexity for nonmonotonic reasoning systems has a long tradition for most formalisms such us default logic, autoepistemic logic, circumscription, abduction and logic programming [9, 21], This may be explained in part by the fact that, historically, implementations of argumentation systems have been limited to areas with no real time response restriction (see [58, 7, 53]). The appearance of several recent applications, for instance, multiagent systems, medical systems and autonomous vehicle among others [38, 43, 34, 45, 20, 51] has changed this. Scalability and robustness of such approaches heavily depend on the computational properties of the underlying algorithms, making their complexity analysis necessary.

We aim in fill this gap, by compiling and reviewing existing complexity results of DeLP, extending some of them, and identifying some interesting open research questions. First, we present some decision problems from the point of view of argumentation systems. Since DeLP builds the arguments from a defeasible logic program, we focus on two questions. The first is whether a set of defeasible rules is an argument for a literal in a given program. We show that this problem is **P**-complete. The second question is the existence of an argument for a literal under a given defeasible logic program, which has been proved to be in **NP**-complete. Afterwards, we analyse when an agent answer YES to a query considering whether a dialectical tree for the literal is marked undefeated, which has been proved **PSPACE**-complete and **AP**-complete. We also analyse when the agent has no information about a query, which has been proved co-**NP**-complete.

Finally, we introduce a novel concept: DeLP as a query language. In this context, we define data, expression and combined complexity [57, 48, 21] of DeLP, in order to evaluate the efficiency of DeLP implementations with very large data instances that share a fixed set of defeasible rules. *Data complexity* is the complexity of evaluating a specific query in the language, when the query is fixed, and we study the complexity of applying this query to arbitrary databases; the complexity is thus given as a function of the size of the database. *Program or Expression complexity* appears when a specific database is fixed, and we study the complexity of applying queries represented by arbitrary expressions in the language; the complexity is given as a function of the length of the expression. *Combined complexity* considers both query and database instance as input variables.

In particular, we study data and combined complexity of query answering to assess DeLP applications over database technologies. As far as we know data and combined complexity has not been introduced in the context of argumentation systems.

Although different computational complexity results [23, 26, 2, 35, 56, 59] have been presented on argumentation abstract framework [8, 25], based on admissibility and preferability semantics, those results do not apply directly to DeLP because of their quite different semantics. Another notable study of the computational complexity of defeasible systems has been done in [41]. But, defeasible theory analysed in this work greatly differs from DeLP in several points, such as knowledge representation (facts and strict rule, defeasible and defeaters rules) and their proof theories. In [36], the authors have presented a novel study of the computational complexity of problems that arise in abstract argumentation in the context of dynamic argumentation, minimal change, and aggregation. Even though their results covers several different semantics, this work does not apply DeLP semantics.

The paper is structured as follows. In the following section we briefly outline the fundamentals of DeLP, and describe the declarative game-based semantics $\mathcal{GS}$. Then, we discuss DeLP through $\mathcal{GS}$ semantics pointing out the decision problems that are of central importance. Afterwards, we give complexity results on the existence of an argument for a literal L under a defeasible logic program $\mathcal{P}$, on the decision problem of whether a subset of defeasible rules is an argument for a literal L under $\mathcal{P}$. and on the decision problem of the existence of a game won by the proponent. In section 5, we introduce a novel concept: DeLP as a query language. Next, we define data, expression and combined complexity in the context of DeLP and we analyse data and combined complexity for DeLP, and we present complexity results for two decision problems on entailment. In the last section, we summarise the main contributions of this work, and we present our conclusions and future research lines.

2 DeLP overview

2.1 Syntax and proof theory

We will start by introducing some DeLP's basic concepts for knowledge representation and some details of its proof theory. For complete details on DeLP see [30, 31].

In the language of DeLP a literal L is a atom A or a negated atom $\sim A$, where $\sim$ represents the strong negation in the logic programming sense. The complement of a literal L, denoted as $\overline{L}$, is defined as follows: $\overline{L} = \sim A$, if L is an atom, otherwise if L is a negated atom, $\overline{L} = A$. Let X be a set of literals, $\overline{X}$ is the set of the complement of every member in X.

Definition 1. *A* fact *is a ground literal and it is written as L. A* strict rule *is an ordered pair, denoted "Head $\leftarrow$ Body", where "Head" is a ground literal,* and "Body" is a finite set of ground literals. A strict rule with head L_0 and body $\{L_1, \ldots L_n, n > 0\}$ is written as $L_0 \leftarrow L_1, \ldots L_n$. Strict rules correspond syntactically to basic rules in Logic Programming[37]. A defeasible rule is an ordered pair, denoted "Head $\prec$ Body", where "Head" is a ground literal, and "Body" is a finite, non-empty set of ground literals. A defeasible rule with head L_0 and body $\{L_1, \ldots L_n, n > 0\}$ is written as $L_0 \prec L_1, \ldots L_n$.

A defeasible logic program $\mathcal{P}$, abbreviated de.l.p., is a set of facts, strict rules and defeasible rules. We will distinguish the subsets Π_F of facts, Π_R of strict rules, $\Pi = \Pi_F \cup \Pi_R$ and the subset Δ of defeasible rules.

We denote by *Lit* the set of all the ground literals that can be generated considering the underlying signature of a de.l.p. an we denote by Lit^+ the set of all the atoms in *Lit*. Intuitively, whereas Π is a set of certain and exception-free knowledge, Δ is a set of defeasible knowledge, i.e., tentative information that could be used, whenever nothing is posed against it. Facts are used for representing information that is considered to hold in the application domain. As a de.l.p. is required to be representationally coherent, it cannot contain two complementary facts.

By definition a de.l.p. may be an infinite set of strict and defeasible rules but for complexity analysis we restrict ourselves to finite defeasible logic programs. Several approaches to defeasible reasoning have introduced a defeasible rule with an empty body[30, 42, 46], which is called a *presumption*. However in the context of this work, we do not consider this class of defeasible rule.

DeLP proof theory is based on developments in non monotonic argumentation systems [50, 55]. An *argument for a literal* L is a minimal subset of Δ that together with Π consistently entails L. Note that an argument for a literal uses the complete set Π in order to *consistently* entails L. Therefore, for any program we assume that no pair of contradictory literals can be derived from Π.

The notion of entailment corresponds to the usual SLD derivation used in logic programming, performed by backward chaining on both strict and defeasible rules, where negated atoms are treated as a new atom in the underlying signature. Thus, an agent can explain a literal L, throughout this argument.

In order to determine whether a literal L is supported from a de.l.p. a dialectical tree for L is built. An argument for L represents the root of the dialectical tree, and every other node in the tree is a defeater argument against its parent. At each level, for a given a node we must consider all the arguments against that node. Thus every node has a descendant for every defeater. A comparison criteria is needed for determining whether an argument defeats another. Even though there exist several preference relations considered in the literature[30, 52, 28, 44], in this approach we will abstract away from that issue.

We will say that a literal L is warranted if there is an argument for L, and in the dialectical tree each defeater of the root is itself defeated. Recursively, this leads to a marking procedure of the tree that begins by considering the fact that leaves of the dialectical tree are undefeated arguments as a consequence of having no defeaters. Finally, an agent will believe in a literal L, if L is a warranted literal. The intuitive meaning of a literal being warranted is that it is supported by an argument that has no counterargument still standing against it.

Given a *de.l.p.*, there exist four possible answers for a query L: YES if L is warranted, NO if $\overline{L}$ is warranted (i.e., the complement of L is warranted), UNDECIDED if neither L nor $\overline{L}$ are warranted, and UNKNOWN if L is not in the underlying signature of the program.

2.2 $\mathcal{GS}$: A Game Semantics for DeLP

Games have an analogy with a dispute and, therefore, that analogy extends to argument-based reasoning. A dispute can be seen as a game where in an alternating manner, the player P, the proponent, starts with an argument for a literal. The player O, the opponent, attacks the previous argument with a counterargument strong enough to defeat it. The dispute could continue with a counterargument of the proponent, and so on. When a player runs out of moves, i.e., that player cannot find a counterargument for any of his adversary's arguments, the game is over. If the proponent's argument has not been defeated then she has won the game.

The semantics $\mathcal{GS}$ is a declarative trivalued game-based semantics for DeLP that links game-semantics [1] and model theory. Soundness and completeness of $\mathcal{GS}$ with respect to DeLP proof theory have been proved in [15]. In the following we present some notions of $\mathcal{GS}$, for more details see [15].

Let X be a set and $\{x_1, \ldots, x_n\} \subseteq X$, X^* is the set of finite sequences over X and $[x_1 \ldots x_n]$ denotes the sequence of the elements $x_1, \ldots, x_n$. We write $|s|$ for the length of a finite sequence and s_i for the ith element of s, $1 \leq i \leq |s|$. Concatenation of sequences is indicated by juxtaposition. If $t = su$ for some sequences t, s, u, then we say that s is a prefix of t. Let $Pref(S)$ be a set of prefix of S, then S is prefix closed if $S = Pref(S)$. The set of ground literals X is *rigorously closed under a de.l.p.* $\mathcal{P}$, if for every strict rule $Head \leftarrow Body$ of $\mathcal{P}$, $Head \in X$ whenever $Body \subseteq X$, and for every defeasible rule $Head' \prec Body'$ of $\mathcal{P}$, $Head' \in X$ whenever $Body' \subseteq X$. The set X is *consistent* if there is no literal L such that $\{L, \overline{L}\} \subseteq X$. Otherwise, we will say that X is *inconsistent*. We say that X is *logically closed* if it is consistent or it is equal to *Lit*.

Let's introduce game concept and $\mathcal{GS}$ semantics.

Definition 2. *Let* $\mathcal{P} = (\Pi, \Delta)$ *be a de.l.p.,* L *a literal and* $\langle \mathcal{A}, L \rangle$ *an argument structure for* L. *A game for* $\langle \mathcal{A}, L \rangle$ *with respect to* $\mathcal{P}$, *that we denote* $G(\langle \mathcal{A}, L \rangle, \mathcal{P})$, *is a structure*

$$(M_{G(\langle \mathcal{A}, L \rangle, \mathcal{P})}, J_{G(\langle \mathcal{A}, L \rangle, \mathcal{P})}, P_{G(\langle \mathcal{A}, L \rangle, \mathcal{P})})$$

where

- $M_{G(\langle \mathcal{A}, L \rangle, \mathcal{P})}$ *is a set of argument structure.*

- $J_{G(\langle \mathcal{A}, L \rangle, \mathcal{P})} : M_{G(\langle \mathcal{A}, L \rangle, \mathcal{P})} \times I \to \{P, O\}$ where I is an enumerable index;
- $P_{G(\langle \mathcal{A}, L \rangle, \mathcal{P})} \subseteq M^*_{G(\langle \mathcal{A}, L \rangle, \mathcal{P})}$, where $P_{G(\langle \mathcal{A}, L \rangle, \mathcal{P})}$ is a non-empty, prefix-closed set.

  Each sequences s of $P_{G(\langle \mathcal{A}, L \rangle, \mathcal{P})}$ satisfy:
  1. $s = [\langle \mathcal{A}, L \rangle]s'$, s' possibly empty.
  2. For all i, $1 < i \leq |s|$

  $$J_{G(\langle \mathcal{A}, L \rangle, \mathcal{P})}(s_1, 1) = \mathsf{P}$$
  $$J_{G(\langle \mathcal{A}, L \rangle, \mathcal{P})}(s_i, i) = \overline{J_{G(\langle \mathcal{A}, L \rangle, \mathcal{P})}(s_{i-1}, i-1)}$$

  $\overline{\mathsf{P}} = \mathsf{O}$ and $\overline{\mathsf{O}} = \mathsf{P}$.
  3. If $s \in P_{G(\langle \mathcal{A}, L \rangle, \mathcal{P})}$, then for each argument structure $\langle \mathcal{A}_2, L_2 \rangle$ that is a legal move for $s_{|s|}$, there exists a sequence $t \in P_{G(\langle \mathcal{A}, L \rangle, \mathcal{P})}$, such that $t = s[\langle \mathcal{A}_2, L_2 \rangle]$.
  4. No other sequence belongs to $P_{G(\langle \mathcal{A}, L \rangle, \mathcal{P})}$.

Movements in a game are arguments. A legal move in the game over a sequence s is an argument $\mathcal{A}$ such that strictly defeats $s_{|s|}$ or defeats non strictly $s_{|s|}$ and $s_{|s|}$ strictly defeats $s_{|s|-1}$. Furthermore, such legal move $\mathcal{A}$ cannot be part of another argument in s, ie we cannot introduce more than once an argument neither for nor against the first move. Finally, this move must be consistent with every move made by the same player in the sequence s.

For every argument $\mathcal{A}$ for a literal L we can built a game whose first move is $\langle \mathcal{A}, L \rangle$. Thus, a family of games, denoted as $\mathcal{F}(L, \mathcal{P})$, will be obtained considering all the arguments for L in a de.l.p. $\mathcal{P}$, $\langle \mathcal{A}_1, L \rangle, \ldots, \langle \mathcal{A}_n, L \rangle$:

$$\mathcal{F}(L, \mathcal{P}) = \{G(\langle \mathcal{A}_1, L \rangle, \mathcal{P}), G(\langle \mathcal{A}_2, L \rangle, \mathcal{P}), \ldots, G(\langle \mathcal{A}_n, L \rangle, \mathcal{P})\}$$

Definition 3. *Let a be the first proponent movement in the game. A sequence s is complete if $s = [a]s_1$, with s_1 potentially empty, then there is no movement $b \in M_{G(\langle \mathcal{A}, L \rangle, \mathcal{P})}$ such that $[a]s_1[b] \in P_{G(\langle \mathcal{A}, L \rangle, \mathcal{P})}$. A sequence s is preferred if each opponent movement has a proponent answer. In other words, a sequence s is preferred if $|s|$ is odd.*

Definition 4. *A strategy over a game G is a set of sequences S, such that for all sequence $s \in S$, either:*

- *s is preferred; or*
- *there exists other sequence $s' \in S$, such that s' is preferred and s and s' has a prefix t, $|t| = n$, n is even and $s_{n+1} \neq s'_{n+1}$.*

Definition 5. *Let $\mathcal{P}$ be a de.l.p., $L \in Lit$ and $G(\langle \mathcal{A}, L \rangle, \mathcal{P}) \in \mathcal{F}(L, \mathcal{P})$. We say that P wins the game $G(\langle \mathcal{A}, L \rangle, \mathcal{P})$ or that $G(\langle \mathcal{A}, L \rangle, \mathcal{P})$ is won by P, if the set of complete sequences of $P_{G(\langle \mathcal{A}, L \rangle, \mathcal{P})}$ is an strategy. Otherwise, we say that O wins the game or that $G(\langle \mathcal{A}, L \rangle, \mathcal{P})$ is won by O.*

A player can win a game even though he does not win every complete sequence in such game. In [52] the authors have developed an argument-based extended logic programming system which differs from DeLP in its winning rule: a player wins a dialogue tree if and only if he wins all the branches of the tree.

Definition 6. *Let $\mathcal{P}$ be a de.l.p.. A game-based interpretation for $\mathcal{P}$, or G-Interpretation for $\mathcal{P}$ for short, is a tuple $\langle T, F \rangle$, such that T and F are subsets of atoms of the underlying signature of $\mathcal{P}$ and $T \cap F = \varnothing$.*

In the previous definition T stands for true while F stands for false. The set of atoms UNDECIDED is defined as the set $U = Lit^+ - \{T \cup F\}$. Each game can finish in two possible ways: won by the proponent P or won by the opponent O. There is no possibility for a draw. As the first move is made by the P, we are interested in those games won by this player.

Definition 7. *Let $\mathcal{P}$ be a de.l.p., h an atom of the underlying signature of $\mathcal{P}$, $\mathcal{F}(h, \mathcal{P})$ the game family for h and $\mathcal{F}(\overline{h}, \mathcal{P})$ the game family for $\overline{h}$ under a de.l.p. $\mathcal{P}$. A game-based model for $\mathcal{P}$, that we name G-Model of $\mathcal{P}$, is a G-interpretation $\langle T, F \rangle$ such that:*

- *If there exists a game $G(\langle \mathcal{A}, h \rangle$ in the family $\mathcal{F}(h, \mathcal{P})$ won by P, then h belongs to T.*
- *If there exists a game $G(\langle \mathcal{A}, \overline{h} \rangle, \mathcal{P})$ in the family $\mathcal{F}(\overline{h}, \mathcal{P})$ won by P, then h belongs to F.*

Since we only consider literals under the signature of de.l.p., the G-model definition does not contemplate the answer UNKNOWN. The minimal G-model defines a sound and complete semantics $\mathcal{GS}$ for DeLP. We will say that $\mathcal{GS}$ entails a literal L from a de.l.p. $\mathcal{P}$, denoted by $\mathcal{P} \models_{\mathcal{GS}} L$, whenever $L \in T$ or $\overline{L} \in F$, being $\langle T, F \rangle$ the minimal G-model of $\mathcal{P}$.

The following theorem relates proof theory and game-based semantics, showing soundness and completeness.

Theorem 1. *Let $\mathcal{P}$ be a de.l.p. and L a literal. L is warranted under $\mathcal{P}$ if and only if L belongs to the set T or $\overline{L}$ belongs to the set F of the minimal G-models $\langle T, F \rangle$ of $\mathcal{P}$ under $\mathcal{GS}$ semantics. [15]*

We have briefly presented the DeLP declarative game-based semantics $\mathcal{GS}$. Now, we will be able to analyse the system and study some complexity properties.

3 Some DeLP Decision Problems

DeLP is a formalism that offers an argumentation based knowledge representation and a computational reasoning system. In order to exploit this system in concrete problem solving, there are some reasoning issues related to its proof theory that have both theoretical and practical importance from a complexity point of view. Hence, we start with listing relevant reasoning decision problems in the context of a *de.l.p.* $\mathcal{P} = (\Pi, \Delta)$:

- ARGUMENTSAT: Given a literal α and a set $\mathcal{A} \in \Delta$, deciding whether $\langle \mathcal{A}, \alpha \rangle$ is an argument in the context of $\mathcal{P}$.

- ARGUMENTEXISTENCE: Given a literal α, deciding whether there exists an argument for α in the context of $\mathcal{P}$.

- GAMESAT: Given a literal α and an argument $\mathcal{A} \in \Delta$, deciding whether the game $G(\langle \mathcal{A}, \alpha \rangle, \mathcal{P})$ is won by the proponent in the context of $\mathcal{P}$.

- PWINGAME: Given a literal α, deciding whether there is a game for α won by the proponent in the context of $\mathcal{P}$.

- LITERALUNDECIDABILITYINGAME: Given a literal α, deciding whether there is no game for α neither for its complement won by the proponent in the context of $\mathcal{P}$.

We will denote NOARGUMENTEXISTENCE to ARGUMENTEXISTENCE's complement, i.e., , deciding whether there is no argument for a given literal α in the context of a *de.l.p.* $\mathcal{P}$. This decision problem can be expressed equivalently as determining whether the Family Game for the literal is empty. GAMESAT problem involves checking whether the set of complete sequences is a strategy. It is closed related with the DeLP marking procedure for dialectical trees[14]. A positive PWINGAME answer for a given *de.l.p.* $\mathcal{P}$ and a literal L implies that $L \in T$, being $\langle T, F \rangle$ the minimal G-model, i.e., $\mathcal{P} \models_{\mathcal{GS}} L$. A positive PWINGAME answer for $\overline{L}$ means that $L \in F$ in the minimal G-model, i.e., $\mathcal{P} \models_{\mathcal{GS}} \overline{L}$.

In order to capture LITERALUNDECIDABILITYINGAME, it is necessary to find all the games for the literal and for its complement, and to establish that none of them is won by the proponent. The LITERALUNDECIDABILITYINGAME decision problem for a *de.l.p.* $\mathcal{P}$ and a literal L is equivalent to determining if given the minimal G-model $\langle T, F \rangle$ of $\mathcal{P}$, $L \in Lit^+ - \{T \cup F\}$, i.e., $\mathcal{P} \not\models_{\mathcal{GS}} L$ and $\mathcal{P} \not\models_{\mathcal{GS}} \overline{L}$.

With regards to LITERALUNDECIDABILITYINGAME, three situations about the agent uncertainty knowledge can be contemplated, establishing the followings decision problems[13]:

(P1) *Whether there is no game for a literal L, neither for its complement $\overline{L}$. The game families for a literal L and for its complement $\overline{L}$, $\mathcal{F}(L, \mathcal{P})$ and $\mathcal{F}(\overline{L}, \mathcal{P})$ respectively, are empty. L has no argument neither for nor against it. Therefore, the agent has no information about such query.*

(P2) *Whether there is no game for a literal L, and the non empty set of all games in the family of its complement $\overline{L}$ are won by the opponent.* The game family for a literal L, $\mathcal{F}(L, \mathcal{P})$, is empty and only games won by the opponent are in the non empty family $\mathcal{F}(\overline{L}, \mathcal{P})$. L has no argument for and all the arguments for its complement are defeated. Therefore, the agent has no information for L, and he cannot defend its complement. In a similarly way, we can define the case where the agent cannot defend a literal L, and has no information about its complement.

(P3) *Whether all games in the non empty families for L and for its complement $\overline{L}$ are won by the opponent.* $\mathcal{F}(L, \mathcal{P})$ and $\mathcal{F}(\overline{L}, \mathcal{P})$ are non empty set and all the argument are defeated. The agent cannot defend any argument neither for nor against the literal L.

While PWinGame decision problem is close related to the agent answers YES or NO, the above problems are of great interest because they characterise agent's information indecision about a literal. In this cases, the agent answers UNDECIDED, because even the literal is in the signature, she cannot warrant neither the literal nor the complement- all the arguments that exists are defeated.

4 Complexity Results

Arguments and counterarguments are the movements in a game, and hence the core of DeLP. Dung's formalism [25] and some extensions that have been developed [5, 6, 2], offer a powerful tool for the abstract analysis of defeasible reasoning. However, these approaches operate with arguments and their attack and defeat relation at an abstract level, avoiding to deal with the underlying logical language used to structure the arguments. On the other hand DeLP does construct the arguments and analyses the defeater relationship, making the ArgumentSat problem core to the argumentation system.

Theorem 2. ArgumentSat *is **P**-complete.[13]*

Now, our aim is to determine the complexity of computing the set of all the arguments under a *de.l.p.*. This is motivated in that PWinGame and LiteralUndecidabilityInGame require for playing a game to compute every argument that defeats each argument introduced in a previous move. A subset $A \subseteq \Delta$ may be a potential argument of different literals in the language. Thus, the maximum number of checks for potential arguments that depends on the size of the set of defeasible rules and on the size of Lit, is $|Lit| * 2^{|\Delta|}$.

Lemma 1. *Let **ASP** be the polynomial time needed for the decision problem* ArgumentSat. *Then, the upper bound time for computing all the arguments is $|Lit| * 2^{|\Delta|} * ASP$.[13]*

Even though we must verify whether every subset of Δ is an argument for every literal in the language of the de.l.p., because the consistency condition in the definition of argument, $A \subseteq \Delta$ cannot be an argument for a literal and for its complement, so we will consider only $\frac{|Lit|}{2} * 2^{|\Delta|} = |Lit| * 2^{|\Delta|-1}$ potential arguments in order to play a game or equivalently to build the dialectical tree. This upper bound could be improved by considering minimality over the arguments, i.e., no $A_1 \subseteq \Delta$ would be an argument for a literal L if A_2 is an argument of L and $A_2 \subseteq A_1$.

Finally, we consider the argument existence decision problem.

Theorem 3. ARGUMENTEXISTENCE *is* ***NP**-complete.*

Proof. Membership is shown in [13]. For the hardness we shall reduce 3-SAT to ARGUMENTEXISTENCE. Given a formula in conjunctive normal form $F = C_1 \wedge \ldots \wedge C_n$, where each clause $C_i = L_i^1 \vee L_i^2 \vee L_i^3$ has exactly 3 literals, we build a de.l.p. $\mathcal{P}$ as follows: the language $\mathcal{L}$ is the set of all the propositional variables in F and the new symbols F' representing the formula F, $C_1', ..., C_n'$ representing each clause in F and a special symbol T representing the value true.

Let $\Pi_F = \{T\}$ and Π_R be the set of the following strict rules:

$$\Pi_R : \begin{cases} F' \leftarrow C_1', \ldots, C_n' \\ C_1' \leftarrow L_1^1 \\ C_1' \leftarrow L_1^2 \\ C_1' \leftarrow L_1^3 \\ \vdots \\ C_n' \leftarrow L_n^1 \\ C_n' \leftarrow L_n^2 \\ C_n' \leftarrow L_n^3 \end{cases}$$

Finally, define Δ to be the set of defeasible rules of the form $L \prec T$, for every literal (propositional variable or the negation of a propositional variable) L in the language of F. In order to obtain $\mathcal{P}$ it is required to go once through the complete formula F. Thus, $\mathcal{P}$ is built in lineal time over the length of F. Therefore, the reduction can be done in polynomial time.

We must now prove that F is satisfiable if and only if there exists an argument for F' in $\mathcal{P}$. As F is satisfiable there exists a model $\mathcal{M}$ for F. Thus, for each clause $C_i \in F, 1 \leq i \leq n$ there exists at least one literal L_i^j, $1 \leq j \leq 3$ such that $\mathcal{M} \models L_i^j$ and therefore $\mathcal{M} \models C_i$. Define a set $\mathcal{A}$ containing the defeasible rules $L \prec T$, where L is every literal such that $\mathcal{M} \models L$. Since each clause C_i has three strict rules associated, one for each literal, $\{C_i' \leftarrow L_i^j\} \cup \mathcal{A} \vdash C_i'$. Then $\Pi \cup \mathcal{A} \vdash F'$. Let $\mathcal{A}'$ be the minimal subset of $\mathcal{A}$, such that $\Pi \cup \mathcal{A}' \vdash F'$ Finally, in order to prove that $\mathcal{A}'$ is an argument for F' we shall demonstrate that no pair of complementary literals can be derived from $\Pi \cup \mathcal{A}'$. Suppose that a pair of complementary literals $L, \bar{L}$ can be derived. We know that Π is consistent, so inconsistency arises from the literals in head of defeasible rules $L \prec T$ and

$\bar{L} \rightarrowtail T$. But these rules are added to $\mathcal{A}$ iff $\mathcal{M} \models L$ and $\mathcal{M} \models \bar{L}$, and this is impossible since $\mathcal{M}$ is a model. Then $\mathcal{A}'$ is an argument for F in $\mathcal{P}$.

Conversely, suppose there exists an argument $\mathcal{A}$ for F. We can construct an interpretation $\mathcal{M}$ making true all the literals that are consequents of the defeasible rules in $\mathcal{A}$, and assigning arbitrary truth value to those propositional variables that do not appear among those literals. $\mathcal{M}$ is consistent by the consistency of $\mathcal{A}$, and moreover $\mathcal{M} \models F$ since $\Pi \cup \mathcal{A} \vdash F$.

We have then showed that 3-SAT $\leq_p$ ARGUMENTEXISTENCE, and then that this latter problem is **NP**-complete.

These results contrast with those of [49], where determining whether there is an argument for a formula h is Σ_2^P-complete. Even thought there are some similarities between argument definitions, they differ in the underlying logic. While in DeLP an argument is a subset of defeasible rules, and the inference mechanism is logic programming based, an argument in the formalism described in [49] is a subset of formulas of a propositional language, and $\vdash$ stands for classical inference. This clear difference in computational complexity makes DeLP systems more keen to knowledge representation applications.

The next result indicates that checking *whether the Family Game of a literal is empty* is not a tractable problem, unless **P** = **NP**.

Corollary 1. NOARGUMENTEXISTENCE *is co-**NP**-complete.*

Proof. Direct consequence of the theorem 3.

The decision problem (P1) expresses the situation when the agent has no information at all about a literal, ie, the literal has no argument neither for nor against it. The corollary above is the core for the next result.

Theorem 4. *The decision problem (P1) is co-**NP**-complete.*

Proof. (P1) is defined when the game families for the literal L and for its complement $\bar{L}$, $\mathcal{F}(L, \mathcal{P})$ and $\mathcal{F}(\bar{L}, \mathcal{P})$ respectively, are empty. So we must check twice the decision problem NOARGUMENTEXISTENCE : one for L and the other for $\bar{L}$. Thus, as a direct consequence of the corollary 1 (P1) is co-**NP**, and by theorem 3 it is also co-**NP**-complete.

In this point, we must analyse whether given a game, it is won by the proponent. This decision problem called GAMESAT and it is closed related to the marking process of the dialectical tree.

Theorem 5. *The decision problem* GAMESAT *is* **PSPACE**-*complete.*

Proof. GAMESAT is defined as determining whether a game is won by the proponent. Thus, by Theorem 1, it is equivalent to marking the dialectical tree and checking if the label is undefeated, which has been proved to be **PSPACE**-complete[14]. Thus, GAMESAT is **PSPACE**-complete.

It is possible to characterise some decision problems through its membership into other complexity classes, because of the similarity between DeLP proof theory and the interaction among two players.

Theorem 6. *Given a dialectical tree, determinate whether the root is labeled undefeated is* **AP**-*complete.*

Proof. For the membership in **AP**, in [14] we have proved that DeLP marking procedure for dialectical trees is **PSPACE**-complete. The set of languages decided in polynomial time with an ATM is the class **AP** $= ATIME[n^k]$ that is equal to the class **PSPACE**[16, 3]. Since the decision problem is in **PSPACE**, then it is in **AP**.

For the hardness, QSAT is **AP**-complete[47]. In [14], we have presented a polynomial reduction from QSAT to the marking process of the dialectical tree.

Corollary 2. *The decision problem* GAMESAT *is* **AP**-*complete.*

The result above states an lower bound for the decision problems PWINGAME, LITERALUNDECIDABILITYINGAME, (P2) and (P3), since they include, among others, determining whether the dialectical tree is won by the proponent.

5 DeLP as a query language

Exploring and querying the data explicitly represented or inferred is essential when analysing a knowledge representation and reasoning system. In this direction, we propose to define DeLP as a query language in order to evaluate the system through different approaches borrow from databases theory.

Making an analogy with database concepts, Π_F represents the input databases, also called the *extensional part*, and $\Pi_R \cup \Delta$ are the inference rules, called the *intensional part* of the database. Similarly, in the ontology context, we can compare DeLP to Description Logics[4]: Π_F represents the $\mathcal{A}_{box}$ and $\Pi_R \cup \Delta$ represents the $\mathcal{T}_{box}$.

For methodological issues, it is important to distinguish in a *de.l.p.* the input data from the inference rules. Thus, hereafter, we will denote $\mathcal{P} = \langle \Pi_F, \Pi_R \cup \Delta \rangle$, where Π_F is a finite set of ground facts, and $\Pi_R \cup \Delta$ is a finite set of ground strict and defeasible rules.

Definition 8. *A general query is a finite set of strict and defeasible rules together with a literal L_{l_i} with arity l_i.*

Intuitively, the answer for a general query is the set of all the tuples that holds the relation A_{l_i} and can be entailed from the DeLP semantics $\mathcal{GS}$.

Note that in a general query, some or all the relation arguments L_{l_i} can be instantiated. The main reason is to capture queries where some elements are fixed and some are variable. Thus, the answer should not contain fixed elements.

Definition 9. *A boolean query is a finite set of strict and defeasible rules together with a ground literal L.*

The intended intuitive meaning of defining such query is the following: we want to know whether a literal L is entailed by $\mathcal{GS}$ from $\Pi_R \cup \Delta$ together with the database Π_F.

Example 1. The following *de.l.p.* is adapted from [29]:

$$\Pi_F : \{duck(lucas), hen(tina), scared(tina)\}$$

$$\Pi_R : \left\{ \begin{array}{c} bird(lucas) \leftarrow duck(lucas) \\ bird(tina) \leftarrow hen(tina) \end{array} \right\}$$

$$\Delta : \left\{ \begin{array}{c} \sim flies(tina) \prec hen(tina) \\ flies(tina) \prec hen(tina), scared(tina) \end{array} \right\}$$

The general query *which are the flying objects?* is

$$\Big(\Pi_R, \Delta, flies\,(x) \Big)$$

The general query *which objects are birds?* is

$$\Big(\Pi_R, \Delta, bird(x) \Big)$$

The boolean query *Is Lucas a bird?* is:

$$\Big(\Pi_R, \Delta, bird\,(lucas) \Big)$$

The definitions below explain query semantics based on $\mathcal{GS}$.

Definition 10. *Let $\mathcal{P} = \langle \Pi_F, \Pi_R \cup \Delta \rangle$ be a de.l.p.. The database universe, denoted $\mathcal{U}$ is the Herbrand's universe of de.l.p., denoted $\mathcal{H}(\mathcal{P})$. Let $Q = \langle \Pi_R, \Delta, A_{l_i} \rangle$ be a general query. The evaluation function AQ of a general query is defined as follows:*

$$AQ(Q) = \left\{ \begin{array}{l} \langle u_1, ..., u_{l_i} \rangle \,|\, u_j \in \mathcal{U},\ 1 \leq j \leq l_i \text{ and exists a game won by the} \\ \text{proponent that begins with an argument for } A_{li}\,(u_1, ..., u_{l_i}) \end{array} \right\}$$

Thus, the answer to a general query is the set of tuples, such that $A_{li}\,(u_1, ..., u_{l_i})$ is warranted.

Definition 11. *Let $\mathcal{P} = \langle \Pi_F, \Pi_R \cup \Delta \rangle$ be a de.l.p. and let $Q_B = \langle \Pi_R, \Delta, L \rangle$ be a boolean query. The evaluation function AQ_B of a boolean query is defined as follows:*

$$AQ_B(Q_B) = \left\{ \begin{array}{l} \text{YES: if exists a game won by the proponent that begins} \\ \text{with an argument for the literal } L \text{ (ie, } L \text{ is warranted)} \\ \\ \text{NO: if exists a game won by the proponent that begins} \\ \text{with an argument for the literal } \bar{L} \text{ (ie, } \bar{L} \text{ is warranted)} \end{array} \right\}$$

Example 2. Let's consider the *de.l.p.* of Example 1. The answer for the general query bird(x) is $\{\langle Lucas \rangle, \langle Tina \rangle\}$. The answer for the general query flies(x) is $\{\langle Tina \rangle\}$. The answer for the boolean query bird(lucas) is YES.

Considering that unary relations BIRD, DUCK,HEN, FLIES and SCARED have just one attribute: the name; attributes domain is string and the universe is $\{Lucas, Tina\}$, we can obtain from Π_F and different queries the following table:

BIRD	DUCK	HEN	FLIES	SCARED
Name	Name	Name	Name	Name
Lucas	Lucas	Tina	Tina	Tina
Tina				

We have presented the characterisation of DeLP as a query language. We proceed now to analyse its data and combined complexity which are the basis for determining the expressive power.

6 Data and Combined Complexity of DeLP

In the direction of determining the computational complexity of some of the decision problems introduced in section 3, we have defined two approaches to study DeLP as a query language: data and combined complexity. When answering queries from a *de.l.p.*, it is relevant to analyse how query evaluation scales to large amounts of instance data. Since the size of the data typically dominates both the sizes of the program and the query, the more important measure for such scalability is data complexity.

The notion of data complexity is borrowed from relational database theory [57] and it is used in the emerging area, called ontology-mediated querying [40, 11]. The idea of using ontologies as a conceptual view over data repositories is becoming nowadays more and more popular. Furthermore, as an ontology provides an enriched vocabulary for querying, it serves as an interface between the query and the data, and it allows the derivation of implicit facts. In this direction, some research has been done merging DeLP and ontologies[33, 32] and linking DeLP and relational databases[22].

Data complexity is the complexity of query evaluation where only the data is considered to be an input, but both the literal in the query and the set of rules are fixed. Data complexity allows us to study DeLP as a query language measuring the complexity of query evaluation focus on the size of the data, using defeasible and strict rules for reasoning purpose. Data complexity is a key measure to determine the efficiency of argumentation system implementations based on database technologies.

On the other hand, combined complexity of a fragment of logic programming has been defined and used in [21]:

Complexity of (some fragment of) logic programming: is the complexity of checking if for variable programs $\mathcal{P}$ and variable ground atoms A, $\mathcal{P} \models A$.

Following the principle and notions above, in the context of DeLP we will define data, program and combined complexity as follows.

Definition 12. *Let Ω be any of the decision problems introduced above, $\mathcal{P} = \langle \Pi_F, \Pi_R \cup \Delta \rangle$ and $(\Pi_R \cup \Delta, L)$ a general or a boolean query:*

- *The* data complexity *of Ω is the complexity of Ω when Π_F varies and the query is fixed, i.e., parameters $\Pi_R \cup \Delta$ and L are fixed.*
- *The* program or expression *complexity of Ω is the complexity of Ω when the database instance is fixed, i.e., the parameter Π_F is fixed, and the query varies.*
- *The* combined complexity *of Ω is the complexity where every parameter Π_F, $\Pi_R \cup \Delta$ and L vary.*

Expression and combined complexity are quite close and they are rarely differentiated. For this reason we will only discuss data and combined complexity.

In this work we prioritise the focus on data complexity because the complexity of managing a large size of data is of great interest and requires reconsider existing approaches to face it.

In order to determine the upper bound for the data complexity of the decision problems PWINGAME and LITERALUNDECIDABILITYINGAME, we have analysed the dialectical tree structure over the size of the facts and the strict and defeasible rules.

The dialectical tree is explored in a complete depth first way, like in minimax. If the maximum depth of the tree is m, there are b legal movements at each point, then the time complexity will be $O(b^m)$[54]. If we implement the technique alpha-beta pruning, and we consider that successors are examined in random order, then the time complexity will be roughly $O(b^{\frac{3m}{4}})$[54]. The maximum depth of a dialectical tree for an argument under a de.l.p. with $|\Delta|$ defeasible rules is $2^{|\Delta|}$, i.e., we can consider every potential argument in one branch of the tree. Any argument can appear more than once in the tree but at most once in every branch, because of the acceptable argumentation line definition. What about branch factor: there exists $|Lit|/2$ literals that can be in conflict with the last argument. These literals may have at most $2^{|\Delta|}$ potential arguments. So our branching factor is in the worst case $|Lit|/2 * 2^{|\Delta|}$. Thus, exploring the dialectical tree like in minimax has an upper bound of $O((|Lit| * 2^{|\Delta|-1})^{2^{|\Delta|}})$.

Every time we must insert a neighbour node B of a node A in the tree structure or equivalently, when a player makes a move, we must check if it is a legal move in the game, i.e., if B attacks and defeats A, and if B does not introduce inconsistency. In order to determine whether B is a defeater of A, we must take into account the preference criterion between arguments. Any preference criterion defined among arguments could be used in DeLP. For this reason, the complexity class of the following decision problem "whether an argument is a legal move over a sequence in the game" will be left parameterized in the class C.

Theorem 7. *Let C be the complexity class for the decision problem:"whether an argument is a legal move over a sequence in the game". The upper bound for data complexity of* PWINGAME *is* $\mathbf{NP}^C$.*[13]*

Proof. For fixed $\Pi_R \cup \Delta$, the size of the dialectical tree for an argument $\langle \mathcal{A}, L \rangle$ is polynomial in the size of the literals in Π_F. Furthermore, computing each argument is in **P**, and determining whether each argument is a legal move in the game is in C. In order to decide whether a literal L belongs to the set T of the minimal G-model, we guess for an argument of L such that the game played from this argument is won by the Proponent. The number of arguments is polynomial when $\Pi_R \cup \Delta$ is fixed, and determining whether the game is won by the Proponent can be done with a C oracle and with constant space when query is fixed. This proves membership in $\mathbf{NP}^C$.

Since LITERALUNDECIDABILITYINGAME is a conjunction of PWINGAME complements an immediate corollary to the result above follows naturally.

Corollary 3. *Let C be the complexity class for the decision problem:"'whether an argument is a legal move over a sequence in the game". The upper bound for data complexity of* LITERALUNDECIDABILITYINGAME *is co-$\mathbf{NP}^C$.[13]*

Even though we have not analysed in depth the complexity for computing the preference criterion, we illustrate this concept with two different cases.

In [17, 18], the authors use specificity [55] as a syntax-based criterion among conflicting arguments, preferring those arguments which are more informed or more direct, in order to assess natural language usage based on the web corpus and to evaluate and rank search results, respectively. Computing specificity depends strongly on the set $2^{|Lit|}$.

Other DeLP implementations use a static preference relation [19]. In this case, the preference criterion is computed by comparing arguments values. Such values are obtained through different mathematical formulas applied to the certainty of a formula in the language. Computing such preference criterion involves just a comparison between two certainty values. However, an extra cost is considered in the argument construction procedure, since the certainty value is computed keeping a trace of all uncertain information used to derive a goal.

When analysing combined complexity we fix the size of the data Π_F, and we vary the general o boolean query. In this case, if we consider constructing the greatest dialectical tree and exploring it as minimax does $(O(((|Lit|*2^{|\Delta|})^{2^{|\Delta|}}))$, then the combined complexity is upper bound in Triple-Exponential time with Linear Exponent, i.e., 3ExpTime. In this analysis, we have not take into account several issues that can decrease the size of the dialectical tree, such as the inability of using an argument that does not defeat previous in a branch.

Considering the results of measuring the complexity of answering a query over an ontology, DeLP results come as no surprise. Data Complexity on $\mathcal{ALCI}$, $\mathcal{SH}$, $\mathcal{SHIQ}$ is co-**NP**-complete[24, 10, 11], while combined complexity is 2ExpTime-complete[12, 39]. Similarly with DeLP, in the OBDA setting, the size of the data largely dominates the size of the conceptual layer and of the query. This results make us conclude that increasing data size is not a problem at all when representing knowledge. In the context of DeLP the problem is not the size of Π_F but the large number of possible arguments we can construct considering the strict and defeasible rules schema. Therefore, DeLP is a system that can be scaled in the size of the data.

7 Conclusion and Future Work

We have review the complexity of DeLP through the $\mathcal{GS}$ semantics, pointing out some relevant decision problems. We extended previous results on complexity of these problems. Particularly interesting is the completeness of ARGUMENTEXISTENCE problem.

We also introduced the notion of DeLP as a query language. In this context, we have defined data, expression and combined complexity. As far as we know, argumentation systems have not been studied yet in this sense, and, therefore, there is no previous data and combined complexity analysis for defeasible reasoning. Table 1 summarises the problems studied and the main complexity results obtained.

Decision Problem	**Complexity**
ARGUMENTSAT	**P**-complete
ARGUMENTEXISTENCE	**NP**-complete
NOARGUMENTEXISTENCE	co-**NP**-complete
(P1) Agent has no information	co-**NP**-complete
GAMESAT	**PSPACE**-complete **AP**-complete
PWINGAME	Data Complexity $\mathbf{NP}^C$ Combined Complexity 3ExpTime (upper bound)
LITERALUNDECIDABILITYINGAME	Data Complexity $co - \mathbf{NP}^C$

Table 1. Decision problems and the main complexity results obtained.

As DeLP do not assume as input the argument set, the first results that has been established where related to arguments, the movements of a game. We have focused on the existence of an argument in order to play a game, and on verifying whether a set is an argument.

We state an exponential upper bound for the set of all the arguments. Because of the underpinning logic of DeLP our complexity results are a bit better than those based on classical logic.

Furthermore, we have analysed the decision problem GAMESAT and we have shown that it is **PSPACE**-complete and **AP**-complete.

Data complexity results on the decision problems PWINGAME and LITERALUNDECIDABILITYINGAME give a guideline for determining expressive power for DeLP. Since our results are parameterized, we can state a lower bound on **NP**, otherwise known as Σ_1^1, which coincides with the class of properties of finite structures expressible in existential second-order logic [27].

We also analysed combined complexity of the decision problem PWINGAME without taking into account restrictions on the dialectical tree size. Thus, we could express an upper bound in 3ExpTime.

Data and combined complexity results show that the intractability of DeLP is not on the data, but mainly on the number of argument we can build. In this sense, the formalisms supports query answering over very large Π_F. Unfortunately, the results over combined complexity (upper bound) are not tailored towards obtaining efficient implementations, and more research on this is needed.

We are currently following several directions to continue the work reported in this paper. First, when analysing data complexity we have fixed the query and we have parameterized the preference criteria. An interesting topic for future research is to study to what extent this results can be applied to others rule-based argumentation systems whose theory proof is rather similar. Also, although here we focused mainly on data complexity, we are also working on characterizing the complexity of query answering with respect to the size of the query, expression complexity and with respect to combined complexity. Finally, we are interested on determining which kind of queries can be express in DeLP and its limitation, circumscribing the properties that DeLP is enable to express. By studying the expressive power of DeLP, we compare this system with other non monotonic formalisms.

References

1. Samson Abramsky and Guy McCusker. Game Semantics. In H. Schwichtenberg and U. Berger, editors, *Logic and Computation: Proceedings of the 1997 Marktoberdorf Summer School*. Springer-Verlag, 1997.
2. Leila Amgoud and Claudette Cayrol. A reasoning model based on the production of acceptable arguments. *Annals of Math and Artificial Intelligence*, 34:197–215, 2002.

3. Sanjeev Arora and Boaz Barak, editors. *Computational Complexity: a modern approach*. Cambridge University Press, New York, 2009.
4. Franz Baader, Diego Calvanese, Deborah L. McGuinness, Daniele Nardi, and Peter F. Patel-Schneider, editors. *The Description Logic Handbook: Theory, Implementation, and Applications*. Cambridge University Press, New York, NY, USA, 2003.
5. Trevor J. M Bench-Capon. Value-based argumentation frameworks. In *NMR 2002*, pages 443–454, 2002.
6. Trevor J. M Bench-Capon. Persuasion in Practical Argument Using Value Based Argumentation Frameworks. *Journal of Logic and Computation*, 13(3):429–448, 2003.
7. Floris Bex, Henry Prakken, Tom M. van Engers, and Bart Verheij. Introduction to the special issue on artificial intelligence for justice (AI4J). *Artif. Intell. Law*, 25(1):1–3, 2017.
8. A. Bondarenko, P.M. Dung, R Kowalski, and F. Toni. An Abstract, Argumentation-Theoretic Approach to Default Reasoning. *Artificial Intelligence*, 93(1-2):63–101, 1997.
9. M. Cadoli and M. Schaerf. A survey of complexity results for nonmonotonic logics. *Journal of Logic Programming*, 17:127–160, 1993.
10. Diego Calvanese, Giuseppe De Giacomo, Domenico Lembo, Maurizio Lenzerini, and Riccardo Rosati. Data complexity of query answering in description logics. In *Proceedings, Tenth International Conference on Principles of Knowledge Representation and Reasoning, Lake District of the United Kingdom, June 2-5, 2006*, pages 260–270, 2006.
11. Diego Calvanese, Giuseppe De Giacomo, Domenico Lembo, Maurizio Lenzerini, and Riccardo Rosati. Data complexity of query answering in description logics. *Artif. Intell.*, 195:335–360, 2013.
12. Diego Calvanese, Giuseppe De Giacomo, and Maurizio Lenzerini. Conjunctive query containment and answering under description logic constraints. *ACM Trans. Comput. Log.*, 9(3):22:1–22:31, 2008.
13. Laura A. Cecchi, Pablo R. Fillottrani, and Guillermo R. Simari. On the complexity of DeLP through game semantics. In J. Dix and A. Hunter, editors, *XI International Workshops on Nonmonotonic Reasoning*, pages 386–394, Clausthal University, 2006.
14. Laura A. Cecchi and Guillermo Simari. *DeLP marking procedure for dialectical trees is PSPACE-complete*. EDULP - Editorial de la Universidad de La Plata, 2012.
15. Laura Andrea Cecchi. *Una semántica basada en juegos para la programación lógica rebatible*. PhD thesis, Universidad Nacional del Sur, 2011.
16. Ashok K. Chandra, Dexter Kozen, and Larry J. Stockmeyer. Alternation. *J. ACM*, 28(1):114–133, 1981.
17. C. Chesñevar and A. Maguitman. An Argumentative Approach to Assessing Natural Language Usage based on the Web Corpus. In *Proc. of the European Conference on Artificial Intelligence (ECAI) 2004*, pages 581–585, Valencia, Spain, August 2004.
18. C. Chesñevar and A. Maguitman. ARGUENET: An Argument-Based Recommender System for Solving Web Search Queries. In *Proc. of the 2nd IEEE Intl. IS-2004 Conference*, pages 282–287, Varna, Bulgaria, June 2004.

19. Carlos Chesñevar, Guillermo Simari, Teresa Alsinet, and Lluís Godo. A Logic Programming Framework for Possibilistic Argumentation with Vague Knowledge. In *Proc. of the UAI-2004*, pages 76–84, 2004.
20. Carlos Iván Chesñevar, Ana Gabriela Maguitman, Elsa Estevez, and Ramón F. Brena. Integrating argumentation technologies and context-based search for intelligent processing of citizens' opinion in social media. In *6th International Conference on Theory and Practice of Electronic Governance, ICEGOV '12, Albany, NY, USA, October 22-25, 2012*, pages 166–170, 2012.
21. Evgeny Dantsin, Thomas Eiter, Georg Gottlob, and Andrei Voronkov. Complexity and expressive power of logic programming. *ACM Computing Surveys (CSUR)*, 33(3):374 – 425, September 2001.
22. Cristhian Ariel David Deagustini, Santiago Emanuel Fulladoza Dalibón, Sebastian Gottifredi, Marcelo Alejandro Falappa, Carlos Iván Chesñevar, and Guillermo Ricardo Simari. Defeasible argumentation over relational databases. *Argument & Computation*, 8(1):35–59, 2017.
23. Yannis Dimopoulos, Bernhard Nebel, and Francesca Toni. On the Computational Complexity of Assumption-based Argumentation for Default Reasoning. *Artificial Intelligence*, 141(1):57–78, 2002.
24. Francesco M. Donini, Maurizio Lenzerini, Daniele Nardi, and Andrea Schaerf. Deduction in concept languages: From subsumption to instance checking. *J. Log. Comput.*, 4(4):423–452, 1994.
25. Phan M. Dung. On the acceptability of arguments and its fundamental role in nonmonotonic reasoning and logic programming and n-person games. *Artificial Intelligence*, 77:321–357, 1995.
26. Paul E. Dunne and Trevor J. M. Bench-Capon. Complexity in value-based argument systems. In *Logics in Artificial Intelligence, 9th European Conference, JELIA 2004, Lisbon, Portugal, September 27-30, 2004, Proceedings*, pages 360–371, 2004.
27. Ronald Fagin. Generalized first-order spectra and polynomial-time recognizable sets. In R. Karp, editor, *Complexity of Computation. SIAM-AMS Proceedings*, volume 7, pages 43–73, 1974.
28. Edgardo Ferretti, Luciano H. Tamargo, Alejandro Javier García, Marcelo Luis Errecalde, and Guillermo Ricardo Simari. An approach to decision making based on dynamic argumentation systems. *Artif. Intell.*, 242:107–131, 2017.
29. Alejandro J. García. *Programación en Lógica Rebatible: Lenguaje, Semántica Operacional y Paralelismo*. PhD thesis, Universidad Nacional del Sur, 2000.
30. Alejandro J. García and Guillermo R. Simari. Defeasible Logic Programming: An Argumentative Approach. *Theory and Practice of Logic Programming*, 4(1):95–138, 2004.
31. Alejandro Javier García and Guillermo Ricardo Simari. Defeasible logic programming: Delp-servers, contextual queries, and explanations for answers. *Argument & Computation*, 5(1):63–88, 2014.
32. Sergio Alejandro Gómez, Carlos Iván Chesñevar, and Guillermo Ricardo Simari. An Argumentative Approach to Reasoning with Inconsistent Ontologies. In Thomas Meyer and Mehmet A. Orgun, editors, *Proc. of the Knowledge Representation in Ontologies Workshop (KROW 2008)*, volume CPRIT 90, pages 11–20, Sydney, Australia, 2008.
33. Sergio Alejandro Gómez, Carlos Iván Chesñevar, and Guillermo Ricardo Simari. Ontoarg: A decision support framework for ontology integration based on argumentation. *Expert Syst. Appl.*, 40(5):1858–1870, 2013.

34. Kathrin Grosse, María Paula González, Carlos Iván Chesñevar, and Ana Gabriela Maguitman. Integrating argumentation and sentiment analysis for mining opinions from twitter. *AI Commun.*, 28(3):387–401, 2015.
35. Eun Jung Kim, Sebastian Ordyniak, and Stefan Szeider. Algorithms and complexity results for persuasive argumentation. *Artif. Intell.*, 175(9-10):1722–1736, 2011.
36. Eun Jung Kim, Sebastian Ordyniak, and Stefan Szeider. The complexity of repairing, adjusting, and aggregating of extensions in abstract argumentation. *CoRR*, abs/1402.6109, 2014.
37. Vladimir Lifschitz. Principles of knowledge representation. chapter Foundations of Logic Programming, pages 69–127. Center for the Study of Language and Information, Stanford, CA, USA, 1996.
38. L. Longo and P. Dondio. Defeasible reasoning and argument-based systems in medical fields: An informal overview. In *2014 IEEE 27th International Symposium on Computer-Based Medical Systems (CBMS)*, volume 00, pages 376–381, May 2014.
39. Carsten Lutz. Inverse roles make conjunctive queries hard. In *Proceedings of the 2007 International Workshop on Description Logics (DL2007), Brixen-Bressanone, near Bozen-Bolzano, Italy, 8-10 June, 2007*, 2007.
40. Carsten Lutz and Frank Wolter. The data complexity of description logic ontologies. *Logical Methods in Computer Science*, 13(4), 2017.
41. Michael J. Maher. Propositional defeasible logic has linear complexity. *Theory and Practice of Logic Programming*, 1(6):691–711, 2001.
42. Maria Vanina Martinez, Alejandro Javier García, and Guillermo Ricardo Simari. On the use of presumptions in structured defeasible reasoning. In *Computational Models of Argument - Proceedings of COMMA 2012, Vienna, Austria, September 10-12, 2012*, pages 185–196, 2012.
43. Maria Vanina Martinez and Sebastian Gottifredi. Query answering in the semantic social web: An argumentation-based approach. In *Encyclopedia of Social Network Analysis and Mining*, pages 1441–1455. 2014.
44. Sanjay Modgil and Henry Prakken. Corrigendum to "a general account of argumentation with preferences" [artif. intell. 195 (2013) 361-397]. *Artif. Intell.*, 263:107–110, 2018.
45. Ana Lucía Nicolini, Ana Gabriela Maguitman, and Carlos Iván Chesñevar. Argp2p: An argumentative approach for intelligent query routing in P2P networks. In *Theory and Applications of Formal Argumentation - Third International Workshop, TAFA 2015, Buenos Aires, Argentina, July 25-26, 2015, Revised Selected Papers*, pages 194–210, 2015.
46. Donald Nute. Defeasible reasoning and decision support systems. *Decision Support Systems*, 4(1):97–110, 1988.
47. Christos Papadimitriou. *Computational Complexity*. Addison-Wesley Publishing Company, 1994.
48. Christos H. Papadimitriou and Mihalis Yannakakis. On the complexity of database queries (extended abstract). In *PODS '97: Proc of the 16th ACM SIGACT-SIGMOD-SIGART Symp on Principles of Database Systems*, pages 12–19, New York, NY, USA, 1997. ACM Press.
49. Simon Parsons, Michael Wooldridge, and Leila Amgoud. Properties and complexity of some formal inter-agent dialogue. *Journal of Logic and Computation*, 13(3):347–376, 2003.

50. John Pollock. Defeasible Reasoning. *Cognitive Science*, 11:481–518, 1987.
51. Henry Prakken. On the problem of making autonomous vehicles conform to traffic law. *Artif. Intell. Law*, 25(3):341–363, 2017.
52. Henry Prakken and Giovanni Sartor. Argument-based extended logic programming with defeasible priorities. *Journal of Applied Non-Classical Logics*, 7(1):25–75, 1997.
53. Henry Prakken, Adam Z. Wyner, Trevor J. M. Bench-Capon, and Katie Atkinson. A formalization of argumentation schemes for legal case-based reasoning in ASPIC+. *J. Log. Comput.*, 25(5):1141–1166, 2015.
54. Stuart Russell and Peter Norvig. *Artificial Intelligence: A modern approach.* Prentice Hall, New Jersey, second edition, 2003.
55. Guillermo R. Simari and Ronald P. Loui. A mathematical treatment of defeasible reasoning and its implementation. *Artificial Intelligence*, 53:125–157, 1992.
56. Hannes Strass and Johannes Peter Wallner. Analyzing the computational complexity of abstract dialectical frameworks via approximation fixpoint theory. *Artif. Intell.*, 226:34–74, 2015.
57. Moshe Y. Vardi. The complexity of relational query languages. In *Proc of the 14th Annual ACM Symposium on Theory of Computing, STOC'82*, pages 137–146, New York, NY, USA, May 1982. ACM Press.
58. Bart Verheij, Enrico Francesconi, and Anne Gardner. ICAIL 2013: The fourteenth international conference on artificial intelligence and law. *AI Magazine*, 35(2):81–82, 2014.
59. Johannes Peter Wallner, Andreas Niskanen, and Matti Järvisalo. Complexity results and algorithms for extension enforcement in abstract argumentation. *J. Artif. Intell. Res.*, 60:1–40, 2017.

Argumentation as Information Input

Dov Gabbay[1] and Michael Gabbay[2]

[1] Bar Ilan University, King's College London, University of Luxembourg
[2] University of Cambridge

Abstract. Given a network (S, R), with $R \subseteq S^2$, we view the nodes of S as containing information and view xRy as x transmitting information to y. We argue that such networks provide a more general account of attack and defense and general arguments exchange on issues between participants in Facebook and Twitter, as well as being able to simulate the traditional Dung approach. We define a general semantics for such networks.

Keywords: argumentation, information input, bipolar networks

1 Background and Orientation

1.1 Introducing information input

The traditional view of an argumentation system (S, R) is set theoretical and static. S is a non-empty set of atomic elements and R is a static subset of S^2 and we are looking for subsets E of S, which we call complete extensions, satisfying certain properties involving R. Although we view xRy as x 'attacks' y, we do not use the dynamic idea of x actually 'sending' something to y. The dynamic idea exists however, for example if we view (S, R) as an Ecology, S as a set of species and R as a predator-prey relation, and view the complete extensions as possible groups of species in equilibrium, see [13]. This view also works for S being a set of arguments, when xRy means that x is a counter argument to y, (in logic this would be x 'proves' $\neg y$, see [14, 15]). In the informational view, we look at xRy differently, we take xRy to mean that x is actually sending information to y. This information might change y or even 'kill'/contradict it.

We shall soon explain this informational idea a bit more formally, but first we must present the traditional notion of what we shall refer to as "argumentation as attack". There are two ways to present the semantics for argumentation as attack, the traditional set theoretical approach and the Caminada labelling approach. For the mapping connections between the two approaches see [17].

Let us briefly quote the traditional set theoretic approach:

1. We begin with a pair (S, R) where S is a nonempty set of points (arguments) and R is a binary relation on S (the "attack" relation).
2. Given (S, R), a subset E of S is said to be conflict free if for no x, y in E do we have xRy.

3. E protects an element $a \in S$, if for every x such that xRa, there exists a $y \in E$ such that yRx holds.
4. E is admissible if E protects all of its elements.
5. E is a complete extension if E is admissible and contains every element which it protects.

Various different semantics (types of extensions) can be defined by identifying different properties of E. For example we might define that E is a *stable extension* if E is a complete extension and for each $y \notin E$ there exists $x \in E$ such that xRy or we might look at the grounded extension as the unique minimal extension or we might consider a preferred extension, being a maximal (with respect to set inclusion) complete extension. The above properties give rise to corresponding semantics (stable semantics, grounded semantics and preferred semantics).

We can also present the complete extensions of (S,R), using the Caminada labelling approach, see [17].

A Caminada labelling of S is a function $\lambda : S \mapsto \{\text{in}, \text{out}, \text{und}\}$ such that the following holds

(C1) $\lambda(x) = \text{in}$, if for all y attacking x, $\lambda(y) = \text{out}$.
(C2) $\lambda(x) = \text{out}$, if for some y attacking x, $\lambda(y) = \text{in}$.
(C3) $\lambda(x) = \text{und}$, if for all y attacking x, $\lambda(y) \neq \text{in}$, and for some z attacking x, $\lambda(z) = \text{und}$.

A consequence of (C1) is that if x is not attacked at all, then $\lambda(x) = \text{in}$.

Any Caminada labelling yields a complete extension and vice versa. Any $\{\text{in}, \text{out}\}$ Caminada labelling (i.e. with no "und" value) yields a stable extension and vice versa. Set theoretic minimality or maximality conditions on extensions E correspond to the respective conditions on the "in" parts of the corresponding Caminada labellings. See [17].

We now want to continue and introduce our ideas about argumentation as information input. It would be helpful to have three useful stories in mind.

Story 1, The Party: We are planning a party and we have a set S which is the maximal set of all relatives, friends, colleagues etc. who can be invited to the party. The problem is that some of them do not get along/hate some others. So we have a relation R, where xRy (which we might denote by $x \twoheadrightarrow y$) means that if x is invited y must not be invited. We get here a basis for traditional argumentation network with attack relation R, provided we assume some social conventions such as that we should try to invite as many relatives as possible while respecting R.

Story 2, The Debate: We have a group of people S discussing a certain topic. The discussion proceeds by each person $x \in S$ providing extra information to a subset of the other members of S. Any $x \in S$ may pass information on to y while receiving some additional information from some (or many) other z.

x may then think about this received information and then choose to send more to y. For simplicity we assume that every member of S only ever passes on information to the same subset of S; that x passes the same information to each of its chosen subset; and that when information is passed, it is done simultaneously throughout S.

Let us denote this information input relation by xRy, which means that y is one of the elements of S to whom x will send information. We begin up with a system $(S, R, \mathbf{f}, \tau)$, where $\mathbf{f}$ is a function assigning some information I_x to each $x \in S$ (it does not matter what exactly 'information' is here). τ is a function from $x \in S$ and information I to information $\tau(x, I)$ (in the story $\tau(x, I)$ is the information x chooses to pass to all y such that xRy when x has previously received the information I). We then have a sequence of systems representing the cycles of information.

- Let $\mu(x) = \mathbf{f}(x)$ and let $S_0 = (S, R, \mu_0, \tau)$.
- Let $\mu_{n+1}(x) = \mu_n(x) \cup \bigcup_{zRx} \tau(z, \mu_m(z))$ and let $S_{n+1} = (S, R, \mu_{m+1}, \tau)$.

So each S_i represents a stage in the potentially infinite cycle of the $x \in S$ receiving information, processing it and passing some function of that information on to others. The sequence of S_i could be made more subtle by allowing R to depend on μ and τ.

Story 3, The security agency: We have a group of security agents involved in collecting information, say about possible terrorist threats. The relation xRy on S means that x reports to y. This story is different from the debate in Story 2 in that the relation R is substantially well-founded following the hierarchy of the agency. We may have an agent a and another agent b responsible for spying on a foreign country. The agent a employs a local agent y, who in turn employs several other locals say $x_1 \ldots x_n$. We note three properties of this information network:

1. R is fixed, and is external and independent of the information involved.
2. If $x_i R y$ and yRa hold then y waits for all the information from all x_i to arrive, and then y processes it and only afterwards passes it on to a. Note that this is a special case of the Story 1 with recursive iterations, where the node a sends nothing/wait at each iteration until it gets info from all x such that xRa and then sends info in the nest iteration just once and stops.
3. If we have aRb and bRa, then this means that a and b cooperate and share information and we need to determine how they do that.
4. The simplest most practical view of cycles is to consider the cycle $\{a, b\}$ as a single composite node. This means that mathematically we can assume (S, R) to be acyclic for the case of information input networks.
5. The view of this story has an additional geometrical (i.e. conditions on R) consequence. Suppose the nodes x and y satisfy the following condition

$$\{z \mid zRx\} = \{z \mid zRy\}$$

If we take the view that x and y depend only on the information collected by their agents, then we can assume that x and y have the same information and therefore we can add the geometrical condition $x = y$. However, it may be that x and y also collect information themselves.

6. There is also the question of how the local subagent reports his information to his master. Does he just send verified facts or also recommendations and interpretations? Mathematically the question is, if agent x collects info $\mathbf{f}(x)$, and xRy holds what does x send to y? Does x send all the information?, Some of it?, Only facts and his conclusions/assumptions?

We now continue more formally with some key motivating examples. We shall abuse conceptual sensitivity and call the relation "attack" relation instead of "information input" relation.

Consider the simple network of Figure 1. In this network b attacks c and a

Fig. 1. a attacks b which attacks c

attacks b. On standard Caminada labelling, a is "in" as it is free from attack, and so it negates the attack on c by b and so c is also free from attack. So we have one grounded extension, namely a = in, b = out and c = in.

The meaning of the attack relation above is taken basically as:

$$x \twoheadrightarrow y \text{ means that if } x \text{ is "in" then } y \text{ is "out"} \qquad (*)$$

This meaning $(*)$ corresponds to Story 1: the party. It is set theoretical in its nature. We need to define a subset E of S satisfying certain conditions, and $(*)$ is one of them. There is no dynamics involved in the concept of the extension E, except possibly in the case when we give an algorithm for finding E (such an argument resides in the meta-level, from our point of view outside the object level (S, R)), in which case there will be inductive steps by step "dynamics". This "dynamics" is external to the argumentation conceptual framework. For example when there are no cycles in (S, R), a directional inductive algorithm can be given to calculate the unique complete (ground) extension of (S, R). This algorithm is in the meta-level and has no meaning from the point of view of the object level (S, R) and its extensions. But in the case of information flow , the direction of the induction corresponds to the direction of the information flow. The $(*)$ interpretation can be formalised (in the meta-level) in a variety of logics, all meaning basically the above. It can also be instantiated/explained in a variety of ways, for example, in the case of Story 1, instantiating/explaining would mean that instead of just listing abstractly who cannot get along with whom, we can collect statements expressing the reasons for their not getting along and such statements can be used, again, in a manner based on the above.

As another example, we may have, in a propositional logic with the language containing $\{\to, \wedge, \neg\}$, that $x = A \to e$ and $y = B \wedge \neg e \to d$ and so we have that $x \twoheadrightarrow y$, where the attack considerations are conducted in the specific logic of $\{\to, \wedge, \neg\}$, where we view y as an argument for d relying on $\neg e$ as an assumption and x is an argument for e, and therefore x attacks y.

In comparison, Stories 2 and 3 involve the transmission of information and not necessarily attack. At this junction we must immediately clarify a methodological point. The perceptive reader might ask, if we indeed view (S, R) as information transmitting network, then what is the connection with argumentation? An information network is more like an electrical network/grid distributing electricity or a water network distributing water , and here we distribute information. So, again, what is the connection with argumentation? We admit that there are many networks sending information along arrows, this is not new, but we claim we do need to model the sending information in the context of argumentation. Our evidence for this is twofold.

1. On the one hand, in many debates and arguments in many cases the response to an attacking argument is to give more information to deflect the attack. This is a fact of life.
   In fact, issue came up several times during the COMMA 2016 conference, see for example the following COMMA 2016 papers:
   – Paper by Tom Bosc, Elena Cabrio and Serena Villata [4], modelling Twitter argumentation sequences: 30% of the political tweets are information, which could be either attack or defence.
   – Keynote lecture of Jens Allwood where he clearly demonstrated that some arguments can be either attack or support, depending on context.
   – Paper by Cyras, Sato & Toni [8], where there is an example of buying a certain car, where some information (that the car is red), is not known whether it is attack or support (to the argument for buying this car), and was therefore left unmarked in their system.
   – The keynote lecture by Professor Marie-Francine Moens [24], where she repeatedly stressed the need for additional Context information (to go along with the arguments). She quoted the paper by Saint Dizier [27], in which he said "Knowledge is a major bottleneck to argument mining, given a controversial issue and a set of texts in which arguments can be found". Indeed Saint-Dizier, in his paper attaches additional knowledge to arguments.
   – The paper [3] by Tom Blount et. al. on information input in Social media.
2. On the other hand, from the technical point of view, the idea of information input can also be used to actually attack. If we transmit certain information from x to y, then the additional information transmitted may render the information in y as unacceptable. Note further in this connection that our paper [16], suggested a different type of instantiation, using non-monotonic logic and in the context of non-monotonicity, information

input can serve as attack. Let $\vdash$ be a non-monotonic consequence relation. The non-monotonicity property allows for the following for a theory Δ:

$$\Delta \vdash A, \text{ but } \Delta \cup \Delta' \not\vdash A.$$

Thus if x is instantiated by a theory Δ_x and y by Δ_y, then we may have that $\Delta_y \vdash A$ but $\Delta_x \cup \Delta_y \not\vdash A$. So if we view being "in" as proving a fixed A, being out as proving "$\neg A$" and being undecided as neither, then we may define the attack of x on y as the input of Δ_x into Δ_y, to form $\Delta_x \cup \Delta_y$ and thus cause A no longer to be derivable.

We thus have a new meaning for the "attack" relation:

$$x \twoheadrightarrow y \text{ means that } x \text{ adds some information } \tau(x) \text{ to the information of } y \quad (**)$$

Remark 1. Note the following about this "attack" relation:

1. The attacker x and its target y can co-exist in the sense that x and y together can be consistent. In fact, x attacking y just changes y into a new y'. So the "attack" passing of information can even actually be "support".
2. The traditional notions of re-instatement, admissibility, extensions etc, need no longer apply. We get a new game here.
3. Consider the attack of x on itself, $x \twoheadrightarrow x$. According to the $(*)$ reading, x wants itself to be out and so the Caminada labelling can only give x the value out or undecided, it cannot give x the value in. The reading $(**)$ on the other hand, lets x join its information to itself, which can be harmless, or can alter what is derivable via x, depending on the logic governing the information. If we are dealing with a resource logic, for example, such as linear logic, we do have $x \vdash x$ but not $x, x \vdash x$. Of course we can let x send its opposite $\neg x$, in which case x would be mounting a traditional attack on itself, but then x would be sending false information (from x's point of view).
4. We can go further and generalise and understand $x \twoheadrightarrow y$ as $(***)$ below:

$$x \text{ sends an algorithm which revises } y \text{ to a new } y'. \quad (***)$$

If y is information which yields a conclusion A, x can be additional information which now yields the conclusion $\neg A$, this is $(**)$. However, in practice, one (the supporter of x) may "tell" (the supporter of) y that he (the supporter of) y has not gathered the information correctly and actually y should be replaced by y', and the conclusion is actually $\neg A$. This is $(***)$.
5. There is a surprising connection is with bipolar argumentation networks. See for example [7, 12, 1]. Bipolar networks have the form (S, R'', R'), where S is a non-empty set and R'' and R' are two disjoint binary relations on S. $xR''y$ means x attacks y (our notation, $x \twoheadrightarrow y$) and $xR'y$ means x supports

y (notation, $x \to y$). The meaning of attack is the traditional one, as (∗) above. As for the meaning of support, there are various approaches almost all compatible with our informational Story 2 and Story 3. There is a lot of discussion and approaches in the literature on how to define extensions for networks with both attack and support. See for example [7, 12, 1]. The interest from our point of view is that the "information input" transmission can be either attack or support, depending on what information is being sent, and so we have an opportunity to connect and contribute to the bipolar debate.

6. We elaborate more on the connection with bipolar argumentation. Consider again the geometrical network of $S = \{a, b, c\}$ and $R = \{(a, b), (b, c)\}$. As informational network each element pair (x, y) in R might be attack or support. We cannot tell what it is, because it depends on the information sent from x to y, (in our example, from a to b and from b to c). So depending on the information being sent, we may have a traditional case of $aR''b$ and $bR''c$, or a case of $aR''b$ and $bR'c$. Now the challenge for us is to develop machinery for defining "informational extensions" in the abstract for (S, R) in such a way that what we get in the abstract for the case of $aR''b$ and $bR''c$ will agree with the traditional approach to it, and what we get for the case of $aR''b$ and $bR'c$ will turn out to be the same as one of the known Bipolar approaches to it.

1.2 Some examples of information input

We have already mentioned that the role of information input as "attack" is not completely alien to the field of non-monotonic reasoning. The paper [20] studies practical instances of the use of expert opinion in court and in such cases argumentation attacks involve informational input. Let us give some examples:

Example 1. This example relates to Figure 1, consider the following real (Israeli) court dialogue (from [20]):[3]

    c = Sex offenders who attack male victims are more dangerous to society than those attacking female victims
    b = This is clearly prejudiced judgement between males and females. You see a man attacking another man as sick and therefore you make him more dangerous.
    a = There is no prejudice here. The observation is based on statistical data.

This court example has the same structure as the famous "Tweetie" example from the non-monotonic reasoning literature, is the following:

---

[3] The (Israeli) courts may call a senior expert on the risk assessment of sex offenders as part of the process of deciding what to do with a convicted offender. All the arguments involved between the expert and the defence attorneys are of informational nature. In fact in [20] we list the types of such arguments and claim that they are templates for any expert witness testimony interactions.

c = Tweetie does not fly.
b = Tweetie is a bird and birds fly.
a = Tweetie is a Penguin and Penguins do not fly.

Fig. 2. x_0 attacks x_1 attacks x_2 attacks x_3

The mechanism for propagating the attacks works differently for $(*)$ and $(**)$.

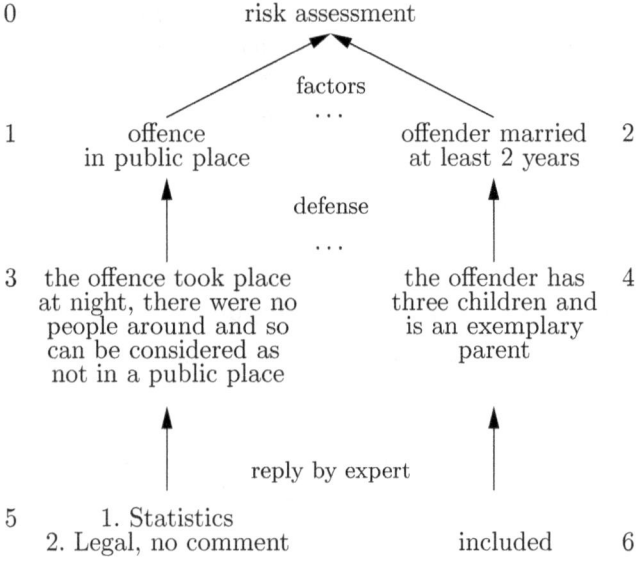

Fig. 3.

Example 2. The two figures Figure 3 and Figure 4 illustrate the idea information input.

Figure 3 is from a real life sex offender Israeli court case. The top argument at node 0 is that the sex offender presents a serious risk to society. The nodes 1 and 2 contain information from an expert on the case. There are about 40 factors the expert has to address before the Judge, we list only two of them. As you can see, node 1 is support to the risk statement while node 2 is attack. The second layer, nodes 3 and 4 are arguments / information by the defence attorney. His job is to attack the attack and support the support nodes of the first layer. Indeed node 3 makes a legal point attacking node 1 and node 4

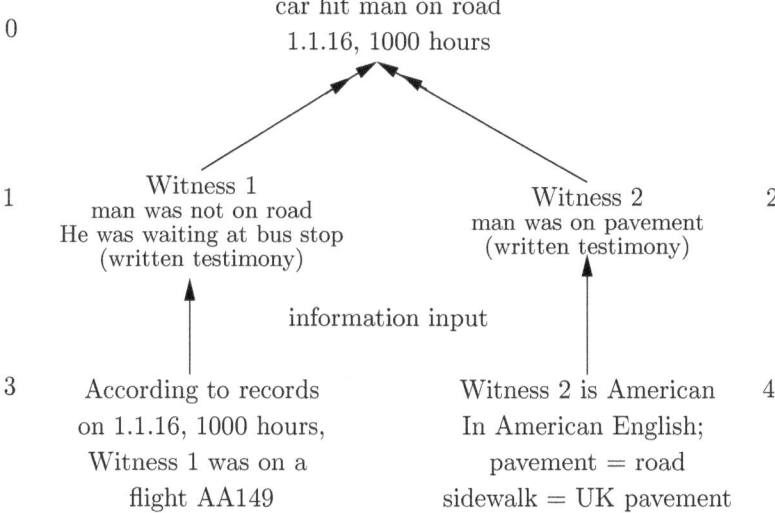

Fig. 4.

gives further information. The expert replies in layer 3 and this concludes the proceedings. Node 5 can be two possible answers.

One is that the legal point is for the Judge to decide. Another possible answer is that the risk assessment of this factor is based on statistical data and the statistical cases include both day and night attacks in public places. Node 6 simply says the presence of children is included in the statistics.

Figure 4 is a constructed story to illustrate the point of information input. The prosecution of the driver of the car wants to show the man/victim was hit while waiting at a bus stop or otherwise certainly not on the road. This is bad for the driver. Nodes 1 and 2 are statements of 2 witnesses. The defence for the driver attack nodes 1 and 2 by supplying more information. Node 3 states that witness 1 was on a flight at the time of the accident. This leaves witness 2 who categorically wrote that the victim was on the pavement (i.e. not on the road). However, node 4 says the witness is an American, and in American English pavement means road, and 'otherwise not on the road' means 'on the sidewalk'.

We assume the court case takes place in the UK..

Example 3. As a simple technical illustration of the (∗∗) meaning, consider the chain of Figure 2. Let us define a simple non-monotonic consequence relation $\vdash$ on a language of atomic formulas $A_0, A_1 \ldots$ by:

$$\Gamma \vdash A \text{ iff } A \in \Gamma \text{ and } |\Gamma| \text{ is odd}$$

where Γ is a finite set and $|\Gamma|$ is its cardinality.[4] Now, let us assign "initial" values $\mathbf{f}(x_i)$ and "final" values $\mu(x_i)$ to to each of $x_0 \ldots x_3$ as follows:

$$\mathbf{f}(x_i) = \{A_i\}$$
$$\mu(x_0) = \mathbf{f}(x_0)$$
$$\mu(x_{n+1}) = \mu(x_n) \cup \mathbf{f}(x_{n+1})$$

The attack relation so defined has the meaning of $(**)$. Formally here $\tau(x)$ is the same as $\mu(x)$. This is a simple example of Story 3, the security agency, where information is transmitted from one agent to the next agent. For some reason an agent can divulge his information only if the number of information pieces he has is odd. Call this superstition. Whatever one thinks of this behaviour, one fact is clear; there is no concept of agent x_i being "in" or "out". We can however, mathematically introduce this concept, for example we can stipulate arbitrarily that x_i is "in" if $\mu(x_i) \vdash A_i$.

Then we have that:

$$\begin{aligned}
\mu(x_0) &= \{A_0\} & &\vdash A_0 \\
\mu(x_1) &= \{A_0, A_1\} & &\nvdash A_1 \\
\mu(x_2) &= \{A_0, A_1, A_2\} & &\vdash A_2 \\
\mu(x_3) &= \{A_0, A_1, A_2, A_3\} & &\nvdash A_3
\end{aligned}$$

and so x_0 and x_2 are "in" and x_1 and x_3 are "out".

This illustrates how we may view the graph of Figure 2 as a sequence of inputs generated backwards from x_3:

x_3: claim A_3.

$x_2 \twoheadrightarrow x_3$: Adds A_2 as information which casts doubt on A_3.

$x_1 \twoheadrightarrow x_2 \twoheadrightarrow x_3$: Adds A_1 which resolves all doubts.

$x_0 \twoheadrightarrow x_1 \twoheadrightarrow x_2 \twoheadrightarrow x_3$ Adds A_0 which creates a new (global) doubt.

So instead of regarding an attack as updating premises so as to move from a monotonic consequence $\Gamma \vdash A$ to $\Delta' \vdash \neg A$ we regard it as adding new premises to move from a non-monotonic $\Gamma \vdash A$ to $\Gamma, \Delta \nvdash A$.

Note that this illustration is imperfect it does not distinguish between a chain of attacks (as in a graph $\{x_0 \twoheadrightarrow x_1, x_1 \twoheadrightarrow x_2, x_2 \twoheadrightarrow x_3\}$) and a number of attacks on the same target ($\{x_0 \twoheadrightarrow x_3, x_1 \twoheadrightarrow x_3, x_2 \twoheadrightarrow x_3\}$). But it suffices as an initial illustration of the idea: in particular notice that traditional concepts of argumentation theory are absent, e.g. that of an extension subset $E \subseteq \{x_0, x_1, x_2, x_3\}$.

1.3 Informational input compared with Dung attacks

If we want to show that informational R'' "attack" networks (to be defined later in Section 2 for the acyclic case and in Section 4 for the general case)

---

[4] Note that this relation is transitive in the sense that $\Gamma \vdash A$ and $A, \Delta \vdash B$ implies $\Gamma, \Delta \vdash B$. But it is not reflexive, as although we have $A \vdash A$, we do not have $A, B \vdash A$.

can simulate the traditional Dung networks, then to succeed in simulating a given abstract network (S, R), we need some specific consequence relation $\vdash$ and some specific initial theories (using S as being included in the atomic propositions of the language) and some specific correspondence theorems which will yield the connection. We use Logic Programming and the idea is illustrated in Example 7.

To further analyse what is needed, consider the traditional attack of x on y and its meaning $(*)$:

$$x \twoheadrightarrow y \text{ means that if } x \text{ is "in" then } y \text{ is "out"} \qquad (*)$$

When x is in we may also say x is "alive" or is "in the game" or "active" and when y is out we may also say y is "dead" or is "out of the game" or "inactive".

In informational terms this means that adding the information that x sends to y deletes or destroys y or renders useless its informational content. Let us understand for simplicity that the notation "x" is both the node x and the information that the node x has. So x sends some information to y which practically deletes y. To do this neatly, without simply sending the negation $\neg x$ to y (which is sending false information from the point of view of x), we must use logics which can "delete via addition". D. Gabbay [18, 19] has identified several options for such logics.

1. We can use negation as failure in logic programming (see Appendix B): Given a clause $A \to a$, give it a name, a unique new literal $\mathbf{n}(A)$ and write, instead of $A \to a$, the clause $A' = \neg \mathbf{n}(A) \wedge A \to a$. Now the addition of $\mathbf{n}(A)$ to A' actually deletes it for all computational purposes.
2. We use the constant $\mathbf{e}$, the nill unit in resource logic, which satisfies for any wff A, $A \oplus \mathbf{e} = A$. Now, if we send and add to A the wff $A \multimap \mathbf{e}$ ($\multimap$ denotes resource implication), then A together with $A \multimap \mathbf{e}$ will be reduced via modus ponens to $\mathbf{e}$, in other words, A is deleted.

Let us consider some examples

Example 4. Let us seek an example which can be viewed from both the traditional $(*)$ approach and from the informational $(**)$ approach. For this purpose we regard any letter x as having a dual Categorial role, it can be both an argument element x in S and also an atomic proposition in a logic. So for example the element x in S can contain the information x in the logic.

Consider our Story 1 above, the problem of deciding whom, out of a volatile collection of friends, to invite to a party. We want the party to run smoothly, but for some individuals $s \in S$ who would be the focus of a scene if certain others $x \in L(s) \subseteq S$ are also present. These are the individuals x who do not want to see s in the party.

We now have both an attack relation $R \subseteq S \times S$ and an information input $\tau(s)$ from s to any $x \in L(s)$. We have:

- yRx iff $y \in L(x)$

- $\mathbf{f}(x) =_{\text{def}} x \leftrightarrow \bigwedge_{y \in L(x)} \neg y$
- $\tau(x) = x$

Let us check whether we get some correspondence between the two approaches. In other words, if $x \in L(s)$ would cause a scene if s were also at the party, then we may think of this as x attacking s (i.e. xRs).

Now, we want to invite as many guests as we can, and so we should take into account that we definitely invite all those to whom nobody objects (i.e. all those not 'attacked' by anyone).

So if we apply the above to the graph of Figure 2 and take a standard (monotonic) logic of implication and negation. We can let:

$$\mathbf{f}(x_0)=\{A_0\}$$
$$\mathbf{f}(x_{n+1})=\{A_n \leftrightarrow A_{n+1}\}$$
$$\mu(x_0)=\mathbf{f}(x_0)$$
$$\mu(x_{n+1})=\mu(x_n) \cup \mathbf{f}(x_{n+1})$$

The final values for the nodes are then:

$$\mu(x_0)= \quad \{A_0\} \quad \vdash A_0$$
$$\mu(x_1)= \quad \{A_0, A_0 \leftrightarrow \neg A_1\} \quad \vdash \neg A_1$$
$$\mu(x_2)= \quad \{A_0, A_0 \leftrightarrow \neg A_1, A_1 \leftrightarrow \neg A_2\} \quad \vdash A_2$$
$$\mu(x_3)=\{A_0, A_0 \leftrightarrow \neg A_1, A_1 \leftrightarrow \neg A_2, A_2 \leftrightarrow \neg A_3\} \vdash \neg A_3$$

we may then set:

$$x_i = \text{in iff } \mu(x_i) \vdash x_i$$

This sets the tone for a more general theorem.

Example 5. There is another way of looking at Figure 2. We use the Equational Approach of [10]. The information we pass along the attacks are numbers and the theory is $\{0,1\}$ multiplication. On this approach each arrow represents an equation that applies to the initial value assigned to each member of S to produce its final value, in this case either 1 or 0. If the final value is 1 we say the node is "in", otherwise it is "out". We capture the idea that un-attacked nodes should be "in" by assigned all nodes with the initial value 1.

$$\mathbf{f}(x_i)=1$$
$$\mu(x_0)=\mathbf{f}(x_0)$$
$$\mu(x_{n+1})=(1-\mu(x_n)) \cdot \mathbf{f}(x_{n+1})$$

The final values for the nodes are then:

$$\mu(x_0)=1$$
$$\mu(x_1)=0$$
$$\mu(x_2)=1$$
$$\mu(x_3)=0$$

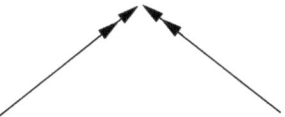

Fig. 5. Informational attack

1.4 Examples from Logic Programs

Example 6. Let us tentatively consider the situation in Figure 5.

In this figure we assume that x, y and z are pieces of information and that y and z attack x. We understand the attack of any a on b, written as $a \twoheadrightarrow b$, to mean that we update b with information $\tau(a)$ sent from a to b. $\tau(a)$ is part of the information a, which is sent to attack b. We could have $\tau(a) = a$. We obtain a new piece of information $b \oplus \tau(a)$ (in many cases we have $\tau(x) = x$ and $\oplus = \cup$ and so the result is $b \cup a$). So the result of z and y passing information to x can be the information $z \cup y \cup x$.

Depending on the information sent, the transmission can be attack or can be support. To see the usefulness of our approach consider the bipolar Example 3 of paper [7, p. 386]. We have the same situation as in Figure 5, with y attacking x and z supporting x. The paper [17] tries to define extensions to such abstract network. According to [17] there is a unique stable extension $\{z, y\}$. The point is that if we, in the current paper, have a reasonable information transmission machinery, we can instantiate x, y, z as informational pieces, with the y information attacking x and the z information supporting x, and see what we get and compare with [17]. [17] says the attacking y overrides the supporting z. Is this indeed always the case? See Example 7 for an instantiation.

Example 7. 1. Consider Figure 6

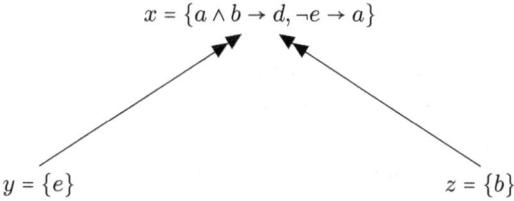

Fig. 6. A logic programming example

The nodes in Figure 6 are logic programming databases where "$\neg$" is negation by failure, "$\wedge$" is conjunction, and "$\to$" is the logic programming implication (see Appendix B). a, b, e and d are atoms. The database x can derive a, we write $x \vdash a$. If x gets attacked by y alone, then it gets the input e and so $\neg e$ no longer succeeds from $x \oplus \{e\}$, and so $x \oplus y$ cannot derive a. If x is attacked by z alone, then we get that x gets the input b *alone* so we have $x \oplus z$, which derives d as well as e.

If x is attacked by both $y = \{e\}$ and $z = \{b\}$, then it becomes $x \oplus y \oplus z$ which cannot derive a and can derive just d.

So far Figure 6 gives us no more and no less than a geometrical network (S, R) of nodes with which information bases are associated as well as the informational "attacks" flow along R as described in item 1. above. If we want to talk about nodes being "in" or "out" or "und" in the Dung sense, we need to allow for a projection function which will give these values for each node. Let $\alpha(x) = a$, $\alpha(y) = b$ and $\alpha(z) = e$.

Define in general that any node u is "in" if the theory at u can prove $\alpha(u)$. According to the projection defined by α, we have that both y and z are "in", y supports x (because the information it sends to x strengthens it) while z attacks x.

2. To further our understanding, note that the databases $x' = \{\neg e \to a, a \wedge b \to d\}$ and $x = \{a, a \wedge b \to d\}$ are not the same, even though both derive $\{a, a \wedge b \to d\}$, because $x \oplus \{e\}$ derives a while $x' \oplus \{e\}$ does not derive a.

3. Consider now Figure 7. In this figure the node u attacks the node v. The

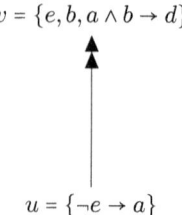

Fig. 7. Second logic programming example

question we ask is what information does u send to v?

On the one hand u can derive a. So if it sends $\{a\}$ (i.e. $\tau(u) = \{a\}$), it will enable $v \oplus \tau(u)$ to derive d, because v derives $a \wedge b \to d$.

But if it sends itself, namely $\tau(u) = u$, then $v \oplus u$ cannot derive a and so also cannot derive d.

The correct choice to send is the entire information u. The reason is that u derives a on the non-monotonic defeasible assumption $\neg e$. By sending $\{a\}$ only, u is sending misleading information. v can derive e and therefore needs to know that a is derived on the basis of $\neg e$. The reader should note that sending only a is not a case of sending some information but it is the case of sending misleading information.

4. We can therefore recommend that if the underlying information logic is non-monotonic, the entire information should be sent and not just what can be derived.

Remark 2. Let us summarise our options of what kind of information $\tau(x)$ a node x (itself being an information carrying node) can send to a target node y (i.e. when $x \twoheadrightarrow y$). Note that x may be a database in some logic $\vdash$.

1. x can send part of itself or all of itself.
2. x can send part of what it can prove or all of what it can prove.[5]
3. x can send misleading information which x itself knows is false, for example x sends $\neg x$.
4. x can send information which is consistent with itself, but is not derivable from itself. It is information which it can consistently pretend to have as true.

1.5 Examples from substructural logics

Remark 3. This remark and the following two Examples 8 and 9 are intended for readers familiar with substructural logics [26].

The idea of information update and the notation $\oplus$ is not new. It appears in the context of the logic and semantics for substructural logics, in connection with the semantical condition and the substructural implication $\multimap$.

We envisage a language with atoms and $\multimap$ and with semantical models of the form $(S, \oplus, h)$, where S is a set of pieces of information and $\oplus$ is an associative and commutative binary operation and where h is the assignment to the atoms (e.g. $h(q) \subseteq S$) and we let, for $x \in S$, satisfaction $\vDash$ to be defined by

- $x \vDash q$, if $x \in h(q)$, for q atomic
- $x \vDash A \multimap B$ iff for every $y \vDash A$ we have $x \oplus y \vDash B$.

Here $\tau(a) = a$, i.e. the node a sends the complete/entire information it has to its target. In practice (witness cases) only some part or a combination $\tau(a)$ derived from a is sent.

Example 8. To give examples from substructural logics (see [26]), consider the following possibilities.

1. If S is a family of multi sets and $\oplus$ is multi set union then $\multimap$ becomes linear implication.
2. If S is a family of sets and $\oplus$ is set union then $\multimap$ becomes relevance implication.
3. If we interpret $\oplus$ as equality $=$ then $\multimap$ becomes S5 strict implication.

---
[5] Note that this is not the same as sending itself or part of itself, as item 3 of Example 7 shows.

Thus we can use $\oplus$ to explain the attacks in Figure 5 to mean that $x = \tau(y) \oplus \tau(z)$ or $x = y \oplus z$ if we let $\tau(u) = u$.

We need to give some examples of the use of $\oplus$ in attack.

Example 9. Let us go through example 4 again, this time using linear logic. Consider the basic situation of x being attacked by $L(x)$. $y \in L(x)$ are all the people who do not like x and might cause a scene if they see that x is present at the party with them. So we have:

- x is invited if and only if none of the $y \in L(x)$ is invited.
- If we have not decided yet whether to invite some particular $y \in L(x)$ and none of the other elements $z \in L(x)$ are definitely invited, then x is not yet decidedly invited.

Let us use linear logic with $\multimap$ and the null element $\mathbf{e}$ to assign information to each person $x \in S$. Using the set theory of multisets:

- $\mathbf{f}(x) = \{y \multimap \mathbf{e} \mid y \in L(x)\} \cup \{\left(\bigoplus_{y \in L(x)} y \multimap \mathbf{e}\right) \multimap x\}$
- $\mu(x) = \bigcup_{y \in L(x)} \tau(y) \cup \mathbf{f}(x)$.
- $\tau(x) = \begin{cases} \{x\} & \text{if } \mu(x) \vdash x \\ \varnothing & \text{otherwise} \end{cases}$

According to this linear logic interpretation, what x is saying is as follows:

$\mathbf{f}(x)$ says "all those y who attack x (i.e. all $y \in L(x)$) amount to nothing $(y \multimap \mathbf{e})$, and since they are all nothing, I am 'alive' " (that is $\left(\bigoplus_{y \in L(x)} y \multimap \mathbf{e}\right) \multimap x$).

To fully appreciate the Linear Logic informational attack view, we need a major example, but first let us do a quick example, and then move to our next major example .

Applying the linear logic view (of Example 9) to Figure 2 we get:

$\mathbf{f}(x_0) = \{x_0\}$
$\mathbf{f}(x_1) = \{x_0 \multimap \mathbf{e}, (x_0 \multimap \mathbf{e}) \multimap x_1\}$
$\mathbf{f}(x_2) = \{x_1 \multimap \mathbf{e}, (x_1 \multimap \mathbf{e}) \multimap x_2\}$
$\mathbf{f}(x_3) = \{x_2 \multimap \mathbf{e}, (x_2 \multimap \mathbf{e}) \multimap x_3\}$

Therefore we get:

$\mu(x_0) = \{x_0\}$
$\mu(x_1) = \{x_0, x_0 \multimap \mathbf{e}, (x_0 \multimap \mathbf{e}) \multimap x_1\}$
$\mu(x_2) = \mathbf{f}(x_2)$
$\mu(x_3) = \{x_2\} \cup \mathbf{f}(x_3)$

And we have that $\mu(x_0) \vdash x_0$ but $\mu(x_1) \nvdash x_1$, because either (i) the resources $x_0 \multimap \mathbf{e}$ and x_0 are 'used up' by a Modus Ponens to $\mathbf{e}$ and then $\{\mathbf{e}, (x_0 \multimap \mathbf{e}) \multimap x_1\} = \{(x_0 \multimap \mathbf{e}) \multimap x_1\} \nvdash x_1$; or else (ii) the resources $x_0 \multimap \mathbf{e}$ and $(x_0 \multimap \mathbf{e}) \multimap x_1$ are 'used up' by a Modus Ponens to x_1, but x_0 remains and $\{x_0, x_1\} \nvdash x_1$ in linear logic. Similarly, $\mathbf{f}(x_2) \vdash x_2$ and $\{x_2\} \cup \mathbf{f}(x_3) \nvdash x_3$.

2 Formal presentation of the acyclic informational networks

We are now ready for the formal machinery. We first define the traditional argumentation networks for finite acyclic graphs. These have only the traditional grounded extension. This will allow us to appreciate the next definition, that of informational argumentation network for finite acyclic graphs.

We have no problems or technical issues with the presence of cycles in the context of Argumentation as information input. We know how to handle them. Our treating the acyclic case first is mainly for reasons of effective exposition. We want to highlight certain points in stages and starting with acyclic graphs is advantageous.

We remarked in the previous section that for the acyclic case of a traditional Dung network (S,R), we can use an inductive algorithm for calculating the unique grounded Dung extension of (S,R). The direction of the algorithm (i.e. the direction of the induction) corresponds to the direction of information flow, when we view (S,R) as information network.

This directional point of view can be enlisted to help in the case of cycles in an informational network. Consider a cycle of the form (S,R), with $S = \{a,b\}$ and R being $R = \{(a,b),(b,a)\}$, i.e. the cycle is aRb and bRa. As a Dung traditional network, there are two non-empty extensions, namely $\{a\}$ and $\{b\}$. There is no notion of a starting point in the cycle, because there is no direction. As informational network, we can add a starting point, as an object level part of the cycle and use it to our advantage (see also item 2 of Remark 4). It is for such reasons that we want to address the acyclic case first in this section. In fact, since information flow is directional, one can argue that there is no place for cycles in Information Networks. We shall discuss the place of cycles in information networks in the Methodological Section 4.

We show that informational attacks can simulate traditional attacks for acyclic graphs. When we have cycles we have problems as the meanings of the attack diverge, as seen in item 3 of Remark 1. We shall also see later that for the case of finite acyclic graphs there is better correspondence between the "dynamics" aspects of the traditional attack argumentation and informational argumentation. This will be clarified in Remark 4 below.

Definition 1. *1. Let S be a non-empty set and let $R \subseteq S \times S$ be a binary relation on S. We say R is acyclic iff there does not exist a sequence $(x_1,\ldots,x_n)$ in $S, n \geq 1$ such that*

$$x_1 R x_2, x_2 R x_3, \ldots, x_n R x_1.$$

2. x is said to be a source point if there is no y s.t. yRx. x is an endpoint if there is no y such that xRy.

Proposition 1. *Let (S,R) be a finite acyclic graph. Then for some $x \in S$ we have that x is a source point.*

Proof. Assume that there are no source points. Let $x_1 \in S$, then for some $x_2, x_2 R x_1$. Similarly for some $x_3, x_3 R x_2$. We carry on and get a sequence $x_1, x_2, x_3, \ldots, x_n, \ldots$ such that $x_{n+1} R x_n$. Since S is finite, for some $m, n, m \neq n, m < n$ we have $x_m = x_n$. Thus gives us a cycle $(x_m, \ldots, x_n)$.

Definition 2. *Let (S, R) be finite acyclic. We define a Caminada $\{0, 1\}$ labelling λ on S as follows.*

Step 1. Let all source points x be labelled $\lambda(x) = 1$.

Step 2. Let y be any point such that for some $x, x R y$ and $\lambda(x)$ is defined in Step 1 and $\lambda(x) = 1$. Let $\lambda(y) = 0$.

⋮

Step $2n + 1$. Let x be any point such that $\lambda(x)$ is not yet defined but for all z, such that $z R x$, we have that $\lambda(z)$ is defined and $\lambda(z) = 0$. Let $\lambda(x) = 1$.

Step $2n + 2$. Let y be such that $\lambda(y)$ is not yet defined but for some $z, \lambda(z)$ is defined and $\lambda(z) = 1$. Let $\lambda(y) = 0$.

Let S_λ be all points x such that $\lambda(x)$ is defined at any n.

Proposition 2. 1. *For the S_λ of Definition 2 we have that $S_\lambda = S$.*
2. *The function λ thus defined is unique and satisfies the conditions (C1)-(C3) of the Caminada labelling.*

Proof. 1. Let $x_1 \in S_\lambda - S$. Consider the set of all z such that $z R x$.
  (a) If the set is empty then $\lambda(x) = 1$ by Step 1 of Definition 2.
  (b) If the set is not empty and all its members are in S_λ, then if for some $z, z R x_1$ and $\lambda(z) = 1$ then at some stage $\lambda(x_1)$ is defined and $\lambda(x) = 0$. On the other hand, if for all z such that $z R x_1$ we have $\lambda(z) = 0$, then at some stage $\lambda(x_1)$ is defined and $\lambda(x_1) = 1$.
  (c) Therefore there exists an element z such that $z R x_1$ and $\lambda(z)$ is not defined. Call this element x_2.
  Assume by induction that we have $(x_m, x_{m-1}, \ldots, x_1)$ such that

$$x_m R x_{m-1}, \ldots, x_{m-1} R x_{m-2}, \ldots, x_2 R x_1$$

and $\lambda(x_j)$ are all undefined. Repeat our reasoning for x_m and get x_{m+1} for which $\lambda(x_{m+1})$ is undefined.
  Since S is finite, we have that for some $m_1 < m_2$ we have $x_{m_1} = x_{m_2}$. This gives us a cycle. A contradiction. So $S_\lambda = S$.
2. It is clear from the construction that λ is unique and satisfies the conditions (C1)-(C3) of the Caminada labelling.

Definition 3. 1. *An information system is a set I with a binary associative and commutative operation $\oplus$ on I. Let $\mathbf{e} \in I$ be empty information with $a \oplus \mathbf{e} = a$ for all a. Assume further that for each $a \in I$, there is associated another piece of information $\tau(a)$. (if a is a set of wffs of a logic with consequence relation $\vdash$, then usually $\tau(a)$ is either a itself or the set $\{b \mid a \vdash b\}$, which might be different from a).*

2. Let (S, R) be a finite acyclic graph as in Definition 2 and let $\mathbf{f}$ be a function giving for each point x in S a value $\mathbf{f}(x) \in \mathbf{I}$. We call the system $(S, R, \mathbf{f}, \mathbf{I}, \oplus)$ an acyclic information network.
We define an information label $\mu(x)$, for each $x \in S$, in steps, using $\mathbf{f}$.

Step 1. Let x be any source point. Define $\mu(x)$ to be $\mathbf{f}(x)$. Declare the source point x as having permission to send information to its target points (i.e. points y such that xRy).[6]

Step $n+1$. Let x be any point such that $\mu(x)$ is not yet defined but assume that for all z such that $zRx, \mu(z)$ is defined. Further assume that all such points z have now permission to send information to its target points such as x. Assume the information the node z can send is $\tau(z)$. Then let $\mu(x)$ be defined as

$$\mu(x) = \mathbf{f}(x) \oplus \tau(z_1) \oplus \ldots \oplus \tau(z_m)$$

where $z_1, \ldots, z_m$ are all the points in S such that zRx holds. Declare x to be now as having permission to send information to its targets and let $\tau(x)$ be defined as $\tau(\mu(x))$.

Let S_μ be the set of all $x \in S$ for which $\mu(x)$ is defined.

Proposition 3. *We have for S_μ of Definition 3 that $S_\mu = S$ and that μ is unique.*

Proof. Similar to the proof of Proposition 2.

Remark 4. 1. We need to discuss and compare the construction of the Caminada labelling of Definition 2 with the construction of the information labelling of Definition 3.
The construction of the Caminada labelling in Definition 2 is just an algorithm for finding the labelling λ, defining a set theoretic object, a subset E_λ of S:

$$E_\lambda = \{s \in S \mid \lambda(s) = 1\}.$$

The inductive steps involved in the construction of λ and hence in the construction of the set E_λ, have no conceptual meaning in the context of the network (S, R). If we consider the party example of Story 1, all we have is a relation xRy, where x says he does not want to come to the party if y is there, and we are looking for a maximal set of invitees to the party satisfying certain conditions. The step by step construction of such a set is not part of the concept of the story of the party. By comparison, if we

---

[6] This declaration is done by virtue that x itself is a source point, and therefore is not expecting any information from any other point (i.e. from any z such that zRx, because there are no such z). Define $\tau(x)$ to be $\tau(\mathbf{f}(x))$.
Think of Story 3, the security agency. In this story xRy means that x reports to y. So x will wait to process all the information from his informants and then process it and then report to y.

interpret (S,R) according to the security agency story 3, the relation xRy, meaning that x passes information to y, also implies a hierarchy, that if $xRy \land yRz$ then y cannot pass any information to z until it gets the full information from all x_i such that $x_i Ry$. Thus the step by step construction of μ in Definition 3, does have a conceptual meaning in the context of information transmission.

2. We can also look at the construction of Definition 2 as a case of Story 3, a case of information transmission, and try to simulate/create the function λ of Caminada. This may be possible to do (see Example 11) but the simulation may not be of any conceptual value and only of mathematical value.

3. The difference between Story 1 and Story 3 emerges in the case of cycles. If we have two nodes only $\{a,b\}$ with aRb and bRa, the set theoretical approach will look for set theoretical complete extensions and so would seek to extend the algorithm of Definition 2 to the case of cycles to yield the three extensions $\{a\}$, $\{b\}$ and $\varnothing$. The informational approach would have to look at $\{a,b\}$ as cooperating nodes with respect to information and would require extending the algorithm of Definition 3 to yield a solution where $\mu(a) = \mu(b)$. There is no guarantee that the algorithm of one approach will be meaningful when viewed by the other approach.

Example 10. We now give a useful example of an information system labelling. Let (S, R) be acyclic graph. Regard the elements of S as atoms for constructing a logic programming information system. For each $x \in S$ and for $y_1, \ldots, y_m$ being *all* the nodes y such that yRx, construct the clause $C_x =_{\text{def}} \neg y_1 \land \ldots \land \neg y_m \to x$. If x is is a source point then let $C_x =_{\text{def}} x$. Consider as pieces of information as finite sets of clauses of the form C_x. Let $\oplus$ be defined on such pieces of information as set union $\cup$. Consider the function $\mathbf{f}$ on S, defined for $x \in S$ by
$$\mathbf{f}(x) = C_x.$$
We can now consider an information system I whose elements are sets of clauses $\{C_x\}$, with $\oplus$ taken as union $\cup$. Consider μ defined from $\mathbf{f}$ as in Definition 3. We now got a special information system labelling for any acyclic (S, R). To fully appreciate the role of this example we need to know more about logic programming. The next definitions and theorem will tell us a bit more.

Definition 4. 1. *Let $(I, \oplus)$ be an information system. Let Q be a set of atomic sentences. A function $\pi : Q \times I \mapsto \{0,1\}$ is called a consequence function. For $q \in Q$ and $x \in I$ we also write $x \vdash_\pi q$ when $\pi(q,x) = 1$.[7]*

2. *Assume we are given an acyclic information network $(S, R, \mathbf{f}, \mathbf{I}, \oplus)$ containing the information system $(\mathbf{I}, \oplus)$ and a consequence function π defined on it as defined in item 1 above. Let us consider $S = Q$ as a set of distinct atomic sentences and let us therefore consider π as defined on $S \times \mathbf{I}$. Under these conditions, we can get a $\{in, out, und\}$ valued function λ_μ for (S,R),*

---

[7] There can be various restrictions on $\vdash$, but we need not go into the details here.

by using π and the information label $\mu(x)$, for each $x \in S$, as defined in Definition 3, as follows:
- $\lambda_\mu(x) = 1$, if $\mu(x) \vdash x$,
- $\lambda_\mu(x) = 0$, if $\mu(x) \vdash \neg x$.
- If neither $\mu(x) \vdash x$ nor $\mu(x) \vdash \neg x$ holds we say that the value of λ_μ is undecided.

Note that the projection λ_μ we just defined may not be a legitimate Caminada labelling. Its values depend on the information network we use.
See Example 11 for a choice of information network which does yield a legitimate Caminada labelling.

Example 11. Let us go back to the logic programming information system labelling of Example 10, defined for any acyclic finite graph (S, R). Consider now $\mu(x)$, for $x \in S$. $\mu(x)$ is a set of logic programming clauses. Let $\vdash$ be the logic programming consequence relation. We are not defining this here, we assume it is know to the reader). Let λ_μ be defined on S by

- $\lambda_\mu(x) = 1$, if $\mu(x) \vdash x$,
- $\lambda_\mu(x) = 0$, if $\mu(x) \vdash \neg x$.
- If neither $\mu(x) \vdash x$ nor $\mu(x) \vdash \neg x$ holds we say that the value of λ_μ is undecided.

The above considerations introduced a very specific $\{0, 1\}$ labelling λ_μ on (S, R) obtained from a very specific logic programming information system labelling for (S, R). The reader can appreciate that the kind of labelling function λ_μ that we get depends on the information involved. In fact we might get a function indicating support , rather than attack, or a mixed bipolar network with both support and attack, or nothing in particular-just a three valued function..

We now want to compare the very specific λ_μ function that we have defined with the traditional Caminada labelling λ for (S, R), introduced in Definition 2. In fact we can prove the following Proposition:

$$\lambda = \lambda_\mu$$

This shows that the informational approach can simulate the traditional approach at least for the case of acyclic networks.

3 Examples with cycles

3.1 Orientation discussion about cycles

We begin this section with a methodological orientation discussion, taking stock of what we have so far and what is left for us to do.

1. Section 1 motivated the idea of information input.

2. Section 2 gave a precise definition of the concept of information network for the case of finite acyclic graph. This means that we start with finite graph (S, R), and an information system $(\mathbf{I}, \oplus)$ and a function $\mathbf{f}$ associating information $\mathbf{f}(x) \in \mathbf{I}$, for $x \in S$. We then defined the propagation of information along the network in the direction of R, namely we defined the function μ.
3. Item 2 above describes the general concept of acyclic information network. This is a very general concept. Having reached this definition we wanted to connect it with the traditional Dung extensions on (S, R). So from the informational side we have (S, R) and an additional $(\mathbf{I}, \oplus)$ and a μ and on the traditional Dung side we need a legitimate Caminada labelling λ. How do we connect the information system with a Caminada labelling? This we do by choosing a particular very specific information system and projecting it (using the information function μ and a projection π) to obtain a three valued labelling function λ_μ.
4. The reader should realise that the notion of information system is much more general. We are not doing here Traditional Caminada labelling under the guise of information systems. To make the situation crystal clear, consider the notion of Assumption Based Argumentation (ABA, see [28]). In this case the declared aim of the ABA community is to provide a different but equivalent system to traditional argumentation. We are not doing this , we are investigating a more general notion of information system.
5. So our aim in Section 3 is to extend our general definition of Information Networks to include cycles. The problem is that to get an idea of how to generalise the notion we need examples and the best examples come from Logic Programming or from some special non monotonic logics. We ask the reader to be clear on this point. Although in the preceding Section 2 we used some specific information systems based on Logic Programming to simulate traditional dung argumentation, in this section we use such examples to try and get ideas and motivate how to handle cycles in general information networks. We simply do not have other examples for this purpose. We are not initially seeking to simulate traditional argumentation networks that contain cycles, we are just trying to look at examples to give us ideas of how to deal with cycles (though later we shall simulate bipolar three valued argumentation networks, but this is not our immediate problem).

This section provides preparatory examples for networks with cycles. The problem with cycles such as odd or even loops, is that we do not have source points and so we do not yet have procedures for giving permissions to send information to targets. We may also have that all elements of the cycle have the same information under the $(**)$ interpretation, and so there may be a problem to distinguish between them in terms of in, out, undecided labelling. Furthermore, in the presence of cycles, we will not be able to find a way to simulate the traditional extensions in terms of information as we did in the acyclic case, in Example 11.

We stress again that the idea of informational network comes from widespread applications and our attempts to connect with traditional Dung networks is only for the benefit of the community familiar with Dung extensions and not by way of justification of the informational approach. We do not mind if we cannot simulate the traditional Dung approach for networks with cycles. In fact this is tolerable to us because in practice x passes information to y only after x sees y and decides to respond, and so we do not get cycles in such practice (where the (∗∗) meaning plays a role). Nevertheless we would like to know what happens formally when we have a network (S, R) with cycles under the (∗∗) interpretation. We can reduce (S, R) to its SCC components network $(S^\approx, R^\approx)$, which is acyclic (see [2] and Appendix A), but still we need to address what happens in each cycle.

3.2 Some useful examples of cycles

We shall progress by looking at examples with cycles. Each example we use shall illustrate a point we need to consider.

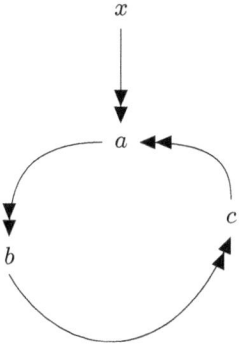

Fig. 8. A cyclic graph

Example 12. Consider Figure 8 which will help us make a few explanatory remarks. We considered on acyclic graphs (S, R) and compared the Caminada extension labelling with the information system labelling. The graph of Figure 8 is cyclic and so is not acceptable to the machinery of Section 2. It does, however, have a unique grounded extension with Caminada labelling $\lambda(a) = 0$, $\lambda(x) = 1$, $\lambda(b) = 1$, $\lambda(c) = 0$, and so we suspect that the machinery of Section 2 might work. The cycles however, could present a problem for the informational labelling because of the way these are defined. So let us analyse the informational labelling approach of Definition 3 when applied to Figure 8. We use the same **f** as given in Example 10. Thus we have:

- $\mathbf{f}(x) = x$
- $\mathbf{f}(a) = \neg x \wedge \neg c \to a$

- $\mathbf{f}(b) = \neg a \to b$
- $\mathbf{f}(c) = \neg b \to c$

We now define μ.

- $\mu(x) = x$
- $\mu(a) = \mathbf{f}(a) \oplus \mu(x) \oplus \mu(c)$
  $= \{\neg x \wedge \neg c \to a\} \cup \{x\} \cup \mu(c)$
- $\mu(b) = \mathbf{f}(b) \oplus \mu(a)$
  $= \{\neg a \to b\} \cup \mu(a)$
- $\mu(c) = \mathbf{f}(c) \oplus \mu(b)$
  $= \{\neg b \to c\} \cup \{\neg a \to b\} \cup \mu(a)$

Therefore

($*$) $\mu(c) = \{\neg b \to c\} \cup \{\neg a \to b\} \cup \{\neg x \wedge \neg c \to a, x\} \cup \mu(c)$

This can happen only if

($**$) $\{\neg b \to c, \neg a \to b, \neg x \wedge \neg c \to a, x\} \subseteq \mu(c)$

This actually happens only if we have equality in ($**$) because all the players' wff are participating. We therefore get that:

($***$) $\mu(a) = \mu(b) = \mu(c) = \{\neg b \to c, \neg a \to b, \neg x \wedge \neg c \to a, x\}$, and $\mu(x) = x$

Logic programming calculations will show that

$$\mu(x) \vdash x$$
$$\mu(a) \vdash \neg a$$
$$\mu(b) \vdash b$$
$$\mu(c) \vdash \neg c$$

Therefore $\lambda_\mu = \lambda$, in this case also.

Let us now view the network of Figure 8 according to Story 3, the security agency. According to this view agent x reports $\mathbf{f}(x)$ to agent a and then all of the agents in the cycle $\{a, b, c\}$ share the information they all have, namely, agent a has information $\mathbf{f}(a)$ and $\mathbf{f}(x)$ and agent b has $\mathbf{f}(b)$ and agent c has $\mathbf{f}(c)$. So when they all share we have that

$$\mu(a) = \mu(b) = \mu(c) = \{\neg b \to c, \neg a \to b, \neg x \wedge \neg c \to a, x\}$$

which is the same as before. If this is not a coincidence, it suggests an inductive algorithm for computing μ for general networks with cycles. We consider the acyclic graph of SCC cycle (for the definition of SCC, see Appendix A) and propagate. We leave this to Section 4.

Example 13. The cycle in Figure 8 was broken by an external attack, $x \twoheadrightarrow a$. Let us examine a cycle which is not broken. Consider the four cycle graph of Figure 9. Let us see what extensions we get using the logic programming informational model. As before we let:

Argumentation as Information Input   169

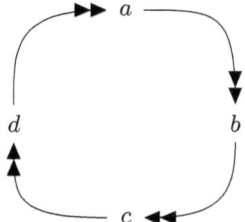

Fig. 9. A four cycle

- **f**(a) = ¬d → a
- **f**(b) = ¬a → b
- **f**(c) = ¬b → c
- **f**(d) = ¬c → d

The function μ can be calculated as before. We get:

- $\mu(a) = \{\neg d \to a\} \cup \mu(d)$
- $\mu(b) = \{\neg a \to b\} \cup \mu(a)$
- $\mu(c) = \{\neg b \to c\} \cup \mu(b)$
- $\mu(d) = \{\neg c \to d\} \cup \mu(c)$

Solving these set equations we get:

$$\mu(d) = \{\neg c \to d, \neg b \to c, \neg a \to b, \neg d \to a\} \cup \mu(d)$$

Therefore
$$\mu(a) = \mu(b) = \mu(c) = \mu(d) = \Delta$$

where $\Delta = \{\neg d \to a, \neg a \to b, \neg b \to c, \neg c \to d\}$. Since

$$\Delta \nvdash a \qquad \Delta \nvdash \neg a$$
$$\Delta \nvdash b \qquad \Delta \nvdash \neg b$$
$$\Delta \nvdash c \qquad \Delta \nvdash \neg c$$
$$\Delta \nvdash d \qquad \Delta \nvdash \neg d$$

we get the whole undecided extension.

However there are two answer set models for Δ (for answer set semantics for Logic Programs see [22]). These are $\{a,c\}$ and $\{b,d\}$ and these give the other two extensions. We ask ourselves is answer set semantics for logic programs a useful instrument for dealing with cycles in the informational case (where the information is expressed in Logic Programming)? Again, we leave this question to Section 4.

Example 14. The examples used so far used Logic Programming as the information systems. Let us use a different logic this time. Let us use Linear Logic

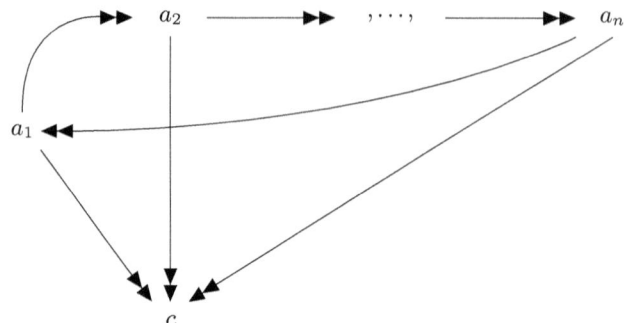

Fig. 10. An n cycle on the attack.

We consider an example of a cycle which will help us further understand the linear logic informational view of Example 9.

Consider Figure 10

The information associated with each node is as follows (using the linear logic view):

$$\mathbf{f}(a_1) = \{a_n \multimap \mathbf{e}, (a_n \multimap \mathbf{e}) \multimap a_1\}$$
$$\mathbf{f}(a_{i+1}) = \{a_i \multimap \mathbf{e}, (a_i \multimap \mathbf{e}) \multimap a_{i+1}\} \quad \text{for } 1 \leq i < n$$
$$\mathbf{f}(c) = \{a_i \multimap \mathbf{e} \mid 1 \leq i \leq n\} \cup \{\left(\bigoplus_{i=1}^n (a_i \multimap \mathbf{e})\right) \multimap c\}$$

We have that $\mathbf{f}(a_i) \vdash a_i$ for $1 \leq i \leq n$ and $\mathbf{f}(c) \vdash c$.

We define an information equilibrium function μ as any function $\mu(x)$ into the wffs of the logic which satisfies for any x:

$$\mu(x) = \mathbf{f}(x) \cup \bigcup_{y \twoheadrightarrow x} \{z \mid \mu(y) \vdash z\} \qquad (\#)$$

This μ corresponds to the traditional notion of complete extension, or for the case of the equational approach it corresponds to the solution to the equations associated with the network [10].

We use $\mu(x)$ to define whether x is to be considered "in" or "out" or undecided. In the equational approach the solution to the equations is a numerical function μ with values in $[0, 1]$. So $\mu(x) = 1$ means $x =$ in, and $\mu(x) = 0$ means $x =$ out and otherwise x is undecided. In the informational case, $\mu(x)$ is an informational database. We are free to define in some reasonable way the conditions of when to consider x to be in or out or neither, based on the information at x. Mathematically such a decision is an arbitrary projection. We find it useful to agree that if $\mu(x) \vdash x$ then we say x is in. If negation (of any kind) is in the language, (say "¬"), we can use it to say x is out if $\mu(x) \vdash \neg x$ and undecided otherwise. If we do not have any negation, we can say x is out iff $\mu(x) \not\vdash x$. In this case we do not have a notion of being undecided. We stress that the notions of in and out are conceptually marginal to the informational view. We use them only to simulate the traditional view. What is important is only what $\mu(x)$ can derive for each x.

Remark 5. Following the previous Example 14, we shall examine our options for loops more closely.

Let us return to the cycle of Figure 10. We need to deal with it as an informational network. What we need here (this is a design stage for us) is a procedure for the nodes to send information to their targets. We are all agreed that nodes which are not attacked, or such that none of their attackers are capable of sending information, should be allowed to send information themselves. So this principle alone allowed us to deal with the network of Figure 5, as we did in Example 6. However, Figure 10 starts with a cycle, so what do we do with the cycle?

Let us examine how traditional argumentation handles this cycle. Actually, it depends on whether the cycle is odd or even. If the cycle is even, say $n = 2m$, then the cycle can be resolved in (or it has) two extensions

$$E_1 = \{a_1, a_3, \ldots, a_{2m-1}\}$$
$$E_2 = \{a_2, a_4, \ldots, a_{2m}\}$$

and the elements of the, extension (i.e. those elements which are "in") continue to propagate the attack. If the cycle is odd, say $n = 2m+1$, then the only extension has all members of the cycle "undecided" and the undecided value is propagated. In [2] and [11], various methods of handling cycles are discussed. [2] resolves an odd cycle by taking maximal conflict-free sets of elements from the cycle $\{a_1, \ldots, a_{2m+1}\}$, while [11], for example, uses annihilators to resolve cycles (external attackers with "kill" elements in the cycle). We note that whatever method $\mathbb{R}$ of resolving cycles we use, its net function is to suggest or extract several sets $\mathbb{E}_1, \ldots, \mathbb{E}_k$ of elements from the cycle which are to be deemed "in" and are allowed to propagate the attack to their targets. The traditional Dung $\mathbb{R}_{\text{Dung}}$ approach is to take extensions, while the $CF2$ semantics $\mathbb{R}_{CF2}$ (see [2]) takes maximal conflict free sets. There are other Gabbay methods in [11], for resolving cycles.

So in this paper we need to offer our own method, $\mathbb{R}_{\text{info}}$, of resolving cycles, telling us which points in the cycle are allowed to send information to their targets within the informational paradigm outlined above. We first observe that in traditional "attack" argumentation, the resolution of geometrical cycles (i.e. cycles in (S, R)) is external to the argumentation networks. The basic concept of traditional argumentation is that of semantics and extensions, satisfying some set theoretic conditions. When we propose algorithms for computing these extensions we encounter problems with cycles in the computation because of the geometrical cycles and we seek methods for dealing with them. Alternatively, if we believe that odd cycles prevent us from getting enough legal set theoretic extensions, again we look for ways to resolve the geometric cycles. However in comparison in the informational case geometrical cycles simply mean sharing of information and so we should be able to deal with them more easily and in a more tolerant way. Nevertheless to the extent that inductive algorithms are involved, we do need to address (external) resolution of cycles also in the informational case to facilitate our algorithms. See also the discussion in Remark 4 above. Let us consider two extreme options for the external handling of cycles and cycles.

The first option is to choose one a_i to send information, and the second option is to let all the a_i attack (by sending information). Let us see what happens in the case of $n = 6$. For convenience we redraw the network in Figure 11. We note the following:

1. If we allow a_i alone to attack successfully we get E_1 or E_2 depending on i, and of course we get that c = out.
2. If we let all of the a_i to attack successfully, we get that every node a_i, as well as c, are out.
3. If we let $\{a_1, a_2, a_4, a_5\}$ to attack successfully, we get the maximal conflict free $CF2$ extension $\{a_1, a_4\}$ of the cycle, and of course, c is out.

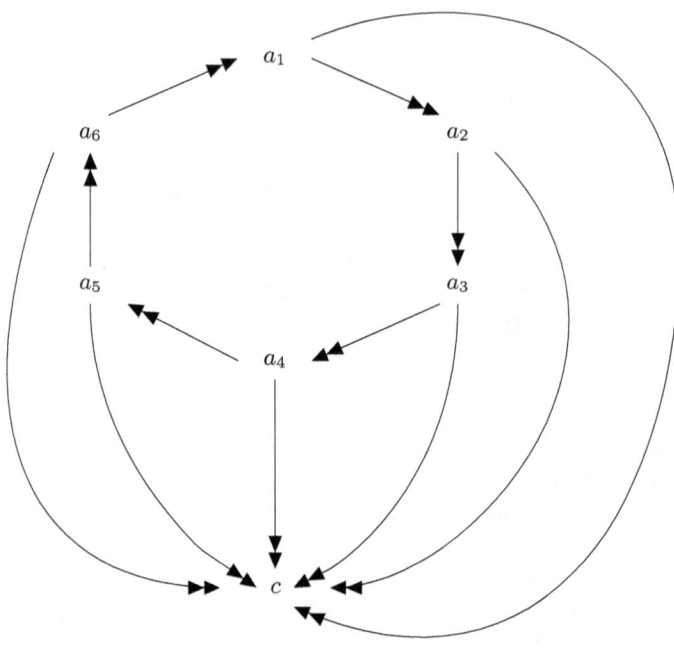

Fig. 11. A cycle of 6

Example 15. Let us continue out explanatory example with Figure 12. This is a cycle with a $CF2$ extension $\{a_1, a_4\}$ which cannot be simulated by the informational approach of Example 14. In this figure we have:

$\mathbf{f}(a_i) = \{a_{i-1} \multimap \mathbf{e}, (a_{i-1} \multimap \mathbf{e}) \multimap a_i\}$ $i = 2, 3, 5, 6$
$\mathbf{f}(a_1) = \{a_6 \multimap \mathbf{e}, (a_6 \multimap \mathbf{e}) \multimap a_1\}$
$\mathbf{f}(a_4) = \{a_7 \multimap \mathbf{e}, a_3 \multimap \mathbf{e}, ((a_7 \multimap \mathbf{e}) \oplus (a_3 \multimap \mathbf{e})) \multimap a_4\}$
$\mathbf{f}(a_7) = \{a_3 \multimap \mathbf{e}, (a_3 \multimap \mathbf{e}) \multimap a_7\}$

Traditional Dung extensions for this figure are either to have $E_1 = \{a_3, a_5, a_1\}$ or to have everything undecided.

We also have the $CF2$ extension of the maximal conflict free set $\{a_1, a_4\}$. This set cannot be obtained by the informational approach $\mathbb{R}_{\text{info}}$ as outlined in example 14. There is no subset E of $\{a_1, \ldots, a_7\}$ which can be chosen to be allowed to attack and it will result in exactly $\{a_1, a_4\}$ being live. To get such a set, none of the attackers of a_4 can be allowed to attack. So a_3 cannot successfully attack, but then a_7 cannot be taken out, and so it will attack a_4.

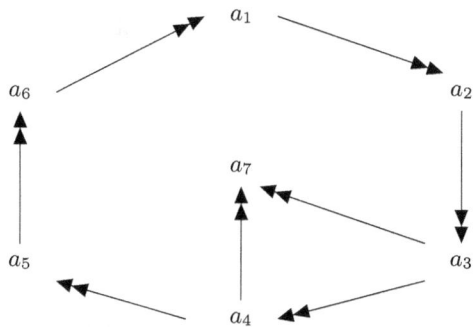

Fig. 12. A cycle distinguishing $CF2$ from the informational approach

Example 16. We need one more comparative example to fully appreciate the informational approach. Consider Figure 13.

The left side (1) is a 3-cycle. We compare 3 methods of handling this cycle.

Gabbay's annihilator method This method (see (2)) uses an external annihilator node h to take out one of the elements in the cycle (in this case a_1). We get at $a_1 = $ out, $a_2 = $ in and $a_3 = $ out.

CF2 semantics This takes a maximal conflict free subset of the cycle. In this case we can take $\{a_2\}$.

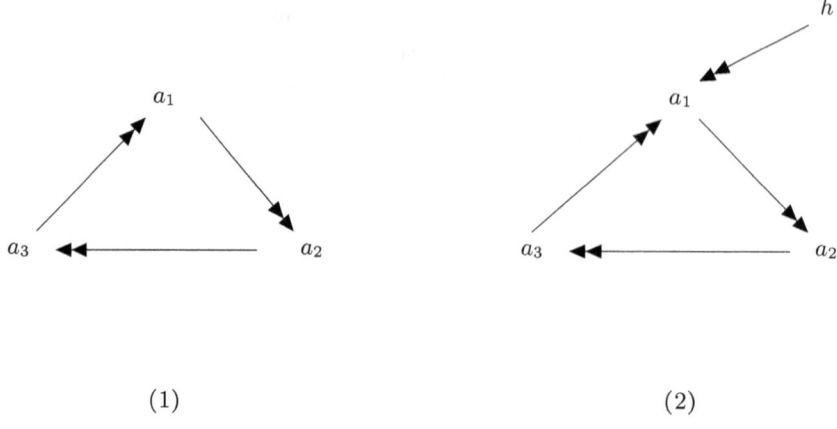

Fig. 13. cycles of three

The informational approach Let $\mathbf{f}(a_i)$ be as in Example 14 (note that item (1) of Figure 13 is a case $n = 3$ of Figure 10). The informational approach for resolving cycles uses the device of picking some element of the cycle to 'start' firing information. We note (as we shall see) that to 'start' with, say a_3 does not imply that we have ultimately that $a_3 =$ in. Let us check what we have in item (1) of Figure 13.

$$\mathbf{f}(a_1) = \{a_3 \multimap \mathbf{e}, (a_3 \multimap \mathbf{e}) \multimap a_1\}$$
$$\mathbf{f}(a_2) = \{a_1 \multimap \mathbf{e}, (a_1 \multimap \mathbf{e}) \multimap a_2\}$$
$$\mathbf{f}(a_3) = \{a_2 \multimap \mathbf{e}, (a_2 \multimap \mathbf{e}) \multimap a_3\}$$

Suppose we allow a_3 to start sending its information.

1. Since $\mathbf{f}(a_3) \vdash a_3$, we send a_3 to node a_1. Now the information at a_1 becomes:

$$\mu(a_1) = \{a_3, a_3 \multimap \mathbf{e}, (a_3 \multimap \mathbf{e}) \multimap a_1\}$$

a_1 no longer can prove anything in linear logic. So it cannot sent anything to node a_2. a_2 is thus free from attack and it has information to sent.
2. Thus $\mu(x_2) = \mathbf{f}(x_2)$ and so $\mathbf{f}(a_2) \vdash a_2$ we get that a_2 is sent to a_3.
3. Node a_3 now has the information

$$\mu(a_3) = \{a_2, a_2 \multimap \mathbf{e}, (a_2 \multimap \mathbf{e}) \multimap a_3\}$$

$\mu(a_3)$ does not prove anything and so the cycle of sent information stops. We end up with only $\mu(a_2) \vdash a_2$ (i.e. $a_2 =$ in) the others, a_1 and a_3 are out.

Example 17. 1. The previous examples discussed single cycles and how to resolve them. We now want to show what to do when we have a network with cycles containing nodes attacking nodes of other cycles. This will prepare us

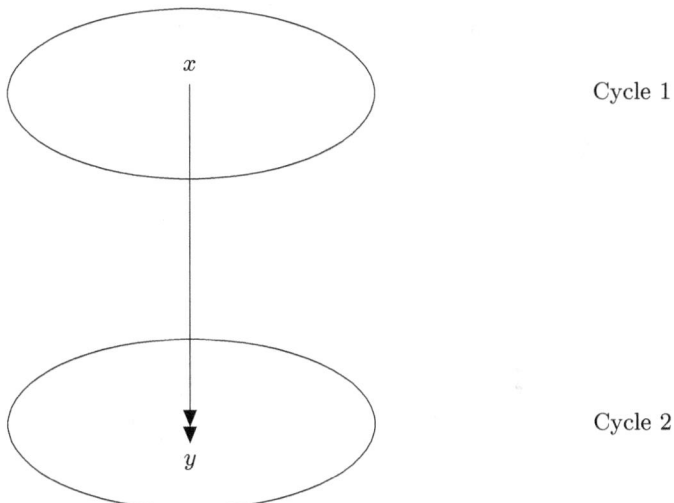

Fig. 14. Cycle 1 containing x which attacks y in Cycle 2

for the next formal section. We first explain the machinery involved. Consider Figure 14. In this figure Cycle 1 attacks Cycle 2 through the point x attacking y. Note that we assume no point in Cycle 2 attacks any point in Cycle 1. Let **f** be the data function for the network. We resolve Cycle 1 first according to the principles mentioned in item (2) of Example 14 using the given function **f**. As a result, node x will either send or not send information to node y in Cycle 2. The new information in node y will be $\mathbf{f}'(y)$. Once we have finished the handling of Cycle 1 and emerge with a new **f**′ for Cycle 2 we begin handling Cycle 2, but with the new information function **f**′.

This is the process we use. We work our way down the acyclic SCC (see [2] and Appendix A) of cycles. We shall define this is Section 4. Meanwhile, let us do a key example from [5]. It is their Example 1 intended to illustrate the connection between Logic Programs and argumentation networks. The Logic Program P has clauses $r_1, \ldots, r_5$ and the arguments are built up from it, as illustrated in Figure 15. To conform with the formal definition of an information network, we shall augment the usual Logic Programming inference engine with the following (u):

(u) If we have a clause $\neg b \wedge A \to x$ and we succeed in deriving b, we say that we also have derived the negation of the clause.

Actually we have derived $\neg x$, but formally it is convenient to say that $\neg(\neg b \wedge A \to x)$. In fact, b actually 'kills' (or 'deletes') the clause!

2. We now continue with the example from [5]. The clauses are $r_1, \ldots, r_5$ below.

$$r_1 : \neg c \to c:$$
$$r_2 : \neg b \to a:$$
$$r_3 : \neg a \to b:$$
$$r_4 : \neg a \wedge \neg c \to c:$$
$$r_5 : \neg g \wedge \neg b \to c:$$

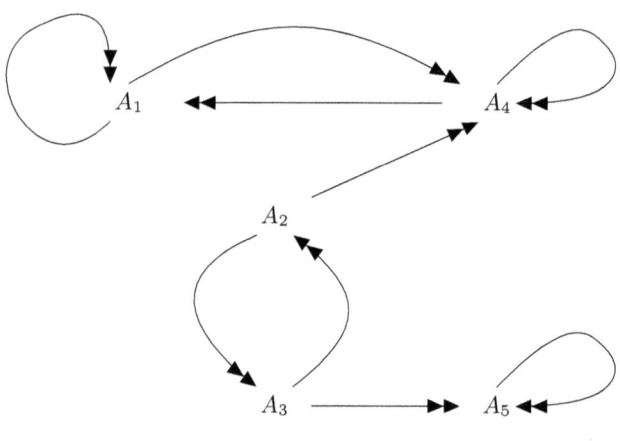

Fig. 15.

The nodes of the argumentation network which is built up from $\{r_1, \ldots, r_5\}$ are denoted by $A_1, \ldots, A_5$ respectively. Note that the attack relation in the network $\{A_1, \ldots, A_5\}$ is determined and is forced on us by the meaning of the clauses. We get Figure 15 which is the same as Figure 1 of [5], the example is attributed in [5] to Wolfgang Dvořák, who later co-authored [6].

Let us apply out information input machinery to this example, what we get, for $i = 1, \ldots 5$ is $\mathbf{f}(A_i) = r_i$. The network of Figure 15 has three cycles: one top cycle $\{A_2, A_3\}$, attacking two bottom cycles $\{A_1, A_4\}$ and $\{A_5\}$. So using our informational procedures as discussed in part (2) of Example 14 we need to select either node A_2 or A_3 (or both) to start sending information.

Case 1 We let node A_2 begin. A_2 will send the literal a to A_3 (which is what can be proved from r_2). Now $r_2 \oplus A$ can derive $\neg B$ and the cycle ends up with A_3 = out, A_2 = in. A_2 also sends the information a to A_4. So we get that the information at A_4 is now $r_4 \oplus a$ and so A_4 can prove $\neg c$, so A_4 is out. Also A_5 is now not being attacked by A_3 (which cannot prove anything) and so it is free to send information to itself, again it can send nothing. So we have so far that A_5 is a node which can prove nothing, in particular neither g nor $\neg g$, but its information sending cycle is finished. So A_5 is undecided. Now, A_4 can prove $\neg c$ but it has already received the information from its attacked A_2. So we now have the cycle of Figure 16, which needs to be resolved.

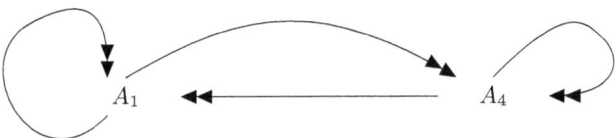

Fig. 16. The information at A_1 is r_1, the information at A_4 is $r_4 \oplus a$

We need to say which node in Figure 16 is to start sending information. In this case it does not matter because A_1 is undecided (it cannot prove c or $\neg c$) and A_4 is out (it proves $\neg c$). This completes case 1.

Case 2 We let A_3 start sending information. The result is that A_2 and A_5 will be out, and A_1 and A_4 will be undecided.

Case 3 Both A_2 and A_3 start firing together. Here we get that the information at A_2 becomes $\{b, \neg b \to a\}$ and at A_3 becomes $\{a, \neg a \to b\}$. Thus A_3 proves $\neg b$ and A_2 proves $\neg a$ and so both out out. We get also in view of that, that A_5, A_1, A_4 are all undecided.

We must compare what we have obtained with the traditional view of this network. This helps us to see the connection between the two. Regarding the network traditionally, we get the following extensions:

E_1 All undecided.
E_2 A_3 = in, A_2 = A_5 = out.
E_3 A_2 = in, A_3 = A_4 = out and A_1 = A_5 = und.

We observe that the family of extensions we obtained for the informational interpretation of Figure 15 is different from what we get for the traditional Dung interpretation (or semantics). For the informational case we did not get the 'all undecided' extension but instead we got the extension $\{A_2 = A_3 =$ out, $A_5 = A_1 = A_4 =$ undecided$\}$. This should not surprise us. The information $\mathbf{f}(x)$ attached to the nodes $x \in \{A_1, \ldots, A_5\}$ is responsible for that. The next Example 18 will give a different $\mathbf{f}$ to the nodes and there will be a complete match.

Example 18. Consider the network of Figure 15. Define a function **g** from nodes to Logic Programs as follows:

$$\mathbf{g}(A_1) = \neg A_1 \wedge \neg A_4 \to A_1$$
$$\mathbf{g}(A_2) = \neg A_3 \to \neg A_2$$
$$\mathbf{g}(A_3) = \neg A_2 \to \neg A_3$$
$$\mathbf{g}(A_4) = \neg A_4 \wedge \neg A_2 \wedge \neg A_1 \to A_4$$
$$\mathbf{g}(A_5) = \neg A_3 \wedge \neg A_5 \to A_5$$

In general given (S, R) the definition of any $\mathbf{g}(x)$ for $x \in S$ is

$$C_x = \neg y_1 \wedge, \ldots, \neg y_k \to x$$

where $\{y_1, \ldots, y_k\}$ are all the attackers y of x (all y s.t. yRx). The set of clauses $\Delta_{S,R} = \{C_x \mid x \in S\}$ satisfies the following properties.

1. For each $x \in S$, there exists a unique clause in $\Delta_{(S,R)}$ with head x.
2. The clauses all have bodies built up from negated literals.

In fact, given a set of clauses Δ satisfying (1) and (2), (S, R) can be defined from Δ by:

(a) $S = \{x \mid x \text{ is a head of a clause in } \Delta\}$
(b) $R = \{(x, y) \mid \neg x \text{ is in the body of the clause with head } y\}$

Let us now calculate the informational extensions corresponding to Figure 15 with **g** above. The top cycle is $\{A_3, A_2\}$ with A_3RA_2 and A_2RA_3. We have three ways of resolving the cycle according to $\mathbb{R}_{\text{info}}$: start with A_2; start with A_3; start with both. Let us adopt a convention that, when resolving a cycle, we only ever start with a single node, call this $\mathbb{R}^1_{\text{info}}$. So following this route we get the extensions

$$A_1 = \text{in}, A_4 = \text{out}, A_1 = A_5 = \text{undecided}$$
$$A_3 = \text{in}, A_5 = A_2 = \text{out}, A_1 = A_4 = \text{undecided}$$

We still do not get the solution where all nodes are undecided. We will address the problem of simulating the traditional approach in Section 4.

4 Formal presentation of general informational networks

In this section we give general definitions for abstract informational networks (with cycles) and also show that such networks can simulate the traditional Dung networks, when we choose some specific Logic Programming information.

To define an abstract information network, we must agree on what we are modelling, i.e. what model of information passing and information flow we have in mind. This will influence the abstract definitions and their semantics. Let us start by looking at Story 3, where we have nodes S and the relation R says that if xRy holds then x passes information to y. We need an agreed view of how this works and then use this view to define the abstract model. The view needs to be general and flexible. Let us start with Figure 1. Think of French Intelligence (node a) passing terrorist threat information to the British (node b) who in turn pass information to the Americans (node c).

1. We assume each node x can collects information independently and let $\mathbf{f}(x)$ be the information x has collected on its own. Now in Figure 1 the nodes a, b, c have independent information $\mathbf{f}(a), \mathbf{f}(b)$ and $\mathbf{f}(c)$ respectively.
2. The node a is an initial node of the network so $\mathbf{f}(a)$ is all the information it has. Denote the total information a node x has after receiving everything from its informants ($\{y \mid yRx\}$) by $\mu(x)$. So for node a we have that $\mathbf{f}(a) = \mu(a)$.
   Node a needs to report/send information to node b. What does it send? Does it send all the information it has or does it process it first (using a process function τ) and then sends say $\tau(\mu(a))$? To be general, let us use τ which may be identity, if we want.
3. Now node b has the information $\mu(b) = \big(\mathbf{f}(b) + \tau(\mu(a))\big)$, and so it will send to node c the information $\tau\big(\mathbf{f}(b) + \tau(\mu(a))\big) = \tau(\mu(b))$.
4. Node c has now the total information $\mu(c) = \mathbf{f}(c) + \tau\big(\mathbf{f}(b) + \tau(\mathbf{f}(a)))\big)$.
5. Suppose we want to ask a query from the network, which node will answer? We would expect perhaps node c, but only if all other nodes passed on all the information they had, only then node c has the full information. Otherwise we need to ask all nodes.
6. We have to decide how we deal with loops and how to answer questions in the general case if nodes withhold information or if there is no single minimal node. Also note that with the three countries, we will still get three different answers to our question. So really we should ask all nodes all our questions.

Our strategy is to start with a general abstract network (S, R), and move to the network $(S^\approx, R^\approx)$ of SCC cycles as defined in [2], see Appendix A. Given (S, R) and a logic with the information function $\mathbf{f}$ from S into wffs of the logic, we use $(S^\approx, R^\approx)$ to propagate the information in three steps.

1. Describe how to propagate the information for a single SCC.
2. Propagate from one SCC to another following the structure of $(S^\approx, R^\approx)$.
3. Use the result of (1) and (2) to define extensions.

The scenario we have in mind is that of Story 3. the security agency. So we assume that we have a non-empty set S of information agents and a relation xRy meaning agent x is contracted to pass information to agent y. We have a logic **L** in which information is expressed and a function $\mathbf{f}(x), x \in S$, describing the initial information available to agent x. $\mathbf{f}(x)$ is a set of wffs of **L**.[8] We assume a

---

[8] We compare the use of the logic **L** with our Definitions 3 and 4. The set **I** of 3 is simply the set of wffs of the logic **L** and the operation $\oplus$ is whatever the logic **L**

dynamic view, where agents y receive information from their contracted agents x_i ($x_i R y, i = 1, 2, \ldots$) and once they receive all the information they pass it on to any z such that yRz. This dynamic process stabilises and we end up with a function $\mu(x)$, giving the final information which each agent possesses at the end of the dynamic process.

In earlier sections we also used a function $\tau(x)$, which describes what information agent x sends to any y such that xRy. The options were whether x sends all the information which he has or only some of it. For the sake of generality we want to assume that any x sends everything to all y such that xRy (i.e. $\tau(x) = \mu(x)$). We adopt this following the advice of items (3) and (4) of Example 7.

We also agree that if we have a cycle then some sort (to be determined) of cooperation should exist between the agents of the cycle. The exact nature of such co-operation is to be formalised after we consider some additional examples.

Let us now start with some technical definitions in the spirit of the above discussion.

Let us now start with some technical definitions in the spirit of the above discussion.

Definition 5. *Let* **L** *be a propositional language for a logic. Assume this language contains connectives to define wffs which include classical conjunction and a negation symbol* ¬. *Also assume that the language has semantics which allows theories* Θ *(i.e. sets of wffs) to have models. In a model any wffs can be true, false or undecided. There are many possibilities for such logics and semantics, such as Logic Programming, Answer Set Semantics, Default Logics, Kraus-Lehman-Magidor semantics and many more. We use the logic to express our informational theories.*

Definition 6. 1. *Let* (S, R) *be a finite network with* $S \neq \emptyset$, $R \subseteq S \times S$. *Assume a logic* **L** *and semantics for it and/or a proof theory system for it. Assume the elements of* S *do not appear in* **L**.
2. *Let* **f** *be a function giving for each* $t \in S$ *a theory* $\mathbf{f}(t) = \Delta_t$ *of* **L**. *Let* $\tau(t)$ *and* $\mu(t)$ *be other theories of* **L** *as defined in the next Definition 7. Let* $\alpha(t)$ *be another functions on* S *giving to each* t *an atom* $\alpha(t)$ *of* **L** *such that* $t \neq s \Rightarrow \alpha(t) \neq \alpha(s)$. *We use the notation* $\tau(t)$ *to denote the information that node* t *transmits and we use* $\mu(t)$ *to denote the final information at node* t *after all information passing is over and stable.* τ *and* μ *are defined in the next Definition 7.*
3. *We consider the system* $(S, R, \mathbf{f}, \tau, \mu, \alpha)$ *as an information system with a test function as follows:*
    (a) (S, R) *is the information flow geometric system.*

---

uses to combine theories, for example set or multi-set union. Further if we let the set Q in Definition 4 to be S itself then the projection π of this definition can be simplified to a projection α as defined below, in Definition 6.

(b) $\mathbf{f}(t), t \in S$, is the information collected by the node t on its own, and $\mu(t)$ is the total information note t has after it received the information submitted to it from parent nodes.

(c) $\tau(t)$ is the information which node t passes to any node s such that tRs holds.

(d) Note that the simplest τ is $\tau = \mathbf{i}$.

(e) $\alpha(t)$ is a projection function for any node t. Its role is to connect the information system with a possible bipolar argumentation network $\mathbb{B}(\alpha)$ based on (S, R) and defined using $\mathbf{f}, \tau, \mu$ and α.

Definition 7. Let $\mathcal{A} = (S, R, \mathbf{f}, \tau, \mu, \alpha)$ be as in Definition 6.

1. Let $\mathbf{L}$ be a logic. Let τ be a universal (transmission) function of the logic $\mathbf{L}$ such that for any theory Δ of $\mathbf{L}$, $\tau(\Delta)$ is a (possibly empty) subset of all $\mathbf{L}$-consequences of Δ (i.e. every element of $\tau(\Delta)$ is a logical consequence of Δ according to $\mathbf{L}$).
   For each $t \in S$, we may have a different universal transmission function τ_t, which node t uses to decide what to transmit to other nodes.
2. Let $\mu(t)$, for $t \in S$ be a function defined implicitly as a solution (if (S, R) is acyclic then a unique solution exists, otherwise we choose one arbitrarily if there is one at all) to the equation $\mu(t) = \mathbf{f}(t) \cup \bigcup_{sRt} \tau(\mu(s))$.
3. Let $\mathbf{m}$ be a function associating with each $t \in S$ a model $\mathbf{m}_t$ of the theory $\mu(t)$. We require the restriction on the models that if tRs holds and $\mu(t) \subset \mu(s)$ then $\mathbf{m}_t$ is a restriction of $\mathbf{m}_s$ to the language of $\mu(t)$.
4. Let $E_\mathbf{m} = \{t \mid \alpha(t) \text{ holds in } \mathbf{m}(t) \text{ or equivalently } \mu(t) \text{ proves } \alpha(t)\}$.

We say that $E_\mathbf{m}$ is an informational extension of $\mathcal{A}$. The meaning of $E_\mathbf{m}$ depends on the underlying logic $\mathbf{L}$ and its models $\mathbf{m}$. It may not be a Dung complete extension of (S, R).

Remark 6. 1. Note that Definition 7 is really very general. A more practical definition is to assume that all parties pass all the information they have along the arrow and that extensions are obtained from models of the union of all information.
   So in this case given (S, R), let R^* be the transitive and reflexive closure of R, and we have that $\mu(t) = \bigcup_{sR*t} \mathbf{f}(s)$
2. Note that the informational system concept is much more general than the traditional [9] Dung argumentation concept and its notion of complete extensions. In the literature there are papers describing networks with nodes sending information, papers such as [23, 25], but not under the context of argumentation and attack. We further stress that α is not a central informational concept and is used only to be able to define technically "informational extensions" and use them to show that informational attacks and support can simulate bipolar argumentation.

The following Definition 8 and Theorem 1 show that traditional argumentation machinery can be simulated by informational networks.

Definition 8. Let (S, R) be a finite network, we now define a very specific associated informational network. The Language **L** is the logic programming language based on the atoms S. The semantics is the Answer Set Semantics (see [22]).

1. For each $x \in S$ such that $\{y_1, \ldots, y_n\}$ are all the attackers of x (i.e. $\{y_1, \ldots, y_n\} = \{y \mid yRx\}$), let $C_x = \neg y_1 \wedge \ldots \neg y_n \to x$. Note that the empty conjunction is considered as $\top$, so if x is not attacked by any other node then $C_x = x$. Let $\mathbf{f}(x) = \tau(x) = \{C_x\}$. Let $\alpha(x) = x$. Thus **f** translates from S into the language **LP** of Logic Programs (Recall that in this language, "$\neg$" is negation as failure, see Example 7).[9]
2. Define Δ to be $\{C_x \mid x \in S\}$

Theorem 1. Let (S, R) be a finite network then the informational network of Definition 8 can simulate the traditional complete extensions for (S, R).

Proof. A very long proof is in Appendix B. The proof of the current theorem follows from Theorem 3. We show that the informational extensions, which are in this case the same as all the answer set Programming with loop (ASPL) models of Δ of item 2 of Definition 8 are the same as all the Dung extensions of (S, R).

Remark 7. Looking more closely at the proof 1 we see that this machinery can also deal with (or simulate) bipolar networks with attack and support, see the Appendix C where we discuss support. See also [7, 12, 1].[10]

5 Conclusion

We have shown that information input networks exist in practical argumentation (see for example [20]) and can simulate traditional attack or support or explanation, all depending on the nature of the information. We need to study

---

[9] Note that this translation has two properties.
(a) Each $x \in S$ has exactly one clause of which it is the head.
(b) The elements in the body of each clause are all negated atoms.

[10] Appendix C contains a general discussion of support. This is not central to Information Networks. We are talking about bipolar networks (S, R'', R'), with R'' attack and R' support, where we understand support to behave according to the rules:

1. x is in, if all of its attackers are out and all of its supporters are in.
2. x is out, if one attacker is in or one supporter is out.
3. Otherwise x is und.

This view lumps an argument as a conjunction with all of its supporters and their supporters, and their supporters' supporters etc., all into one big set; and if any one of these supporters is "out" then the original argument is "out".

carefully and extensively how information input is used in real debate and arguments. This is the real challenge here. On the technical side we can compare with logic programming, ASPIC and Assumption Based Argumentation (ABA), but this is to be done in subsequent papers.

A Definition of Strongly Connected Components

In this appendix we define the notion of an *SCC*, taken from [2].

Definition 9. *Let (S, R) be an argumentation network with $S \neq \emptyset$ and $R \subseteq S \times S$.*

1. *We say that a sequence $(x_1, \ldots, x_n)$ of elements of S is a cycle of length n if we have*
$$x_1 R x_2, x_2 R x_3, \ldots, x_{n-1} R x_n, x_n R x_1.$$
2. *Define a relation $x \approx y$ on S by setting $x \approx y$ iff $x = y$ or x and y share a cycle. That is, $x \approx y$ iff there is a cycle $(z_1, \ldots, z_n)$ of length n such that $x = z_i$ and $y = z_j$ for some $1 \leq i, j \leq n$.*
3. *Since $\approx$ is an equivalence relation on S (see [2]) let*
   - *$x^\approx = \{y \mid x \approx y\}$, for any $x \in S$.*
   - *$S^\approx = \{x^\approx \mid x \in S\}$.*
   - *$R^\approx = \{(x^\approx, y^\approx) \mid a R b \text{ for some } a \in x^\approx, b \in y^\approx\}$*

   $R^\approx$ is well defined and is an antisymmetric relation (see [2]).
4. *Let $<$ be the transitive-reflexive closure of $R^\approx$. That is, $x^\approx < y^\approx$ iff either $x^\approx = y^\approx$ or for some $z_1^\approx \ldots z_k^\approx$, $k \geq 1$ we have $x^\approx R^\approx z_1^\approx, z_1^\approx R^\approx z_2^\approx, \ldots, z_{k-1}^\approx R^\approx z_k^\approx$ and $z_k^\approx = y^\approx$.*

Then $<$ is a partial ordering on $S^\approx$.

B Goal directed Logic Programming computation

This appendix shows a connection between traditional Dung argumentation and some specific information networks based on Logic Programming.

Consider a logic programming language with the connectives $\neg$ for negation as failure, $\to$ for implication and $\land$ for conjunction. We need the following concepts.

Definition 10. 1. *Let S be a finite set of propositional atoms $\{x_1, x_2 \ldots\}$, or literals.*
2. *By a clause C_q we mean an expression of the form*
$$C_q : \quad \left(\bigwedge \{\neg x_i \mid i \in I\} \land \bigwedge \{y_j \mid j \in J\}\right) \to q$$
where I and J are possibly empty index sets and $\{x_i \mid i \in I\} \cup \{y_j \mid i \in J\} \cup \{q\} \subseteq S$.

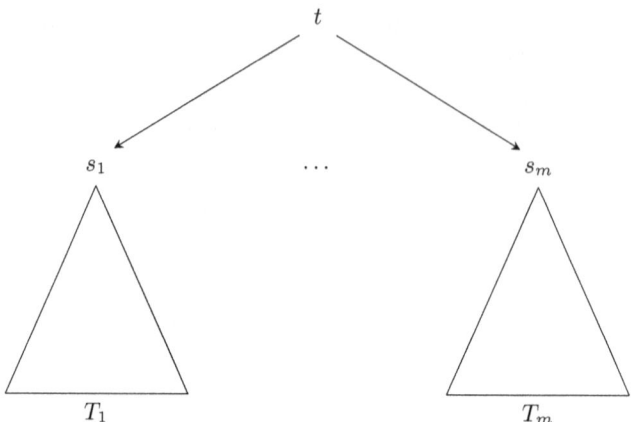

Fig. 17. Grafting trees

3. We say that q is the head of C_q and $\bigwedge_{i \in I} \neg x_i \wedge \bigwedge_{j \in J} y_j$ is the body of C_q. We say the x_i occur negatively and the y_j occur positively in the body of C_q.
4. A database Δ is a set of clauses. We say Δ is based on a set of literals S if every clause in Δ contains only literals in S.
5. We also define a relation R_Δ between literals. We say $uR_\Delta z$ if for some clause C_z in Δ, u appears in the body of C (where z is the head of C_z).
6. A query Q has the form $?q = 0$ or $?q = 1$ where $q \in S$.
7. A history H is sequence $(Q_1, \ldots, Q_n)$ of distinct queries.
8. A decoration D has the form (Q, H, ζ) where Q is a query, H is a history and ζ is one of the following expressions: "success", "failure", "loop". We also sometimes refer to ζ as decorating (Q, H).
9. Say that a decoration is positive if it is either of the form $(?q = 1, H, success)$ or of the form $(?q = 0, H, fail)$.
   Say that a decoration is negative if it is either of the form $(?q = 0, H, success)$ or of the form $(?q = 1, H, fail)$.
10. Say that a decoration is 1-positive if it is of the form $(?q = 1, H, success)$
    Say that a decoration is 1-negative if it is of the form $(?q = 1, H, fail)$
    Say that a decoration is 1-looping if it is of the form $(?q = 1, H, loop)$
11. A tree $\boldsymbol{T}$ has the form $(T, >, t)$, where T is finite and $>$ is a partial ordering of T with $t \in T$ as the head element (that is, there is no $s \in T$ such that $s > t$ and for every $s \in T$ $t > s$).
    If $t > s$ and there is no r s.t. $t > r > s$ then call s and immediate (downwards) successor of t and write $t >_0 s$.
12. If the trees $(T_i, >_i, s_i)$, for $i \in I = \{1, \ldots, m\}$ are distinct and disjoint trees and $t \notin T_i$ for any i, we can graft the trees (see in Figure 17). We set $T = \bigcup_i T_i \cup \{t\}$ and $>= \bigcup_{i \in I} >_i \cup \{(t, s) | s \in T_i \text{ for all } i \in I\}$ and our grafted tree is $(T, >, t)$.

13. Given a tree $(T,>,t)$ with the $(T_i,>_i,s_i)$ as in Figure 17, we also view $(T_i,>_i,s_i)$ as the tree obtained from $(T,>,t)$ by truncating at s_i. We may also denote each $(T_i,>_i,s_i)$ as $(T\upharpoonright s_i,>\upharpoonright s_i,s_i)$.
   Thus we may think of $(T,>,t)$ as being grafted from the $(T_i,>_i,s_i)$ (and t), or we may view each $(T_i,>_i,s_i)$ as having been obtained by truncating $(T,>,t)$.
14. A tree with decoration has the form $(T,>,t,\delta)$ where $\delta: T \mapsto \mathbf{D}$ where $\mathbf{D}$ is a set of decorations D as defined in item 8 above. For $s \in T$ we write $s:(Q,H,\zeta)$ to mean that $\delta(s) = (Q,H,\zeta)$.

Definition 11. *Let Δ be a finite database based on S. Assume for simplicity that for each $p \in S$ there is at most one clause $C_p \in \Delta$.*
   Suppose $\mathbf{T} = (T,>,t,\delta)$ is a decorated tree and that $\delta(t) = (Q,H,\zeta)$, where Q is $?q = 1$ or $?q = 0$. Then, by induction on $(T,>)$, we define $\mathbf{T}$ to be legitimate as follows.

Base Case Suppose T is a single node, i.e. $T = \{t\}$. Then $\mathbf{T}$ is legitimate if either $q \in \Delta$ and $\delta(t)$ is positive; or there is no $C_q \in \Delta$ and $\delta(t)$ is negative; or $Q \in H$ and ζ = loop.

Induction Case Suppose T contains many nodes. Then t has some immediate downwards successors. Then $\mathbf{T}$ is legitimate when:

1. $(T\upharpoonright s,>\upharpoonright s,s,\delta)$ is legitimate for all s such that $t >_0 s$.
2. $\delta(s)$ is either 1-positive, 1-negative or 1-looping for all s such that $t >_0 s$.
3. There is a $C_q \in \Delta$, which by Definition 10, must have the form

$$\left(\bigwedge\{\neg x_i \mid i \in I\} \wedge \bigwedge\{y_j \mid j \in J\}\right) \to q.$$

4. There is a bijection $f: \{\neg x_i \mid i \in I\} \cup \{y_j \mid j \in J\} \longmapsto \{s \mid t >_0 s\}$.
5. Q is not in H and each s such that $t >_0 s$, the history $H_s \in \delta(s)$ is the sequence (Q,H).[11]
6. Either (a) or (b) holds:
   (a) $\zeta \neq$ loop and either:
      i. $\delta(t)$ is positive and $\delta(f(x_i))$ is 1-negative and $\delta(f(y_j))$ is 1-positive for each $i \in I, j \in J$.
      ii. $\delta(t)$ is negative and either $\delta(f(x_i))$ is 1-positive for some $i \in I$ or $\delta(f(y_j))$ 1-negative for some $j \in I$.
   (b) ζ = loop and both:
      i. $\delta(f(x_i))$ is not 1-positive and $\delta(f(y_j))$ is not 1-negative for each $i \in I, j \in J$.
      ii. there is an $k \in I$ and $l \in J$ such that either $\delta(f(x_k))$ or $\delta(f(y_l))$ is 1-looping.

---

[11] That is, if $H = (Q_1,\ldots,Q_n)$ then $H_s = (Q, Q_1, \ldots, Q_n)$. Remember that a history must be a non repeating sequence.

Lemma 1.

1. Let $\mathbf{T}$ be a decorated tree with top decoration $(?q = n, H, success)$, and let $\mathbf{T'}$ be an otherwise identical decorated tree except its top node is decorated by $(?q = 1-n, H, fail)$ (where $n \in \{0,1\}$. Then, for any Δ and q, $\mathbf{T}$ is legitimate iff $\mathbf{T'}$ is legitimate.
2. Let $\mathbf{T}$ be a decorated tree with top decoration $(?q = n, H, loop)$, and let $\mathbf{T'}$ be an otherwise identical decorated tree except its top node is decorated by $(?q = 1-n, H, loop)$ (where $n \in \{0,1\}$. Then, for any Δ and q, $\mathbf{T}$ is legitimate iff $\mathbf{T'}$ is legitimate.

Proof. Both parts follow from the inductive definition of 11. For the first part, depending on n, the decorations at the top nodes of $\mathbf{T}$ and $\mathbf{T'}$ are either both positive or both negative (Definition 10) and so are subject to the same clause of 11. The second part is equally straightforward.

Lemma 2. *For each S, Δ, each query Q and each history H there exists a legitimate tree $\mathbf{T'} = (T', >', t', \delta')$ such that $\delta'(t') = (Q, H, \zeta')$. Moreover, $\mathbf{T'}$ is unique in the sense that for any legitimate $\mathbf{T} = (T, >, t, \delta)$ where $\delta(t) = (Q, H, \zeta)$*

1. $(T, >, t)$ is isomorphic to $(T', >', t')$.
2. $\zeta = \zeta'$

Proof. First notice that since S is finite there are only a finite number of histories based on S. This is because a history is a non-repeating sequence of queries and a query is an element of $S \times \{0,1\}$. If $|S|$ is the cardinality of S, then l_H, the length of H, is no longer than $2 \cdot |S|$. We argue by induction on $n = 2 \cdot |S| - l_H$.

Base case Suppose that $n = 0$, i.e. $l_H = 2 \cdot |S|$.

Then H is of maximal length and so must contain Q. Now from Definition 11 the only legitimate trees $(T, >, t, \delta)$ for which $\delta(t) = (Q, H, \zeta)$ are single node trees $(\{t\}, \emptyset, t, \delta)$ where $\delta(t) = (Q, H, loop)$. Clearly both conditions of the lemma must hold.

Induction case Suppose $n > 0$ and let Q be $?q = n$ for $n \in \{1, 0\}$.

If $Q \in H$ and either $q \in \Delta$ or there is no $C_q \in \Delta$, then as with the base case, the only legitimate trees meeting the condition of the lemma are single node trees, and the result follows as above. So suppose $Q \notin H$, $q \notin \Delta$ and there is a $C_q \in \Delta$. Then C_q must have the form:

$$\left(\bigwedge\{\neg x_i \mid i \in I\} \wedge \bigwedge\{y_j \mid j \in J\}\right) \to q.$$

where $I \cup J \neq \emptyset$. Then by the induction hypothesis, for each $i \in I$ and $j \in J$, there are legitimate trees $\mathbf{T}_i = (T_i, >_i, t_i, \delta_i)$ and $\mathbf{T}_j = (T_j, >_j, t_j, \delta_j)$ such that $\delta_i(t_i) = (?x_i = 1, (Q, H), \zeta_i)$ and $\delta_j(t_j) = (?x_i = 1, (Q, H), \zeta_j)$; and all these trees are unique in the sense of the lemma (without loss of generality, we may also assume that the T_i are disjoint).

Now (looking to Figure 17) pick a new object t' and set $\mathbf{T}' = (T', >', t', \delta')$ where

$T' = \bigcup_{k \in I \cup J} T_k$
$>' = \bigcup_{k \in I \cup J} >_k \cup \{(t', s) \mid s \in T_k \text{ for some } k \in I \cup J\}$
$\delta' = \bigcup_{k \in I \cup J} \delta_k \cup \{(t', (Q, H, \zeta'))\}$

and where ζ' is the unique choice (given by the inductive clause of Definition 11) that makes $\mathbf{T}'$ a legitimate tree. The uniqueness of $\mathbf{T}'$ now follows from the uniqueness of the T_i and T_j and the fact that ζ' is determined by Definition 11 given the ζ_i the ζ_j and the construction of $\mathbf{T}'$.

Definition 12. Let (S, R'', R') be an argumentation network with $R'' \subset S \times S$ is the attack relation and $R' \subset S \times S$ is the support relation and R'' and R' are disjoint. Let $\Delta = \{C_x \mid x \in S\}$, where $C_x = \bigwedge_{y R'' x} \neg y \wedge \bigwedge_{z R' x} z \to x$. Let $E \subseteq S$ and let $\Delta_E = \{C_x \mid \text{no } e \in E \text{ appears negatively in the body of } C_x\}$.

Consider the query $?q = 1$ for $q \in S$ and the history $\varnothing$. Then by Lemma 2, for $\Delta = \Delta_E$, there exists a unique (in the sense of the lemma) legitimate tree with top decoration $(?q = 1, \varnothing, \zeta)$.

Let $E' = \{q \mid \text{there is a legitimate tree with top decoration } (?q = 1, \varnothing, \text{success})\}$. If $E' = E$ we say that E is an answer set with loops (ASPL) for Δ.

Lemma 3. Let (S, R), Δ and Δ_E be as in Definition 12.

1. Assume E is ASPL for Δ. For any q, let $\mathbf{T}_q$ be the unique tree (by Lemma 2) for the query $?q = 1$ and history $\varnothing$.
   Now define λ_E so that for any q:
   (a) $\lambda_E(q) = \text{in}$ if $\mathbf{T}_q$ is decorated by $(?q = 1, \varnothing, \text{success})$ at its top node.[12]
   (b) $\lambda_E(q) = \text{out}$ if $\mathbf{T}_q$ is decorated by $(?q = 1, \varnothing, \text{fail})$ at its top node.
   (c) $\lambda_E(q) = \text{und}$ if $\mathbf{T}_q$ is decorated by $(?q = 1, \varnothing, \text{loop})$ at its top node.
   Then $\lambda_E(q)$ is a legitimate Caminada labelling on (S, R).
2. Let λ be a legitimate Caminada labelling for (S, R). Let E_λ be the set of all points x in S such that $\lambda(x) = 1$. Then E_λ is ASPL for Δ.

Proof. 1. Assume E is ASPL for Δ. We show that λ_E is a Caminada labelling. Consider the basic configuration in (S, R) for a node q where $s_1, \ldots, s_m$ are all of its attackers. The clause C_q is

$$C_q: \quad \bigwedge_i \neg s_i \to q.$$

(a) First, note that since E is the set of all nodes q such that the top node of the tree for $?q = 1, \varnothing$ obtained by Lemma 2 is decorated with $(?q = 1, \varnothing, \text{success})$, then all nodes in E are "in".

---
[12] Given Lemma 1 this is equivalent to the condition that $\mathbf{T}$ is decorated by $(?q = 1, \varnothing, \text{fail})$ at its top node.

(b) Let q be a node such that for some $x \in E$, xRq. Then $\neg x$ appears in the body of C_q, and so the clause C_q is not in Δ_E. Therefore $\mathbf{T}_q$ of the lemma has decoration ($?q = 1, \emptyset, \text{fail}$), since there is no clause in Δ_E with head q. So $\lambda_E(q)$ is out.[13]

(c) If all the attackers of q are "out", then the tree $\mathbf{T}_q$ must be decorated by ($?q = 1, \emptyset, \text{success}$) at its top node. This is because all the truncations $\mathbf{T}_q \upharpoonright s_i$ are decorated by ($?s_i = 1, \emptyset, \text{fail}$) at their top nodes, and so $\lambda_E(q) = \text{in}$.

(d) Assume q is such that its tree $\mathbf{T}_q$ is decorated by ($?q = 1, \emptyset, \text{loop}$) at its top node. Then none of its immediate truncated sub-trees is decorated by ($?s_i = 1, (?q = 1), \text{success}$) for each s_i. The immediate truncated sub-trees are all decorated either by ($?s_i = 1, (?q = 1), \text{fail}$) for some s_i or by ($?s_i = 1, (?q = 1), \text{loop}$) for some s_i; and at least one is decorated with ($?s_i = 1, (?q = 1), \text{loop}$) for some s_i.

Let us now examine each case (for arbitrary i):

i. Suppose an immediate sub-tree T (of T_q) is decorated by ($?s_i = 1, (?q = 1), \text{fail}$) at its top node. But then by Lemma 2 this tree is unique in the sense that there is no legitimate tree decorated by (($?s_i = 1, (?q = 1), \text{success}$) at its top node.
Thus by case (a) in the definition of λ_E, $\lambda_E(s_i) \neq \text{in}$.

ii. Suppose an immediate subtree T (of T_q) is decorated by ($?s_i = 1, (?q = 1), \text{loop}$) at its top node. We claim that we must also have a legitimate tree for ($?s_i = 1, \emptyset, \text{loop}$), i.e. T_{s_i} loops. Otherwise, T_{s_i} is decorated either by ($?s_i = 1, \emptyset, \text{fail}$) or by ($?s_i = 1, \emptyset, \text{success}$) at its top node. Now since T is decorated by ($?s_i = 1, (?q = 1), \text{loop}$), and since the only difference between T_{s_i} and T is the addition of $?q = 1$ to its history, it follows that there is a subtree T' of T_{s_i} which is decorated by ($?q = 1, H', \zeta'$) for some H' and ζ'. Furthermore $\zeta' = \text{success}$ or $\zeta' = \text{fail}$, for otherwise T_{s_i} would be decorated with loop. But this is contrary to our initial assumption of this case (d), that the tree $\mathbf{T}_q$ is decorated by ($?q = 1, \emptyset, \text{loop}$) (as if the query $?q = 1$ loops with history H, then it will loop with history $H' \supseteq H$).

Either way, we get that condition (C3) for Caminada labelling holds.

2. Let λ be a Caminada labelling for (S, R), we show that E_λ is *ASPL* for Δ. Consider $\Delta_\lambda = \Delta_{E_\lambda}$, we prove for any q that:

(a) If $\lambda(q) = \text{in}$, then there is a (unique) legitimate tree for $?q = 1$ and $\emptyset$ and its top node is decorated by ($?q = 1, \emptyset, \text{success}$).[14]

(b) If $\lambda(q) = \text{out}$, then for any history H such that $(?q = 1) \notin H$ there is a (unique) legitimate tree for $?q = 1$ and H and its top node is decorated by ($?q = 1, \emptyset, \text{failure}$).[15]

---

[13] In fact, the tree for any query of the form ($?q = 1, H, \text{fail}$) (i.e. with that decoration at its top node) will be legitimate as long as $?q = 1 \notin H$

[14] Equivalently, ($?q = 0, \emptyset, \text{fail}$) decorates the tree for $?q = 0$ and $\emptyset$.

[15] Equivalently, using Lemma 1, ($?q = 0, \emptyset, \text{success}$) decorates the tree for $?q = 0$ and H, for ($?q = 0$) $\notin H$.

(c) If $\lambda(q) = $ und, then for any history H there is a (unique) legitimate tree for $?q = 1$ and H and its top node is decorated by $(?q = 1, \emptyset, \text{loop})$.[16]

We prove the above in turn.

(a) Assume $\lambda(q) = $ in. Then all attackers y of q are out. We construct a legitimate tree for Δ_E, with its top node decorated by $(?q = 1, \emptyset, \text{success})$. First we observe that since all attackers of q are out we have that C_q is in Δ_E.

If $C_q = q$, then there is a one point legitimate tree decorated by $(?q = 1, \emptyset, \text{success})$.

If C_q has a body $\wedge \neg y_i$, where, for all i, $q \neq y_i$ and all y_i are out. Therefore, by next item (b) below, each of the y_i has a one point legitimate tree decorated by $(?y_i = 1, (?q = 1), \text{fail})$. We can graft all these one point trees together into a tree for $(?q = 1, \emptyset, \text{success})$.

(b) Assume $\lambda(q) = $ out. Then for some $x \in E$ we have xRq. This means that $\neg x$ appears in C_q and so $C_q \notin \Delta_E$. Therefore, since $?q = 1 \notin H$ the legitimate tree for $?q = 1$ and H is decorated by $(?q = 1, H, \text{failure})$ at its top node.

(c) Assume $\lambda(q) = $ und. Consider all the attackers of q. Since $\lambda(q) = $ und, we must have that none of the attackers is "in" and at least one is "und". For any q let $O_q = \{x \in S \mid xRq \text{ and } \lambda(x) = \text{out}\}$ and let $U_q = \{x \in S \mid xRq \text{ and } \lambda(x) = \text{und}\}$. If $\lambda(q) = $ und then U_q must be nonempty, but O_q could be empty.

We must now construct a legitimate tree $\mathbf{T} = (T, >, t, \delta)$ such that $\delta(t) = (?q = 1, \emptyset, \text{loop})$. The nodes of $\mathbf{T}$ will be sequences of queries on elements from S ordered by length.

Stage 0: set $t_0 = \delta_0 = \emptyset$

Stage 1: Set $T_1 = \{?q = 1\}$ and set $\delta_1(?q = 1) = (?q = 1, t_0, \text{loop})$.

Stage n+1 ($n \geq 1$) Take all points $t \in T_n$ such that $\delta_n(t_n) = (?p = 1, t_{n-1}, \text{loop})$ for some p.
- For each $x \in O_p$ create the node $s_x = (t_n, ?x = 1)$ and set $\delta_{n+1}(s_x) = (?x = 1, t_n, \text{fail})$.
- For each $x \in U_p$ such that $?x = 1 \notin H$ create the node $r_x = (t_n, ?x = 1)$ and set $\delta_{n+1}(r_x) = (?x = 1, t_n, \text{loop})$.

Set $T_{n+1} = T_n \cup \{s_x \mid x \in O_p\} \cup \{r_x \mid x \in O_p\}$.

Now let $T = \bigcup_{n>0} T_n$ and $\delta = \bigcup_n \delta_n$. Finally let $x > y$, for $x, y \in T$, when x is a subsequence of y.

The top node of of $\mathbf{T}$ is clearly decorated by $\delta(t) = (?q = 1, \emptyset, \text{loop})$. So we must verify, by induction on n that $\mathbf{T}$ is a legitimate tree.

First note that there is a k such that $T \subseteq \bigcup_{m<k} T_m$. For otherwise there is an infinite sequence $t_1 > t_2 \ldots$ where each $t_i \in T_i$ and $\delta_i(t_{i+1}) = (?p_{i+1} = 1, t_i, \text{loop})$ for some p_i. But since S is finite there are only

---

[16] Equivalently, $(?q = 0, \emptyset, \text{loop})$ decorates the tree for $?q = 0$ and H.

finitely many p_i, and so for some i, $p_{i+1} \in t_i$, and in such a case there is no new node constructed from t_i at stage T_{i+1}.

Now we can argue by induction on the subtrees (from the endpoints up) of **T** that **T** is legitimate.

Base Case: Consider an endpoint s in **T**. Then $s = t_{n+1} \in T_{n+1}$ for some n and no supersequence of queries is added to T_{n+2}. Then $\delta(t_{n+1}) = (?x = 1, t_n, \text{fail})$ or $\delta(t_{n+1}) = (?x = 1, t_n, \text{loop})$, in the former case $\mathbf{T} \upharpoonright t_(n+1)$ is a legitimate tree by 2b, and in the latter case $?x = 1 \in t_n$ – for otherwise, since U_x is nonempty, a supersequence of t_{n+1} would be added to t_{n+2} and t_{n+1} would not be an endpoint – but then again $\mathbf{T} \upharpoonright t_(n+1)$ is a legitimate tree.

Induction Case: Let $s_1 \ldots s_m$ be the immediate successors of $t \in \mathbf{T}$, and suppose by the induction hypothesis that the $\mathbf{T} \upharpoonright s_i$ are legitimate trees. Then $\mathbf{T} \upharpoonright t$ is legitimate if it meets the condition of Definition 11. Now $t \in T_n$ for and the $s_i \in T_{n+1}$ for some $n > 0$, and examining the inductive case of the construction of **T** above, we see it matches clause 6b of Definition 11. We conclude that $\mathbf{T} \upharpoonright t$ is legitimate.

This completes the induction and the proof.

Remark 8. We need to show that our results in this Appendix conform with the requirements of Definition 7. For this purpose we take a look at Definition 12.

Given a geometric network (S, R) let R^* be the reflexive and transitive closure of R. For each x in S let S_x be the set $\{y \mid yR^*x\}$ and let R_x be the restriction of R to S_x. Assume that xR^*y holds. We claim that all the answer set models of the theory $\Delta_x = \{C_z \mid z \in S_x\}$ are exactly all the restrictions to S_x of of all the answer set models of the theory $\Delta_y = C_z \mid z \in S_y$.

C Bipolar networks with attack and support

We begin this Appendix by stating that we need an approach to Bipolar network compatible with information input. In fact we already have an operational view, the view dictated by the translation into Logic Programming clauses, which we can do in a similar way to Definition 8. This is Definition 13 below. We just need to understand what this definition does in terms of Argumentation way of thinking and proceed from there. So let us first give the definition.

Definition 13. *Let (S, R'', R') be a finite bipolar network with attack relation R'' and a disjoint support relation R'. We now define a very specific associated informational network, and use the informational semantics for this associated network to endow argumentation semantics to the bipolar network. The Language L is the logic programming language based on the atoms S. The semantics is the Answer Set Semantics (see [22] and Appendix B).*

1. For each $x \in S$ let $\{y_1, \ldots, y_n\}$ be all the attackers of x (i.e. $\{y_1, \ldots, y_n\} = \{y \in S \mid yR''x\}$) and let $\{z_1, \ldots, z_k\}$ be all the (R') supporters of x.
   Let $C_x = (\neg y_1 \wedge \cdots \wedge \neg y_n \wedge z_1 \wedge \cdots \wedge z_k) \to x$.
   Note that the empty conjunction is considered as ⊤. So if x is neither attacked nor supported by any other node then $C_x = x$.
   Let $\mathbf{f}(x) = \tau(x) = \{C_x\}$. Let $\alpha(x) = x$.
   Thus **f** translates from S into the language LP of Logic Programs (Recall that in this language, "¬" is negation as failure, see Example 7).
2. Define Δ to be $\{C_x \mid x \in S\}$

Given a finite Bipolar network (S, R'', R'), we want to define for it our own view of semantics. We consider the theory Δ of Definition 13, based on the atoms of S and proceed to build the annotated computation trees based on Δ and each $x \in S$ as done in Appendix B, and define the answer sets with loop for S as in Definition 12. This defines for us semantics for the Bipolar network (S, R'', R').

Remark 9. To avoid confusion we shall write in this remark ↪ for the arrow of a logic program clause and → for the arrow of argumentation support and ↠ for argumentation attack.

Note that Δ of Definition 13 is a Logic programming theory with a special structure. Every atom x appearing in S has exactly one clause with head x. Thus the logic program $\{a \hookrightarrow b\}$ does not have the right form. It is not the Δ theory of the the bipolar network $(\{a, b\}, \emptyset, \{(a, b)\})$, (i.e. a network with nodes a, b in which a supports b). The translation of this bipolar network is the logic program $\{a, a \hookrightarrow b\}$. If we want to be able to obtain the logic program $\{a \hookrightarrow b\}$ as the translation of a bipolar network we must allow for ⊤ to be an argument and take the network ⊤ ↠ $a \to b$.

Our translation gives an object level approach of connecting Logic Programming and Abstract Argumentation. The reader interested in this area can also consult [21, 6, 5].

We need not formulate the analog of Lemma 3 for our case for two reasons:

1. There is no agreed semantics for Bipolar networks with respect to which we can formulate a lemma.
2. Our business in this paper is to study informational networks and not necessarily join the debate about Bipolar Networks

It is therefore sufficient for us to discuss properties and do comparisons using examples to give the reader a feel of what our semantic options are. Perhaps we pursue the Logic Programming Bipolar view in a subsequent paper.

To understand the way we view bipolar networks (S, R'', R') with attack R'' denoted by ↠ and support R' denoted by →, we use Figure 18.

We use the network of this figure to explain the view which is compatible with the answer set loop semantics of Logic Programming.

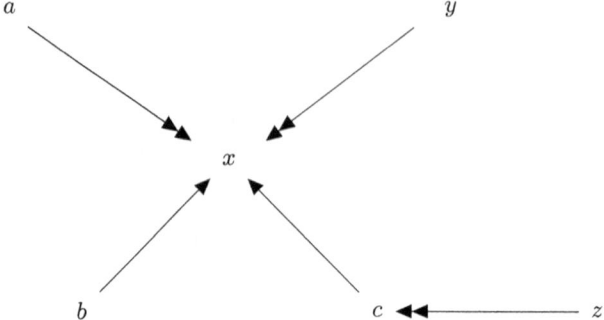

Fig. 18. We do not show what other nodes attack a, b, y, z, but the nodes x and c are attacked exactly by what is shown

The node x is attacked by a and y and is supported by b and c. c is attacked by z. The figure may be a local part of a larger network. So we do not show what other nodes attack a, b, y, z, but the nodes x and c are attacked exactly by what is shown in the figure.

The logic Programming clauses for x and c as heads are therefore fully known. These are:

1. $\neg a \wedge \neg y \wedge b \wedge c \hookrightarrow x$
2. $\neg z \hookrightarrow c$
3. Body1(not shown) $\hookrightarrow a$
4. Body2(not shown) $\hookrightarrow y$
5. Body3(not shown) $\hookrightarrow b$

In our paper [12] we presented several options for understanding and re-writing the attacks in such a figure so as to define the extensions.

Option 1 Add attacks from attackers of x to its supporters by closing the network under the rule:

$$\alpha \twoheadrightarrow \beta \,\&\, \gamma \to \beta \quad \Rightarrow \quad \alpha \twoheadrightarrow \gamma$$

So since $a \twoheadrightarrow x$ we would add $a \twoheadrightarrow b$ and $a \twoheadrightarrow c$.

Option 2 Add attacks from attackers of supporters of x (e.g. c in the figure) to x by closing the network under the rule:

$$\alpha \twoheadrightarrow \beta \,\&\, \beta \to \gamma \quad \Rightarrow \quad \alpha \twoheadrightarrow \gamma$$

Now consider option 2. This can read: x is "out" if either a is "in" or y is "in" or b is "out" or c is "out". This is in confirmation of the clause for x. Compare with footnote 10.

In fact if we follow the usual logic programming computation, we get from clauses 1 and 2 the clause below, which confirms Option 2:

6. $(\neg a \wedge \neg y \wedge b \wedge \neg z) \hookrightarrow x$

Remark 10. The approach of [12] to support is a variation on Option 2. Consider the network with just $\{a\}$, and no attacks, i.e the network $A_1 = (\{a\}, \varnothing, \varnothing)$. The complete extension for this network is $a =$ "in". It is reasonable, when support is available, to have that a supports itself, i.e. $a \to a$. So we expect that the network A_1 is essentially the same as the network $A_2 = (\{a\}, \varnothing, \{(a, a)\})$. The Logic Programming approach/translation, will have a loop for a and therefore will not yield $a =$ "in" but $a =$ "loop/undecided". Similarly if we have the mutual support group $\{a, b\}$ with $a \to b$ and $b \to a$, we get value "loop/undecided" for both a and b. However, another way to look at this is that the group $\{a, b\}$ is not attacked in any way, and therefore all its members should be in.

Note that this problem connects with the question of how to interpret loops in information systems. If $a \to b$ is read and information input from a to b, then the loop $\{a \to b, b \to a\}$, is just a group all sharing the same information and therefore can be considered as one node in the system. This view would entail that we should consider only acyclic networks for information input systems.

References

1. L. Amgoud, C. Cayrol, M. C. Lagasquie-Schiex, and P. Livet. On bipolarity in argumentation frameworks. *International Journal Of Intelligent Systems*, 23:1062–1093, 2008. doi:10.1002/int.20307.
2. Pietro Baroni, Massimiliano Giacomin, and Giovanni Guida. Scc-recursiveness: a general schema for argumentation semantics. *Artificial Intelligence*, 168(1-2):162–210, 2005.
3. Tom Blount, David E. Millard, and Mark J. Weal. An ontology for argumentation on the social web: Rhetorical extensions to the aif 119. In Tatjana Scheffler Pietro Baroni, Thomas F. Gordon and Manfred Stede, editors, *COMMA 2016, Computational Models of Argument*, volume 287 of *Frontiers in Artificial Intelligence and Applications*, pages 119–127. IOS press, 2016.
4. Tom Bosc, Elena Cabrio, and Serena Villata. Tweeties squabbling: Positive and negative results in applying argument mining on social media 21. In Tatjana Scheffler Pietro Baroni, Thomas F. Gordon and Manfred Stede, editors, *COMMA 2016, Computational Models of Argument*, volume 287 of *Frontiers in Artificial Intelligence and Applications*, pages 21–33. IOS press, 2016.
5. Martin Caminada, Samy Sá, and João Alcântara. On the equivalence between logic programming semantics and argumentation semantics. In *ECSQARU 2013*, volume 7958 of *LNAI*, pages 97–108. Springer, 2013.
6. Martin Caminada, Samy Sá, Wolfgang Dvořák, and Jo ao Alcântara. On the equivalence between logic programming semantics and argumentation semantics. *International Journal of Approximate Reasoning*, 58:87–111, 2015.
7. C. Cayrol and M.C. Lagasquie-Schiex. On the acceptability of arguments in bipolar argumentation frameworks. In L. Godo, editor, *ECSQARU 2005*, volume 3571 of *LNAI*, page 378389. Springer, 2005.
8. Toni F. Cyras K., Satoh K. Explanation for case-based reasoning via abstract argumentation. In Tatjana Scheffler Pietro Baroni, Thomas F. Gordon and Manfred Stede, editors, *COMMA 2016, Computational Models of Argument*, volume

287 of *Frontiers in Artificial Intelligence and Applications*, pages 243–254. IOS press, 2016.
9. Phan Minh Dung. On the acceptability of arguments and its fundamental role in non-monotonic reasoning, logic programming and n-person games. *Artificial Intelligence*, 77(2):321–357, 1995.
10. D. M. Gabbay. An equational approach to argumentation networks. *Argument & Computation*, 3(2-3):87–142, 2012.
11. D. M. Gabbay. The handling of loops in argumentation networks. *Journal of Logic and Computation*, 26:1065–1148, 2016. Special issue on Loops. First published online February 20, 2014 doi:10.1093/logcom/exu007.
12. D. M. Gabbay. Logical foundations for bipolar argumentation networks. *Journal of Logic and Computation*, 26(1):247–292, 2016. Special issue in honour of Arnon Avron. First published online: July 22, 2013, doi: 10.1093/logcom/ext027.
13. D. M. Gabbay, H. Barringer, and J. Woods. Temporal dynamics of support and attack networks: from argumentation to zoology. In D. Hutter and W. Stephan, editors, *Mechanising Mathematical Reasoning*, volume 2605 of *LNCS*, pages 59–98. Springer, 2005. Volume Dedicated to Joerg Siekmann.
14. D. M. Gabbay and M. Gabbay. The attack as strong negation, part 1. *Logic Journal of the IGPL*, 23(6):881–941, 2015. doi: 10.1093/jigpal/jzv033. First published online September 28, 2015.
15. D. M. Gabbay and M. Gabbay. The attack as intuitionistic negation. *Logic Journal of the IGPL*, 24:807–838, 2016. Article ID: JIGPAL-jzw012.
16. D. M. Gabbay and A. Garcez. Logical modes of attack in argumentation networks. *Studia Logica*, 93(2-3):199–230, 2009.
17. D. M. Gabbay and Caminada M. A logical account of formal argumentation. *Studia Logica*, 93(109), 2009.
18. D. M. Gabbay, O. Rodrigues, and J. Woods. Belief contraction, anti-formulas, and resource overdraft: Part I. *Logic Journal of the IGPL*, 10:601–652, 2002.
19. D. M. Gabbay, O. Rodrigues, and J. Woods. Belief contraction, anti-formulas, and resource overdraft: Part II. In J. Symons D. M. Gabbay, S. Rahman and J-P van Bendegem, editors, *Logic, Epistemology and the Unity of Science*, pages 291–326. Kluwer, 2004.
20. D. M. Gabbay and G. Rozenberg. Reasoning schemes, expert opinion and critical questions: Sex offenders case study. *IFCoLog Journal of Logics and their Applications*, 4(6):1687–1789, 2017.
21. D. M. Gabbay, Y. Wu, and M. Caminada. Complete extensions in argumentation coincide with 3-valued stable models in logic programming. *Studia Logica*, 93(1-2):383–403, 2009. Special issue: new ideas in argumentation theory.
22. Michael Gelfond. Answer sets. In Frank van Harmelen, Vladimir Lifschitz, and Bruce Porter, editors, *Handbook of Knowledge Representation*, pages 285–316. Elsevier, 2008.
23. Umberto Grandi, Emiliano Lorini, and Laurent Perrussel. Propositional opinion diffusion. In Elkind Bordini and Yolum Weiss, editors, *Proceedings of the 14th International Conference on Autonomous Agents and Multiagent Systems (AAMAS 2015)*. 2015. Istanbul, Turkey.
24. Marie-Francine Moens. Argumentation mining: How can a machine acquire world and common sense knowledge., 2016. Keynote Lecture, COMMA 2016.
25. Gauvain Bourgne Nicolas Schwind, Katsumi Inoue and Sebastien Konieczny. *Belief Revision Games*. Association for the Advancement of Artificial Intelligence (www.aaai.org), 2015.

26. Greg Restall. *Introduction to substructural logics*. Routledge, 2000.
27. Patrick Saint-Dizier. Challenges of argument mining: Generating an argument synthesis based on the qualia structure, 2016. Paper presented at The 9th International Natural Language Generation, Edinburgh, Scotland, September 5-8, doi: 10.18653/v1/W16-6613.
28. F. Toni. A tutorial on assumption-based argumentation. *Argument & Computation*, 5:89–117, 2014. Special Issue: Tutorials on Structured Argumentation.

On the Functional Completeness of Argumentation Semantics

Massimiliano Giacomin[1], Thomas Linsbichler[2] and Stefan Woltran[2]

[1] Department of Information Engineering
University of Brescia, Italy
[2] Institute of Logic and Computation
TU Wien, Vienna, Austria

Abstract. Abstract argumentation frameworks (AFs) are one of the central formalisms in AI; equipped with a wide range of semantics, they have proven useful in several application domains. We contribute to the systematic analysis of semantics for AFs by connecting two recent lines of research – the work on input/output frameworks and the study of the expressiveness of semantics. We do so by considering the following question: given a function describing an input/output behaviour by mapping extensions (resp. labellings) to sets of extensions (resp. labellings), is there an AF with designated input and output arguments realizing this function under a given semantics? For the major semantics we give exact characterizations of the functions which are realizable in this manner.

1 Introduction

Dung's argumentation frameworks (AFs) have been extensively investigated, mainly because they represent an abstract model unifying a large variety of specific formalisms ranging from nonmonotonic reasoning to logic programming and game theory [16]. After the development and analysis of different semantics [28, 10, 2], recent attention has been drawn to their expressive power, i.e. determining which sets of extensions [17] and labellings [19] can be enforced in a single AF under a given semantics. Such results have recently been facilitated in order to express AGM-based revision in the context of abstract argumentation [14].

In [1] it has been shown that an AF can be viewed as a set of partial interacting sub-frameworks, each characterized by an input/output behavior, i.e. a semantics-dependent function which maps each labelling of the "input" arguments (the external arguments affecting the sub-framework) into the set of labellings prescribed for the "output" arguments (the arguments of the sub-framework affecting the external ones). It turns out that under the major semantics, i.e. complete, grounded, stable and (under some mild conditions) preferred semantics, sub-frameworks with the same input/output behavior can be safely exchanged, i.e. replacing a sub-framework with an equivalent one does not affect the justification status of the arguments not involved in the exchange: semantics of this kind are called *transparent* [1].

As a simple example, consider an argumentation framework including a chain of 4 arguments $a_1, \ldots, a_4$ where for $i \in \{2, 3, 4\}$, a_{i-1} attacks a_i and a_i does not receive other attacks, and a_1 is unattacked. This chain can be seen as a sub-framework with input argument a_1 and output argument a_4, which under any transparent semantics can be replaced with any even-length chain without affecting the justification status of the arguments outside the sub-framework.

Then, somewhat resembling functional completeness of a specific set of logic gates, a natural question concerns the expressive power of transparent semantics in the context of an interacting sub-framework: given a so-called *I/O specification*, i.e. a function describing an input/output behaviour by mapping extensions (resp. labellings) to sets of extensions (resp. labellings), is there an AF with designated input and output arguments realizing this function under a given semantics?

Turning to the example above, the sub-framework including the 4-length chain *realizes* the mapping where $\{a_1\}$ is mapped to $\emptyset$ (i.e. if a_1 belongs to an extension then a_4 does not belong to it) and $\emptyset$ is mapped to $\{a_4\}$ (i.e. if a_1 does not belong to an extension then a_4 belongs to it): we call this kind of mapping a two-valued I/O specification. On the other hand, one may want to distinguish between *out* arguments (i.e. attacked by an extension) and *undecided* arguments (i.e. neither belonging to the extension nor attacked by it). Considering the sub-framework above, if a_1 is accepted then a_4 is *out*, if a_1 is *out* then a_4 is accepted, if a_1 is undecided then a_4 is undecided too. We call this kind of mapping a three-valued I/O specification. As it will be shown in the following, not all three-valued I/O specifications are realizable, e.g. there is no sub-framework realizing the variant of the mapping above where a_1 undecided yields a_4 accepted.

In this paper, we answer the question of realizability as follows:

- For the stable, preferred, semi-stable, stage, complete, ideal, and grounded semantics we exactly characterize all realizable two-valued I/O specifications.
- For the preferred and grounded labellings we exactly characterize all realizable three-valued I/O specifications. Moreover, we give sufficient conditions for realizability for semi-stable and ideal labellings and some observations for complete labellings.

Answering this question is essential in many aspects. First, it adds to the analysis and *comparison of semantics* (see e.g. [9, 3]), by providing an absolute characterization of their functional expressiveness, which holds independently of how the abstract argumentation framework is instantiated. Second, it lays foundations towards a theory of *dynamic and modular argumentation*. More specifically, a functional characterization provides a common ground for different representations of the same sub-framework, as in metalevel argumentation [24] where meta-level arguments making claims about object-level arguments allow for equivalent characterizations of the same framework at different levels of abstraction, or to devise a summarized version of a sub-framework in order to simplify a given argumentation framework. One may also translate a different formalism to an AF or vice versa, e.g. to express a logical system as an AF or provide an argument-free representation of a given AF for human/computer interaction issues. In all of these cases, it is important to know whether an input/output behavior is realizable under a given argumentation semantics. Finally, our results are important in the dynamic setting of *strategic argumentation*, where a player may exploit the fact that for some set of arguments certain labellings are achievable (or non achievable) independently of the labelling of other arguments, or more generally she/he may exploit knowledge on the set of realizable dependencies. For example, an agent may desire to achieve some goal, i.e., ensure that a certain argument is justified. Considering arguments brought up by other agents as input arguments, our results enable the agent to verify whether the goal is achievable and provide one particular way for the agent to bring up further arguments in order to succeed.

The paper is organized as follows. After providing the necessary background in Section 2, Section 3 introduces the notion of an I/O-gadget to represent a sub-framework, and tackles the above problem with extension-based two-valued specifications. Labelling-based three-valued specifications are investigated in Section 4 while Section 5 considers partial specifications where for some inputs the output is not specified. Section 6 concludes the paper.

This article is an extended version of a conference version [20]. Additional contributions concern results on three-valued I/O specifications with respect to complete, semi-stable and ideal labellings as well as comparison of the different semantics in terms of signatures [17]. Furthermore, additional examples and full proofs are provided. The additional material is taken from the 2nd author's PhD thesis [21] but is unpublished elsewhere.

Relation to Guillermo's Work. Besides his pioneering work in the field of computational argumentation in general, Guillermo always insisted that the dynamic aspects of argumentation and dialogue have to be taken into account properly and need formal treatment[1]. This is not only witnessed by his work on temporal argumentation frameworks [7, 8, 12, 13, 23] and dynamics in defeasible logic programming [25], but also by his restless activities in organizing events that focus on the topics of argumentation and belief change, in particular the "Madeira Workshop on Belief Revision and Argumentation" series or the Dagstuhl Seminar on "Belief Change and Argumentation in Multi-Agent Scenarios" [15]. Guillermo's ideas and considerations were very inspiring for all three of us and his tireless support for the argumentation community cannot

---

[1] Input/output frameworks which we study here provide a means to analyse whether changes within a particular subpart of an argumentation framework result in an overall change of the acceptability of certain arguments. Hence, they play a particular role to analyse dynamics in argumentation.

be appreciated highly enough. We are thus very happy and honoured to contribute to this Festschrift dedicated to Guillermo Simari on the occasion of his 70th Birthday. *Ad multos annos!*

2 Background

We assume a countably infinite domain of arguments $\mathfrak{A}$. An *argumentation framework* (AF) is a pair $F = (A, R)$ where $A \subseteq \mathfrak{A}$ and $R \subseteq A \times A$. We assume that A is non-empty and finite. For an AF $F = (A, R)$ and a set of arguments $S \subseteq A$, we define $S_F^+ = \{a \in A \mid \exists s \in S : (s, a) \in R\}$, $S_F^\oplus = S \cup S_F^+$, and $S_F^- = \{a \in A \mid \exists s \in S : (a, s) \in R\}$.

Given $F = (A, R)$, a set $S \subseteq A$ is *conflict-free* (in F), if there are no arguments $a, b \in S : (a, b) \in R$. An argument $a \in A$ is *defended* (in F) by a set $S \subseteq A$ if $\forall b \in A : (b, a) \in R \Rightarrow b \in S_F^+$. A set $S \subseteq A$ is *admissible* (in F) if it is conflict-free and defends all of its elements. We denote the set of conflict-free and admissible sets in F as $cf(F)$ and $ad(F)$, respectively.

An extension-based *semantics* σ associates to any $F = (A, R)$ the (possibly empty) set $\sigma(F) \subseteq 2^A$ of subsets of A called σ-extensions, where 2^A denotes the powerset of A. In this paper we focus on complete, grounded, preferred, ideal, stable, stage and semi-stable semantics, with extensions defined as follows:

- $S \in co(F)$ iff $S \in ad(F)$ and $a \in S$ for all $a \in A$ defended by S;
- $S \in gr(F)$ iff S is the least (wrt. $\subseteq$) element in $co(F)$;
- $S \in pr(F)$ iff $S \in ad(F)$ and $\nexists T \in ad(F)$ s.t. $T \supset S$;
- $S \in id(F)$ iff $S \in ad(F)$, $S \subseteq \bigcap pr(F)$ and $\nexists T \in ad(F)$ s.t. $T \subseteq \bigcap pr(F)$ and $T \supset S$;
- $S \in st(F)$ iff $S \in cf(F)$ and $S_F^\oplus = A$;
- $S \in sg(F)$ iff $S \in cf(F)$ and $\nexists T \in cf(F)$ s.t. $T_F^\oplus \supset S_F^\oplus$;
- $S \in se(F)$ iff $S \in ad(F)$ and $\nexists T \in ad(F)$ s.t. $T_F^\oplus \supset S_F^\oplus$.

Given $F = (A, R)$ and a set $O \subseteq A$, the restriction of $\sigma(F)$ to O, denoted as $\sigma(F)|_O$, is the set $\{E \cap O \mid E \in \sigma(F)\}$.

Given a set of arguments A, a *labelling* v is a function assigning each argument $a \in A$ exactly one label[2] among $\mathbf{t}$, $\mathbf{f}$ and $\mathbf{u}$, i.e. $v : A \mapsto \{\mathbf{t}, \mathbf{f}, \mathbf{u}\}$. If the arguments $A = \{a_1, \ldots, a_n\}$ are ordered, then we denote a labelling of A as a sequence of labels, e.g. the labelling **tuf** of arguments $\{a_1, a_2, a_3\}$ maps a_1 to $\mathbf{t}$, a_2 to $\mathbf{u}$, and a_3 to $\mathbf{f}$. We denote the set of all possible labellings of A as $\mathcal{V}(A)$. Likewise, given an AF F, we denote the set of all possible labellings of (the arguments of) F as $\mathcal{V}(F)$. Given a labelling v and an argument a, $v(a)$ denotes the labelling of a wrt. v; finally $v^\mathbf{t}$, $v^\mathbf{f}$, and $v^\mathbf{u}$ denotes the arguments labeled to $\mathbf{t}$, $\mathbf{f}$, and $\mathbf{u}$ by v, respectively. By $\neg v$ we denote the inverse labelling of v, i.e. $(\neg v)^\mathbf{t} = v^\mathbf{f}$, $(\neg v)^\mathbf{f} = v^\mathbf{t}$, and $(\neg v)^\mathbf{u} = v^\mathbf{u}$.

---

[2] Note that in the original work [11], labels are among *in*, *out*, and *undec*.

Definition 1. *Given a set of arguments A and labellings v_1, v_2 thereof, $v_1 \leq_i v_2$ iff $v_1^t \subseteq v_2^t$ and $v_1^f \subseteq v_2^f$. As usual, $v_1 <_i v_2$ holds iff $v_1 \leq_i v_2$ but $v_2 \not\leq_i v_1$. Moreover, we call v_1 and v_2*

- *comparable if $v_1 \leq_i v_2$ or $v_2 \leq_i v_1$, and*
- *compatible if $v_1^t \cap v_2^f = v_1^f \cap v_2^t = \emptyset$.*

Note that if v_1 and v_2 are comparable then they are also compatible, while the reverse does not hold.

For each semantics there is also a three-valued (or labelling-based) version of the semantics, giving a more fine-grained view of the acceptance status of arguments The three-valued version σ_3 of a semantics σ associates to F a set $\sigma_3(F) \subseteq \mathcal{V}(F)$, where any labelling $v \in \sigma_3(F)$ corresponds to an extension $S \in \sigma(F)$ as follows.

Definition 2. *Given an AF $F = (A, R)$, let $S \subseteq A$. The labelling corresponding to S is called $e2l(S)$ and is defined as $(e2l(S))^t = S$, $(e2l(S))^f = S_F^+ \setminus S$, and $(e2l(S))^u = A \setminus S_F^\oplus$. For a semantics σ, its three-valued (or labelling-based) version σ_3 is defined as $\sigma_3 = \{e2l(E) \mid E \in \sigma(F)\}$.*

In words, the labelling corresponding to an extension is such that an argument is labelled **t** if it is contained in the extension, **f** if it is attacked by an argument which is contained in the extension, and **u** otherwise.

Finally, given F and a set $O \subseteq A$, the restriction of $\sigma_3(F)$ to O, denoted as $\sigma_3(F)|_O$, is the set $\{v \cap (O \times \{\mathbf{t}, \mathbf{f}, \mathbf{u}\}) \mid v \in \sigma_3(F)\}$.

The following well-known result states that preferred labellings are the maximal (wrt. $\leq_i$) complete labellings, and can be deduced e.g. from the semantics account given in [11].

Proposition 1. *For an AF F, $pr_3(F) = \max_{\leq_i}(co_3(F))$.*

3 Extension-based I/O-gadgets

An I/O-gadget represents a (partial) AF where two sets of arguments are identified as input and output arguments, respectively, with the restriction that input arguments do not have any ingoing attacks[3].

Definition 3. *Given a set of input arguments $I \subseteq \mathfrak{A}$ and a set of output arguments $O \subseteq \mathfrak{A}$ with $I \cap O = \emptyset$, an I/O-gadget is an AF $F = (A, R)$ such that $I, O \subseteq A$ and $I_F^- = \emptyset$.*

---

[3] Differently from an I/O-gadget, the notion of *argumentation multipole* in [1] assumes a fixed set of incoming and outgoing attacks rather than of input and output arguments. However, for the purposes of the present paper the two notions are equivalent insofar as input (output) arguments of an I/O-gadget are identified with the sources (destinations) of incoming (outgoing) attacks of the corresponding argumentation multipole.

The injection of a set $J \subseteq I$ to an I/O-gadget F simulates the input J in the way that all arguments in J are accepted (none of them has ingoing attacks since F is an I/O-gadget) and all arguments in $(I \setminus J)$ are rejected (each of them is attacked by the newly introduced argument z, which has no ingoing attacks).

Definition 4. *Given an I/O-gadget $F = (A, R)$ and a set of arguments $J \subseteq I$, the* injection *of J to F is the* AF

$$\triangleright(F, J) = (A \cup \{z\}, R \cup \{(z, i) \mid i \in (I \setminus J)\}),$$

where z is a newly introduced argument.

An I/O-specification describes a desired input/output behaviour by assigning to each set of input arguments a set of sets of output arguments.

Definition 5. *A two-valued[4] I/O-specification consists of two sets $I, O \subseteq \mathfrak{A}$ and a total[5] function $\mathfrak{f} : 2^I \mapsto 2^{2^O}$.*

In order for an I/O-gadget F to satisfy $\mathfrak{f}$ under a semantics σ, the injection of each $J \subseteq I$ to F must have $\mathfrak{f}(J)$ as its σ-extensions restricted to the output arguments. So, informally, with input J applied the set of outputs under σ should be exactly $\mathfrak{f}(J)$.

Definition 6. *Given $I, O \subseteq \mathfrak{A}$, a semantics σ and an I/O-specification $\mathfrak{f}$, an I/O-gadget F* realizes *$\mathfrak{f}$ under σ iff $\forall J \subseteq I : \sigma(\triangleright(F, J))|_O = \mathfrak{f}(J)$.*

The following example illustrates these basic concepts.

Example 1. Consider the sets $I = \{a, b\}$ and $O = \{c, d\}$. An exemplary I/O-specification is given by the function $\mathfrak{f} : 2^I \mapsto 2^{2^O}$ such that

$$\mathfrak{f}(\emptyset) = \{\{d\}\}$$
$$\mathfrak{f}(\{a\}) = \{\{c, d\}\}$$
$$\mathfrak{f}(\{b\}) = \{\{c\}, \{d\}\}$$
$$\mathfrak{f}(\{a, b\}) = \{\{c, d\}, \{c\}\}$$

Considering, for instance, the case of input $\{b\}$, the intended meaning of $\mathfrak{f}$ is that if b is accepted and a is not, then either c or d, but not both, should be accepted. On the other hand, in the case of input $\{a\}$, i.e. a is accepted and b is not, both c and d should be accepted. The AF F in Figure 1 represents an I/O-gadget with dedicated input arguments $\{a, b\}$ and output arguments $\{c, d\}$. It turns out that F realizes $\mathfrak{f}$ under stable semantics. In order to show this we have to check, for each $J \subseteq I$, whether the

---

[4] In the following we omit this specification.
[5] The case of a partial function will be discussed in Section 5.

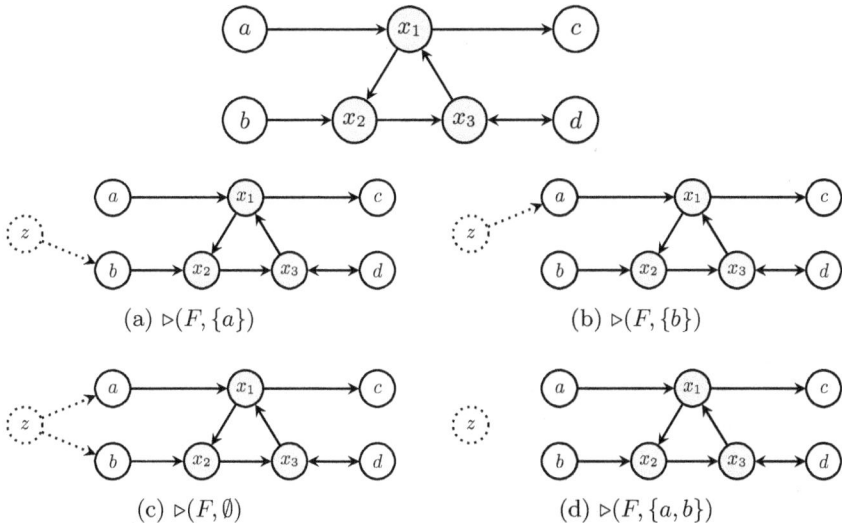

Fig. 1: I/O-gadget F with $I = \{a, b\}$ and $O = \{c, d\}$ realizing the I/O-specification given in Example 1 under $\{st, pr, se, sg\}$ on top and the injections of all possible input assignments in (a) – (d).

injection of J to F, $\triangleright(F, J)$, has exactly $\mathfrak{f}(J)$ as stable extensions restricted to O, i.e. $st(\triangleright(F, J))|_O = \mathfrak{f}(J)$. Considering $J = \emptyset$ (cf. Figure 1c), we have that $\triangleright(F, \emptyset)$ is F together with a new argument z attacking both a and b. Hence we have $st(\triangleright(F, \emptyset)) = \{\{z, x_1, d\}\}$, meaning that $st(\triangleright(F, \emptyset))|_O = \{\{d\}\}$, which is as specified by $\mathfrak{f}$. Moreover we get $st(\triangleright(F, \{a\})) = \{\{z, a, x_2, c, d\}\}$ (cf. Figure 1a), $st(\triangleright(F, \{b\})) = \{\{z, b, x_3, c\}, \{z, b, x_1, d\}\}$ (cf. Figure 1b), and $st(\triangleright(F, \{a, b\})) = \{\{z, a, b, c, d\}, \{z, a, b, x_3, c\}\}$ (cf. Figure 1d). This shows that for all possible inputs, the extensions restricted to the output arguments are as specified by $\mathfrak{f}$, hence F realizes $\mathfrak{f}$ under stable semantics.

While it is easy to verify that F also realizes $\mathfrak{f}$ under preferred, semi-stable and stage semantics, it does not realize $\mathfrak{f}$ under grounded, ideal and complete semantics. For $J = \{b\}$ we have that $gr(\triangleright(F, \{b\}))|_O = id(\triangleright(F, \{b\}))|_O = \{\emptyset\}$ and $co(\triangleright(F, \{b\}))|_O = \{\emptyset, \{c\}, \{d\}\}$, being not in line with $\mathfrak{f}$ (recall that $\mathfrak{f}(b) = \{\{c\}, \{d\}\}$). ◇

The question we want to address is the following: which conditions must $\mathfrak{f}$ fulfill in order to be realizable by some I/O-gadget and how can such an I/O-gadget be constructed? The following generic AF will be the key concept for the forthcoming characterization results.

Definition 7. *Given an I/O-specification* $\mathfrak{f}$, *let* $Y = \{y_i \mid i \in I\}$ *and* $X = \{x_J^S \mid J \subseteq I, S \in \mathfrak{f}(J)\}$. *The* canonical I/O-gadget *(for* $\mathfrak{f}$*) is defined as*

$$\mathfrak{C}(\mathfrak{f}) = (I \cup O \cup Y \cup X \cup \{w\},$$
$$\{(i, y_i) \mid i \in I\} \cup$$
$$\{(y_i, x_J^S) \mid x_J^S \in X, i \in J\} \cup \{(i, x_J^S) \mid x_J^S \in X, i \in (I \setminus J)\} \cup$$
$$\{(x, x') \mid x, x' \in X, x \neq x'\} \cup \{(x, w) \mid x \in X\} \cup \{(w, w)\} \cup$$
$$\{(x_J^S, o) \mid x_J^S \in X, o \in (O \setminus S)\}).$$

Besides the dedicated input and output arguments, $\mathfrak{C}(\mathfrak{f})$ consist of a copy of each input argument, an argument for each combination of input and output given by $\mathfrak{f}$, as well as the argument w. Intuitively, the argument x_J^S shall enforce output S for input J. It does so by attacking all other arguments in X and all output arguments except S. Moreover, w ensures that any stable extension of (an injection to) $\mathfrak{C}(\mathfrak{f})$ must contain at least one argument of X, making it possible to realize under stable semantics I/O-specifications that for some input do not prescribe any output.

The following theorem shows that any I/O-specification is realizable under stable semantics.

Theorem 1. *Every I/O-specification* $\mathfrak{f}$ *is realized by* $\mathfrak{C}(\mathfrak{f})$ *under st.*

Proof. Let $I, O \subseteq \mathfrak{A}$ and $\mathfrak{f}$ be an arbitrary I/O-specification. We have to show that $st(\triangleright(\mathfrak{C}(\mathfrak{f}), J))|_O = \mathfrak{f}(J)$ holds for any $J \subseteq I$. Consider such a $J \subseteq I$.

First let $S \in \mathfrak{f}(J)$. We show that[6] $E = \{z\} \cup J \cup \{y_i \mid i \in (I \setminus J)\} \cup \{x_J^S\} \cup S \in st(\triangleright(\mathfrak{C}(\mathfrak{f}), J))$, thus $S \in st(\triangleright(\mathfrak{C}(\mathfrak{f}), J))|_O$. E is conflict-free in $\triangleright(\mathfrak{C}(\mathfrak{f}), J)$ since z only attacks the arguments in $(I \setminus J)$ (and thus outside E), any argument $i \in J$ only attacks arguments y_i with $i \in J$ (thus $y_i \notin E$) and arguments $x_{J'}^{S'}$ with $i \notin J'$ (thus $x_{J'}^{S'} \notin E$), any y_i with $i \in (I \setminus J)$ only attacks arguments $x_{J'}^{S'}$ with $i \in J'$ (thus $x_{J'}^{S'} \notin E$), x_J^S only attacks other arguments $x \in X$, w, and arguments in $(O \setminus S)$ (all of them not belonging to E), and the arguments in S do not attack any argument. E is stable in $\triangleright(\mathfrak{C}(\mathfrak{f}), J)$ since x_J^S attacks w, all other $x \in X$ and all $o \in (O \setminus S)$; z attacks all $i \in (I \setminus J)$; each y_j with $j \in J$ is attacked by j.

It remains to show that there is no $S' \in st(\triangleright(\mathfrak{C}(\mathfrak{f}), J))|_O$ with $S' \notin \mathfrak{f}(J)$. Towards a contradiction assume there is some $S' \in st(\triangleright(\mathfrak{C}(\mathfrak{f}), J))|_O$ with $S' \notin \mathfrak{f}(J)$. Hence there must be some $E' \in st(\triangleright(\mathfrak{C}(\mathfrak{f}), J))$ with $S' \subset E'$. Since w attacks itself, $w \notin E'$, thus by construction of $\mathfrak{C}(\mathfrak{f})$ there must be some $x_{J'}^{S'} \in (X \cap E')$ attacking w, and $x_{J'}^{S'}$ must attack all $o \in (O \setminus S')$. Since $S' \notin \mathfrak{f}(J)$ by assumption, it must hold that $J' \neq J$. Now note that $z \in E'$ and $j \in E'$ for all $j \in J$, since they are not attacked by construction of $\triangleright(\mathfrak{C}(\mathfrak{f}), J)$. Now if $J' \subset J$ then there is some $j \in (J \setminus J')$ attacking $x_{J'}^{S'}$, a contradiction to conflict-freeness of E'. On the other hand if $J' \not\subseteq J$ there is some $j' \in (J' \setminus J)$ which is attacked by z. Therefore also $y_{j'} \in E'$, which attacks $x_{J'}^{S'}$, again a contradiction.

---

[6] Recall that the argument z is introduced by the injection.

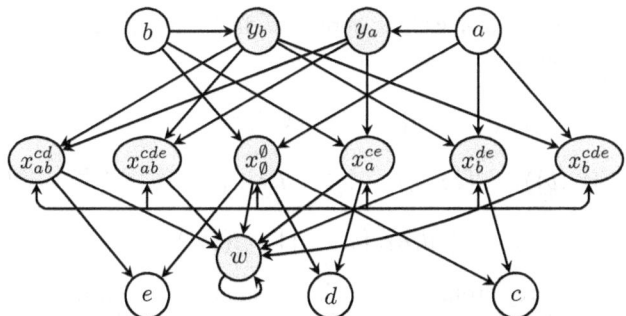

Fig. 2: I/O-gadget realizing the I/O-specification given in Example 2 under $\{st, pr, se, sg\}$.

As to preferred, semi-stable and stage semantics, any I/O-specification is realizable, provided that an output is prescribed for any input.

Proposition 2. *Every I/O-specification $\mathfrak{f}$ such that $\forall J \subseteq I : \mathfrak{f}(J) \neq \emptyset$ is realized by $\mathfrak{C}(\mathfrak{f})$ under $\sigma \in \{pr, se, sg\}$.*

Proof. We show that for all $J \subseteq I$ stable, preferred, stage and semi-stable extensions coincide in $\triangleright(\mathfrak{C}(\mathfrak{f}), J)$, thus the result follows from Theorem 1.

First, according to the hypothesis and Theorem 1, there exists a stable extension of $\triangleright(\mathfrak{C}(\mathfrak{f}), J)$ for each $J \subseteq I$, thus stable, stage and semi-stable extensions coincide. As to preferred semantics, we know that any stable extension is also preferred, and we show that the reverse also holds in $\triangleright(\mathfrak{C}(\mathfrak{f}), J)$. By construction of $\triangleright(\mathfrak{C}(\mathfrak{f}), J)$, it is easy to see that for any preferred extension E it holds that $E = \{z\} \cup J \cup \{y_i \mid i \in (I \setminus J)\} \cup \{x_J^S\} \cup S$, where S is a set among $\mathfrak{f}(J)$ and exists by the hypothesis. E is stable, since x_J^S attacks w, all other $x \in X$ and all $o \in (O \setminus S)$, z attacks all $i \in (I \setminus J)$, and each y_j with $j \in J$ is attacked by j.

Theorem 2. *An I/O-specification $\mathfrak{f}$ is realizable under $\sigma \in \{pr, se, sg\}$ iff $\forall J \subseteq I : \mathfrak{f}(J) \neq \emptyset$.*

Proof. The if-direction is a direct consequence of Proposition 2.

The only-if-direction follows directly by the fact that in every AF, particularly in any injection of some set of arguments to an I/O-gadget, a σ-extension exists.

Example 2. Consider the I/O-specification $\mathfrak{f}$ with $I = \{a, b\}$ and $O = \{c, d, e\}$ defined as follows:

$$\mathfrak{f}(\emptyset) = \{\emptyset\}$$
$$\mathfrak{f}(\{a\}) = \{\{c, e\}\}$$
$$\mathfrak{f}(\{b\}) = \{\{c, d, e\}, \{d, e\}\}$$
$$\mathfrak{f}(\{a, b\}) = \{\{c, d, e\}, \{c, d\}\}$$

The canonical I/O-gadget $\mathfrak{C}(\mathfrak{f})$ is depicted[7] in Figure 2. Let σ be a semantics among $\{st, pr, se, sg\}$. One can verify that for every possible input $J \subseteq I$, the injection of J to $\mathfrak{C}(\mathfrak{f})$ has exactly $\mathfrak{f}(J)$ as σ-extensions restricted to O. As an example let $J = \{b\}$. $\triangleright(\mathfrak{C}(\mathfrak{f}), \{b\})$ adds to $\mathfrak{C}(\mathfrak{f})$ the argument z attacking a. Now

$$\sigma(\triangleright(\mathfrak{C}(\mathfrak{f}), \{b\})) = \{\{z, b, y_a, x_{\{b\}}^{\{d,e\}}, d, e\}, \{z, b, y_a, x_{\{b\}}^{\{c,d,e\}}, c, d, e\}\},$$

hence $\sigma(\triangleright(\mathfrak{C}(\mathfrak{f}), \{b\}))|_O = \{\{d, e\}, \{c, d, e\}\} = \mathfrak{f}(\{b\})$. ◇

Also for complete, grounded and ideal semantics we are able to identify a necessary and sufficient condition for realizability. While we show sufficiency of these conditions in more detail, their necessity is by the well-known facts that the intersection of all complete extensions is always a complete extension too, and ideal and grounded semantics always yield exactly one extension. We define the former property for I/O-specifications:

Definition 8. *An I/O-specification $\mathfrak{f}$ is* closed *iff for each $J \subseteq I$ it holds that $\mathfrak{f}(J) \neq \emptyset$ and $\bigcap \mathfrak{f}(J) \in \mathfrak{f}(J)$.*

Example 3. Again considering the I/O-specification $\mathfrak{f}$ given in Example 2, we observe that $\mathfrak{f}$ is closed, since $\bigcap \mathfrak{f}(J) \in \mathfrak{f}(J)$ for each $J \subseteq \{a, b\}$. For instance, $\bigcap \mathfrak{f}(\{b\}) = \{d, e\}$ and, indeed, $\{d, e\} \in \mathfrak{f}(\{b\})$. ◇

Proposition 3. *Every closed I/O-specification $\mathfrak{f}$ is realized by $\mathfrak{C}(\mathfrak{f})$ under co.*

Proof. Let $J \subseteq I$. By construction of $\triangleright(\mathfrak{C}(\mathfrak{f}), J)$, $E^* = \{z\} \cup J \cup \{y_i \mid i \in (I \setminus J)\}$ is contained in all complete extensions, while the elements of $(I \setminus J) \cup \{y_i \mid i \in J\}$ are attacked by E^* and thus by all complete extensions. All $x_{J'}^{S'}$ with $J' \neq J$ are attacked by J or some y_i with $i \in (I \setminus J)$, thus they are attacked by E^*, while all x_J^S with $S \in \mathfrak{f}(J)$ attack each other, and the other attacks they receive come from elements attacked by E^*. Two cases can then be distinguished. If $|\mathfrak{f}(J)| = 1$ then by construction of $\triangleright(\mathfrak{C}(\mathfrak{f}), J)$ there is just one x_J^S defended by E^*, thus the only complete extension is $E^* \cup \{x_J^S\} \cup S$. If, on the other hand, $|\mathfrak{f}(J)| > 1$, any x_J^S with $S \in \mathfrak{f}(J)$ can be included, giving rise to the complete extension $E^* \cup \{x_J^S\} \cup S$, or none of x_J^S can be included, giving rise to the complete extension $E^* \cup \bigcap \mathfrak{f}(J)$ since an x_J^S attacks all $o \in (O \setminus S)$. Taking into account that $\bigcap \mathfrak{f}(J) \in \mathfrak{f}(J)$, in both cases we have that $co(\triangleright(\mathfrak{C}(\mathfrak{f}), J))|_O = \mathfrak{f}(J)$.

The attentive reader might have already noticed that, for an I/O-specification $\mathfrak{f}$ and an input $J \subseteq I$, there is not necessarily a one-to-one correspondence between $co(\triangleright(\mathfrak{C}(\mathfrak{f}), J))$ and $\mathfrak{f}(J)$. There can be more than one complete extensions of the injection of J to $\mathfrak{C}(\mathfrak{f})$ corresponding to a single output given by $\mathfrak{f}(J)$. In particular, for $S = \bigcap \mathfrak{f}(J)$, there are two distinct complete extensions $\{z\} \cup J \cup \{y_i \mid i \in (I \setminus J)\} \cup S$ and $\{z\} \cup J \cup \{y_i \mid i \in (I \setminus J)\} \cup \{x_J^S\} \cup S$. The restriction of the complete extensions to the output arguments takes care of the one-to-one correspondence to $\mathfrak{f}(J)$.

---

[7] In the figure, argument names such as $x_{\{a,b\}}^{\{c,d\}}$ are abbreviated by x_{ab}^{cd}.

Theorem 3. *An I/O-specification $\mathfrak{f}$ is realizable under co iff $\mathfrak{f}$ is closed.*

Proof. The if-direction is a consequence of Proposition 3.
The only-if-direction follows directly by the fact that in any AF, particularly in any injection of some extension to an *I/O*-gadget, the intersection of all complete extensions is always a complete extension too.

Proposition 4. *Every I/O-specification $\mathfrak{f}$ with $|\mathfrak{f}(J)| = 1$ for each $J \subseteq I$ is realized by $\mathfrak{C}(\mathfrak{f})$ under gr and id.*

Proof. Let $J \subseteq I$. By construction of $\triangleright(\mathfrak{C}(\mathfrak{f}), J)$, $E^* = \{z\} \cup J \cup \{y_i \mid i \in (I \setminus J)\}$ is contained in all complete extensions, while the elements of $(I \setminus J) \cup \{y_i \mid i \in J)\}$ are attacked by E^* and thus by all complete extensions. All $x_{J'}^{S'}$ with $J' \neq J$ are attacked by J or some y_i with $i \in (I \setminus J)$, thus they are attacked by E^*, while the unique x_J^S with $S \in \mathfrak{f}(J)$ is defended by E^*. As a consequence, there is only one complete extension, i.e. $E^* \cup \{x_J^S\} \cup S$, which is consequently also the grounded and ideal extension. The result directly follows.

Theorem 4. *An I/O-specification $\mathfrak{f}$ is realizable under gr and id iff $|\mathfrak{f}(J)| = 1$ for each $J \subseteq I$.*

Proof. The if-direction is a consequence of Proposition 4.
The only-if-direction follows directly by the fact that in any AF, particularly in any injection of some extension to an *I/O*-gadget, the grounded and ideal extension are uniquely defined.

In order to compare the expressiveness of semantics in terms of *I/O*-realizability, we define the notion of an *I/O*-signature.

Definition 9. *Let σ be a semantics. The (two-valued) I/O-signature of σ consists of all I/O-specifications that are realizable under σ:*

$$\Sigma_{AF\triangleright}^{\sigma} = \{\sigma_{\triangleright}(F) \mid F \text{ is an } I/O\text{-gadget}\},$$

where $\sigma_{\triangleright}$ is the I/O-version of σ, defined as a function mapping I/O-gadgets to I/O-specifications such that, given an I/O-gadget F, $\sigma_{\triangleright}(F) = J \subseteq I \mapsto \sigma(\triangleright(F, J))|_O$.

We summarize the presented results on *I/O*-realizability in the following theorem.

Theorem 5. *In accordance with Figure 3 it holds that*

$$\Sigma_{AF\triangleright}^{gr} = \Sigma_{AF\triangleright}^{id} \subset \Sigma_{AF\triangleright}^{co} \subset \Sigma_{AF\triangleright}^{pr} = \Sigma_{AF\triangleright}^{se} = \Sigma_{AF\triangleright}^{sg} \subset \Sigma_{AF\triangleright}^{st}$$

Proof. The relations follow from Theorems 1, 2, 3, and 4.

Note that we have disregarded *I/O*-realizability of admissible semantics. This is because the concept of an injection enforcing a certain acceptance status of input arguments is not applicable to admissible semantics (e.g. $\emptyset$ is always admissible and does not include the arguments of J if $J \neq \emptyset$).

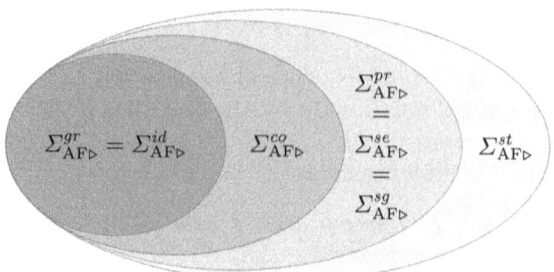

Fig. 3: A Venn diagram illustrating the I/O-signatures of grounded, ideal, complete, semi-stable, stage, preferred, and stable semantics.

4 Labelling-based I/O-gadgets

Until now we have dealt with realizing I/O-specifications mapping extensions to sets of extensions. As explained in Section 2, there are two reasons why an argument does not belong to an extension, namely either because it is attacked by the extension (i.e. it is assigned **f** in the three-valued version of the semantics) or because it is undecided due to insufficient justification (i.e. it is assigned **u**). This distinction impacts on the justification status of arguments, since attacks from undecided arguments can prevent attacked arguments from belonging to an extension, while attacks from arguments labelled **f** are ineffective. Therefore, a full description of the behaviour of a module's interaction within a larger AF has to take into account this distinction. In order to do so, we first provide a three-valued counterpart of the corresponding notions introduced in Section 3.

First, 3-valued I/O-specifications map labellings of input arguments to sets of labellings of output arguments.

Definition 10. *A 3-valued I/O-specification consists of two sets $I, O \subseteq \mathfrak{A}$ and a total function* $\mathfrak{f} : \mathcal{V}(I) \mapsto 2^{\mathcal{V}(O)}$.

The 3-valued injection now distinguishes between an input argument being **f** or **u**. In the first case, it is attacked by the new argument z and in the second case, it attacks itself and remains otherwise unattacked.

Definition 11. *Given an I/O-gadget $F = (A, R)$ and a labelling $v \in \mathcal{V}(I)$, the 3-valued injection of v to F is the* AF

$$\blacktriangleright(F, v) = (A \cup \{z\}, R \cup \{(z, a) \mid v(a) = \mathbf{f}\} \cup \{(b, b) \mid v(b) = \mathbf{u}\}),$$

where z is a newly introduced argument.

Definition 12. *Given $I, O \subseteq \mathfrak{A}$, a semantics σ_3 and a 3-valued I/O-specification $\mathfrak{f}$, the I/O-gadget F realizes $\mathfrak{f}$ under σ_3 iff*

$$\forall v \in \mathcal{V}(I) : \sigma_3(\blacktriangleright(F, v))|_O = \mathfrak{f}(v).$$

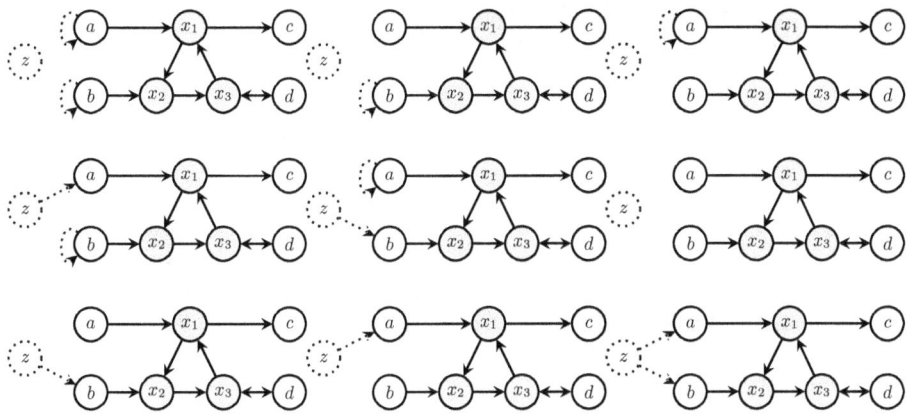

Fig. 4: 3-valued injections to the I/O-gadget from Figure 1, as explained in Example 4.

The following example illustrates these concepts.

Example 4. A possible 3-valued I/O-specification for $I = \{a, b\}$ and $O = \{c, d\}$ is the function $\mathfrak{f} : \mathcal{V}(I) \mapsto 2^{\mathcal{V}(O)}$ such that:

$\mathfrak{f}(\mathbf{uu}) = \{\mathbf{ut}\}$ $\mathfrak{f}(\mathbf{tu}) = \{\mathbf{tt}\}$ $\mathfrak{f}(\mathbf{ut}) = \{\mathbf{ut}, \mathbf{tf}\}$
$\mathfrak{f}(\mathbf{fu}) = \{\mathbf{ft}\}$ $\mathfrak{f}(\mathbf{uf}) = \{\mathbf{ut}\}$ $\mathfrak{f}(\mathbf{tt}) = \{\mathbf{tt}, \mathbf{tf}\}$
$\mathfrak{f}(\mathbf{tf}) = \{\mathbf{tt}\}$ $\mathfrak{f}(\mathbf{ft}) = \{\mathbf{tf}, \mathbf{ft}\}$ $\mathfrak{f}(\mathbf{ff}) = \{\mathbf{ft}\}$

Recall our notation for labellings: a sequence of truth values denotes the labelling mapping the ith argument to the ith value in the sequence. That is, the labelling **fu** for I is short for $\{a \mapsto \mathbf{f}, b \mapsto \mathbf{u}\}$.

Inspecting $\mathfrak{f}$ we observe that, for instance, setting a to **f** and b to **u** shall have the effect that c evaluates to **f** and d evaluates to **t**. Setting both input arguments to **t** shall have two possible outputs, namely one where both output arguments are accepted and one with c accepted and d rejected. It turns out that the I/O-gadget F depicted in Figure 1 realizes $\mathfrak{f}$ under preferred semantics. The 3-valued injections to F are depicted in Figure 4. Consider, for instance, the injection of **fu**, i.e. $\blacktriangleright(F, \mathbf{fu})$ (second row, left). Since $pr(\blacktriangleright(F, \mathbf{fu})) = \{\{z, x_1, d\}\}$ it indeed holds that $pr_3(\blacktriangleright(F, \mathbf{fu}))|_O = \{\mathbf{ft}\} = \mathfrak{f}(\mathbf{fu})$.   ◇

By definition of the stable semantics it is clear that in order to be realized under st_3, a 3-valued I/O-specification must have empty output for all inputs including an argument assigned to **u** and, as can be derived from the two-valued case, no output argument assigned to **u** for inputs with each argument assigned to **t** or **f**.

Theorem 6. *A 3-valued I/O-specification $\mathfrak{f}$ is realizable under st_3 iff for each $v \in \mathcal{V}(I)$ it holds that*

- if $\exists i \in I : v(i) = \mathbf{u}$ then $\mathfrak{f}(v) = \emptyset$, and
- otherwise $w(o) \neq \mathbf{u}$ for all $w \in \mathfrak{f}(v)$ and $o \in O$.

Proof. For the only-if-direction consider the case where $\exists i \in I : v(i) = \mathbf{u}$. Then for any I/O-gadget F, $\blacktriangleright(F,v)$ contains a self-attacking argument otherwise unattacked, hence $st_3(\blacktriangleright(F,v)) = \emptyset$. In the other case, by definition of the stable semantics it is clear that each $o \in O$ must be assigned either $\mathbf{t}$ or $\mathbf{f}$ by any stable labelling.

For the if-direction we get that if $\exists i \in I : v(i) = \mathbf{u}$ then $st_3(\blacktriangleright(\mathfrak{C}(\mathfrak{f}),v)) = \emptyset$. Otherwise the 3-valued injection coincides with the injection from the two-valued case and the result follows from Theorem 1.

In order to characterize those 3-valued I/O-specifications which are realizable under the other semantics we need the following concept of monotonicity.

Definition 13. *A 3-valued I/O-specification $\mathfrak{f}$ is* monotonic *if for all v_1 and v_2 such that $v_1 \leq_i v_2$ it holds that*

$$\forall w_1 \in \mathfrak{f}(v_1) \, \exists w_2 \in \mathfrak{f}(v_2) : w_1 \leq_i w_2.$$

The intuitive meaning of monotonicity is the following: if w_1 is an output for input v_1, then for every input which is more committed than v_1 there must be an output more committed than w_1.

Example 5. Consider again the 3-valued I/O-specification $\mathfrak{f}$ from Example 4. We check whether $\mathfrak{f}$ is monotonic. Note that is suffices to check the condition for the direct successor wrt. $\leq_i$.

- For $\mathbf{ut} \in \mathfrak{f}(\mathbf{uu})$ there is $\mathbf{tt} \in \mathfrak{f}(\mathbf{tu})$, $\mathbf{ut} \in \mathfrak{f}(\mathbf{ut})$, $\mathbf{ft} \in \mathfrak{f}(\mathbf{fu})$, $\mathbf{ut} \in \mathfrak{f}(\mathbf{uf})$.
- For $\mathbf{tt} \in \mathfrak{f}(\mathbf{tu})$ there is $\mathbf{tt} \in \mathfrak{f}(\mathbf{tt})$, $\mathbf{tt} \in \mathfrak{f}(\mathbf{tf})$.
- For $\mathbf{ut} \in \mathfrak{f}(\mathbf{ut})$ there is $\mathbf{tt} \in \mathfrak{f}(\mathbf{tt})$, $\mathbf{ft} \in \mathfrak{f}(\mathbf{ft})$.
- For $\mathbf{tf} \in \mathfrak{f}(\mathbf{ut})$ there is $\mathbf{tf} \in \mathfrak{f}(\mathbf{tt})$, $\mathbf{tf} \in \mathfrak{f}(\mathbf{ft})$.
- For $\mathbf{ft} \in \mathfrak{f}(\mathbf{fu})$ there is $\mathbf{ft} \in \mathfrak{f}(\mathbf{ft})$, $\mathbf{ft} \in \mathfrak{f}(\mathbf{ff})$.
- For $\mathbf{ut} \in \mathfrak{f}(\mathbf{uf})$ there is $\mathbf{tt} \in \mathfrak{f}(\mathbf{tf})$, $\mathbf{ft} \in \mathfrak{f}(\mathbf{ff})$.

Therefore we conclude that $\mathfrak{f}$ is monotonic. ◇

Coming to necessary conditions for 3-valued I/O-specifications we start with rather obvious observations:

Proposition 5. *For every 3-valued I/O-specification $\mathfrak{f}$ which is realizable under gr_3, $|\mathfrak{f}(v)| = 1$ for all $v \in \mathcal{V}(I)$.*

Proof. This is immediate by the fact that $|gr_3(F)| = 1$ for every AF F.

Proposition 6. *For every 3-valued I/O-specification $\mathfrak{f}$ which is realizable under pr_3, $|\mathfrak{f}(v)| \geq 1$ for all $v \in \mathcal{V}(I)$.*

Proof. This is immediate by the fact that $|pr_3(F)| \geq 1$ for every AF F.

Monotonicity is a necessary condition for grounded and preferred semantics.

Proposition 7. *Every 3-valued I/O-specification which is realizable under gr_3 or pr_3 is monotonic.*

Proof. Let $\mathfrak{f}$ be a 3-valued I/O-specification and suppose it is realized by the I/O-gadget F under gr_3. Moreover let $v_1 \leq_i v_2$ be labellings over I.

gr_3: $\forall w_2 \in co_3(\blacktriangleright(F, v_2))|_O \exists w_1 \in co_3(\blacktriangleright(F, v_1))|_O : w_1 \leq_i w_2$ was shown in [1, Proposition 7]. From this and the fact that the grounded labelling is the least (wrt. $\leq_i$) complete labelling, monotonicity for gr_3 follows.

pr_3: We know, again from [1, Proposition 7], that $\forall w_1 \in co_3(\blacktriangleright(F, v_1))|_O \exists w_2 \in co_3(\blacktriangleright(F, v_2))|_O : w_1 \leq_i w_2$. Now observe that each preferred labelling is also complete and for each complete labelling v of F there exists a preferred labelling w of F such that $v \leq_i w$. Hence the result for pr_3 follows.

In Propositions 5, 6 and 7 we have given necessary conditions for 3-valued I/O-specifications to be realizable under gr_3 and pr_3. In the following we show that these conditions are also sufficient in the sense that we can find a realizing I/O-gadget. The constructions of these I/O-gadgets will depend on the given 3-valued I/O-specification and on the semantics, but they will share the same input and output part. The semantics-specific parts, denoted by $X_{\mathfrak{f}}^{\sigma}$ and $R_{\mathfrak{f}}^{\sigma}$ in the following definition, will be given later.

Definition 14. *Given a 3-valued I/O-specification $\mathfrak{f}$ we define $I' = \{i' \mid i \in I\}$, $O' = \{o' \mid o \in O\}$, $R_I = \{(i, i') \mid i \in I\}$ and $R_O = \{(o', o'), (o', o) \mid o \in O\}$. The 3-valued canonical I/O-gadget for semantics σ_3 and the 3-valued I/O-specification $\mathfrak{f}$ is defined as*

$$\mathfrak{D}_{\mathfrak{f}}^{\sigma} = (I \cup I' \cup X_{\mathfrak{f}}^{\sigma} \cup O' \cup O, R_I \cup R_{\mathfrak{f}}^{\sigma} \cup R_O).$$

with $R_{\mathfrak{f}}^{\sigma} \subseteq ((I \cup I') \times X_{\mathfrak{f}}^{\sigma}) \cup (X_{\mathfrak{f}}^{\sigma} \times X_{\mathfrak{f}}^{\sigma}) \cup (X_{\mathfrak{f}}^{\sigma} \times (O' \cup O)).$

The semantics-independent part of $\mathfrak{D}_{\mathfrak{f}}^{\sigma}$ guarantees that the labelling of I coincides with the injected labelling and the labelling of I' is just the negation.

Lemma 1. *Given an arbitrary 3-valued I/O-specification $\mathfrak{f}$ and a semantics $\sigma_3 \in \{gr_3, id_3, co_3, pr_3, se_3, sg_3\}$ it holds for every $v \in \mathcal{V}(I)$ that*

$$\sigma_3(\blacktriangleright(\mathfrak{D}_{\mathfrak{f}}^{\sigma}, v))|_I = \{v\}, and$$
$$\sigma_3(\blacktriangleright(\mathfrak{D}_{\mathfrak{f}}^{\sigma}, v))|_{I'} = \{\neg v\}.$$

Proof. By the fact that arguments in $I \cup I'$ are not allowed to be attacked by the semantics-specific arguments $X_{\mathfrak{f}}^{\sigma}$, it follows that, in $\blacktriangleright(\mathfrak{D}_{\mathfrak{f}}^{\sigma}, v)$, an argument $a \in I$ is unattacked if $v(a) = \mathbf{t}$, attacked by the unattacked argument z if $v(a) = \mathbf{f}$, and self-attacking and otherwise unattacked if $v(a) = \mathbf{u}$. Hence the result for I follows. The result for I' is then immediate by the fact that each $a' \in I'$ is only attacked by $a \in I$ and therefore has the negated labelling of a.

Now we turn to the semantics-specific constructions. For grounded semantics we need the concept of determining input labellings. A labelling v over the input arguments is determining for output argument o if v is a minimal (w.r.t. $\leq_i$) input labelling where o gets a concrete value ($\mathbf{t}$ or $\mathbf{f}$) according to $\mathfrak{f}$.

With abuse of notation, in the following we may identify a set including a single labelling with the labelling itself.

Definition 15. *Given a 3-valued I/O-specification $\mathfrak{f}$ with $|\mathfrak{f}(v)| = 1$ for all $v \in \mathcal{V}(I)$ and an argument $o \in O$, a labelling v over I is determining for o (in $\mathfrak{f}$), if $\mathfrak{f}(v)(o) \neq \mathbf{u}$ and $\forall v' <_i v : \mathfrak{f}(v')(o) = \mathbf{u}$. We denote the set of labellings which are determining for o (in $\mathfrak{f}$) as $\mathfrak{d}_\mathfrak{f}(o)$.*

Note that for 3-valued I/O-specifications which are monotonic, two different labellings which are determining for a certain output argument cannot be comparable. The following example illustrates the concept of determining labellings.

Example 6. Let $\mathfrak{f}$ be the following 3-valued I/O-specification with $I = \{a, b\}$ and $O = \{c, d\}$:

$$\mathfrak{f}(\mathbf{uu}) = \{\mathbf{uu}\} \qquad \mathfrak{f}(\mathbf{tu}) = \{\mathbf{tu}\} \qquad \mathfrak{f}(\mathbf{ut}) = \{\mathbf{ut}\}$$
$$\mathfrak{f}(\mathbf{uf}) = \{\mathbf{uf}\} \qquad \mathfrak{f}(\mathbf{fu}) = \{\mathbf{uu}\} \qquad \mathfrak{f}(\mathbf{tt}) = \{\mathbf{tt}\}$$
$$\mathfrak{f}(\mathbf{tf}) = \{\mathbf{tf}\} \qquad \mathfrak{f}(\mathbf{ft}) = \{\mathbf{ut}\} \qquad \mathfrak{f}(\mathbf{ff}) = \{\mathbf{tf}\}$$

We have the following sets of determining labellings: $\mathfrak{d}_\mathfrak{f}(c) = \{\mathbf{tu}, \mathbf{ff}\}$ and $\mathfrak{d}_\mathfrak{f}(d) = \{\mathbf{ut}, \mathbf{uf}\}$. Consider, for instance, the input labelling $\mathbf{ff}$. We have $\mathfrak{f}(\mathbf{ff}) = \mathbf{tf}$. In order to check if $\mathbf{ff}$ is determining for c we have to look at all input labelling being less committed than $\mathbf{ff}$. Now we observe $\mathfrak{f}(\mathbf{uf}) = \mathbf{uf}$, $\mathfrak{f}(\mathbf{fu}) = \mathfrak{f}(\mathbf{uu}) = \mathbf{uu}$. In all of these desired output labelling c has value $\mathbf{u}$, so $\mathbf{ff}$ is determining for c. On the other hand $\mathbf{ff}$ is not determining for d, since $\mathfrak{f}(\mathbf{uf})(d) = \mathbf{f}$. ◇

The semantics-specific construction for grounded semantics uses the concept of determining labelling and is defined as follows.

Definition 16. *Given a 3-valued I/O-specification $\mathfrak{f}$ with $|\mathfrak{f}(v)| = 1$ for all $v \in \mathcal{V}(I)$, the gr-specific part of $\mathfrak{D}_\mathfrak{f}^{gr}$ is given by*

$$X_\mathfrak{f}^{gr} = \{x_o^v \mid o \in O, v \in \mathfrak{d}_\mathfrak{f}(o)\}, \text{ and}$$
$$R_\mathfrak{f}^{gr} = \{(i, x_o^v) \mid x_o^v \in X_\mathfrak{f}^{gr}, v(i) = \mathbf{f}\} \cup \{(i', x_o^v) \mid x_o^v \in X_\mathfrak{f}^{gr}, v(i) = \mathbf{t}\} \cup$$
$$\{(x_o^v, o') \mid x_o^v \in X_\mathfrak{f}^{gr}, \mathfrak{f}(v)(o) = \mathbf{t}\} \cup \{(x_o^v, o) \mid x_o^v \in X_\mathfrak{f}^{gr}, \mathfrak{f}(v)(o) = \mathbf{f}\}.$$

For every $o \in O$ and each input labelling v which is determining for o, there is the argument x_o^v. This argument can be assigned $\mathbf{t}$ if v is the labelling of I (recall Lemma 1) and intuitively enforces the labelling of o to be as given by $\mathfrak{f}(v)$.

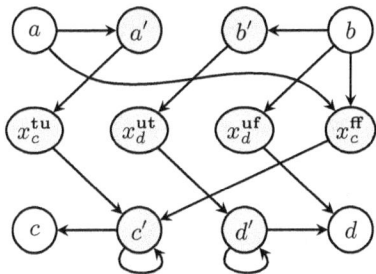

Fig. 5: Canonical I/O-gadget $\mathfrak{D}_{\mathfrak{f}}^{gr}$ for the 3-valued I/O-specification $\mathfrak{f}$ given in Example 6. As discussed in Example 8, it indeed realizes $\mathfrak{f}$ under gr_3.

Example 7. Again consider the 3-valued I/O-specification $\mathfrak{f}$ from Example 6. We have seen the determining labellings there. The I/O-gadget $\mathfrak{D}_{\mathfrak{f}}^{gr}$ is depicted in Figure 5. It can be seen that, for every output argument o, each determining labelling $v \in \mathfrak{d}_{\mathfrak{f}}(o)$ has a corresponding argument x_o^v in $\mathfrak{D}_{\mathfrak{f}}^{gr}$. Depending on the value of $\mathfrak{f}(v)(o)$, x_o^v either attacks argument o or o'. ◇

The next results, requiring two preliminary lemmas, characterize realizability of grounded semantics.

Lemma 2. *Let $\mathfrak{f}$ be a 3-valued I/O-specification which is monotonic and s.t. $|\mathfrak{f}(v)| = 1$ for each $v \in \mathcal{V}(I)$. Moreover let $o \in O$ and $v_1, v_2 \in \mathcal{V}(I)$ be such that $\mathfrak{f}(v_1)(o) = \mathbf{t}$ and $\mathfrak{f}(v_2)(o) = \mathbf{f}$. Then v_1 and v_2 are not compatible.*

Proof. Towards a contradiction assume that v_1 and v_2 are compatible, and let $V \in \mathcal{V}(I)$ such that for each $i \in I$, $V(i) = \mathbf{t}$ iff $v_1(i) = \mathbf{t} \vee v_2(i) = \mathbf{t}$, $V(i) = \mathbf{f}$ iff $v_1(i) = \mathbf{f} \vee v_2(i) = \mathbf{f}$, $V(i) = \mathbf{u}$ iff $v_1(i) = v_2(i) = \mathbf{u}$. Note that V is well-defined since $v_1^{\mathbf{t}} \cap v_2^{\mathbf{f}} = v_1^{\mathbf{f}} \cap v_2^{\mathbf{t}} = \emptyset$. It holds that $v_1 \leq_i V$, thus $\mathfrak{f}(V)(o) = \mathbf{t}$ since $\mathfrak{f}$ is monotonic. However, it also holds that $v_2 \leq_i V$, thus $\mathfrak{f}(V)(o) = \mathbf{f}$, a contradiction.

Lemma 3. *Given a 3-valued I/O-specification $\mathfrak{f}$ which is monotonic and s.t. $|\mathfrak{f}(v)| = 1$ for each $v \in \mathcal{V}(I)$, let $o \in O$ and $v_1, v_2 \in \mathcal{V}(I)$ be such that v_1 is determining for o. Then $gr_3(\blacktriangleright(\mathfrak{D}_{\mathfrak{f}}^{gr}, v_2))(x_o^{v_1})$ is*

1. *$\mathbf{t}$ if $v_1 \leq_i v_2$;*
2. *$\mathbf{f}$ if v_1 and v_2 are not compatible; and*
3. *$\mathbf{u}$ if v_1 and v_2 are compatible but $v_1 \not\leq_i v_2$.*

Proof. Let $g = gr_3(\blacktriangleright(\mathfrak{D}_{\mathfrak{f}}^{gr}, v_2))$ and note that, by Lemma 1, we know that $g|_I = v_2$ and $g|_{I'} = \neg v_2$.

(1) If $v_1 \leq_i v_2$ then, by construction of $\mathfrak{D}_{\mathfrak{f}}^{gr}$ and Lemma 1, all attackers of $x_o^{v_1}$ are $\mathbf{f}$ in g, hence $g(x_o^{v_1}) = \mathbf{t}$.

(2) If v_1 and v_2 are not compatible then there is some $i \in I$ such that either $v_1(i) = \mathbf{t}$ and $v_2(i) = \mathbf{f}$ or $v_1(i) = \mathbf{f}$ and $v_2(i) = \mathbf{t}$. In the first case

$x_o^{v_1}$ is attacked by i' and $g(i') = \mathbf{t}$, in the second case $x_o^{v_1}$ is attacked by i and $g(i) = \mathbf{t}$, both entailing $g(x_o^{v_1}) = \mathbf{f}$.

(3) If v_1 and v_2 are compatible then, by construction of $\mathfrak{D}_\mathfrak{f}^{gr}$ and Lemma 1, all attackers of $x_o^{v_1}$ are either $\mathbf{f}$ or $\mathbf{u}$. Moreover, since $v_1 \not\leq_i v_2$ there is some $i \in I$ with $v_2(i) = \mathbf{u}$ and $v_1(i) \neq \mathbf{u}$. But then $g(i) = g(i') = \mathbf{u}$ and $x_o^{v_1}$ is attacked by either i or i', hence $g(x_o^{v_1}) = \mathbf{u}$.

Proposition 8. *Every 3-valued I/O-specification $\mathfrak{f}$ which is monotonic and s.t. $|\mathfrak{f}(v)| = 1$ for each $v \in \mathcal{V}(I)$, is realized by $\mathfrak{D}_\mathfrak{f}^{gr}$ under gr_3.*

Proof. Consider some input labelling v. We have to show $gr_3(\blacktriangleright(\mathfrak{D}_\mathfrak{f}^{gr}, v))|_O = \mathfrak{f}(v)$. To this end let $o \in O$.

Assume $\mathfrak{f}(v)(o) = \mathbf{u}$. Then, since $\mathfrak{f}$ is monotonic, $\mathfrak{f}(v')(o) = \mathbf{u}$ for all $v' \leq_i v$. Therefore, there is no $v' \leq_i v$ with $v' \in \mathfrak{d}_\mathfrak{f}(o)$. By Lemma 3 we therefore get that for all $v'' \in \mathfrak{d}_\mathfrak{f}(o)$ it holds that $gr_3(\blacktriangleright(\mathfrak{D}_\mathfrak{f}^{gr}, v))(x_o^{v''}) \neq \mathbf{t}$. Since, by construction of $\mathfrak{D}_\mathfrak{f}^{gr}$, such $x_o^{v''}$ with $v'' \in \mathfrak{d}_\mathfrak{f}(o)$ are the only attackers of o and o', $gr_3(\blacktriangleright(\mathfrak{D}_\mathfrak{f}^{gr}, v))(o) = \mathbf{u}$.

Next assume $\mathfrak{f}(v)(o) = \mathbf{t}$. Then there is some $v' \leq_i v$ with $v' \in \mathfrak{d}_\mathfrak{f}(o)$ and $\mathfrak{f}(v')(o) = \mathbf{t}$. By Lemma 3 we get $gr_3(\blacktriangleright(\mathfrak{D}_\mathfrak{f}^{gr}, v))(x_o^{v'}) = \mathbf{t}$. Moreover, $x_o^{v'}$ attacks o', hence $gr_3(\blacktriangleright(\mathfrak{D}_\mathfrak{f}^{gr}, v))(o') = \mathbf{f}$. Towards a contradiction assume there is some $x_o^{v''}$ attacking o with $gr_3(\blacktriangleright(\mathfrak{D}_\mathfrak{f}^{gr}, v))(x_o^{v''}) \in \{\mathbf{t}, \mathbf{u}\}$. Then, by Lemma 3, v'' and v are compatible and as $v' \leq_i v$ also v'' and v' are compatible. However, $x_o^{v''}$ attacking o and $x_o^{v'}$ attacking o' means, by construction of $\mathfrak{D}_\mathfrak{f}^{gr}$, $\mathfrak{f}(v'')(o) = \mathbf{f}$ and $\mathfrak{f}(v')(o) = \mathbf{t}$, respectively. But then, by Lemma 2, v'' and v' are not compatible, a contradiction. Therefore all attackers of o are labelled $\mathbf{f}$ by $gr_3(\blacktriangleright(\mathfrak{D}_\mathfrak{f}^{gr}, v))$, hence $gr_3(\blacktriangleright(\mathfrak{D}_\mathfrak{f}^{gr}, v))(o) = \mathbf{t}$.

Finally assume $\mathfrak{f}(v)(o) = \mathbf{f}$. Then there is some $v' \leq_i v$ with $v' \in \mathfrak{d}_\mathfrak{f}(o)$ and $\mathfrak{f}(v')(o) = \mathbf{f}$. By Lemma 3 we get $gr_3(\blacktriangleright(\mathfrak{D}_\mathfrak{f}^{gr}, v))(x_o^{v'}) = \mathbf{t}$. Moreover, $x_o^{v'}$ attacks o, hence $gr_3(\blacktriangleright(\mathfrak{D}_\mathfrak{f}^{gr}, v))(o) = \mathbf{f}$.

Example 8. Again consider the 3-valued I/O-specification $\mathfrak{f}$ from Example 6 and the corresponding I/O-gadget $\mathfrak{D}_\mathfrak{f}^{gr}$ depicted in Figure 5. Consider, for instance, the 3-valued injection of $\mathbf{fu}$ to $\mathfrak{D}_\mathfrak{f}^{gr}$, which adds the additional arguments z attacking a as well as a self-attack of b to $\mathfrak{D}_\mathfrak{f}^{gr}$. We get $gr(\blacktriangleright(\mathfrak{D}_\mathfrak{f}^{gr}, \mathbf{fu})) = \{z, a'\}$, hence $gr_3(\blacktriangleright(\mathfrak{D}_\mathfrak{f}^{gr}, \mathbf{fu}))|_O = \mathbf{uu}$, being in line with $\mathfrak{f}$. One can check that this holds for all possible 3-valued injections, hence $\mathfrak{D}_\mathfrak{f}^{gr}$ realizes $\mathfrak{f}$ under the grounded semantics. ◇

Theorem 7. *A 3-valued I/O-specification $\mathfrak{f}$ is realizable under gr_3 iff $\mathfrak{f}$ is monotonic and for each $v \in \mathcal{V}(I)$, $|\mathfrak{f}(v)| = 1$.*

Proof. The if-direction is a direct consequence of Proposition 8. The only-if-direction follows by Propositions 5 and 7.

Now we present the part of the 3-valued canonical I/O-gadget which is specific to the preferred semantics.

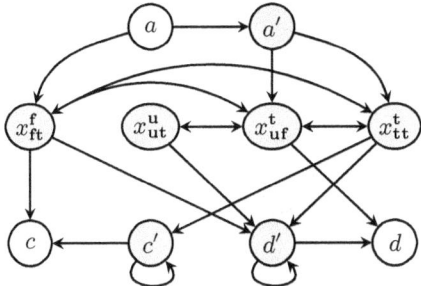

Fig. 6: 3-valued canonical I/O-gadget $\mathfrak{D}_{\mathfrak{f}}^{pr}$ for the 3-valued I/O-specification $\mathfrak{f}$ given in Example 9.

Definition 17. *Given a 3-valued I/O-specification $\mathfrak{f}$, the pr-specific part of $\mathfrak{D}_{\mathfrak{f}}^{pr}$ is given by*

$X_{\mathfrak{f}}^{pr} = \{x_w^v \mid v \in \mathcal{V}(I), w \in \mathfrak{f}(v)\}$, and
$R_{\mathfrak{f}}^{pr} = \{(i, x_w^v) \mid x_w^v \in X_{\mathfrak{f}}^{pr}, v(i) = \mathbf{f}\} \cup \{(i', x_w^v) \mid x_w^v \in X_{\mathfrak{f}}^{pr}, v(i) = \mathbf{t}\} \cup$
$\{(x_w^v, o') \mid x_w^v \in X_{\mathfrak{f}}^{pr}, w(o) = \mathbf{t}\} \cup \{(x_w^v, o) \mid x_w^v \in X_{\mathfrak{f}}^{pr}, w(o) = \mathbf{f}\} \cup$
$\{(x_w^v, x_{w'}^{v'}) \mid \neg(v <_i v' \wedge w \leq_i w') \wedge \neg(v' <_i v \wedge w' \leq_i w)\}.$

Every combination of an input labelling v and a corresponding output labelling w is represented by an argument x_w^v in $\mathfrak{D}_{\mathfrak{f}}^{pr}$. The way the input arguments are linked to x_w^v makes sure that, with input labelling v injected, x_w^v is not attacked by any argument among $I \cup I'$ which can be $\mathbf{t}$ in a preferred labelling (cf. Lemma 1). Therefore each such argument x_w^v can act as a representative for a preferred labelling, enforcing output labelling w. The attacks among arguments $X_{\mathfrak{f}}^{pr}$ are symmetric such that x_w^v attacks all $x_{w'}^{v'}$ except those where either $v <_i v'$ and $w \leq_i w'$ or $v' <_i v$ and $w' \leq_i w$.

Example 9. Consider the 3-valued I/O-specification for $I = \{a\}$ and $O = \{c, d\}$ given by

$$\mathfrak{f}(\mathbf{u}) = \{\mathbf{ut}\} \qquad \mathfrak{f}(\mathbf{t}) = \{\mathbf{tt}, \mathbf{uf}\} \qquad \mathfrak{f}(\mathbf{f}) = \{\mathbf{ft}\}.$$

It is easy to see that $\mathfrak{f}$ is monotonic, since $\mathbf{ut} \in \mathfrak{f}(\mathbf{u})$ has a successor wrt. $\leq_i$ in both $\mathfrak{f}(\mathbf{t})$ (namely $\mathbf{tt}$) and $\mathfrak{f}(\mathbf{f})$ (namely $\mathbf{ft}$).

The AF in Figure 6 depicts the 3-valued canonical I/O-gadget $\mathfrak{D}_{\mathfrak{f}}^{pr}$. Observe the symmetric attacks between arguments $x_w^v, x_{w'}^{v'} \in X_{\mathfrak{f}}^{pr}$ whenever $v = v'$ ($x_{\mathbf{uf}}^{\mathbf{t}}$ and $x_{\mathbf{tt}}^{\mathbf{t}}$), v and v' are not comparable (e.g. $x_{\mathbf{ft}}^{\mathbf{f}}$ and $x_{\mathbf{uf}}^{\mathbf{t}}$), or $v <_i v'$ but $w \not\leq_i w'$ ($x_{\mathbf{ut}}^{\mathbf{u}}$ and $x_{\mathbf{uf}}^{\mathbf{t}}$). However, there is no attack if both $v <_i v'$ and $w \leq_i w'$ holds, as for instance between $x_{\mathbf{ut}}^{\mathbf{u}}$ and $x_{\mathbf{tt}}^{\mathbf{t}}$. To see the motivation behind this consider the injection of $\mathbf{t}$ to $\mathfrak{D}_{\mathfrak{f}}^{pr}$. We get $pr(\blacktriangleright(\mathfrak{D}_{\mathfrak{f}}^{pr}, \mathbf{t})) = \{\{z, a, x_{\mathbf{ut}}^{\mathbf{u}}, x_{\mathbf{tt}}^{\mathbf{t}}, c, d\}, \{z, a, x_{\mathbf{uf}}^{\mathbf{t}}\}\}$, giving rise to $pr_3(\blacktriangleright(\mathfrak{D}_{\mathfrak{f}}^{pr}, \mathbf{t}))|_O = \{\mathbf{tt}, \mathbf{uf}\}$, therefore realizing $\mathfrak{f}$ under pr_3. A (symmetric) attack between x_{ut}^u and x_{tt}^t would give **ut** as output labelling, which is not as specified by $\mathfrak{f}(\mathbf{t})$. ◇

In the following we formally show that $\mathfrak{D}_\mathfrak{f}^{pr}$ realizes $\mathfrak{f}$ under preferred semantics, given that $\mathfrak{f}$ is monotonic and there is at least one output labelling for each input labelling. We begin with a technical lemma, giving sufficient conditions on the status of the arguments in $X_\mathfrak{f}^{pr}$ to get the desired labelling of the output arguments.

Lemma 4. *Given a 3-valued I/O-specification $\mathfrak{f}$ and an input labelling $v \in \mathcal{V}(I)$, it holds for each preferred labelling $p \in pr_3(\blacktriangleright(\mathfrak{D}_\mathfrak{f}^{pr}, v))$ that $p|_O = w$ for some $w \in \mathfrak{f}(v)$ if*

- $p(x_w^v) = \mathbf{t}$ *and*
- *for all $x_{w'}^{v'} \in X_\mathfrak{f}^{pr}$,*
  - $w <_i w'$ *implies $p(x_{w'}^{v'}) \neq \mathbf{t}$ and*
  - w *and* w' *being not comparable implies $p(x_{w'}^{v'}) = \mathbf{f}$.*

Proof. Consider some $w \in \mathfrak{f}(v)$ and an arbitrary $o \in O$. We show that $p(o) = w(o)$.

First assume $w(o) = \mathbf{u}$. By the hypothesis $p(x_{w'}^{v'}) \neq \mathbf{t}$ for all $w' \not\leq_i w$. Moreover, for all $w' \leq_i w$ we have that $w'(o) = \mathbf{u}$ since $w(o) = \mathbf{u}$, thus by construction of $\mathfrak{D}_\mathfrak{f}^{pr}$, $x_{w'}^{v'}$ attacks neither o nor o'. Summing up, neither o nor o' is attacked by an argument which is $\mathbf{t}$ in p. Hence $p(o) = \mathbf{u}$.

Next let $w(o) = \mathbf{t}$. Since $p(x_w^v) = \mathbf{t}$ we must have that $p(o') = \mathbf{f}$. Besides that, o is attacked by all $x_{w'}^{v'}$ with $w'(o) = \mathbf{f}$. But this means that w and w' are not comparable, hence $p(x_{w'}^{v'}) = \mathbf{f}$ by assumption. Now we know that all attackers of o are $\mathbf{f}$ in p, therefore $p(o) = \mathbf{t}$.

Finally let $w(o) = \mathbf{f}$. Since $p(x_w^v) = \mathbf{t}$ and x_w^v attacks o we get that $p(o) = \mathbf{f}$.

We proceed by showing that every monotonic function $\mathfrak{f}$ assigning at least one output labelling to each input labelling is realized by $\mathfrak{D}_\mathfrak{f}^{pr}$ under the preferred semantics.

Proposition 9. *Every 3-valued I/O-specification $\mathfrak{f}$ which is monotonic and s.t. $|\mathfrak{f}(v)| \geq 1$ for each $v \in \mathcal{V}(I)$ is realized by $\mathfrak{D}_\mathfrak{f}^{pr}$ under pr_3.*

Proof. Consider an arbitrary input labelling $v \in \mathcal{V}(I)$. In the following we show that $pr_3(\blacktriangleright(\mathfrak{D}_\mathfrak{f}^{pr}, v))|_O = \mathfrak{f}(v)$.

By construction of $\mathfrak{D}_\mathfrak{f}^{pr}$, those $x_{w'}^{v'} \in X_\mathfrak{f}^{pr}$ with $v' \leq_i v$ are the only arguments in $X_\mathfrak{f}^{pr}$ which can be $\mathbf{t}$ in a preferred labelling of $\blacktriangleright(\mathfrak{D}_\mathfrak{f}^{pr}, v)$, since their attackers in $I \cup I'$ are all $\mathbf{f}$, while the other arguments in $X_\mathfrak{f}^{pr}$ are attacked by an argument $\mathbf{t}$ or $\mathbf{u}$ of $I \cup I'$ (recall also Lemma 1). The arguments x_w^v with $w \in \mathfrak{f}(v)$ (there is at least one such argument by the hypothesis) form a clique in $\blacktriangleright(\mathfrak{D}_\mathfrak{f}^{pr}, v)$. Moreover each of these x_w^v defends itself from all other $x_{w'}^{v'}$, hence there is a preferred labelling of $\blacktriangleright(\mathfrak{D}_\mathfrak{f}^{pr}, v)$ for each $w \in \mathfrak{f}(v)$ identified by x_w^v. Let p_w

be the preferred labelling with $p_w(x_w^v) = \mathbf{t}$ where $w \in \mathfrak{f}(v)$. All $x_{w'}^{v'}$ with $w' \not\leq_i w \wedge w \not\leq_i w'$ are attacked by x_v^w, hence $p_w(x_{w'}^{v'}) = \mathbf{f}$. Assume $w <_i w'$. If $v \not<_i v'$, then $x_{w'}^{v'}$ is again attacked by x_v^w and $p_w(x_{w'}^{v'}) = \mathbf{f}$. If $v <_i v'$, $p_w(x_{w'}^{v'}) \neq \mathbf{t}$ since it is attacked by some argument among $I \cup I'$ which is $\mathbf{u}$ in p_w. Therefore, by Lemma 4, $p_w|_O = w$.

It remains to show that there is no other preferred labelling besides these p_w with $w \in \mathfrak{f}(v)$. Towards a contradiction, assume that there is a preferred labelling p' where no x_w^v with $w \in \mathfrak{f}(v)$ is $\mathbf{t}$. By our initial considerations, those $x_{w'}^{v'}$ with $v' <_i v$ are the only arguments of $X_{\mathfrak{f}}^{pr}$ which can be $\mathbf{t}$ in p'. It cannot be the case that none of them is $\mathbf{t}$, since p' would not be preferred, being less committed (wrt. $\leq_i$) than p_w with $w \in \mathfrak{f}(v)$, of which there exists at least one. Therefore there is at least one $x_{w'}^{v'}$ which is $\mathbf{t}$ in p', with $v' <_i v$, and without loss of generality we can assume that there is no $x_{w''}^{v''}$ which is $\mathbf{t}$ and $v' <_i v''$. Now, since $\mathfrak{f}$ is monotonic there has to be a $w \in \mathfrak{f}(v)$ such that $w' \leq_i w$. We prove that no argument in $X_{\mathfrak{f}}^{pr}$ attacking x_w^v is $\mathbf{t}$ in p'.

First, the only arguments in $X_{\mathfrak{f}}^{pr}$ that can be $\mathbf{t}$ are those $x_{w''}^{v''}$ with $v'' <_i v$. Note that, according to Definition 17, $x_{w'}^{v'}$ does not attack x_w^v, since $v' <_i v$ and $w' \leq_i w$. If an attacker $x_{w''}^{v''}$ is attacked in turn by $x_{w'}^{v'}$ then it is $\mathbf{f}$, otherwise either $v'' <_i v' \wedge w'' \leq_i w'$ or $v' <_i v'' \wedge w' \leq_i w''$. The first case is impossible, since we would have $v'' <_i v \wedge w'' \leq_i w$, entailing that $x_{w''}^{v''}$ does not attack x_w^v. In the other case, by the assumption on $x_{w'}^{v'}$ it holds that $x_{w''}^{v''}$ is not $\mathbf{t}$.

Now, x_w^v defends itself against all arguments in $X_{\mathfrak{f}}^{pr}$ and none of them is $\mathbf{t}$, moreover by construction of $\blacktriangleright(\mathfrak{D}_{\mathfrak{f}}^{pr}, v)$, all attackers from I and I' are $\mathbf{f}$. But then, consider the labelling p'' obtained from p' by assigning to x_w^v the label $\mathbf{t}$, and by assigning to all the attackers of x_w^v the label $\mathbf{f}$. p'' is admissible and $p' \leq_i p''$, contradicting the maximality of p'.

Theorem 8. *A 3-valued I/O-specification $\mathfrak{f}$ is realizable under pr_3 iff $\mathfrak{f}$ is monotonic and such that $|\mathfrak{f}(v)| \geq 1$ for each $v \in \mathcal{V}(I)$.*

Proof. The if-direction was shown in Proposition 9 while the only-if-direction directly follows from Propositions 6 and 7.

Comparing the capabilities of stable and preferred semantics in realizing 3-valued I/O-specifications we find that they are incomparable. Taking into account Theorems 6 and 8, while we can realize 3-valued I/O-specifications which demand the empty set of output labellings for certain input labelling under st_3 but not under pr_3, we cannot realize anything involving output labels $\mathbf{u}$ under st_3.

For the remaining semantics we have to leave the exact characterizations of realizable 3-valued I/O-specifications open. However, we can show that realizability of a 3-valued I/O-specification under gr_3 is a sufficient condition for realizability under id_3, as well as that realizability under pr_3 is sufficient for se_3. First we show that Proposition 8 also applies to ideal semantics since for each labelling v the grounded labelling of $\blacktriangleright(\mathfrak{D}_{\mathfrak{f}}^{gr}, v)$ coincides with the ideal labelling.

Proposition 10. *Every 3-valued I/O-specification $\mathfrak{f}$ which is monotonic and s.t. $|\mathfrak{f}(v)| = 1$ for each $v \in \mathcal{V}(I)$, is realized by $\mathfrak{D}_{\mathfrak{f}}^{gr}$ under id_3.*

Proof. Let $\mathfrak{f}$ be an 3-valued *I/O*-specification which is monotonic and s.t. $|\mathfrak{f}(v)| = 1$ for each $v \in \mathcal{V}(I)$. Consider some input labelling v, and let $F = \blacktriangleright(\mathfrak{D}_{\mathfrak{f}}^{gr}, v)$. We show that $id(F) = gr(F)$.

To this end assume, towards a contradiction, that there is some $E \in ad(F)$ such that $E \supset gr(F)$, i.e. $E \setminus gr(F) \neq \emptyset$. Let $E' = E \setminus gr(F)$. First of all, $z \in gr(F)$, hence $z \notin E'$. Next, consider some $i \in I$. If $v(i) = \mathbf{t}$ then $i \in gr(F)$, if $v(i) = \mathbf{f}$ then $i' \in gr(F)$ (cf. Lemma 1); either way, $i, i' \in gr(F)^{\oplus}$, hence $i, i' \notin E'$. If $v(i) = \mathbf{u}$ then i is self-attacking and not attacked otherwise in F, hence neither i nor i' can be included in an admissible set, i.e. $i, i' \notin E'$. Now consider $x_o^{v'} \in X_{\mathfrak{f}}^{gr}$. By Lemma 3 it follows that if $v' \leq_i v$ or v and v' are not compatible, then $x_o^{v'} \in gr(F)^{\oplus}$, hence $x_o^{v'} \notin E'$. If v and v' are compatible but $v' \not\leq_i v$ then $x_o^{v'} \notin gr(F)^{\oplus}$, but there is some $i \in I$ with $v'(i) \neq \mathbf{u}$ and $v(i) = \mathbf{u}$. Consequently, $x_o^{v'}$ is attacked by either i or i', which, as discussed before, can both not be defended in F, hence also $x_o^{v'} \notin E'$. Finally, consider $o \in O$ and assume by contradiction that $o \notin gr(F)^{\oplus}$. For $o \in E'$, there must be some $x \in X_{\mathfrak{f}}^{gr}$ with $x \in E'$. But this cannot be the case, as just shown.

We have shown that there is no $E \in ad(F)$ with $E \supset gr(F)$. Hence $id(F) = gr(F)$. Since, by Proposition 8, $\blacktriangleright(gr_3(\mathfrak{D}_{\mathfrak{f}}^{gr}), v)\big|_O = \mathfrak{f}(v)$, we can infer that also $\blacktriangleright(id_3(\mathfrak{D}_{\mathfrak{f}}^{gr}), v)\big|_O = \mathfrak{f}(v)$. Therefore the result follows.

We cannot apply the result from Proposition 9 directly to semi-stable semantics, since it is not guaranteed that each preferred extension of the 3-valued injection of some input labelling to $\mathfrak{D}_{\mathfrak{f}}^{pr}$ has $\subseteq$-maximal range, hence we might "lose" some elements of $\mathfrak{f}(v)$ under se_3. On the other hand, the well-known translation of arbitrary AFs from preferred to semi-stable semantics, due to [18], makes sure that the semi-stable extensions of the AF obtained from the translation coincide with the preferred extensions of the original AF. We use the idea of this translation to define the semi-stable specific part of the 3-valued canonical *I/O*-gadget.

Definition 18. *Given a 3-valued I/O-specification $\mathfrak{f}$, the se-specific part of $\mathfrak{D}_{\mathfrak{f}}^{se}$ is given by*

$$X_{\mathfrak{f}}^{se} = \{x, x' \mid x \in X_{\mathfrak{f}}^{pr}\}, \text{ and}$$
$$R_{\mathfrak{f}}^{se} = R_{\mathfrak{f}}^{pr} \cup \{(x, x'), (x', x), (x', x') \mid x \in X_{\mathfrak{f}}^{pr}\}$$

The idea of the translation from preferred to semi-stable semantics is to make the range of every argument incomparable to the range of every other argument, hence making also the range of preferred extensions pairwise incomparable. This is achieved by adding, for each argument a, a self-attacking argument a' which is in symmetric attack with a. Since any preferred extension of an injection to $\mathfrak{D}_{\mathfrak{f}}^{pr}$ can be uniquely identified by some $x \in X_{\mathfrak{f}}^{pr}$, it suffices to add the primed arguments for each element of $X_{\mathfrak{f}}^{pr}$.

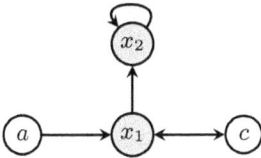

Fig. 7: I/O-gadget realizing a 3-valued I/O-specification under semi-stable (resp. ideal) semantics which is not realizable under preferred (resp. grounded) semantics.

Proposition 11. *Every 3-valued I/O-specification $\mathfrak{f}$ which is monotonic and s.t. $|\mathfrak{f}(v)| \geq 1$ for each $v \in \mathcal{V}(I)$ is realized by $\mathfrak{D}_\mathfrak{f}^{se}$ under se_3.*

Proof. Let $\mathfrak{f}$ be a monotonic 3-valued I/O-specification. Consider an arbitrary input labelling $v \in \mathcal{V}(I)$. Inspecting the proof of Proposition 9 we observe that $pr_3(\mathfrak{D}_\mathfrak{f}^{pr}) = \{p_w \mid w \in \mathfrak{f}(v)\}$, where each p_w contains x_w^v and no other $x_{w'}^v$ with $w' \neq w$. Further observe that $pr(\mathfrak{D}_\mathfrak{f}^{pr}) = pr(\mathfrak{D}_\mathfrak{f}^{se})$. By the construction of $\mathfrak{D}_\mathfrak{f}^{se}$ now every p_w has incomparable range by uniquely attacking $x_{w'}^v$. Therefore $pr(\mathfrak{D}_\mathfrak{f}^{se}) = se(\mathfrak{D}_\mathfrak{f}^{se})$.

Having established $pr(\mathfrak{D}_\mathfrak{f}^{pr}) = se(\mathfrak{D}_\mathfrak{f}^{se})$, we conclude from $\blacktriangleright(pr_3(\mathfrak{D}_\mathfrak{f}^{pr}), v)|_O = \mathfrak{f}(v)$ (cf. Proposition 9) that $\blacktriangleright(se_3(\mathfrak{D}_\mathfrak{f}^{se}), v)|_O = \mathfrak{f}(v)$. Therefore the result follows.

While Propositions 10 and 11 give sufficient conditions for 3-valued I/O-realizability under ideal and semi-stable semantics, respectively, they do not allow us to derive complete characterizations, since there are 3-valued I/O-specifications, which are not monotonic but realizable under ideal and semi-stable semantics, respectively. The following example illustrates this fact.

Example 10. Consider the AF F in Figure 7 which represents an I/O-gadget with $I = \{a\}$ and $O = \{c\}$. The semi-stable and ideal extensions of the various 3-valued injections are $se(\blacktriangleright(F, \mathbf{u})) = id(\blacktriangleright(F, \mathbf{u})) = \{\{c\}\}$, $se(\blacktriangleright(F, \mathbf{t})) = id(\blacktriangleright(F, \mathbf{t})) = \{\{a, c\}\}$, and $se(\blacktriangleright(F, \mathbf{f})) = \{\{z, x_1\}\} \neq id(\blacktriangleright(F, \mathbf{f})) = \{\{z\}\}$. This means that F realizes the 3-valued I/O-specification given by $\mathbf{u} \mapsto \{\mathbf{t}\}$, $\mathbf{t} \mapsto \{\mathbf{t}\}$, and $\mathbf{f} \mapsto \{\mathbf{u}\}$ under ideal semantics, and the one given by $\mathbf{u} \mapsto \{\mathbf{t}\}$, $\mathbf{t} \mapsto \{\mathbf{t}\}$, and $\mathbf{f} \mapsto \{\mathbf{f}\}$ under semi-stable semantics which are both clearly not monotonic. ◊

An exact characterization of 3-valued I/O-specifications which are realizable under semi-stable and ideal semantics, respectively, therefore requires weaker notions of monotonicity. Complete semantics, on the other hand, imposes necessary conditions which are more restrictive. The following is a direct consequence of [1, Proposition 7].

Proposition 12. *Every 3-valued I/O-specification $\mathfrak{f}$ which is realizable under co_3 is monotonic and for all v_1 and v_2 such that $v_1 \leq_i v_2$ it holds that $\forall w_2 \in \mathfrak{f}(v_2) \exists w_1 \in \mathfrak{f}(v_1) : w_1 \leq_i w_2$.*

Example 11. Once more consider the 3-valued I/O-specification $\mathfrak{f}$ from Example 4. We have seen that $\mathfrak{f}$ is monotonic and therefore realizable under pr_3 in Example 5. It is, however, not realizable under complete semantics. To see this let $v_1 = \mathbf{tu}$ and $v_2 = \mathbf{tt}$. We have $w_2 = \mathbf{tf} \in \mathfrak{f}(v_2)$ but there is no $w_1 \in \mathfrak{f}(v_2)$ such that $w_1 \leq_i w_2$. Therefore the condition given in Proposition 12 is violated and $\mathfrak{f}$ is not realizable under co_3. ◇

Exact characterizations of realizable 3-valued I/O-specifications for complete, semi-stable and ideal semantics are subject of future work. With the following definition of the three-valued I/O-signature we can state some relations between semantics.

Definition 19. *Let σ be a semantics. The three-valued I/O-signature of σ_3 consists of all 3-valued I/O-specifications that are realizable under σ_3:*

$$\Sigma_{AF\blacktriangleright}^{\sigma_3} = \{\sigma_\blacktriangleright(F) \mid F \text{ is an } I/O\text{-gadget}\},$$

where $\sigma_\blacktriangleright$ is the three-valued I/O-version of σ, defined as a function mapping I/O-gadgets to 3-valued I/O-specifications such that, given an I/O-gadget F, $\sigma_\blacktriangleright(F) = v \in \mathcal{V}(I) \mapsto \sigma_3(\blacktriangleright(F,v))|_O$.

Theorem 9. *The following relations hold:*

1. $\Sigma_{AF\blacktriangleright}^{gr_3} \subset \Sigma_{AF\blacktriangleright}^{co_3} \subset \Sigma_{AF\blacktriangleright}^{pr_3} \subset \Sigma_{AF\blacktriangleright}^{se_3}$;
2. $\Sigma_{AF\blacktriangleright}^{gr_3} \subset \Sigma_{AF\blacktriangleright}^{id_3}$;
3. $\Sigma_{AF\blacktriangleright}^{id_3} \setminus \Sigma_{AF\blacktriangleright}^{pr_3} \neq \emptyset$ *and* $\Sigma_{AF\blacktriangleright}^{pr_3} \setminus \Sigma_{AF\blacktriangleright}^{id_3} \neq \emptyset$.

Proof. (1) $\Sigma_{AF\blacktriangleright}^{gr_3} \subseteq \Sigma_{AF\blacktriangleright}^{co_3}$ holds by the observation that a 3-valued I/O-specification $\mathfrak{f}$ which is monotonic and has $|\mathfrak{f}(v)| = 1$ is realized by $\mathfrak{D}_\mathfrak{f}^{gr}$ also under complete semantics, since $gr_3(\blacktriangleright(\mathfrak{D}_\mathfrak{f}^{gr}, v)) = co_3(\blacktriangleright(\mathfrak{D}_\mathfrak{f}^{gr}, v))$. This can be read off the proof of Proposition 10, where it is shown that there is no $E \in ad(\blacktriangleright(\mathfrak{D}_\mathfrak{f}^{gr}, v))$ with $E \supset gr(\blacktriangleright(\mathfrak{D}_\mathfrak{f}^{gr}, v))$. Properness is, for instance, by the I/O-gadget $F = (\{a,b,c\}, \{(a,b),(b,a),(b,b),(b,c)\})$ with $I = \{a\}$ and $O = \{c\}$, which has $co_3(\blacktriangleright(F, \mathbf{u}))|_O = \{\mathbf{u}\}$, $co_3(\blacktriangleright(F, \mathbf{t}))|_O = \{\mathbf{u}, \mathbf{t}\}$, and $co_3(\blacktriangleright(F, \mathbf{f}))|_O = \{\mathbf{u}\}$.

$\Sigma_{AF\blacktriangleright}^{co_3} \subseteq \Sigma_{AF\blacktriangleright}^{pr_3}$ is by Theorem 8 and Proposition 12. Properness of the relation was discussed in Example 11.

Finally, $\Sigma_{AF\blacktriangleright}^{pr_3} \subseteq \Sigma_{AF\blacktriangleright}^{se_3}$ is by Theorem 8 and Proposition 11, while properness was shown in Example 10.

(2) $\Sigma_{AF\blacktriangleright}^{gr_3} \subseteq \Sigma_{AF\blacktriangleright}^{id_3}$ follows from Theorem 7 and Proposition 10. Properness of the relation was shown in Example 10.

(3) $\Sigma_{AF\blacktriangleright}^{id_3} \setminus \Sigma_{AF\blacktriangleright}^{pr_3} \neq \emptyset$ is by the I/O-gadget in Figure 7, which realizes a 3-valued I/O-specification under id_3 which is not monotonic and therefore not realizable under pr_3 (cf. Example 10). $\Sigma_{AF\blacktriangleright}^{pr_3} \setminus \Sigma_{AF\blacktriangleright}^{id_3} \neq \emptyset$ is, for instance, by the 3-valued I/O-specification discussed in Example 9, which has more than one output labelling assigned to the input labelling $\mathbf{t}$.

5 Partial I/O-specifications

Until now we have restricted our considerations to total I/O-specifications, where the output is defined for each input. It is however natural to think of situations where we do not care about the output for some inputs, i.e. where we are only interested in realizability of a partial function.

Definition 20. *A partial 2-valued (resp. 3-valued) I/O-specification consists of two sets $I, O \subseteq \mathfrak{A}$ and a partial function $\mathfrak{f} : 2^I \mapsto 2^{2^O}$ (resp. $\mathfrak{f} : \mathcal{V}(I) \mapsto 2^{\mathcal{V}(O)}$).*

An I/O-gadget F realizes $\mathfrak{f}$ under a semantics σ iff for all $J \subseteq I$ (resp. $v \in \mathcal{V}(I)$) such that $\mathfrak{f}$ is defined for J (resp. v), $\sigma(\triangleright(F, J))|_O = \mathfrak{f}(J)$ (resp. $\sigma_3(\blacktriangleright(F, v))|_O = \mathfrak{f}(v)$).

The results provided in Theorems 1, 2, 3, and 4 for the 2-valued case can be directly exploited to handle partial I/O-specifications in the 2-valued case: "don't care"-outputs can be assigned arbitrarily, provided that at least one extension is assigned for pr, se and sg, a single extension for gr and id, and the specification is closed for co. This is because conditions for realizability contain no dependencies between outputs for different inputs. Furthermore, all the proofs also work with a partial I/O-specification by considering only the specified inputs in the definition of the canonical I/O-gadget, i.e. neglecting the inputs with undefined output, yielding a considerable simplification.

Corollary 1. *A partial 2-valued I/O-specification is realizable under semantics $\sigma \in \{pr, se, sg, co, gr, id\}$ iff for each $J \subseteq I$ such that $\mathfrak{f}$ is defined for J it holds that $\mathfrak{f}(J) \neq \emptyset$ and, in case $\sigma = co$, $\bigcap \mathfrak{f}(J) \in \mathfrak{f}(J)$. A partial 2-valued I/O-specification is always realizable under st.*

On the other hand, the following example shows some difficulties in the three-valued case.

Example 12. Consider the partial 3-valued I/O-specification $\mathfrak{f}$ for $I = \{a, b\}$ and $O = \{c\}$ with $\mathfrak{f}(\mathbf{uu}) = \{\mathbf{u}\}$, $\mathfrak{f}(\mathbf{tu}) = \{\mathbf{t}\}$ and undefined, i.e. "don't care", for all other inputs. Clearly, $\mathfrak{f}$ is realized by $\mathfrak{D}_\mathfrak{f}^{gr}$ or, more easily, by the simple I/O-gadget $(\{a, b, x, c\}, \{(a, x), (x, c)\})$. Now note that $\mathfrak{f}$ is not monotonic according to Definition 13, since there is no $w \in \mathfrak{f}(\mathbf{tt})$ with $\mathbf{t} \leq_i w$. It is monotonic for those inputs for which it is defined though.

The same can be observed if we consider $\mathfrak{f}'$ which coincides with $\mathfrak{f}$ on inputs $\mathbf{uu}$ and $\mathbf{tu}$ but also defines $\mathfrak{f}'(\mathbf{ut}) = \{\mathbf{f}\}$. Now one can check that there is no I/O-gadget realizing $\mathfrak{f}'$ under the grounded semantics. The reason for this is that $\mathfrak{f}$ cannot be extended to a total 3-valued I/O-specification $\mathfrak{f}''$ which is still monotonic. In order to be monotonic $\mathfrak{f}''$ extending $\mathfrak{f}'$ would have to fulfill both $\mathbf{t} \leq_i w$ and $\mathbf{f} \leq_i w$ for the unique output $w \in \mathfrak{f}''(\mathbf{tt})$, which is obviously not possible. ◇

This already leads us to the condition for realizability of partial functions, which we state after formally defining what it means to extend a 3-valued I/O-specification.

Definition 21. *Given two (partial) 3-valued I/O-specifications $\mathfrak{f}$ and $\mathfrak{f}'$, we say that $\mathfrak{f}'$ extends $\mathfrak{f}$ iff for all $v \in \mathcal{V}(I)$ such that $\mathfrak{f}(v)$ is defined, $\mathfrak{f}'(v) = \mathfrak{f}(v)$.*

Theorem 10. *A (partial) 3-valued I/O-specification $\mathfrak{f}$ is realizable under semantics σ_3 iff there is a total function $\mathfrak{f}'$ extending $\mathfrak{f}$ which is realizable under σ_3.*

Proof. If $\mathfrak{f}'$ is realized by some I/O-gadget F then F also realizes $\mathfrak{f}$ since $\mathfrak{f}'(v) = \mathfrak{f}(v)$ for all $v \in \mathcal{V}(I)$ such that $\mathfrak{f}(v)$ is defined. If, on the other hand, $\mathfrak{f}$ is realized by some I/O-gadget F' then F' obviously also realizes some total 3-valued I/O-specification $\mathfrak{f}'$ which coincides with $\mathfrak{f}$ on those input labellings which are defined by $\mathfrak{f}$, i.e. $\mathfrak{f}'$ extends $\mathfrak{f}$.

It may be noted that the extension-based case can be viewed as a particular case of 3-valued partial specification where also the output is partially specified, i.e. for those inputs without undecided arguments we specify a set of extensions (i.e. without distinguishing between **f** and **u** arguments).

Another relation can be drawn between three-valued I/O-realizability of stable and preferred semantics. Recalling Theorem 6, in order to be realizable under stable semantics, a 3-valued I/O-specification $\mathfrak{f}$ has to have $\mathfrak{f}(v) = \emptyset$ for each $v \in \mathcal{V}(I)$ where $\exists i \in I : v(i) = \mathbf{u}$. Interpreting the desired output $\emptyset$ as "don't care", we can realize any 3-valued I/O-specification realizable under stable semantics also under preferred semantics.

Proposition 13. *Given a 3-valued I/O-specification $\mathfrak{f}$ which is realizable under st_3, let $\mathfrak{f}'$ be the partial 3-valued I/O-specification with $\mathfrak{f}'(v) = \mathfrak{f}(v)$ if $\mathfrak{f}(v) \neq \emptyset$ and $\mathfrak{f}'(v)$ undefined if $\mathfrak{f}(v) = \emptyset$. It holds that $\mathfrak{f}'$ is realizable under pr_3.*

Proof. First observe that $\mathfrak{f}(v) \neq \emptyset$, and therefore $\mathfrak{f}'(v)$ defined, only if $\nexists i \in I : v(i) = \mathbf{u}$. Let $\mathfrak{f}''$ be the 3-valued I/O-specification which has $\mathfrak{f}''(v) = \mathfrak{f}'(v)$ whenever $\mathfrak{f}'(v)$ is defined and $\mathfrak{f}''(v) = v_\mathbf{u}$, i.e. the labelling mapping all arguments to $\mathbf{u}$, otherwise. Consider $v_1, v_2 \in \mathcal{V}(I)$ with $v_1 \leq_i v_2$. From the first observation it follows that $\mathfrak{f}''(v_1) = \{v_\mathbf{u}\}$, hence every $w_2 \in \mathfrak{f}''(v_2)$ it holds that $w_1 \leq_i w_2$ for all $w_1 \in \mathfrak{f}''(v_1)$ (i.e. for $v_\mathbf{u}$). Since, by definition, $\mathfrak{f}''(v) \neq \emptyset$ for each $v \in \mathcal{V}(I)$, monotonicity of $\mathfrak{f}''$ follows. Hence, by Theorem 8, $\mathfrak{f}''$ is realizable under pr_3 and, by Theorem 10, also $\mathfrak{f}'$ is realizable under pr_3.

6 Conclusions

To the best of our knowledge, this paper and its conference version [20] provide the first characterization of the input/output expressive power of argumentation semantics. In [17], expressiveness has been studied as the capability of enforcing sets of extensions. The problem faced in this paper differs in two aspects: on the one hand, we have to enforce a set of extensions for any input rather than a single set of extensions; on the other hand, we can exploit non-output arguments that are not seen outside a sub-framework. Moreover we also consider labellings besides extensions. A labelling-based investigation exploiting hidden arguments is carried out in [19], but still in the context of an ordinary AF rather than an I/O-gadget. Pührer [26] and Strass [27] have investigated the expressiveness of abstract dialectical frameworks [6] under three-valued and two-valued semantics, respectively, and an algorithmic approach to three-valued realizability [22] has dealt with several generalizations of AFs.

Future work includes the 3-valued I/O-characterization of complete semantics, being the only transparent semantics [1] for which this was left open. Moreover, the investigation of further semantics such as CF2 [4] would be of interest. Another issue is the construction of I/O-gadgets from *compact I/O*-specifications where the function is not explicitly stated but, for instance, described as a Boolean (or three-valued) circuit. We conjecture that I/O-gadgets can then be composed from simple building blocks along the lines of the given circuit. A related question in this direction is the identification of minimal I/O-gadgets satisfying a given specification. Such minimal gadgets would pave the way for systematic simplifications of AFs where sub-frameworks are identified (together with their input/output arguments I and O) and replaced by smaller frameworks realizing the same I/O-specification. Research in this direction has recently been initiated from a slightly different perspective [5] where tailored equivalence notions are proposed for such simplification procedures.

Acknowledgments

This research has been supported by the Austrian Science Fund (FWF) through projects I1102 and I2843.

References

1. Pietro Baroni, Guido Boella, Federico Cerutti, Massimiliano Giacomin, Leendert van der Torre, and Serena Villata. On the Input/Output behaviour of argumentation frameworks. *Artificial Intelligence*, 217:144–197, 2014.
2. Pietro Baroni, Martin Caminada, and Massimiliano Giacomin. An introduction to argumentation semantics. *Knowledge Engineering Review*, 26(4):365–410, 2011.
3. Pietro Baroni and Massimiliano Giacomin. On principle-based evaluation of extension-based argumentation semantics. *Artificial Intelligence*, 171(10-15):675–700, 2007.
4. Pietro Baroni, Massimiliano Giacomin, and Giovanni Guida. SCC-Recursiveness: A general schema for argumentation semantics. *Artificial Intelligence*, 168(1-2):162–210, 2005.
5. Ringo Baumann, Wolfgang Dvořák, Thomas Linsbichler, and Stefan Woltran. A general notion of equivalence for abstract argumentation. In Carles Sierra, editor,

Proceedings of the 26th International Joint Conference on Artificial Intelligence (IJCAI 2017), pages 800–806. ijcai.org, 2017.

6. Gerhard Brewka, Stefan Ellmauthaler, Hannes Strass, Johannes P. Wallner, and Stefan Woltran. Abstract Dialectical Frameworks. An Overview. *IfCoLog Journal of Logics and their Applications*, 4(8):2263–2318, 2017.
7. Maximiliano Celmo Budán, Maria Laura Cobo, Diego C. Marténez, and Guillermo Ricardo Simari. Bipolarity in temporal argumentation frameworks. *International Journal of Approximate Reasoning*, 84:1–22, 2017.
8. Maximiliano Celmo Budán, Mauro Javier Gómez Lucero, Carlos Iván Chesñevar, and Guillermo Ricardo Simari. Modeling time and valuation in structured argumentation frameworks. *Information Sciences*, 290:22–44, 2015.
9. Martin Caminada and Leila Amgoud. On the evaluation of argumentation formalisms. *Artificial Intelligence*, 171(5-6):286–310, 2007.
10. Martin Caminada, Walter A. Carnielli, and Paul E. Dunne. Semi-stable semantics. *Journal of Logic and Computation*, 22(5):1207–1254, 2012.
11. Martin Caminada and Dov M. Gabbay. A logical account of formal argumentation. *Studia Logica*, 93(2):109–145, 2009.
12. Maria Laura Cobo, Diego C. Martínez, and Guillermo Ricardo Simari. On admissibility in timed abstract argumentation frameworks. In Helder Coelho, Rudi Studer, and Michael Wooldridge, editors, *Proceedings of the 19th European Conference on Artificial Intelligence, ECAI 2010*, volume 215 of *Frontiers in Artificial Intelligence and Applications*, pages 1007–1008. IOS Press, 2010.
13. Maria Laura Cobo, Diego C. Martínez, and Guillermo Ricardo Simari. Acceptability in timed frameworks with intermittent arguments. In Lazaros S. Iliadis, Ilias Maglogiannis, and Harris Papadopoulos, editors, *Proceedings of the 12th INNS EANN-SIG International Conference and the 7th IFIP WG 12.5 International Conference on Artificial Intelligence Applications and Innovations, AIAI 2011*, volume 364 of *IFIP Advances in Information and Communication Technology*, pages 202–211. Springer, 2011.
14. Martin Diller, Adrian Haret, Thomas Linsbichler, Stefan Rümmele, and Stefan Woltran. An extension-based approach to belief revision in abstract argumentation. *International Journal of Approximate Reasoning*, 93:395–423, 2018.
15. Jürgen Dix, Sven Ove Hansson, Gabriele Kern-Isberner, and Guillermo Ricardo Simari. Belief change and argumentation in multi-agent scenarios (Dagstuhl Seminar 13231). *Dagstuhl Reports*, 3(6):1–21, 2013.
16. Phan Minh Dung. On the acceptability of arguments and its fundamental role in nonmonotonic reasoning, logic programming and n-person games. *Artificial Intelligence*, 77(2):321–357, 1995.
17. Paul E. Dunne, Wolfgang Dvořák, Thomas Linsbichler, and Stefan Woltran. Characteristics of multiple viewpoints in abstract argumentation. *Artificial Intelligence*, 228:153–178, 2015.
18. Wolfgang Dvořák and Stefan Woltran. On the intertranslatability of argumentation semantics. *Journal of Artificial Intelligence Research*, 41:445–475, 2011.
19. Sjur Kristoffer Dyrkolbotn. How to argue for anything: Enforcing arbitrary sets of labellings using AFs. In Chitta Baral, Giuseppe De Giacomo, and Thomas Eiter, editors, *Proceedings of the 14th International Conference on Principles of Knowledge Representation and Reasoning, KR 2014*, pages 626–629. AAAI Press, 2014.

20. Massimiliano Giacomin, Thomas Linsbichler, and Stefan Woltran. On the functional completeness of argumentation semantics. In Chitta Baral, James P. Delgrande, and Frank Wolter, editors, *Proceedings of the 15th Principles of Knowledge Representation and Reasoning, KR 2016*, pages 43–52. AAAI Press, 2016.
21. Thomas Linsbichler. *Advances in Abstract Argumentation – Expressiveness and Dynamics*. PhD thesis, TU Wien, 2017.
22. Thomas Linsbichler, Jörg Pührer, and Hannes Strass. A uniform account of realizability in abstract argumentation. In Gal A. Kaminka, Maria Fox, Paolo Bouquet, Eyke Hüllermeier, Virginia Dignum, Frank Dignum, and Frank van Harmelen, editors, *Proceedings of the 22nd European Conference on Artificial Intelligence, ECAI 2016*, volume 285 of *Frontiers in Artificial Intelligence and Applications*, pages 252–260. IOS Press, 2016.
23. M. Julieta Marcos, Marcelo A. Falappa, and Guillermo Ricardo Simari. Dynamic argumentation in abstract dialogue frameworks. In Peter McBurney, Iyad Rahwan, and Simon Parsons, editors, *Argumentation in Multi-Agent Systems - 7th International Workshop, ArgMAS 2010. Revised, Selected and Invited Papers*, volume 6614 of *Lecture Notes in Computer Science*, pages 228–247. Springer, 2011.
24. Sanjay Modgil and Trevor J. M. Bench-Capon. Metalevel argumentation. *Journal of Logic and Computation*, 21(6):959–1003, 2011.
25. Martín O. Moguillansky, Nicolás D. Rotstein, Marcelo A. Falappa, Alejandro Javier García, and Guillermo Ricardo Simari. Dynamics of knowledge in DeLP through argument theory change. *Theory and Practice of Logic Programming*, 13(6):893–957, 2013.
26. Jörg Pührer. Realizability of three-valued semantics for Abstract Dialectical Frameworks. In Qiang Yang and Michael Wooldridge, editors, *Proceedings of the 24th International Joint Conference on Artificial Intelligence, IJCAI 2015*, pages 3171–3177. AAAI Press, 2015.
27. Hannes Strass. Expressiveness of two-valued semantics for abstract dialectical frameworks. *Journal of Artificial Intelligence Research*, 54:193–231, 2015.
28. Bart Verheij. Two approaches to dialectical argumentation: admissible sets and argumentation stages. In John-Jules C. Meyer and Linda C. van der Gaag, editors, *Proceedings of the 8th Dutch Conference on Artificial Intelligence, NAIC 1996*, pages 357–368, 1996.

Argumentation, Reasoning, and Belief Revision

Sven Ove Hansson

Division of Philosophy, Royal Institute of Technology (KTH), Stockholm, Sweden
soh@kth.se

Abstract. In order to include argumentation and reasoning in models of belief change, these models have to be adjusted so as to avoid the common assumption that the epistemic agent has already drawn all the valid conclusions from her beliefs that are available to her. A simple model is proposed in which this adjustment has been performed to make room for deductive inference, reasoning, and argumentation. Indications are given of how the model can be extended to cover non-deductive and non-monotonic inferences as well.

Keywords: argumentation; belief base; belief change; faultless inference; inference; inferential belief state; logical omniscience; premise-limited inference; reasoning; spontaneous inference; step-limited inference; subconsequence.

1 Introduction

In our everyday lives, there is a close relationship between argumentation and belief revision. Arguments are the major means by which we try to make other people change their beliefs. Arguments are also the means by which we can achieve the common knowledge that is necessary for human co-operation. But in spite of these close pre-theoretical connections, argumentation and belief revision are seldom included in the same formal model. In the last few decades, an abundance of formal models representing *either* argumentation or belief revision have been presented. ([7]; [3]) However, this has not led to a profusion of models representing *both* argumentation and belief revision. The purpose of this contribution is to suggest one way to construct such combined models.

A major difficulty in such endeavours is the immediacy of logical inference that characterizes the belief state representations in most of the commonly used belief revision models. The predominant representation of an epistemic agent's belief state is a belief set. By this is meant a logically closed set of sentences, i.e. a set K such that $K = \mathrm{Cn}(K)$ for some consequence operation Cn that incorporates all the conclusions that the epistemic agent can ever draw from a set of premises. It follows from this property of the belief state representation that if α follows logically from K, then it is meaningless to tell the agent that α; she is sure to know α already. This property of belief states has often been called "logical omniscience", but that is a somewhat misleading terminology.[1] The operation Cn need not cover all valid inferences that are at all accessible to logical thinking. Furthermore, for Goedelian reasons there may be true statements that are unprovable by the deductive system that Cn represents. But nevertheless: In a standard belief set model, such as AGM ([1]), all conclusions are deductive, and the epistemic agent has already drawn all the valid conclusions that she can ever draw from the information she has access to. Such a model does not have much room for belief changes induced by arguments that do not carry any new factual information.[2]

This deficiency of belief set models does not only affect arguments in the usual sense of patterns of reasoning that are communicated by one agent to another in order to have an impact on the latter's belief state. The assumption that all obtainable valid conclusions are always drawn immediately will also block any meaningful separate representation of the agent's own, internal, reasoning processes. In the standard models of belief change, the agent starts with a belief state that is logically closed, and when revising her beliefs, she simultaneously draws all the logical conclusions that are at all obtainable from new information that she assimilates. This means that there is no room for separate processes of reasoning. Since we normally assume that a rational agent is an agent who reasons, this is a severe limitation in the construction of models of belief change.

The major traditional alternatives to belief set models are models built on belief bases, i.e. sets of sentences that are not logically closed. A belief base B represents the beliefs that are taken by the epistemic agent to have independent justification. For instance, at the moment of writing I believe that Professor Andersson's husband is called "Tom" (a). I also believe that the youngest son of our local grocer is called "Tom" (b). Therefore I also believe that Professor Andersson's husband and the grocer's youngest son have the same first name (s). However, this last-mentioned belief s is a *merely derived* belief, in contrast to a and b, each of which has independent justification.[3] If I receive information that makes me give up a, then my belief in s will have to go as well. The same will happen if I am induced to give up b. This is because s has no justification of its own; all of its justification is lost if I lose my belief in one of the independently justified beliefs on which it is based.

Belief base models extend the expressive power of belief set models in useful ways. Their major advantage is that two different belief bases, B_1 and B_2, can give rise to the same belief set, in other words, we can have $\text{Cn}(B_1) = \text{Cn}(B_2)$ but $B_1 \neq B_2$. Since all changes are performed on the belief base, rather than the belief set, the use of belief bases allows us to introduce important dynamic dis-

---

[1] "Logical immediacy" would be a better term.
[2] See [4] for an exploration of the small scope that is left for such changes.
[3] The distinction between basic and merely derived beliefs is not a property of the beliefs themselves, but depends on how they have been acquired. For instance, I could have a belief state with a and s but not b as basic beliefs. A basic belief in s can result from someone telling me that the grocer's youngest son has the same name as professor Andersson's husband.

tinctions.[4] However, in the way that belief base models are usually constructed, this increased expressive power does not help us to represent reasoning and argumentation. This is because it is assumed in the common belief base models that the beliefs actually held by the epistemic agent are represented by the logical closure of the belief base. In other words, if the agent's belief base is B, then her set of beliefs is $\mathrm{Cn}(B)$. Just as in a belief set model (such as AGM), all valid conclusions have already been drawn, so there is no room for introducing any new ones.

However, a relatively simple adjustment of a belief base model is sufficient to make room for at least a rudimentary representation of reasoning and argumentation. Basically, we have to introduce a distinction in the formal model between those logical consequences of B that the epistemic agent believes in and those that she does not belief in (but which she can be brought to believe in by an argument or an act of reasoning). Once this has been done, it is possible to introduce representations of arguments and their assimilation into a belief state. As was accurately noted by Guillermo Simari and his coworkers, in order to develop the connections between argumentation and belief revision, we need to "start with making the basic steps of reasoning,... more precise, pointing out the way from receiving (new) information to coming up with adequate plausible beliefs on which a decision can be based". ([2], p. 352)

2 A simple deductive model

Both reasoning and argumentation are to a large extent non-deductive and non-monotonic. However, deductive arguments are easier to model, and we can therefore use them as a starting-point for the introduction of reasoning and argumentation into a belief base framework. It should also be observed that many non-deductive arguments can be represented in a deductive format, given certain background beliefs. For instance, given my background knowledge about available transportation facilities, I can conclude from the information that Mario is now in Paris he will not be in time for a meeting in New Delhi that starts one hour from now. Similarly, if told that it is raining heavily where you live, I can conclude that your lawn is wet. Generally speaking, non-deductive inferences can be (roughly) modelled as deductive if it is highly implausible that we will receive any information that overrides them.

2.1 Spontaneous inference

Some inferences require a lot of work and effort, but others require no effort at all. In our everyday lives we continuously make inferences without even noticing that we do so. For instance, since I know that Sophia is a woman, if

---

[4] In our example we can let $B_1 = \{a, b\}$ and $B_2 = \{a, s\}$. We then have $\mathrm{Cn}(B_1) = \mathrm{Cn}(B_2)$. However, with a plausible operation of contraction we also have $B_1 \div a = \{b\}$ and $B_2 \div a = \{s\}$, and consequently $\mathrm{Cn}(B_1 \div a) \neq \mathrm{Cn}(B_2 \div a)$.

you tell me that she is now a parent, then I will conclude that she is a mother, without even thinking of this as an inference. The same phenomenon can also be illustrated with the two examples in the previous paragraph, Mario's travel and the rain on your lawn.[5] For our modelling purposes, we can assume that for any belief base B, there is a set of immediate conclusions which the agent draws from it spontaneously, effortlessly and in principle unavoidably. Together with the belief base B itself, these spontaneous inferences form a set $C(B)$ of immediately available beliefs. It is a superset of B and a subset of $\text{Cn}(B)$:

$$B \subseteq C(B) \subseteq \text{Cn}(B) \text{ (interpolation)} \qquad (1)$$

Typically, we should expect $C(B)$ to be truly intermediate between B and $\text{Cn}(B)$:

$$B \subset C(B) \subset \text{Cn}(B) \text{ (strict interpolation)} \qquad (2)$$

$C(B) \setminus B$ is the set of spontaneous inferences, and $\text{Cn}(B) \setminus B$ the set of inferences from B that are at all obtainable in the logical system under study.

The following property:

$$\text{If } B_1 \subseteq B_2 \text{ then } C(B_1) \subseteq C(B_2) \text{ (monotony)} \qquad (3)$$

is also fairly reasonable, since we are concerned with deductive inferences. If a certain deductive conclusion from B_1 is immediate for the agent, then that must be because there is some subset X of B_1 from which the conclusion follows directly. If $B_1 \subseteq B_2$, then X is also a subset of B_2, which makes it plausible that the same conclusion also follows immediately from B_2. (However, if B_2 is much larger and more difficult to survey than B_1, then this may not hold.)

Thus we can assume that C satisfies two of the three standard properties of consequence operations, namely inclusion ($B \subseteq C(B)$) and monotony. However, the third of these properties:

$$C(C(B)) \subseteq C(B) \text{ (iteration)} \qquad (4)$$

is highly implausible since it would threaten to eliminate the distinction between C and Cn. Suppose that a deduction from B to some element of $\text{Cn}(B)$ can be divided into a series of small steps, such that each of them is immediately available, whereas their combination in a chain of arguments is rather difficult to discover. Since iteration implies all elements in the infinite series:

$$C(C(C(B))) \subseteq C(B)$$
$$C(C(C(C(B)))) \subseteq C(B)$$
$$\ldots$$

---

[5] The spontaneous inferences are parts of the fast System 1 of cognition, as explained by Daniel Kahneman ([6]) and others.

it would follow that $\mathrm{Cn}(B) \subseteq C(B)$ and consequently, due to interpolation, $C(B) = \mathrm{Cn}(B)$.

In summary, operations for spontaneous deductive inference should have the properties listed in the following definition:

DEFINITION 1 *Let* Cn *be a consequence operation. An operation* C *is a* subconsequence operation *based on* Cn *if and only if it satisfies the two properties:*

$B \subseteq C(B) \subseteq \mathrm{Cn}(B)$ *(interpolation) and*
If $B_1 \subseteq B_2$, *then* $C(B_1) \subseteq C(B_2)$ *(monotony)*

The logic of subconsequence operations is much weaker than that of consequence operations, but some properties of interest are obtainable:

OBSERVATION 1 *Let* C *be a subconsequence operation that is based on the consequence operation* Cn. *Then:*

(1) $A \cup C(B) \subseteq C(A) \cup C(B) \subseteq C(A \cup B) \subseteq C(A \cup C(B)) \subseteq C(C(A) \cup C(B))$
(2) If $C(B_1) \subseteq C(B_2)$ *then* $\mathrm{Cn}(B_1) \subseteq \mathrm{Cn}(B_2)$.
(3) $\mathrm{Cn}(C(B)) = \mathrm{Cn}(B)$.

PROOF: *Part 1:* Left to the reader. *Part 2:*
$C(B_1) \subseteq C(B_2)$
$B_1 \subseteq C(B_2)$ (interpolation)
$B_1 \subseteq \mathrm{Cn}(B_2)$ (interpolation)
$\mathrm{Cn}(B_1) \subseteq \mathrm{Cn}(\mathrm{Cn}(B_2))$ (monotony of Cn)
$\mathrm{Cn}(B_1) \subseteq \mathrm{Cn}(B_2)$ (iteration of Cn)
Part 3: Since $B \subseteq C(B)$, it follows from the monotony of Cn that $\mathrm{Cn}(B) \subseteq \mathrm{Cn}(C(B))$. For the other direction:
$C(B) \subseteq \mathrm{Cn}(B)$ (interpolation)
$\mathrm{Cn}(C(B)) \subseteq \mathrm{Cn}(\mathrm{Cn}(B))$ (monotony of Cn)
$\mathrm{Cn}(C(B)) \subseteq \mathrm{Cn}(B)$ (iteration of Cn) □

2.2 Belief states

We can now introduce the states of belief held by an epistemic agent whose independently justified beliefs are represented by the belief base B. Clearly, $\mathrm{Cn}(B)$, the set that traditionally has this role in belief base models, is the upper limit. $C(B)$ contains the beliefs that come "automatically" without any reasoning, and is therefore a lower limit. We can introduce the set I of actual (basic and inferred) beliefs as an *inferential extension* of the belief base, such that:

$$C(B) \subseteq I \subseteq \mathrm{Cn}(B) \tag{5}$$

Since one and the same belief base B can be combined with different inferential extensions, we need to specify both B and I in order to identify a belief state. Formally:

> **DEFINITION 2** *Let C be a subconsequence operation based on the consequence operation* Cn. *A pair $\langle B, I \rangle$ of sets of sentences is an* inferential belief state *with respect to* Cn *and C if and only if $C(B) \subseteq I \subseteq \mathrm{Cn}(B)$.*
>
> *B is the* belief base *and I its* inferential extension *in the inferential belief state $\langle B, I \rangle$.*

Notably, $\mathrm{Cn}(B)$ can contain some logically false sentence without I doing so. Therefore, inferential belief states can be paraconsistent.

Alternatively, we could use either of the pairs $\langle B, I \setminus B \rangle$ or $\langle B, I \setminus C(B) \rangle$ for the same purpose. The option $\langle B, I \rangle$ was chosen because it provides a somewhat more convenient formal structure.

Operations of change on a belief state $\langle B, I \rangle$ can target either B or I, i.e. they can be performed either to modify the set of basic (usually factual) beliefs or the set of inferences that the agent draws from them. The following two subsections are devoted to each of these two classes of operations. We begin with the operations targeting I.

2.3 Operations of inference

Let us consider purely inferential changes on $\langle B, I \rangle$, i.e. changes on I not affecting B. As a starting-point, we can build our model with only one operation for such changes, namely an operation of *inference*, resulting in the extension of I with some new conclusion, but leaving B unchanged. Such an operation can represent either the agent's own reasoning or the receipt of an argument from an external source. Its input can be represented by a pair $\langle X, \alpha \rangle$, where X is a set of premises and α the conclusion propounded by the argument. $\langle X, \alpha \rangle$ is an *inferential input*, and it can be interpreted as the instruction "Since you believe in X, you should conclude that α".

For a start, we can make the simplifying assumption that the agent accepts an inferential input $\langle X, \alpha \rangle$ if and only if it is applicable ($X \subseteq I$) and logically valid ($\alpha \in \mathrm{Cn}(X)$). We can call this an operation of *faultless inference*, and denote it $\oplus$. It can be defined as follows:

> **DEFINITION 3** *A sentential operation $\oplus$ on an inferential belief state $\langle B, I \rangle$ is an operation of* faultless inference *if and only if, for all sentences α and sets X of sentences:*
>
> $\langle B, I \rangle \oplus \langle X, \alpha \rangle = \langle B, I \cup \{\alpha\} \rangle$ *if $X \subseteq I$ and $\alpha \in \mathrm{Cn}(X)$, and*
> $\langle B, I \rangle \oplus \langle X, \alpha \rangle = \langle B, I \rangle$ *if $X \nsubseteq I$ or $\alpha \notin \mathrm{Cn}(X)$.*

Notably, just like the operation of expansion in standard belief change, the operation of faultless inference is purely logical, which means that it can be fully defined without reference to any extra-logical choice mechanism. There is exactly one operation of faultless inference, and it is applicable to all inferential belief states $\langle B, I \rangle$.

This approach to purely inferential changes (changes on I not affecting B) is idealized in at least two important ways. First, as already mentioned, the epistemic agent is assumed to be able to evaluate arguments (inferential inputs) with unerring correctness. Secondly, under the assumption that $\oplus$ is the only operation of change that does not affect B, once a conclusion has been accepted, it cannot be rescinded unless there is first some change in the set B of basic beliefs. These two idealizations are of course closely related. If one of them is removed, then that is a compelling reason to give up the other as well.

2.4 Operations of base change

Let us now turn to changes targeting the belief base, B in $\langle B, I \rangle$. The most important such operation in the standard approach is contraction by some sentence α. When a belief base B is contracted by a sentence α, the outcome is a new belief base $B \div \alpha$, which is a subset of B. Enough elements in B have been removed to ensure that $B \div \alpha$ does not imply α (unless α is a tautology, in which case this cannot be achieved, and $B \div \alpha = B$). The most common construction is partial meet contraction, in which a selection function is applied to the remainder set $B \perp \alpha$, defined as the set of maximal subsets of B not implying α. In formal terms, partial meet contraction can be defined as follows:

DEFINITION 4 *([1]) Let B be a set of sentences and α a sentence. The remainder set $B \perp \alpha$ is the set such that $X \in B \perp \alpha$ if and only if:*

(1) $X \subseteq B$,
(2) $\alpha \notin \mathrm{Cn}(X)$
(3) If $X \subset X' \subseteq B$, then $\alpha \in \mathrm{Cn}(X')$

DEFINITION 5 *([1]) Let B be a belief base. An operation of partial meet contraction on B is an operation $\sim_\gamma$ based on a selection function γ, with $\emptyset \neq \gamma(X) \subseteq X$ for all non-empty subsets X of B, and such that for all sentences α:*

(i) If $B \perp \alpha \neq \emptyset$, then $B \sim_\gamma \alpha = \bigcap \gamma(B \perp \alpha)$.
(ii) If $B \perp \alpha = \emptyset$, then $B \sim_\gamma \alpha = B$.

In parts (2) and (3) of Definition 4, subsets of the belief base are evaluated according to their logical closures (closures under Cn). This is adequate under the usual assumption that the agent has access to all the logical inferences encoded in Cn. However, in the present framework the epistemic agent only has access to a subset I of the valid inferences that can be drawn from her belief base B. It would therefore be strange to assume that when she evaluates a subset X of B, she has access to all the logical inferences from X that are encoded in Cn.

There are two obvious ways to adjust the definition of remainder sets (Definition 4) to represent the agent's actual inferential knowledge. One option is to replace $Cn(X)$ by $C(X)$ in clause (2) of the definition, and similarly $Cn(X')$ by $C(X')$ in clause (3). This corresponds to the assumption that in these evaluations, the agent only has access to the immediately available inferences. The other option is to assume that when evaluating a subset X of B, she has access to all inferences from B that she already knows and which can also be obtained from X.[6] This would lead us to replace $Cn(X)$ by $I \cap Cn(X)$ in clause (2) and $Cn(X')$ by $I \cap Cn(X')$ in clause (3) of Definition 4.

The second of these solutions results, as expected, in a superset of that obtained in the first solution. This can be seen from the following observation:

OBSERVATION 2 *Let Cn be a consequence operation and C a subconsequence operation based on Cn. Furthermore, let $X \subseteq B$ and $C(B) \subseteq I \subseteq Cn(B)$. Then $C(X) \subseteq I \cap Cn(X)$.*

PROOF: It follows by monotony from $X \subseteq B$ that $C(X) \subseteq C(B)$. Thus, $C(X) \subseteq I$. From this it follows that $C(X) \cap Cn(X) \subseteq I \cap Cn(X)$. It follows from $C(X) \subseteq Cn(X)$ that $C(X) \cap Cn(X) = C(X)$. We can conclude that $C(X) \subseteq I \cap Cn(X)$. □

We can now adjust the definition of $\perp$ accordingly, by adjusting clauses (2) and (3) of Definition 4:

DEFINITION 6 *Let B and S be sets of sentences and α a sentence. The S-limited remainder set $B \perp_S \alpha$ is the set such that $X \in B \perp_S \alpha$ if and only if:*

(1) $X \subseteq B$,
(2') $\alpha \notin S \cap Cn(X)$, and
(3') If $X \subset X' \subseteq B$, then $\alpha \in S \cap Cn(X')$

DEFINITION 7 *An inference-limited partial meet contraction is an operation $\sim_\gamma^S$ on belief bases that is based on a selection function γ and a set S of sentences, such that*

(i') If $B \perp_S \alpha \neq \emptyset$, then $B \sim_\gamma^S \alpha = \bigcap \gamma(B \perp_S \alpha)$.
(ii') If $B \perp_S \alpha = \emptyset$, then $B \sim_\gamma^S \alpha = B$.

We can now use these definitions to construct the new inferential belief state that should result after contraction of the inferential belief state $\langle B, I \rangle$ by a sentence α. Clearly, the new belief base should be $B \sim_\gamma^I \alpha$. We also need to find its inferential extension. When we remove sentences from B we also need

---

[6] This assumption is not quite unproblematic. I may contain some inference from B that can also be obtained from X alone, but then only with a much more complex chain of reasoning than what is needed to obtain it from B.

to remove conclusions from I that no longer hold since their premises among the independently justified sentences in B have been lost in $B \sim_\gamma^I \alpha$. Those that are still supported can be identified as the elements of the intersection $I \cap \mathrm{Cn}(B \sim_\gamma^I \alpha)$. Based on this, we can define a partial meet contraction on an inferential belief state as follows:

DEFINITION 8 *The operation $\approx_\gamma$ is a partial meet contraction on the inferential belief state $\langle B, I \rangle$ if and only if:*
$$\langle B, I \rangle \approx_\gamma \alpha = \langle B \sim_\gamma^I \alpha, I \cap \mathrm{Cn}(B \sim_\gamma^I \alpha) \rangle$$

Other partial meet operations, such as consolidation and revision, can be defined in the same style.

3 Models of spontaneous deduction

In the previous section, we have taken the subconsequence operation C as given. For some modelling purposes, it may be useful to have access to simple constructions of C. In this section, two such constructions will be introduced. The first of them is based on the observation that inferences obtainable from a small number of premises tend to be cognitively more accessible than inferences requiring a higher number of premises. The second is based on the observation that inferences obtainable with a small number of proof steps tend to be more easily accessible than those that require a higher number of steps.

3.1 Premise-limited inference

Our first option is to model the spontaneously made inferences as those that can made from a limited number of premises.

DEFINITION 9 *For any set X of sentences, $\#(X)$ is the number of logically non-equivalent elements of X (in a given logic).*

DEFINITION 10 *Let Cn be a consequence operation and k a non-negative integer. The* premise-limited subconsequence operation Cn_k *that is based on Cn is the operation such that for all sets X of sentences:*
$$\mathrm{Cn}_k(X) = \bigcup \{\mathrm{Cn}(X') \mid X' \subseteq X \ \& \ \#(X') \leq k\}$$

This construction has the following properties:

OBSERVATION 3 *Let Cn_k be a premise-limited operation based on Cn. Then:*

1. $\mathrm{Cn}_k(X) \subseteq \mathrm{Cn}(X)$
2. $\mathrm{Cn}_k(X) \subseteq \mathrm{Cn}_{k+1}(X)$
3. *If $k > 0$ then $X \subseteq \mathrm{Cn}_k(X)$.*
4. $\mathrm{Cn}_0(X) = \mathrm{Cn}(\varnothing)$

5. If $X_1 \subseteq X_2$, then $\mathrm{Cn}_k(X_1) \subseteq \mathrm{Cn}_k(X_2)$.
6. If $k = 0$ or $k = 1$, then $\mathrm{Cn}_k(\mathrm{Cn}_k(X)) = \mathrm{Cn}_k(X)$.
7. If $k > 1$ and the language has at least $k + 1$ logical atoms, then it does not hold in general that $\mathrm{Cn}_k(\mathrm{Cn}_k(X)) = \mathrm{Cn}_k(X)$.

PROOF: Parts 1-6 are left to the reader. For part 7, let $X = \{a_1, \ldots a_{k+1}\}$, where all of $a_1, \ldots a_{k+1}$ are atoms. Then $a_1 \& \ldots \& a_{k+1} \notin \mathrm{Cn}_k(X)$. However, since $a_1 \& \ldots \& a_k \in \mathrm{Cn}_k(X)$ and $a_{k+1} \in \mathrm{Cn}_k(X)$, we have $a_1 \& \ldots \& a_{k+1} \in \mathrm{Cn}_k(\mathrm{Cn}_k(X))$. □

It follows from clauses (1), (3), and (5) of Observation 3 that if $k > 0$, then Cn_k is a subconsequence operation based on Cn. It also follows from clauses (2) and (4) that $\mathrm{Cn}(\varnothing) \subseteq \mathrm{Cn}_k(X)$ for all k and X, which is clearly an undesirable property. But on the other hand, this model has the advantage of not requiring any details about the exact construction of proofs. In order to construct Cn_k, it is sufficient that we can determine for any sentence α and any set X of (at most k) sentences whether or not $\alpha \in \mathrm{Cn}(X)$. We do not have to search for proofs with special properties, such as minimal length.

3.2 Step-limited inference

Alternatively, we can construct a subconsequence operation by limiting the number of (elementary) proof steps in a derivation. Provided that a precise proof system has been defined for the logic, this can be done as follows:

DEFINITION 11 *Let* Cn *be a consequence operation and* k *a positive integer.*[7] *The step-limited subconsequence operation* Cn^k *that is based on* Cn *is the operation such that for all sentences α and all sets X of sentences:*

$\alpha \in \mathrm{Cn}^k(X)$ *if and only if there is a proof of α from elements of X in at most k elementary steps.*

This construction has the following properties:

OBSERVATION 4 *Let $k > 0$ and let* Cn^k *be a step-limited operation based on the consequence operation* Cn. *Then:*

1. $X \subseteq \mathrm{Cn}^k(X) \subseteq \mathrm{Cn}(X)$
2. $\mathrm{Cn}^k(X) \subseteq \mathrm{Cn}^{k+1}(X)$
3. $\mathrm{Cn}^1(X) = X$
4. It does not hold in general for any k that $\mathrm{Cn}(\varnothing) \subseteq \mathrm{Cn}^k(X)$.
5. If $X_1 \subseteq X_2$, then $\mathrm{Cn}^k(X_1) \subseteq \mathrm{Cn}^k(X_2)$.
6. It does not hold in general for any $k > 1$ that $\mathrm{Cn}^k(\mathrm{Cn}^k(X)) \subseteq \mathrm{Cn}^k(X)$.

---

[7] We assume that $k > 0$, since a proof in zero steps is not worth any attention.

PROOF: Left to the reader. □

It follows from Definition 1 and clauses (1) and (5) of Observation 4 that Cn^k is a subconsequence operation based on Cn. It follows from clause (4) of Observation 4 that it does not have the problematic property $\text{Cn}(\varnothing) \subseteq \text{Cn}^k(X)$. However, the step-limited operation Cn^k differs from the premise-limited operation Cn_k in requiring detailed information on proof structures.

3.3 Other constructions

Other constructions of subconsequence operations are also possible. For instance, if the belief base is divided into compartments with different subject-matter ([5]), then spontaneous inferences can be assumed always to take place within a single compartment, or with premises from a limited number of compartments. Alternatively, the set of available premises can be limited to those that satisfy some criterion of relevance in relation to the conclusion. Yet another option is to limit the total number of some formal entities, such as atomic sentences, predicates, or variables, in the set of premises (rather than limiting the number of premises).

4 Conclusion

In order to accommodate argumentation and reasoning in a belief revision framework, we need to introduce a representation of arguments and their assimilation into a belief state. In the present contribution, this has been done for a limited class of arguments, which is probably also the class for which it can most easily be done, namely (purely) deductive arguments. In an extension to non-deductive argumentation, much of the formal framework introduced here can be retained, but the properties of subconsequence operations will have to be different. In particular, the monotony property must be given up. We will have to distinguish between (1) the conclusions that can validly be drawn from some subset of the belief base B and (2) the conclusions that can validly be drawn from B as a whole. In deductive inference, the two coincide, but that does not hold in non-monotonic inference ($\vdash$). In other words:

$$\{\alpha \mid (\exists X)(X \subseteq B \ \& \ X \vdash \alpha)\} \subseteq \{\alpha \mid B \vdash \alpha\} \tag{6}$$

does *not* hold in general. This has the important consequence that the operation $\oplus$ of faultless inference cannot be used. We can replace it by an operation of non-monotonic inference, denoted $\circledast$. Then $\langle B, I \rangle \circledast \langle X, \alpha \rangle$ is a new belief state $\langle B, I' \rangle$ such that, notably:

(1) $I \subseteq I'$ does not hold in general, since the assimilation of α can override some previously accepted conclusion that is incompatible with α.

(2) It does not hold in general that $\alpha \in I'$ if $X \subseteq I$ and $\alpha \in \mathrm{Cn}(X)$. Even if α follows from some subset of the agent's belief base, it can be overridden by some previously accepted argument which has more weight.
(3) For essentially the same reason, it does not either hold in general that $\alpha \in I'$ if $X \subseteq I$ and $X \mathrel{\vert\!\sim} \alpha$.

The step from $\oplus$ to $\circledast$ is largely similar to replacing expansion ($+$) by revision ($*$) in standard belief revision, and the same type of formal apparatus (for instance partial meet or kernel operations) can be used to explore it. However, whereas the AGM operation $*$ satisfies the success postulate ($\alpha \in K * \alpha$ for all K and α), $\circledast$ will have to be non-prioritized, i.e. new arguments will not always have precedence over previously adopted inferences.

References

1. Carlos Alchourrón, Peter Gärdenfors, and David Makinson. On the logic of theory change: Partial meet contraction and revision functions. *Journal of Symbolic Logic*, 50:510–530, 1985.
2. Marcelo Alejandro Falappa, Gabriele Kern-Isberner, and Guillermo Ricardo Simari. Belief revision and argumentation theory. In Guillermo Simari and Iyad Rahwan, editors, *Argumentation in Artificial Intelligence*, pages 341–360. Springer US, 2009.
3. Eduardo Fermé and Sven Ove Hansson. *Belief Change. Introduction and Overview*. Springer Briefs in Intelligent Systems. Springer, 2018.
4. Sven Ove Hansson. A dyadic representation of belief. In Peter Gärdenfors, editor, *Belief Revision*, number 29 in Cambridge Tracts in Theoretical Computer Science, pages 89–121. Cambridge University Press, 1992.
5. Sven Ove Hansson and Renata Wassermann. Local change. *Studia Logica*, 70(1):49–76, 2002.
6. Daniel Kahneman. *Thinking, fast and slow*. Farrar, Straus and Giroux, New York, 2011.
7. Iyad Rahwan and Guillermo Simari. *Argumentation in Artificial Intelligence*. Springer, New York, 2009.

Non-monotonic Reasoning in Deductive Argumentation

Anthony Hunter

Department of Computer Science,
University College London,
London, UK

Abstract. Argumentation is a non-monotonic process. This reflects the fact that argumentation involves uncertain information, and so new information can cause a change in the conclusions drawn. However, the base logic does not need to be non-monotonic. Indeed, most proposals for structured argumentation use a monotonic base logic (e.g. some form of modus ponens with a rule-based language, or classical logic). Nonetheless, there are issues in capturing defeasible reasoning in argumentation including choice of base logic and modelling of defeasible knowledge. And there are insights and tools to be harnessed for research in non-monontonic logics. We consider some of these issues in this paper.

1 Introduction

Computational argumentation is emerging as an important part of AI research. This comes from the recognition that if we are to develop robust intelligent systems, then it is imperative that they can handle incomplete and inconsistent information in a way that somehow emulates the human ability to tackle such information. And one of the key ways that humans do this is to use argumentation, either internally, by evaluating arguments and counterarguments, or externally, by for instance entering into a discussion or debate where arguments are exchanged. Much research on computational argumentation focuses on one or more of the following layers: the structural layer (How are arguments constructed?); the relational layer (What are the relationships between arguments?); the dialogical layer (How can argumentation be undertaken in dialogues?); the assessment layer (How can a constellation of interacting arguments be evaluated and conclusions drawn?); and the rhetorical layer (How can argumentation be tailored for an audience so that it is convincing?). This has led to the development of a number of formalisms for aspects of argumentation (for reviews see [10, 40, 5]), and some promising application areas [3]).

Argumentation is related to non-monotonic reasoning. The latter is reasoning that allows for retraction of inferences in the light of new information. Interest in non-monotonic reasoning started with attempts to handle general rules, or defaults, of the form "if α holds, then β normally holds", where α

and β are propositions. It is noteworthy that human practical reasoning relies much more on exploiting general rules (not to be understood as universal laws) than on a myriad of individual facts. General rules tend to be less than 100% accurate, and so have exceptions. Nevertheless it is intuitive and advantageous to resort to such defaults and therefore allow the inference of useful conclusions, even if it does entail making some mistakes as not all exceptions to these defaults are necessarily known. Furthermore, it is often necessary to use general rules when we do not have sufficient information to allow us to specify or use universal laws. For a review of non-monotonic reasoning, see [14].

In using defeasible (or default) knowledge, we might make an inference α on the basis of the information available, and then on the basis of further information, we may want to withdraw α. So with defeasible knowledge, the set of inferences does not increase monotonically with the set of assumptions.

Example 1. Consider the following general statements with the fact `the match is struck`. For this, we infer, `The match lights`. But, if we also have the fact `The match is wet` then we retract `The match lights`.

`A match lights if struck`

`A match doesn't light if struck`

The notion of defeasible knowledge covers a diverse variety of information, including heuristics, rules of conjecture, null values in databases, closed world assumptions for databases, and some qualitative abstractions of probabilistic information. Defeasible knowledge is a natural and very common form of information. There are also obvious advantages to applying the same default a number of times: There is an economy in stating (and dealing with) only a general rule instead of stating (and dealing with) many refined instances of such a general rule.

Even though argumentation and non-monotonic reasoning are widely acknowledged as closely related phenomena, there is a need to clarify the relationship. There is pioneering work such as by Simari *et. al.* [43, 21], Prakken *et. al.* [37, 39, 38], Bondarenko [13], and Toni [44, 45] that looks at aspect of this relationship, but it is potentially valuable to revisit the topic, and in particular investigate it from the point of view of deductive argumentation. So our primary aim in this paper is to investigate how non-monotonic reasoning arises in deductive argumentation, and how that differs from logics for non-monotonic reasoning. Our secondary aim is to investigate how logics for non-monotonic reasoning can be harnessed in deductive argumentation.

We proceed in the rest of the paper as follows: In Section 2, we review deductive argumentation as an example of a framework for structured argumentation; In Section 3, we consider how non-monotonic reasoning arises in structured argumentation; In Section 4, we review default logic, and consider how it can be harnessed in deductive argumentation; In Section 5, we review conditional logics, and consider how they can be harnessed in deductive argumentation; In Section 7, we consider how we can model defeasible knowledge as defeasible rules; And in Section 8, we draw conclusions and consider future work.

2 Deductive argumentation

Deductive argumentation is formalized in terms of deductive arguments and counterarguments, and there are various choices for defining this [10,11]. In the rest of this section, we will investigate some of the choices we have for defining arguments and counterarguments, and for how they can be used in modelling argumentation.

In order to define a specific system for deductive argumentation, we need to use a base logic. This is a logic that specifies the logical language for the knowledge, and the consequence (or entailment) relation for deriving inferences from the knowledge. In this section, we focus on two choices for base logic. These are simple logic (which has a language of literals and rules of the form $\alpha_1 \wedge \ldots \wedge \alpha_n \to \beta$ where $\alpha_1, \ldots, \alpha_n, \beta$ are literals, and modus ponens is the only proof rule) and classical logic (propositional and first-order classical logic).

A **deductive argument** is an ordered pair $\langle \Phi, \alpha \rangle$ where $\Phi \vdash_i \alpha$ holds for the base logic $\vdash_i$. Φ is the support, or premises, or assumptions of the argument, and α is the claim, or conclusion, of the argument. Different deductive argumentation systems can be obtained by imposing constraints (such as minimality or consistency) on the definition of an argument. For an argument $A = \langle \Phi, \alpha \rangle$, the function Support($A$) returns Φ and the function Claim(A) returns α.

A counterargument is an argument that attacks another argument. In deductive argumentation, we define the notion of counterargument in terms of logical contradiction between the claim of the counterargument and the premises or claim of the attacked argument. We will review some of the kinds of counterargument that can be specified for simple logic and classical logic.

2.1 Simple logic

Simple logic is based on a language of literals and simple rules where each **simple rule** is of the form $\alpha_1 \wedge \ldots \wedge \alpha_k \to \beta$ where α_1 to α_k and β are literals. A **simple logic knowledgebase** is a set of literals and a set of simple rules. The consequence relation is modus ponens (i.e. implication elimination).

Definition 1. *The* **simple consequence relation**, *denoted* $\vdash_s$, *which is the smallest relation satisfying the following condition, and where Δ is a simple logic knowledgebase: $\Delta \vdash_s \beta$ iff there is an $\alpha_1 \wedge \cdots \wedge \alpha_n \to \beta \in \Delta$, and for each $\alpha_i \in \{\alpha_1, \ldots, \alpha_n\}$, either $\alpha_i \in \Delta$ or $\Delta \vdash_s \alpha_i$.*

Example 2. Let $\Delta = \{a, b, a \wedge b \to c, c \to d\}$. Hence, $\Delta \vdash_s c$ and $\Delta \vdash_s d$. However, $\Delta \not\vdash_s a$ and $\Delta \not\vdash_s b$.

Definition 2. *Let Δ be a simple logic knowledgebase. For $\Phi \subseteq \Delta$, and a literal α, $\langle \Phi, \alpha \rangle$ is a* **simple argument** *iff $\Phi \vdash_s \alpha$ and there is no proper subset Φ' of Φ such that $\Phi' \vdash_s \alpha$.*

So each simple argument is minimal but not necessarily consistent (where consistency for a simple logic knowledgebase Δ means that for no atom α does $\Delta \vdash_s \alpha$ and $\Delta \vdash_s \neg\alpha$ hold). We do not impose the consistency constraint in the definition for simple arguments as simple logic is paraconsistent, and therefore can support a credulous view on the arguments that can be generated.

Example 3. Let p_1, p_2, and p_3 be the following formulae. Note, we use p_1, p_2, and p_3 as labels in order to make the presentation of the premises more concise. Then $\langle \{p_1, p_2, p_3\}, \mathsf{goodEmployee(John)} \rangle$ is a simple argument.

$p_1 = \mathsf{clever(John)}$
$p_2 = \mathsf{conscientious(John)}$
$p_3 = \mathsf{clever(John)} \land \mathsf{conscientious(John)} \to \mathsf{goodEmployee(John)}$

For simple logic, we consider two forms of counterargument. For this, recall that literal α is the complement of literal β if and only if α is an atom and β is $\neg\alpha$ or if β is an atom and α is $\neg\beta$.

Definition 3. *For simple arguments A and B, we consider the following type of* **simple attack***:*

- *A is a* **simple undercut** *of B if there is a simple rule $\alpha_1 \land \cdots \land \alpha_n \to \beta$ in* Support(B) *and there is an $\alpha_i \in \{\alpha_1, \ldots, \alpha_n\}$ such that* Claim(A) *is the complement of α_i.*
- *A is a* **simple rebut** *of B if* Claim(A) *is the complement of* Claim(B).

Example 4. The first argument A_1 captures the reasoning that the metro is an efficient form of transport, so one can use it. The second argument A_2 captures the reasoning that there is a strike on the metro, and so the metro is not an efficient form of transport (at least on the day of the strike). A_2 undercuts A_1.

$A_1 = \langle \{\mathsf{efficientMetro}, \mathsf{efficientMetro} \to \mathsf{useMetro}\}, \mathsf{useMetro} \rangle$
$A_2 = \langle \{\mathsf{strikeMetro}, \mathsf{strikeMetro} \to \neg\mathsf{efficientMetro}\}, \neg\mathsf{efficientMetro} \rangle$

Example 5. The first argument A_1 captures the reasoning that the government has a budget deficit, and so the government should cut spending. The second argument A_2 captures the reasoning that the economy is weak, and so the government should not cut spending. The arguments rebut each other.

$A_1 = \langle \{\mathsf{govDeficit}, \mathsf{govDeficit} \to \mathsf{cutGovSpending}\}, \mathsf{cutGovSpending} \rangle$
$A_2 = \langle \{\mathsf{weakEconomy}, \mathsf{weakEconomy} \to \neg\mathsf{cutGovSpending}\}, \neg\mathsf{cutGovSpending} \rangle$

So in simple logic, a rebut attacks the claim of an argument, and an undercut attacks the premises of the argument (either by attacking one of the literals, or by attacking the consequent of one of the rules in the premises).

2.2 Classical logic

Classical logic is appealing as the choice of base logic as it better reflects the richer deductive reasoning often seen in arguments arising in discussions and debates.

We assume the usual propositional and predicate (first-order) languages for classical logic, and the usual the **classical consequence relation**, denoted $\vdash$. A **classical knowledgebase** is a set of classical propositional or predicate formulae.

Definition 4. *For a classical knowledgebase Φ, and a classical formula α, $\langle \Phi, \alpha \rangle$ is a* **classical argument** *iff $\Phi \vdash \alpha$ and $\Phi \not\vdash \bot$ and there is no proper subset Φ' of Φ such that $\Phi' \vdash \alpha$.*

So a classical argument satisfies both minimality and consistency. We impose the consistency constraint because we want to avoid the useless inferences that come with inconsistency in classical logic (such as via ex falso quodlibet).

Example 6. The following classical argument uses a universally quantified formula in contrapositive reasoning to obtain the claim about number 77.

$\langle \{\forall \texttt{X.multipleOfTen(X)} \rightarrow \texttt{even(X)}, \neg\texttt{even(77)}\}, \neg\texttt{multipleOfTen(77)} \rangle$

Given the expressivity of classical logic (in terms of language and inferences), there are a number of natural ways that we can define counterarguments. We give some options in the following definition.

Definition 5. *Let A and B be two classical arguments. We define the following types of* **classical attack***.*

A is a **classical undercut** *of B if $\exists \Psi \subseteq \mathsf{Support}(B)$ s.t. $\mathsf{Claim}(A) \equiv \neg \bigwedge \Psi$.*
A is a **classical direct undercut** *of B if $\exists \phi \in \mathsf{Support}(B)$ s.t. $\mathsf{Claim}(A) \equiv \neg \phi$.*
A is a **classical rebuttal** *of B if $\mathsf{Claim}(A) \equiv \neg\mathsf{Claim}(B)$.*

Using simple logic, the definitions for counterarguments against the support of another argument are limited to attacking just one of the items in the support. In contrast, using classical logic, a counterargument can be against more than one item in the support. For example, in Example 7, the undercut is not attacking an individual premise but rather saying that two of the premises are incompatible (in this case that the premises `lowCost` and `luxury` are incompatible).

Example 7. Consider the following arguments. A_1 is attacked by A_2 as A_2 is an undercut of A_1 though it is not a direct undercut. Essentially, the attack is an integrity constraint.

$A_1 = \langle \{\texttt{lowCost}, \texttt{luxury}, \texttt{lowCost} \land \texttt{luxury} \rightarrow \texttt{goodFlight}\}, \texttt{goodFlight} \rangle$
$A_2 = \langle \{\neg\texttt{lowCost} \lor \neg\texttt{luxury}\}, \neg\texttt{lowCost} \lor \neg\texttt{luxury} \rangle$

We give further examples of undercuts in Figure 2.

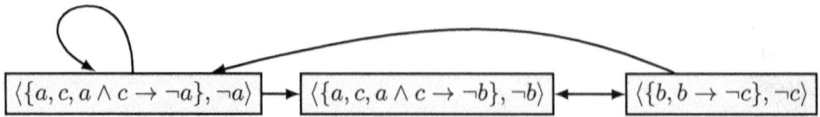

Fig. 1: A generative graph obtained from the simple logic knowlegebase where $\Delta = \{a, b, c, a \wedge c \rightarrow \neg a, b \rightarrow \neg c, a \wedge c \rightarrow \neg b\}$. Note, that this exhaustive graph contains a self cycle, and an odd length cycle.

2.3 Instantiating argument graphs

For a specific deductive argumentation system, once we have a definition for arguments, and for counterarguments (i.e. for the attack relation), we can consider how to use them to instantiate argument graphs. For this, we need to specify which arguments and attacks are to appear in the instantiated argument graph. Two approaches to specifying this are descriptive graphs and generative graphs defined informally as follows.

- **Descriptive graphs** Here we assume that the structure of the argument graph is given, and the task is to identify the premises and claim of each argument. Therefore the input is an abstract argument graph, and the output is an instantiated argument graph. This kind of task arises in many situations: For example, if we are listening to a debate, we hear the arguments exchanged, and we can construct the instantiated argument graph to reflect the debate.
- **Generative graphs** Here we assume that we start with a knowledgebase (i.e. a set of logical formula), and the task is to generate the arguments and counterarguments (and hence the attacks between arguments). Therefore, the input is a knowledgebase, and the output is an instantiated argument graph. This kind of task also arises in many situations: For example, if we are making a decision based on conflicting information. We have various items of information that we represent by formulae in the knowledgebase, and we construct an instantiated argument graph to reflect the arguments and counterarguments that follow from that information. Note, we do not need to include all the arguments and attacks that we can generate from the knowledgebase. Rather we can define a selection function to choose which arguments and attacks to include.

We give an example of generative graph in Figure 1 and an example of a descriptive graph in Figure 2.

For constructing both descriptive graphs and generative graphs, there may be a dynamic aspect to the process. For instance, when constructing descriptive graphs, we may be unsure of the exact structure of the argument graph, and it

Non-monotonic Reasoning in Deductive Argumentation 245

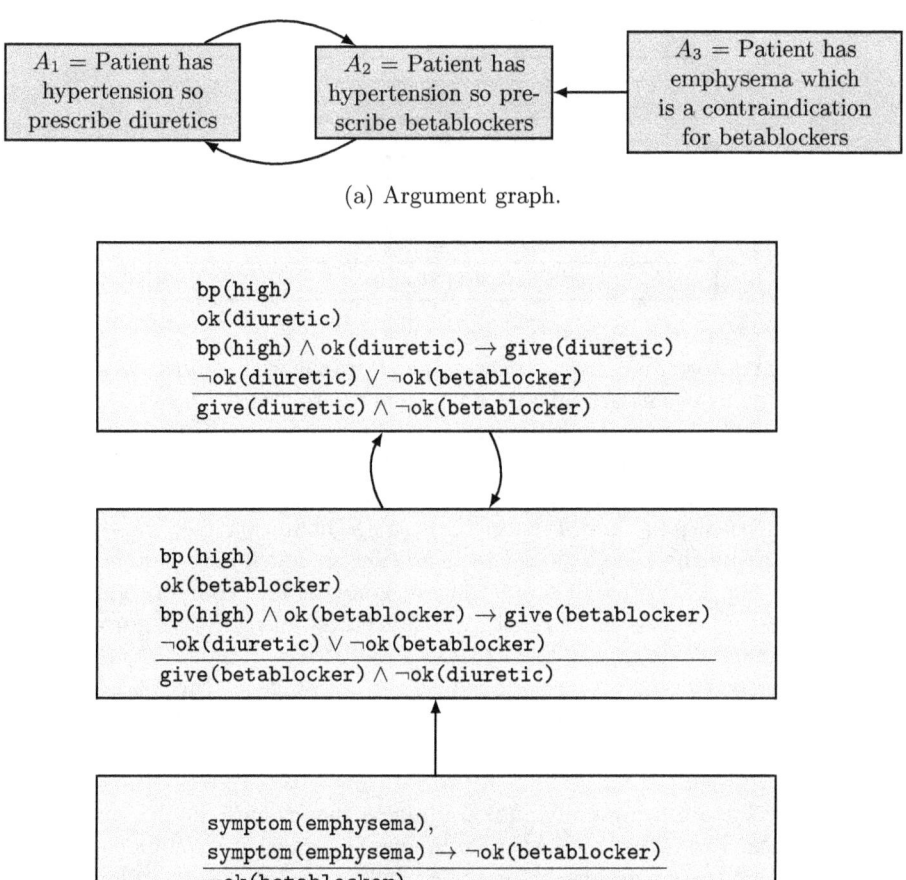

(a) Argument graph.

(b) Descriptive graph representation of the argument graph.

Fig. 2: The abstract argument graph captures a decision making scenario where there are two alternatives for treating a patient, diuretics or betablockers. Since only one treatment should be given for the disorder, each argument attacks the other. There is also a reason to not give betablockers, as the patient has emphysema which is a contraindication for this treatment. The descriptive graph representation of the abstract argument graph is using classical logic. The atom bp(**high**) denotes that the patient has high blood pressure. The top two arguments rebut each other (i.e. the attack is classical defeating rebut). For this, each argument has an integrity constraint in the premises that says that it is not ok to give both betablocker and diuretic. So the top argument is attacked on the premise ok(diuretic) and the middle argument is attacked on the premise ok(betablocker).

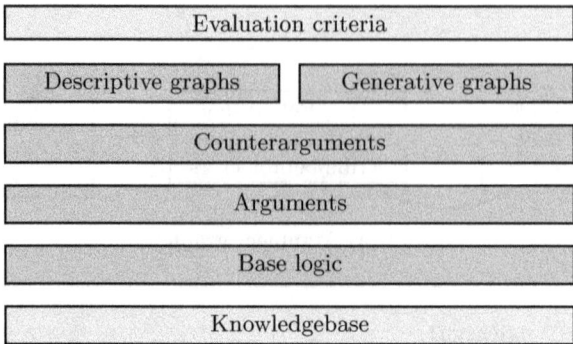

Fig. 3: Framework for constructing argument graphs with deductive arguments: For defining a specific argumentation system, there are four levels for the specification: (1) A base logic is required for defining the logical language and the consequence or entailment relation (i.e. what inferences follow from a set of formlulae); (2) A definition of an argument $\langle \Phi, \alpha \rangle$ specified using the base logic (e.g. Φ is consistent, and Φ entails α); (3) A definition of counterargument specified using the base logic (i.e. a definition for when one argument attacks another); and (4) A definition of which arguments and counterarguments are composed into an argument graph (which is either a descriptive graph or some form of generative graph). Then to use a deductive argumentation system, a knowledgebase needs to be specified in the language of the base logic, and evaluation criteria such as dialectical semantics need to be selected.

is only by instantiating individual arguments that we are able to say whether it is attacked or attacks another argument. As another example, when constructing generative graphs, we may be involved in a dialogue, and so through the dialogue, we may obtain further information which allows us to generate further arguments that can be added to the argument graph.

2.4 Deductive argumentation as a framework

So in order to construct argument graphs with deductive arguments, we need to specify the the base logic, the definition for arguments, the definition for counterarguments, and the definition for instantiating argument graphs. For the latter, we can either produce a descriptive graph or a generative graph. We summarize the framework for constructing argument graphs with deductive arguments in Figure 3.

Key benefits of deductive arguments include: (1) Explicit representation of the information used to support the claim of the argument; (2) Explicit representation of the claim of the argument; and (3) A simple and precise connection between the support and claim of the argument via the consequence relation. What a deductive argument does not provide is a specific proof of the claim from the premises. There may be more than one way of proving the claim from the premises, but the argument does not specify which is used. It is therefore indifferent to the proof used.

There are a number of proposals for deductive arguments using classical propositional logic [16,8,1,24], classical predicate logic [9], description logic [12,34,47,46]) temporal logic [31], simple (defeasible) logic [25,28], conditional logic [7], and probabilistic logic [26,27,29]. These are monotonic logics, though non-monotonic logics can be used as a base logics, as we will investigate in this paper.

There has also been progress in understanding the nature of classical logic in computational argumentation. Types of counterarguments include rebuttals [35,36], direct undercuts [20,19,16], and undercuts and canonical undercuts [8]. In most proposals for deductive argumentation, an argument A is a counterargument to an argument B when the claim of A is inconsistent with the support of B. It is possible to generalize this with alternative notions of counterargument. For instance, with some common description logics, there is not an explicit negation symbol. In the proposal for argumentation with description logics, [12] used the description logic notion of *incoherence* to define the notion of counterargument: A set of formulae in a description logic is incoherent when there is no set of assertions (i.e. ground literals) that would be consistent with the formulae. Using this, an argument A is a counterargument to an argument B when the claim of A together with the support of B is incoherent.

3 Non-monotonic reasoning

For a logic with a consequence relation $\vdash_i$, an important property is the monotonicity property (below) which states that if α follows from a knowledgebase Δ, it still follows from Δ augmented with any additional formula. This is a property that holds for many logics including classical logic, intuitionistic logic, and many modal and temporal logics.

$$\frac{\Delta \vdash_i \alpha}{\Delta \cup \{\beta\} \vdash_i \alpha} \qquad \text{[Monotonicity]}$$

In the following example, we assume a consequence relation $\vdash_i$ for which the monotonicity property does not hold, and show how it reflects a form of defeasible reasoning.

Example 8. Consider the following set of formulae Δ where $\Rightarrow$ is some form of conditional implication symbol.

$$\text{bird}(x) \Rightarrow \text{flyingThing}(x)$$
$$\text{ostrich}(x) \Rightarrow \neg\text{flyingThing}(x)$$
$$\text{ostrich}(x) \Rightarrow \text{bird}(x)$$

Assuming we have an appropriate non-monotonic logic, we would get the following behaviour. So given the rules in Δ and the premise bird(Tweety) we get flyingThing(Tweety), but when we add ostrich(Tweety) to the premises, we no longer get flyingThing(Tweety).

$$\Delta \cup \{\text{bird}(\text{Tweety})\} \vdash_i \text{flyingThing}(\text{Tweety})$$
$$\Delta \cup \{\text{bird}(\text{Tweety}), \text{ostrich}(\text{Tweety})\} \not\vdash_i \text{flyingThing}(\text{Tweety})$$

In the following sections, we will consider specific logics that give us such behaviour.

3.1 Non-monotonicity in deductive argumentation

We now focus on deductive argumentation, and investigate the monotonic and non-monotonic aspects of it.

Argument construction is monotonic In deductive reasoning, we start with some premises, and we derive a conclusion using one or more inference steps in a base logic. Within the context of an argument, if we regard the premises as credible, then we should regard the intermediate conclusion of each inference step as credible, and therefore we should regard the conclusion as credible. For example, if we regard the premises "Philippe and Tony are having tea together in London" as credible, then we should regard that "Philippe is not in Toulouse" as credible (assuming the background knowledge that London and Toulouse are different places, and that nobody can be in different places at the same time). As another example, if we regard that the statement "Philippe and Tony are having an ice cream together in Toulouse" is credible, then we should regard the statement "Tony is not in London" as credible. Note, however, we do not need know that the premises are true to apply deductive reasoning. Rather, deductive reasoning allows us to obtain conclusions that we can regard as credible contingent on the credibility of their premises. This means that we can form arguments monotonically from a knowledgebase: Adding formulae to the knowledgebase allows us to increase the set of arguments.

Argument evaluation is non-monotonic Given a knowledgebase, we use the base logic to construct the arguments and to identify the attack relationships that hold between arguments. We then apply the dialectical criteria to determine which arguments are in an extension according to a specific semantics. If we then add more formulae to the knowledgebase, we may obtain further arguments, but we do not lose any arguments. So when we add formulae to the knowledgebase, we may make additions to the instantiated argument graph but we do not make any deletions. If we then apply the dialectical criteria, we may then lose arguments from our extension. So in this sense, argumentation is non-monotonic.

In the following example, we illustrate how argument construction is monotonic but argument evaluation is non-monotonic.

Non-monotonic Reasoning in Deductive Argumentation 249

Example 9. Let $\Delta = \{a\}$, and so the following is a generative argument graph where A'_1 stands for all arguments with claim implied by a such as $\langle\{a\}, a \vee b\rangle$, $\langle\{a\}, a \vee c\rangle$, etc. Hence, $\{A_1, A'_1, \ldots\}$ is the grounded extension of the graph.

$$A_1 = \langle\{a\}, a\rangle \qquad A'_1 = \langle\{a\}, \ldots\rangle$$

If we add $\neg a$ to Δ, we get the following generative argument graph where A'_2 stands for all arguments with claim implied by $\neg a$ such as $\langle\{\neg a\}, \neg a \vee b\rangle$, $\langle\{\neg a\}, \neg a \vee c\rangle$, etc. Hence, A_1 is no longer in the grounded extension.

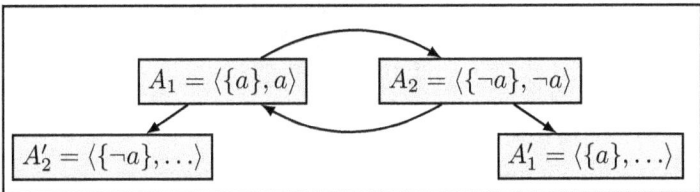

Simple arguments and counterarguments can be used to model defeasible reasoning. For this, we use simple rules that are normally correct but sometimes are incorrect. For instance, if Sid has the goal of going to work, Sid takes the metro. This is generally true, but sometimes Sid works at home, and so it is no longer true that Sid takes the metro, as we see in the next example.

Example 10. The first argument A_1 captures the general rule that if workDay holds, then useMetro(Sid) holds. The use of the simple rule in A_1 requires that the assumption normal holds. This is given as an assumption. The second argument A_2 undercuts the first argument by contradicting the assumption that normal holds

$A_1 = \langle\{\text{workDay}, \text{normal}, \text{workDay} \wedge \text{normal} \to \text{useMetro(Sid)}\}, \text{useMetro(Sid)}\rangle$
$A_2 = \langle\{\text{workAtHome(Sid)}, \text{workAtHome(Sid)} \to \neg\text{normal}\}, \neg\text{normal}\rangle$

If we start with just argument A_1, then A_1 is undefeated, and so useMetro(Sid) is an acceptable claim. However, if we add A_2, then A_1 is a defeated argument and A_2 is an undefeated argument. Hence, if we have A_2, we have to withdraw useMetro(Sid) as an acceptable claim.

So by having appropriate conditions in the antecedent of a simple rule we can disable the rule by generating a counterargument that attacks it. This in effect stops the usage of the simple rule. This means that we have a convention to attack an argument based on the inferences obtained by the simple logic (e.g. as in Example 4 and Example 5), or on the rules used (e.g. Example 10).

This way to disable rules by adding appropriate conditions (as in Example 10) is analogous to the use of abnormality predicates in formalisms such as circumscription (see for example [32]). We can use the same approach to capture defeasible reasoning in other logics such as classical logic (as for example, the use of the ok predicate in the arguments in Figure 2). Note, this does not mean that we turn the base logic into a nonmonotonic logic. Both simple logic and classical logic are monotonic logics. Hence, for a simple logic knowledgebase Δ (and similarly for a classical logic knowledgebase Δ), the set of simple arguments (respectively classical arguments) obtained from Δ is a subset of the set of simple arguments (respectively classical arguments) obtained from $\Delta \cup \{\alpha\}$ where α is a formula not in Δ. But at the level of evaluating arguments and counterarguments, we have non-monotonic defeasible behaviour as illustrated by Example 10.

In the section, we have focused on using simple logic as a base logic. But it is very weak since it only has modus ponens as a proof rule. There is a range of logics between simple logic and classical logic called conditional logics. They are monotonic, and they capture interesting aspects of defeasible reasoning. We will briefly consider one type of conditional logic in Section 5.

3.2 Other approaches to structured argumentation

Other approaches to structured argumentation such as ASPIC+ [38, 33] and assumption-based argumentation (ABA) [44, 45] also have an argument construction process that is monotonic but an argument evaluation process that is non-monotonic. These approaches apply rules to the formulae from the knowledge base, where the rules may be defeasible. These rules are defeasible in the sense they describe defeasible knowledge but the underlying logic is essentially modus ponens (i.e. analogous to simple logic as investigated for deductive argumentation) and so monotonic. Note, in these rule-based approaches, an argument is seen as a tree whose root is the claim or conclusion, whose leaves are the premises on which the argument is based, and whose structure corresponds to the application of the rules from the premises to the conclusion.

Through the dialectical semantics of argumentation, it is possible to formalize non-monotonic logics. Bondarenko *et. al.* showed how assumption-based argumentation subsumes a range of key non-monotonic logics including Theorist, default logic, logic programming, autoepistemic logic, non-monotonic modal logics, and certain instances of circumscription as special cases [13]. The use of assumptions can be introduced to other approaches to structured argumentation (see for example, for ASPIC+ [38]).

In defeasible logic programming (DeLP) [21, 22], another approach to structured argumentation, strict and defeasible rules are used in a form of logic programming. The language includes a default negation (i.e. a form of negation-as-failure) which gives a form of non-monotonic behaviour. Essentially, negation-as-failure means that if an atom cannot be proved to be true, then assume it is false. For example, the following is a defeasible rule in DeLP where the negated atom $\sim$crossRailwayTracks is the consequent, $\prec$ is the defeasible implication symbol, and *not* $\sim$trainIsComing is the condition. Furthermore, $\sim$trainIsComing is a negated atom, and *not* is the default negation operator.

$$\sim\text{crossRailwayTracks} \prec \textit{not} \sim\text{trainIsComing}$$

For the above defeasible rule, if we cannot prove ~trainIsComing (i.e. we cannot show that the train is not coming), then *not* ~trainIsComing holds, and therefore we infer ~crossRailwayTracks (i.e. we shouldn't cross the tracks).

In DeLP, the negation-as-failure is with respect to all the strict knowledge (i.e. the subset of the knowledge that is assumed to be true), rather than just the premises of the argument, and this means that as formulae are added to the knowledgebase, it may be necessary to withdraw arguments. For instance, if there is no strict knowledge, then we can construct an argument that has the above defeasible rule as the premise, and ~crossRailwayTracks as the claim. But, if we then add ~trainIsComing to the knowledgebase as strict knowledge, then we have to withdraw this argument. So unlike deductive argumentation, and the other approaches to structured argumentation, DeLP is not monotonic in argument contruction. However, like deductive argumentation, and the other approaches to structured argumentation, DeLP is non-monotonic in argument evaluation.

4 Default logic

As a basis of representing and reasoning with default knowledge, default logic, proposed by Reiter [41], is one of the best known and most widely studied formalisations of default reasoning. Furthermore, it offers a very expressive and lucid language. In default logic, knowledge is represented as a *default theory*, which consists of a set of first-order formulae and a set of *default rules* for representing default information. Default rules are of the following form, where α, β and γ are classical formulae.

$$\frac{\alpha : \beta}{\gamma}$$

The inference rules are those of classical logic plus a special mechanism to deal with default rules: Basically, if α is inferred, and $\neg\beta$ cannot be inferred, then infer γ. For this, α is called the pre-condition, β is called the justification, and γ is called the consequent.

4.1 Inferencing in default logic

Default logic is an extension of classical logic. Hence, all classical inferences from the classical information in a default theory are derivable (if there is an extension). The default theory then augments these classical inferences by default inferences derivable using the default rules.

Definition 6. *Let (D, W) be a default theory, where D is a set of default rules and W is a set of classical formulae. Let Cn be the function that for a set of formulae returns the set of classical consequences of those formulae. The operator Γ indicates what conclusions are to be associated with a given set E*

of formulae, where E is some set of classical formulae. For this, $\Gamma(E)$ is the smallest set of classical formulae such that the following three conditions are satisfied.

1. $W \subseteq \Gamma(E)$
2. $\Gamma(E) = Cn(\Gamma(E))$
3. For each default in D, where α is the pre-condition, β is the justification, and γ is the consequent, the following holds:

$$\text{if } \alpha \in \Gamma(E), \text{ and } \neg\beta \notin E, \text{ then } \gamma \in \Gamma(E)$$

We refer to E as the satisfaction set, and $\Gamma(E)$ the putative extension.

Once $\Gamma(E)$ has been identified, E is an extension of (D, W) iff $E = \Gamma(E)$. If E is an extension, then the first condition ensures that the set of classical formulae W is also in the extension, the second condition ensures the extension is closed under classical consequence, and the third condition ensures that for each default rule, if the pre-condition is in the extension, and the justification is consistent with the extension, then the consequent is in the extension.

We can view E as the set of formulae for which we are ensuring consistency with the justification of each default rule that we are attempting to apply. We can view $\Gamma(E)$ as the set of putative conclusions of a default theory: It contains W, it is closed under classical consequence, and for each default that is applicable (i.e. the precondition is $\Gamma(E)$ and the justification is satisfiable with E), then the consequent is in $\Gamma(E)$. We ask for the smallest $\Gamma(E)$ to ensure that each default rule that is applied is grounded. This means that it is not the case that one or more default rules are self-supporting. For example, a single default rule is self-supporting if the pre-condition is satisfied using the consequent. The test $E = \Gamma(E)$ ensures that the set of formulae for which the justifications are checked for consistency coincides with the set of putative conclusions of the default theory. If $E \subset \Gamma(E)$, then not all applied rules had their justification checked with $\Gamma(E)$. If $\Gamma(E) \subset E$, then the rules are checked with more than is necessary.

Example 11. Let D be the following set of defaults.

$$\frac{\texttt{bird(X)} : \neg\texttt{penguin(X)} \wedge \texttt{fly(X)}}{\texttt{fly(X)}}$$

$$\frac{\texttt{penguin(X)} : \texttt{bird(X)}}{\texttt{bird(X)}} \qquad \frac{\texttt{penguin(X)} : \neg\texttt{fly(X)}}{\neg\texttt{fly(X)}}$$

For (D, W), where W is $\{\texttt{bird(Tweety)}\}$, we obtain one extension

$$Cn(\{\texttt{bird(Tweety)}, \texttt{fly(Tweety)}\})$$

For (D, W), where W is $\{\texttt{penguin(Tweety)}\}$, we obtain one extension

$$Cn(\{\texttt{penguin(Tweety)}, \texttt{bird(Tweety)}, \neg\texttt{fly(Tweety)}\})$$

Attempt	E	$\Gamma(E)$	Extension?
1	$Cn(\{\text{bird(Tweety)}\})$	bird(Tweety) fly(Tweety)	$E \subset \Gamma(E)$
2	$Cn(\{\text{fly(Tweety)}\})$	bird(Tweety) fly(Tweety)	$E \subset \Gamma(E)$
3	$Cn(\{\text{bird(Tweety)}, \text{fly(Tweety)}\})$	bird(Tweety) fly(Tweety)	$E = \Gamma(E)$
4	$Cn(\{\text{bird(Tweety)}, \text{fly(Tweety)}\})$	bird(Tweety)	$\Gamma(E) \subset E$
5	$Cn(\{\neg\text{bird(Tweety)}, \neg\text{fly(Tweety)}\})$	bird(Tweety)	$\Gamma(E) \not\subseteq E$ & $E \not\subseteq \Gamma(E)$

Fig. 4: A non-exhaustive number of attempts are made for determining an extension. In each attempt, a guess is made for E, and then $\Gamma(E)$ is calculated.

Possible sets of conclusions from a default theory are given in terms of extensions of that theory. A default theory can possess multiple extensions because different ways of resolving conflicts among default rules lead to different alternative extensions. For query-answering this implies two options: in the credulous approach, we accept a query if it belongs to one of the extensions of a considered default theory, whereas in the skeptical approach, we accept a query if it belongs to all extensions of the default theory.

The notion of extension in default logic overlaps with the notion of extension in argumentation. However, there is a key difference between the two notions. In the former, the extensions are derived directly from reasoning with the knowledge, whereas in the latter, the extensions are derived from the instantiated argument graph, and that graph has been generated from the knowledge. As a consequence, default logic suppresses the inconsistency arising within the relevant knowledge whereas argumentation draws the inconsistency out in the form of arguments and counterarguments.

4.2 Using default logic in deductive argumentation

We can use default logic as a base logic. Normally, a base logic is monotonic. However, there is no reason to not use a non-monotonic logic such as default logic. Let $\vdash_d$ be the consequence relation for default logic. For a default theory $\Delta = (D, W)$, and a propositional formula α, $\Delta \vdash_d \alpha$ denotes that α is in a default logic extension of (D, W). So the consequence relation is credulous. We could use a skeptical version as an alternative.

Definition 7. *For a default theory $\Phi = (D, W)$, and a classical formula α, $\langle \Phi, \alpha \rangle$ is a **default argument** iff $\Phi \vdash_d \alpha$ and there is no proper subset Φ' of Φ such that $\Phi' \vdash_d \alpha$.*

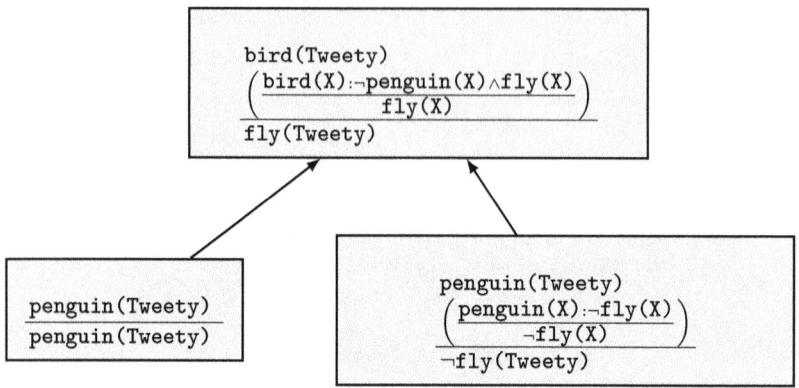

Fig. 5: Example of using default logic as a base logic in deductive argumentation. Here the top argument and the right arguments are default arguments, whereas the left argument is a classical argument. Each argument is presented with the premises above the line, and the claim below the line. Each default rule is given in brackets.

For the above definition, $\Phi' = (D', W')$ is a proper subset of $\Phi = (D, W)$ iff $D' \subseteq D$ and $W' \subset W$ or $D' \subset D$ and $W' \subseteq W$.

We now consider a definition for attack where the counterargument attacks an argument when it negates the justification of a default rule in the premises of the argument. For this we require a function DefaultRules(A) to return the default rules used in the premises of an default argument.

Definition 8. *For arguments A and B, where A is a default argument, and B is either a default argument or a classical argument, B* **justification undercuts** *A if there is $\alpha : \beta/\gamma \in$ DefaultRules(A) s.t. Claim(B) $\vdash \neg\beta$.*

We illustrate justification undercuts in Figure 5. In the figure, the left undercut involves a classical argument, and the right undercut involves a default argument.

Using default logic as a base logic in this way does not affect the argument contruction being monotonic. Adding a formula to the knowledgebase would not cause an argument to be withdrawn. Rather it may allow further arguments to be constructed.

The advantage of using default logic in arguments is that it allows default inferences to be drawn. This means that we use a well-developed and well-understand formalism for representing and reasoning with the complexities of default knowledge. Hence, we can have a richer and more natural representation of defaults. It also means that inferences can be drawn in the absence of reasons to not draw them. For instance, in Figure 5, we can conclude fly(Tweety) from bird(Tweety) in the absence of knowing whether it is a penguin. In other words, we just need to know that it is consistent to believe that it is not a penguin and that it is consistent to believe that it can fly. Furthermore, we just need to do this consistency check within the premises of the argument. Note, this consistency check is different to the consistency check used for default negation in DeLP which involves checking consistency with all the strict knowledge (i.e. the subset of knowledge that is assumed to be correct) [21].

In this paper, we have only given an indication of how default logic can be used to capture aspects of non-monotonic reasoning (for a comprehensive review of default logic, see [6]) in deductive argumentation. There are various ways (e.g. by Santos and Pavão Martins [42]) that default logic could be harnessed in deductive argumentation to give richer behaviour . Also possible definitions for counterarguments could be adapted from an approach to argumentation based on default logic by Prakken [37].

5 KLM logics for non-monotonic reasoning

The KLM logics for non-monotonic reasoning are a family of conditional logics developed by Kraus, Lehmann and Magidor [30] to capture aspects of non-monotonic reasoning and where each logic has a proof theory and a possible worlds semantics. We focus here on a member of this family called System P which has the language composed of formulae of the form $\alpha \Rightarrow \beta$ where α, β are propositional formulae, and it has the the following set of proof rules.

$$(REF) \ \frac{}{A \Rightarrow A} \qquad (CUT) \ \frac{A \Rightarrow B \quad A \wedge B \Rightarrow C}{A \Rightarrow C}$$

$$(LLE) \ \frac{A \equiv B \quad A \Rightarrow C}{B \Rightarrow C} \qquad (RW) \ \frac{\vdash A \rightarrow B \quad C \Rightarrow A}{C \Rightarrow B}$$

$$(AND) \ \frac{A \Rightarrow B \quad A \Rightarrow C}{A \Rightarrow B \wedge C} \qquad (OR) \ \frac{A \Rightarrow C \quad B \Rightarrow C}{A \vee B \Rightarrow C}$$

$$(CM) \ \frac{A \Rightarrow B \quad A \Rightarrow C}{A \wedge B \Rightarrow C}$$

$$(LOOP) \ \frac{A_0 \Rightarrow A_1 \quad A_1 \Rightarrow A_2 \ldots A_{k-1} \Rightarrow A_k \quad A_k \Rightarrow A_0}{A_0 \Rightarrow A_k}$$

We illustrate this proof systems using the following examples of inferences. In the examples, we can see how the reasoning is monotonic in that no inferences (i.e. no formula of the form $\alpha \Rightarrow \beta$) are retracted. However, within these formulae, the defeasible reasoning is encoded. For instance, in Example 12, we have a formula that says "birds fly", and we have the inference that says "birds that are penguins do not fly".

Example 12. From the following statements

$$\text{penguin} \Rightarrow \text{bird} \qquad \text{penguin} \Rightarrow \neg\text{fly} \qquad \text{bird} \Rightarrow \text{fly}$$

we get the following inferences

$$\begin{array}{ll}
\text{penguin} \wedge \text{bird} \Rightarrow \neg\text{fly} & \text{fly} \Rightarrow \neg\text{penguin} \\
\text{bird} \Rightarrow \neg\text{penguin} & \text{bird} \vee \text{penguin} \Rightarrow \text{fly} \\
\text{bird} \vee \text{penguin} \Rightarrow \neg\text{penguin} &
\end{array}$$

We can use System P as a base logic in a deductive argumentation system. We start by giving the following definition of an argument.

Definition 9. *Conditional logic argument For a set of conditional statements Φ and a set of propositional formulae Ψ, if $\Phi \vdash_P \alpha \Rightarrow \beta$ and $\wedge \Psi \equiv \alpha$, then a **preferential argument** is $\langle \Phi \cup \Psi, \alpha \Rightarrow \beta \rangle$.*

Example 13. For premises $\Phi = \{\text{penguin} \Rightarrow \text{bird}, \text{penguin} \Rightarrow \neg\text{fly}\}$ and $\Psi = \{\text{penguin}, \text{bird}\}$, the following is a preferential argument.

$$\langle \Phi \cup \Psi, \text{penguin} \wedge \text{bird} \Rightarrow \neg\text{fly}\rangle$$

Next, we can define a range of attack relations. The following definition is not exhaustive as there are further options that we could consider.

Definition 10. *Some preferential attack relations: Let $A_1 = \langle \Phi \cup \Psi, \alpha \Rightarrow \beta \rangle$ and $A_2 = \langle \Phi' \cup \Psi', \gamma \Rightarrow \delta \rangle$.*

- A_2 is a **rebuttal** of A_1 iff $\delta \vdash \neg \beta$ and $\gamma \vdash \alpha$
- A_2 is a **direct rebuttal** of A_1 iff $\delta \equiv \neg \beta$ and $\gamma \vdash \alpha$
- A_2 is a **undercut** of A_1 iff $\delta \vdash \neg \alpha$
- A_2 is a **canonical undercut** of A_1 iff $\delta \equiv \neg \alpha$
- A_2 is a **direct undercut** of A_1 iff there is $\sigma \in \Psi$ such that $\delta \equiv \neg \sigma$

Example 14. For A_1 and A_2 below, A_2 is a direct rebuttal of A_1, but not vice versa,

- $A_1 = \langle \Phi_1 \cup \Psi_1, \text{bird} \Rightarrow \text{fly}\rangle$
- $A_2 = \langle \Phi_2 \cup \Psi_2, \text{penguin} \wedge \text{bird} \Rightarrow \neg\text{fly}\rangle$

where

- $\Phi_1 = \{\text{bird} \Rightarrow \text{fly}\}$,
- $\Psi_1 = \{\text{bird}\}$,
- $\Phi_2 = \{\text{penguin} \Rightarrow \text{bird}, \text{penguin} \Rightarrow \neg\text{fly}\}$,
- $\Psi_2 = \{\text{penguin}, \text{bird}\}$.

Example 15. For A_1 and A_2 below, A_2 is a direct rebuttal of A_1, but not vice versa,

- $A_1 = \langle \Phi_1 \cup \Psi_1, \texttt{matchIsStruck} \Rightarrow \texttt{matchLights} \rangle$
- $A_2 = \langle \Phi_2 \cup \Psi_2, \texttt{matchIsStruck} \wedge \texttt{matchIsWet} \Rightarrow \neg \texttt{matchLights} \rangle$

Harnessing System P, and the other members of the KLM family, offers the ability to undertake intuitive reasoning with defeasible rules. This allows for plausible consequences from knowledge to be investigated. It also allows for more efficient representation of knowledge to be undertaken since fewer rules would be required when compared with using simple logic. Furthermore, this reasoning can be implemented using automated reasoning [23].

6 Further conditional logics

Conditional logics are a valuable alternative to classical logic for knowledge representation and reasoning. Whilst many conditional logics extend classical logic, the implication introduced is normally more restricted than the strict implication used in classical logic. This means that many knowledge modelling situations, such as for non-monotonic reasoning, can be better captured by conditional logics (such as [18, 30, 2]).

By using conditional logic as a base logic, we have a range of options for more effective modelling complex real-world scenarios. For instance, they can be used to capture hypothetical statements of the form "If α were true, then β would be true". This done by introducing an extra connective $\Rightarrow$ to extend a classical logic language. Informally, $\alpha \Rightarrow \beta$ is valid when β is true in the possible worlds where α is true. Representing and reasoning with such knowledge in argumentation is valuable because useful arguments exist that refer to fictitious and hypothetical situations as shown by Besnard *et. al.* [7].

So we are proposing that we should move beyond simple logic (i.e. modus ponens) that is essentially the logic used for most structured argumentation systems. Even though formalisms such as ASPIC+ and ABA are general frameworks that accept a wide range of proof systems, most exmaples of using the frameworks involve simple rule-based systems (i.e. rules with modus ponens).

Once we accept that we can move beyond simple rule-based systems, we have a wide range of formalisms that we could consider, in particular conditional logics. This then raises the question of what proof theory do we want for reasoning with the defeasible rules. Central to such considerations is whether we want to have contrapositive reasoning.

In considering this issue for argumentation, Caminada recalls two simple examples where the inference of contrapositives is problematic [15].

- Men usually do not have beards. But, this does not imply that if someone does have a beard, then that person is not a man.
- If I buy a lottery ticket, then I will normally not win a prize. But, this does not imply that if I do win a prize, then I did not buy a ticket.

To identify situation where contrapositive reasoning is potentially desirable, Caminada described the following two types of situation.

Epistemic Defeasible rules describe how certain facts hold in relation to each other in the world. So the world exists independently of the rules. Here, contrapositive reasoning may be appropriate.

Constitutive Defeasible rules (in part) describe how the world is constructed (e.g. regulations). So the world does not exist independently of the rules. Here contrapositive reasoning is not appropriate.

The first example below illustrates the first kind of situation, and the second example illustrates the second kind of situation. Note that the knowledge in each example is syntactically identical.

Example 16. The following rules describe an epistemic situation. From these rules, it would be reasonable to infer ¬A from L.

- O ⇒ A - goods ordered three months ago will probably have arrived now.
- A ⇒ C - arrived goods will probably have a customs document.
- L ⇒ ¬C - goods listed as unfulfilled will probably lack customs document.

Example 17. The following rules describe a constitutive situation. From these rules, it would not be reasonable to infer ¬M from P.

- S ⇒ M - snoring in the library is form of misbehaviour.
- M ⇒ R - misbehaviour in the library can result in removal from the library.
- P ⇒ ¬R - professors cannot be removed from the library.

A wide variety of conditional logics have been proposed to capture various aspects of conditionality including deontics, counterfactuals, relevance, and probability [17]. These give a range of proof theories and semantics for capturing different aspects of reasoning with conditions, and many of these do not support contrapositive reasoning.

7 Modelling defeasible knowledge

Another important aspect of non-monotonic reasoning is how we model defeasible knowledge in a meaningful way. A defeasible or default rule is a rule that is generally true but has exceptions and so is sometimes untrue. But this explanation still leaves some latitude as to what we really mean by a defeasible rule. To illustrate this issue, consider the following rule.

$$\text{birds fly} \quad (*)$$

If we take all the **normal** situations (or observations, or worlds or days), and in the majority of these birds fly, then (*) may be equal to the following

$$\text{birds normally fly}$$

Or if we take the set of all birds (or all the birds you have seen, or read about, or watched on TV) and the **majority** of this set fly, then (*) may be equal to the following

> most birds fly

Or perhaps if we take the set of all birds, we know the majority have the **capability** to fly, then (*) could equate with the following. — though this in turn raises the question of what it is to know something has a capability

> most birds have the capability to fly

Or if we take an **idealized** notion of a bird that we in a society are happy to agree upon, then we could equate (*) with the following

> a prototypical bird flies

As another illustration of this issue, consider the following rule which is in the vein as the "birds fly" example but introduces further complications.

> birds lay eggs   (**)

We may take (**) as meaning the following — but on a given day, (or given situation, or observation) a given bird bird will probably not lay an egg.

> birds normally lay eggs

Or we could say (**) means the following — but half the bird population is male and therefore don't lay eggs.

> most birds have the capability lay eggs

Or perhaps we could say (**) means the following — where

> most species of bird reproduce by laying eggs

However, the above involves a lot of missing information to be filled in, and in general it is challenging to know how to formalize a defeasible rule.

Obviously different applications call for different ways to interpret defeasible rules. Specifying the underlying ontology by recourse to description logic may be useful, and in some situations formalization in a probabilistic logic (see for example, [4]) may be appropriate

8   Conclusions

Discrimination between the monotonic and non-monotonic aspects of a deductive argumentation system is important to better understand the nature of deductive argumentation. Each deductive argumentation system is based on a base logic which can be monotonic or non-monotonic. In either case, the construction of arguments and counterarguments is monotonic in the sense that adding knowledge to the knowledgebase may increase the set of arguments and counterargument but it cannot reduce the set of arguments or counterarguments. However, at the evaluation level, deductive argumentation is non-monotonic since adding arguments and counterarguments to the instantiated graph may cause arguments to be withdrawn from an extension, or even for extensions to be withdrawn.

Defeasible formulae (such as defeasible rules) are an important kind of knowledge in argumentation, as in non-monotonic reasoning. They are formulae that are often correct, but sometimes can be incorrect. Depending on what aspect of defeasible knowledge we want to model, there is a wide variety of base logics that we can use to represent and reason with it. This range includes simple logic and classical logic when we use appropriate conventions such abnormality predicates, in which case, we can overturn an argument based defeasible knowledge by recourse counterarguments that attack the assumption of normality. This range of base logics also includes default logic and conditional logics.

References

1. L. Amgoud and C. Cayrol. A reasoning model based on the production of acceptable arguments. *Annals of Mathematics and Artificial Intelligence*, 34:197–215, 2002.
2. H. Arló-Costa and S. Shapiro. Maps between conditional logic and non-monotonic logic. In *Proceedings of the 3rd International Conference on Principles of Knowledge Representation and Reasoning (KR'92)*, pages 553–565. Morgan Kaufmann, 1992.
3. K. Atkinson, P. Baroni, M. Giacomin, A. Hunter, H. Prakken, C. Reed, G. Simari, M. Thimm, and S. Villata. Towards artificial argumentation. *AI Magazine*, 38(3):25–36., 2017.
4. F. Bacchus. *Representing and Reasoning with Probabilistic Knowledge. A Logical Approach to Probabilities*. MIT Press., 1990.
5. P. Baroni, D. Gabbay, and M. Giacomin, editors. *Handbook of Formal Argumentation*, volume 1. College Publications, 2018.
6. Ph. Besnard. *An Introduction to Default Logic*. Springer, 1989.
7. Ph. Besnard, E. Gregoire, and B. Raddaoui. A conditional logic-based argumentation framework. In *Proceedings of the 7th International Conference on Scalable Uncertainty Management (SUM'13)*, volume 7958 of *Lecture Notes in Computer Science*, pages 44–56. Springer, 2013.
8. Ph. Besnard and A. Hunter. A logic-based theory of deductive arguments. *Artificial Intelligence*, 128:203–235, 2001.
9. Ph Besnard and A Hunter. Practical first-order argumentation. In *Proceedings of the 20th National Conference on Artificial Intelligence (AAAI'05)*, pages 590–595. MIT Press, 2005.
10. Ph. Besnard and A Hunter. *Elements of Argumentation*. MIT Press, 2008.

11. Ph. Besnard and A. Hunter. Constructing argument graphs with deductive arguments. *Argument and Computation*, 5(1):5–30, 2014.
12. E. Black, A. Hunter, and J. Pan. An argument-based approach to using multiple ontologies. In *Proceedings of the 3rd International Conference on Scalable Uncertainty Management (SUM'09)*, volume 5785 of *Lecture Notes in Computer Science*, pages 68–79. Springer, 2009.
13. A. Bondarenko, P. Dung, R. Kowalski, and F. Toni. An abstract, argumentation-theoretic approach to default reasoning. *Artificial Intelligence*, 93:63–101, 1997.
14. G. Brewka. *Nonmonotonic Reasoning: Logical Foundations of Commonsense*. Cambridge University Press, 1991.
15. M. Caminada. On the issue of contraposition of defeasible rules. In *Procceings of the International Conference on Computational Models of Argument (COMMA'08)*, pages 109–115, 2008.
16. C. Cayrol. On the relation between argumentation and non-monotonic coherence-based entailment. In *Proceedings of the 14th International Joint Conference on Artificial Intelligence (IJCAI'95)*, pages 1443–1448, 1995.
17. B. Chellas. Basic conditional logic. *Journal of Philosophical Logic*, 4:133–153, 1975.
18. J. Delgrande. A first-order logic for prototypical properties. *Artificial Intelligence*, 33:105–130, 1987.
19. M. Elvang-Gøransson and A. Hunter. Argumentative logics: Reasoning with classically inconsistent information. *Data & Knowledge Engineering*, 16(2):125–145, 1995.
20. M. Elvang-Gøransson, P. Krause, and J. Fox. Acceptability of arguments as 'logical uncertainty'. In *Proceedings of the 2nd European Conference on Symbolic and Quantitative Approaches to Reasoning and Uncertainty (ECSQARU'93)*, volume 747 of *Lecture Notes in Computer Science*, pages 85–90. Springer, 1993.
21. A. García and G. Simari. Defeasible logic programming: An argumentative approach. *Theory and Practice of Logic Programming*, 4:95–138, 2004.
22. A. Garcia and G. Simari. Defeasible logic programming: Delp-servers, contextual queries, and explanations for answers. *Argument and Computatio*, 5(1):63–88, 2014.
23. L. Giordano, V. Gliozzi, N. Olivetti, and G. Pozzato. Analytic tableaux calculi for klm logics of nonmonotonic reasoning. *ACM Transactions on Computational Logic*, 10(3):1–18, 2009.
24. N. Gorogiannis and A. Hunter. Instantiating abstract argumentation with classical logic arguments: Postulates and properties. *Artificial Intelligence*, 175(9-10):1479–1497, 2011.
25. G. Governatori, M. Maher, G. Antoniou, and D. Billington. Argumentation semantics for defeasible logic. *Journal of Logic and Computation*, 14(5):675–702, 2004.
26. R. Haenni. Modelling uncertainty with propositional assumptions-based systems. In *Applications of Uncertainty Formalisms*, volume 1455 of *Lecture Notes in Computer Science*, pages 446–470. Springer, 1998.
27. R. Haenni. Cost-bounded argumentation. *International Journal of Approximate Reasoning*, 26(2):101–127, 2001.
28. A. Hunter. Base logics in argumentation. In *Proceedings of the 3rd International Conference on Computational Models of Argument (COMMA'10)*, volume 216 of *Frontiers in Artificial Intelligence and Applications*, pages 275–286. IOS Press, 2010.

29. A. Hunter. A probabilistic approach to modelling uncertain logical arguments. *International Journal of Approximate Reasoning*, 54(1):47–81, 2013.
30. S. Kraus, D. Lehmann, and M. Magidor. Non-monotonic reasoning, preferential models and cumulative logics. *Artificial Intelligence*, 44:167–207, 1990.
31. N. Mann and A. Hunter. Argumentation using temporal knowledge. In *Proceedings of the 2nd Conference on Computational Models of Argument (COMMA'08)*, volume 172 of *Frontiers in Artificial Intelligence and Applications*, pages 204–215. IOS Press, 2008.
32. J. McCarthy. Circumscription: A form of non-monotonic reasoning. *Artificial Intelligence*, 13(1-2):23–79, 1980.
33. S. Modgil and H. Prakken. The ASPIC+ framework for structured argumentation: a tutorial. *Argument and Computation*, 5(1):31–62, 2014.
34. M. Moguillansky, R. Wassermann, and M. Falappa. An argumentation machinery to reason over inconsistent ontologies. In *Advances in Artificial Intelligence (IBERAMIA 2010)*, volume 6433 of *LNCS*, pages 100–109. Springer, 2010.
35. J. Pollock. Defeasible reasoning. *Cognitive Science*, 11(4):481–518, 1987.
36. J. Pollock. How to reason defeasibly. *Artificial Intelligence*, 57(1):1–42, 1992.
37. H. Prakken. An argumentation framework in default logic. *Annals of Mathematics and Artificial Intelligence*, 9:93–132, 1993.
38. H. Prakken. An abstract framework for argumentation with structured arguments. *Argument and Computation*, 1:93–124, 2010.
39. H. Prakken and G. Sartor. Argument-based extended logic programming with defeasible priorities. *Journal of Applied Non-classical Logic*, 7:25–75, 1997.
40. I. Rahwan and G. Simari, editors. *Argumentation in Artificial Intelligence*. Springer, 2009.
41. R. Reiter. A logic for default reasoning. *Artificial Intelligence*, 13:81–132, 1980.
42. E. Santos and J. Pavão Martins. A default logic based framework for argumentation. In *Proceedings of the European Conference on Artificial Intelligence (ECAI'08)*, pages 859–860, 2008.
43. G. Simari and R. Loui. A mathematical treatment of defeasible reasoning and its implementation. *Artificial Intelligence*, 53:125–157, 1992.
44. F. Toni. A generalized framework for dispute derivation in assumption-based argumentation. *Artificial Intelligence*, 195:1–43, 2013.
45. F. Toni. A tutorial on assumption-based argumentation. *Argument and Computation*, 5(1):89–117, 2014.
46. X. Zhang and Z. Lin. An argumentation framework for description logic ontology reasoning and management. *Journal of Intelligent Information Systems*, 40(3):375–403, 2013.
47. X. Zhang, Z. Zhang, D. Xu, and Z. Lin. Argumentation-based reasoning with inconsistent knowledge bases. In *Advances in Artificial Intelligence*, volume 6085 of *Lecture Notes in Computer Science*, pages 87–99. Springer, 2010.

Defense Semantics of Argumentation: Encoding Reasons for Accepting Arguments

Beishui Liao[1] and Leendert van der Torre[2]

[1] Zhejiang University, China
[2] University of Luxembourg, Luxembourg

Abstract. In this paper we show how the defense relation among abstract arguments can be used to encode the reasons for accepting arguments. After introducing a novel notion of defenses and defense graphs, we propose a defense semantics together with a new notion of defense equivalence of argument graphs, and compare defense equivalence with standard equivalence and strong equivalence, respectively. Then, based on defense semantics, we define two kinds of reasons for accepting arguments, i.e., direct reasons and root reasons, and a notion of root equivalence of argument graphs. Finally, we show how the notion of root equivalence can be used in argumentation summarization.

Keywords: abstract argumentation, defense graph, defense semantics, argumentation equivalence, argumentation summarization

1 Introduction

Abstract argumentation is mainly about evaluating the status of arguments in an argument graph [1–3], which is composed of a set of abstract arguments and a set of attacks between them [4]. In many topics such as equivalence [5–7], summarization [8], and dynamics in argumentation [9, 10], the notion of extensions plays a central role. Since in classical argumentation semantics, an extension is a set of arguments that are collectively accepted, the existing theories and approaches based on this notion are mainly focused on exploiting the status of individual arguments. However, besides the status of individual arguments, in many situations, we need to know the reasons for accepting arguments in terms of a defense relation. The following are two simple examples.

$\mathcal{F}_1:\quad a \rightleftarrows c_1 \rightarrow c_2 \rightarrow b \qquad \mathcal{F}_2:\quad a \leftrightarrow b$
$\qquad\qquad c_4 \leftarrow c_3$

First, consider a, b in $\mathcal{F}_1$ and $\mathcal{F}_2$. In $\mathcal{F}_1$, accepting a is a reason to accept c_2, accepting c_2 is a reason to accept c_3, and accepting c_3 is a reason to accept a. If we allow this relation to be transitive, we find that accepting a is a reason to accept a. Similarly, accepting b is a reason to accept b. Meanwhile, in $\mathcal{F}_2$, we have: accepting a is a reason to accept a, and accepting b is a reason to accept b. So, from the perspective of the reasons for accepting a and b, $\mathcal{F}_2$ is equivalent to $\mathcal{F}_1$, or $\mathcal{F}_2$ is a summarization of $\mathcal{F}_1$.

Second, consider the question when two argument graphs are equivalent in a dynamic setting. For $\mathcal{F}_3$ and $\mathcal{F}_4$ below, both of them have a complete extension $\{a, c\}$. However, the reasons of accepting c in $\mathcal{F}_3$ and $\mathcal{F}_4$ are different. For the former, c is defended by a, while for the latter, c is unattacked and has no defender. In this sense, $\mathcal{F}_3$ and $\mathcal{F}_4$ are not equivalent. For example, in order to change the status of argument c from "accepted" to "rejected", in $\mathcal{F}_3$, one may produce a new argument to attack the defender a, or to directly attack c. However, in $\mathcal{F}_4$ using an argument to attack a cannot change the status of c, since a is not a defender of c.

$$\mathcal{F}_3: \quad a \longrightarrow b \longrightarrow c \qquad \mathcal{F}_4: \quad a \longrightarrow b \qquad c$$

From the above two examples, one question arises: under what conditions, can two argument graphs be viewed as equivalent? The existing notions of argumentation equivalence, including standard equivalence and strong equivalence, are not sufficient to capture the equivalence of the argument graphs in the situations mentioned above. More specifically, $\mathcal{F}_1$ and $\mathcal{F}_2$ are not equivalent in terms of the notion of standard equivalence or that of strong equivalence, but they are equivalent in the sense that the reasons for accepting arguments a and b in these two graphs are the same. $\mathcal{F}_3$ and $\mathcal{F}_4$ are equivalent in terms of standard equivalence, but they are not equivalent in the sense that the reasons for accepting c in these two graphs are different. Although the notion of strong equivalence can be used to identify the difference between $\mathcal{F}_3$ and $\mathcal{F}_4$, conceptually it is not defined from the perspective of reasons for accepting arguments.

Note that the reasons for accepting arguments in the above two examples are depicted in terms of a defense relation, which plays a central role in Dung's concept of admissibility and thus in admissibility based semantics. So, it is natural to define a new semantics in this paper based on a defense relation such that the reasons for accepting arguments can be encoded.

Since the new semantics is defined at the level of abstract argumentation, it can be applied to various structured argumentations systems. In particular, in the field of legal reasoning [20], argumentation can be used to model legal interpretation, dialogue, and deontic reasoning, etc. In all these applications, it is useful to make clear the reasons for accepting arguments in terms of a defense relation. In this paper, we will formulate a defense semantics for abstract argumentation, while its application to various structured argumentation systems is left to future work. The structure of this paper is as follows. In Section 2, we introduce some basic notions of argumentation semantics. In Section 3, we propose the notions of defenses and defense graphs, which lay a foundation of this paper. In Section 4, we formulate defense semantics by applying classical argumentation semantics to defense graphs, and study some properties of this new semantics. In Section 5, we introduce two kinds of reasons for accepting arguments in terms of defense semantics. We conclude in Section 6.

2 Argumentation semantics

An argument graph or argumentation framework (AF) is defined as $\mathcal{F} = (\mathcal{A}, \rightarrow)$, where $\mathcal{A}$ is a finite set of arguments and $\rightarrow \subseteq \mathcal{A} \times \mathcal{A}$ is a set of attacks between arguments [4].

Let $\mathcal{F} = (\mathcal{A}, \rightarrow)$ be an argument graph. Given a set $B \subseteq \mathcal{A}$ and an argument $\alpha \in \mathcal{A}$, B attacks α, denoted $B \rightarrow \alpha$, iff there exists $\beta \in B$ such that $\beta \rightarrow \alpha$. Given an argument $\alpha \in \mathcal{A}$, let $\alpha^{\leftarrow} = \{\beta \in \mathcal{A} \mid \beta \rightarrow \alpha\}$ be the set of arguments attacking α, and $\alpha^{\rightarrow} = \{\beta \in \mathcal{A} \mid \alpha \rightarrow \beta\}$ be the set of arguments attacked by α. When $\alpha^{\leftarrow} = \emptyset$, we say that α is unattacked, or α is an initial argument.

Given $\mathcal{F} = (\mathcal{A}, \rightarrow)$ and $E \subseteq \mathcal{A}$, we say: E is *conflict-free* iff $\nexists \alpha, \beta \in E$ such that $\alpha \rightarrow \beta$; $\alpha \in \mathcal{A}$ is *defended* by E iff $\forall \beta \in \alpha^{\leftarrow}$, it holds that $E \rightarrow \beta$; E is *admissible* iff E is conflict-free, and each argument in E is defended by E; E is a *complete extension* iff E is admissible, and each argument in $\mathcal{A}$ that is defended by E is in E; E is a *grounded extension* iff E is the minimal (w.r.t. set-inclusion) complete extension; E is a *preferred extension* iff E is a maximal (w.r.t. set-inclusion) complete extension; E is a *stable extension* iff E is conflict-free and E attacks each argument that is not in E. We use $\sigma(\mathcal{F})$ to denote the set of argument extensions of $\mathcal{F}$ under semantics σ, where σ is a function mapping each argument graph to a set of argument extensions. We use co, gr, pr and st to denote complete, grounded, preferred and stable semantics respectively. There are some other argumentation semantics (cf. [2] for an overview).

For argument graphs $\mathcal{F}_1 = (\mathcal{A}_1, \rightarrow_1)$ and $\mathcal{F}_2 = (\mathcal{A}_2, \rightarrow_2)$, we use $\mathcal{F}_1 \cup \mathcal{F}_2$ to denote $(\mathcal{A}_1 \cup \mathcal{A}_2, \rightarrow_1 \cup \rightarrow_2)$. The standard equivalence and strong equivalence of argument graphs are defined as follows. For simplicity, when we talk about equivalence of AFs, we mainly consider the cases under complete semantics, while the full-fledged study of equivalence will be presented in an extended version of the present paper.

Definition 1 (Standard and strong equivalence of AFs). *[5] Let $\mathcal{F}$ and $\mathcal{G}$ be two argument graphs, and σ be a semantics. $\mathcal{F}$ and $\mathcal{G}$ are of standard equivalence w.r.t. a semantics σ, in symbols $\mathcal{F} \equiv^{\sigma} \mathcal{G}$, iff $\sigma(\mathcal{F}) = \sigma(\mathcal{G})$. $\mathcal{F}$ and $\mathcal{G}$ are of strong equivalence w.r.t. a semantics σ, in symbols $\mathcal{F} \equiv^{\sigma}_s \mathcal{G}$, iff for every argument graph $\mathcal{H}$, it holds that $\sigma(\mathcal{F} \cup \mathcal{H}) = \sigma(\mathcal{G} \cup \mathcal{H})$.*

Example 1. Consider $\mathcal{F}_1 - \mathcal{F}_4$ in Section 1. In terms of Definition 1, under complete semantics, since $\text{co}(\mathcal{F}_1) \neq \text{co}(\mathcal{F}_2)$, $\mathcal{F}_1 \not\equiv^{\text{co}} \mathcal{F}_2$, which implies that $\mathcal{F}_1 \not\equiv^{\text{co}}_s \mathcal{F}_2$. And, since $\text{co}(\mathcal{F}_3) = \text{co}(\mathcal{F}_4)$, $\mathcal{F}_3 \equiv^{\text{co}} \mathcal{F}_4$. Let $\mathcal{H} = (\{d\}, \{d \rightarrow a\})$. Since $\text{co}(\mathcal{F}_3 \cup \mathcal{H}) \neq \text{co}(\mathcal{F}_4 \cup \mathcal{H})$, $\mathcal{F}_3 \not\equiv^{\text{co}}_s \mathcal{F}_4$.

Given an argument graph $\mathcal{F} = (\mathcal{A}, \rightarrow)$, the kernel of $\mathcal{F}$ under complete semantics, call *c-kernel*, is defined as follows.

Definition 2 (c-kernel of an AF). *[5] For an argument graph $\mathcal{F} = (\mathcal{A}, \rightarrow)$, the c-kernel of $\mathcal{F}$ is defined as $\mathcal{F}^{\text{ck}} = (\mathcal{A}, \rightarrow^{\text{ck}})$, where $\rightarrow^{\text{ck}} = \rightarrow \setminus \{\alpha \rightarrow \beta \mid \alpha \neq \beta, \alpha \rightarrow \alpha, \beta \rightarrow \beta\}$.*

According to [5], it holds that $co(\mathcal{F}) = co(\mathcal{F}^{ck})$, and for any AFs $\mathcal{F}$ and $\mathcal{G}$, $\mathcal{F}^{ck} = \mathcal{G}^{ck}$ iff $\mathcal{F} \equiv_s^{co} \mathcal{G}$.

3 Defenses and defense graph

According to classical argumentation semantics, with respect to an extension E, an argument $\alpha \in E$ is accepted because it is initial or for all $\gamma \in \alpha^\leftarrow$, γ is attacked by an argument in E. So, for all $\alpha, \beta \in E$, if there exists $\gamma \in \mathcal{A} \setminus E$ such that $\alpha \to \gamma$ and $\gamma \to \beta$, we say that accepting α is a (partial) reason to accept β, denoted as $\langle \alpha, \beta \rangle$. And, for all $\beta \in E$ if $\beta^\leftarrow = \emptyset$ (i.e., β is an initial argument), we say that β is accepted without a reason, denoted as $\langle \emptyset, \beta \rangle$ where ø is a symbol denoting an empty position. In this paper, $\langle \alpha, \beta \rangle$ or $\langle \emptyset, \beta \rangle$ is called a *defense*.

Without referring to any specific extension, a defense $\langle \alpha, \beta \rangle$ can be viewed as a relation between α and β satisfying some constraints. Intuitively, there are the following two minimal constraints. First, $\{\alpha, \beta\}$ is conflict-free. Otherwise, they can not be both accepted. Second, there exists $\gamma \in \mathcal{A} \setminus \{\alpha, \beta\}$ such that $\alpha \to \gamma$ and $\gamma \to \beta$, in the sense that α defends β by attacking β's attacker γ. Regarding the defense $\langle \emptyset, \beta \rangle$, the only constraint is that β is initial.

Example 2. Consider $\mathcal{F}_5$ below. $\langle \emptyset, a \rangle$, $\langle a, c \rangle$ and $\langle b, d \rangle$ are defenses. Note that the three defenses do not refer to a specific extension.

$\mathcal{F}_5:$ $a \to b \to c \to d$

Based on the above analysis, we have the following definition.

Definition 3 (Defense). *Let $\mathcal{F} = (\mathcal{A}, \to)$ be an argument graph. For $\alpha, \beta \in \mathcal{A}$, $\langle \alpha, \beta \rangle$ is a defense iff $\{\alpha, \beta\}$ is conflict-free, and $\exists \gamma \in \mathcal{A}$ such that $\alpha \to \gamma$ and $\gamma \to \beta$; $\langle \emptyset, \beta \rangle$ is a defense iff β is initial.*

The set of defenses of $\mathcal{F}$ is denoted as $\mathcal{F}^{\text{DEF}}$. Given a defense $\langle \alpha, \beta \rangle$ or $\langle \emptyset, \beta \rangle \in \mathcal{F}^{\text{DEF}}$, we call α the *defender*, and β the *defendee*, of the defense. Given a set $D \subseteq \mathcal{F}^{\text{DEF}}$, we write defendee$(D) = \{\beta \mid \langle \alpha, \beta \rangle, \langle \emptyset, \beta \rangle \in D\}$ to denote the set of defendees in D, defender$(D) = \{\alpha \mid \langle \alpha, \beta \rangle \in D\}$ to denote the set of defenders in D, and def$(D) = $ defendee$(D) \cup $ defender(D) be the set of defendees and defenders in D. Note that not all arguments of an AF are included in the defenses. Consider the following example.

Example 3. In $\mathcal{F}_6$, $\mathcal{F}_7$ and $\mathcal{F}_8$, $\langle a_2, a_4 \rangle$, $\langle a_3, a_5 \rangle$, $\langle a_4, a_6 \rangle$, $\langle \emptyset, a_7 \rangle$, $\langle a_7, a_9 \rangle$, $\langle \emptyset, a_{11} \rangle$ and $\langle a_{11}, a_{13} \rangle$ are defenses, while some defense-like pairs, for instance (a_1, a_3) and (a_{14}, a_{13}), are not defenses since both $\{a_1, a_3\}$ and $\{a_{14}, a_{13}\}$ are not conflict-free. And, $(\emptyset, a_{10})$, $(\emptyset, a_{14})$ and $(\emptyset, a_{15})$ are not defenses, because they are not initial arguments, but either self-attacked or attacked by a self-attacked argument.

$\mathcal{F}_6:$ $a_1 \to a_2 \to a_3 \to a_4 \to a_5 \to a_6$ $\mathcal{F}_8:$ $a_{11} \to a_{12} \to a_{13}$

$\mathcal{F}_7:$ $a_7 \to a_8 \to a_9 \leftarrow a_{10}$ $a_{14} \to a_{15}$

Given a defense $\langle x, \alpha \rangle$ where $x \in \mathcal{A} \cup \{\emptyset\}$ and $\alpha \in \mathcal{A}$, $\langle x, \alpha \rangle$ can be regarded as a meta-argument. Its status is affected by other defenses and/or other defense-like pairs (cf. (a_1, a_3) and $(\emptyset, a_{10})$ in Example 3). Since the pairs like (a_1, a_3) and $(\emptyset, a_{10})$ are not accepted as a defense, but may be used to hamper the acceptance of some defenses, their behavior is similar to that of *defeaters* in defeasible logic. We call them *defeaters of defenses (DoD)*.

Definition 4 (Defeaters of defenses). Let $\mathcal{F} = (\mathcal{A}, \to)$ be an argument graph. For $\alpha, \beta \in \mathcal{A}$,

– (α, β) is a DoD, iff $\{\alpha, \beta\}$ is not conflict-free, and $\exists \gamma \in \mathcal{A} \setminus \{\alpha, \beta\}$ such that $\alpha \to \gamma$ and $\gamma \to \beta$.
– $(\emptyset, \beta)$ is a DoD, iff β is self-attacked or attacked by a self-attacked argument.

The set of DoDs of $\mathcal{F}$ is denoted as $\mathcal{F}^{\text{DOD}}$.

In this definition, note that (α, β) and $(\emptyset, \beta)$ are not accepted as a defense, and may be used to hamper the acceptance of some defenses. This does not mean that the arguments α and β in the corresponding argument graph can not be accepted, since they may in some defenses at the same time. See the following example.

Example 4. In $\mathcal{F}_9$, $\langle a, c \rangle$, while $(\emptyset, b)$, $(\emptyset, c)$ and (b, d) are DoDs. c is both in the defense $\langle a, c \rangle$ and in the DoD $(\emptyset, c)$. When $\langle a, c \rangle$ is accepted, c is accepted.

$\mathcal{F}_9:$ $a \to b \to c \to d$

Note also that in Definition 4 when β is attacked by a self-attacked argument, it is a DoD. Consider $\mathcal{F}_8$ in Example 3. $\langle a_{11}, a_{13} \rangle$ is a defense. If $(\emptyset, a_{15})$ is not a DoD, then there is no DoD to prevent the acceptance of $\langle a_{11}, a_{13} \rangle$.

Let $\arg(\mathcal{F}^{\text{DOD}}) = \{\alpha, \beta \mid (\alpha, \beta) \in \mathcal{F}^{\text{DOD}}\} \cup \{\beta \mid (\emptyset, \beta) \in \mathcal{F}^{\text{DOD}}\}$ be the set of arguments involved in $\mathcal{F}^{\text{DOD}}$. Let $\text{def}(\mathcal{F}^{\text{DEF}})^{\to}$ be the set of arguments attacked by $\text{def}(\mathcal{F}^{\text{DEF}})$. We have the following proposition.

Proposition 1. Let $\mathcal{F} = (\mathcal{A}, \to)$ be an argument graph. It holds that $\mathcal{A} = \arg(\mathcal{F}^{\text{DOD}}) \cup \text{def}(\mathcal{F}^{\text{DEF}}) \cup \text{def}(\mathcal{F}^{\text{DEF}})^{\to}$.

This proposition states that arguments in $\mathcal{F}$ are equivalent to the union of the arguments in defenses, arguments in defeaters of defenses, and the arguments attacked by the arguments in defenses.

Let $\mathcal{F}^{\text{d}} = \mathcal{F}^{\text{DEF}} \cup \mathcal{F}^{\text{DOD}}$ be the set of defenses and their defeaters. The attack relation between the elements of $\mathcal{F}^{\text{d}}$ can be identified according to the attack relation between the arguments involved. For convenience, we also write $[x, \beta]$ to denote a defense $\langle x, \beta \rangle$ or a defeater of defenses (x, β) where $x \in \mathcal{A} \cup \{\emptyset\}$ and $\beta \in \mathcal{A}$. Formally we have the following definition.

Definition 5 (Attacks between defenses and their defeaters). *For all $[x, \alpha], [y, \beta] \in \mathcal{F}^d$ where $x, y \in \mathcal{A} \cup \{\emptyset\}$ and $\alpha, \beta \in \mathcal{A}$, we say that $[x, \alpha]$ attacks $[y, \beta]$, denoted as $[x, \alpha] \to^d [y, \beta]$ iff $x \to y$, $x \to \beta$, $\alpha \to y$, or $\alpha \to \beta$.*

The set of attacks between defenses and their defeaters is denoted as $\to^d$. Given $D \subseteq \mathcal{F}^d$ and $X \in \mathcal{F}^d$, we use $D \to^d X$ to denote that $\exists Y \in D$ such that $Y \to^d X$.

Since the status of a defense is determined by that of other defenses and affected by defeaters of defenses through the attacks between them, to evaluate the status of normal defenses, one possible way is to use *defense graph*, which is defined as follows.

Definition 6 (Defense graph). *Let $\mathcal{F} = (\mathcal{A}, \to)$ be an argument graph. Let $\mathcal{F}^d = \mathcal{F}^{\text{DEF}} \cup \mathcal{F}^{\text{DoD}}$. A defense graph w.r.t. $\mathcal{F}$, denoted $d(\mathcal{F})$, is defined as follows.*

$$d(\mathcal{F}) = (\mathcal{F}^d, \to^d) \tag{1}$$

A defense graph can be viewed as a kind of meta-argumentation [11].

Example 5. The defense graph of $\mathcal{F}_6$ is as follows.

$$d(\mathcal{F}_6): \quad (a_2, a_1) \leftrightarrow (a_3, a_2) \to (a_2, a_4) \leftrightarrow (a_3, a_5)$$
$$(a_1, a_3) \qquad (a_4, a_6)$$

4 Defense semantics

In a defense graph $d(\mathcal{F}) = (\mathcal{F}^d, \to^d)$, nodes are defenses and/or defeaters of defenses, rather than arguments in the corresponding argument graph $\mathcal{F}$. So, when applying classical semantics to $d(\mathcal{F})$, we get a set of extensions, each of which is a set of defenses. By slightly modifying the definition for classical semantics, defense semantics can be defined as follows.

Definition 7 (Defense semantics). *Defense semantics is a function Σ mapping each defense graph to a set of extensions of defenses. Given a defense graph $d(\mathcal{F}) = (\mathcal{F}^d, \to^d)$ where $\mathcal{F}^d = \mathcal{F}^{\text{DEF}} \cup \mathcal{F}^{\text{DoD}}$, let $D \subseteq \mathcal{F}^{\text{DEF}}$. We have:*

- *D is conflict-free iff $\nexists X, Y \in D$ such that $X \to^d Y$.*
- *$X \in \mathcal{F}^{\text{DEF}}$ is defended by D iff for all $Y \in \mathcal{F}^d$, if $Y \to^d X$, then $\exists Z \in D$ such that $Z \to^d Y$.*
- *D is admissible iff D is conflict-free and each member in D is defended by D.*
- *D is a complete extension of defenses iff D is admissible, and each member in $\mathcal{F}^{\text{DEF}}$ that is defended by D is in D.*

- D is the grounded extension of defenses iff D is the minimal (w.r.t. set-inclusion) complete extension of defenses.
- D is a preferred extension of defenses iff D is a maximal (w.r.t. set-inclusion) complete extension of defenses.
- D is a stable extension of defenses iff D is conflict-free, and $\forall X \in \mathcal{F}^d \setminus D$, $D \to^d X$.

The set of complete, grounded, preferred, and stable extensions of defenses of $d(\mathcal{F})$ is denoted as $\mathrm{CO}(d(\mathcal{F}))$, $\mathrm{GR}(d(\mathcal{F}))$, $\mathrm{PR}(d(\mathcal{F}))$ and $\mathrm{ST}(d(\mathcal{F}))$ respectively.

Note that the notion of defense semantics is similar to that of classical semantics. The only difference is that in a defense graph, we differentiate two kinds of nodes: defenses and defeaters of defenses. The former can be included in extensions, while the latter are only used to prevent the acceptance of some defenses.

Now, let us consider some properties of the defense semantics of an argument graph.

The first property is about the relation between defense semantics and classical semantics. Let $D \in \Sigma(d(\mathcal{F}))$ be a Σ-extension of $d(\mathcal{F})$. Now the question is whether the set of defenders and defendees in D is a σ-extension of $\mathcal{F}$. In order to verify this property, technically, we first present the follow lemma. The lemma states that $\forall \langle \alpha, \beta \rangle \in D$, if α is attacked by an argument $\gamma \in \mathcal{A}$, then $\exists \eta \in \mathrm{def}(D)$ such that η attacks γ.

Lemma 1 *For all $D \in \Sigma(d(\mathcal{F}))$, for all $\langle x, y \rangle \in D$, for all $\gamma \in \mathcal{A}$, if $\gamma \to x$ or $\gamma \to y$, then $\mathrm{def}(D) \to \gamma$.*

Example 6. Consider $d(\mathcal{F}_1)$ below. Under complete semantics, $\mathrm{CO}(d(\mathcal{F}_1)) = \{D_1, D_2, D_3\}$ where $D_1 = \{\}$, $D_2 = \{\langle a, c_2 \rangle, \langle c_2, c_3 \rangle, \langle c_3, a \rangle\}$, $D_3 = \{\langle b, c_4 \rangle, \langle c_1, b \rangle, \langle c_4, c_1 \rangle\}$. Take D_2 and $\langle a, c_2 \rangle$ in D_2 as an example. $\mathrm{def}(D_2) = \{a, c_2, c_3\}$. For a being attacked by c_4, and c_2 being attacked by c_1, it holds that $\mathrm{def}(D_2) \to c_4$ and $\mathrm{def}(D_2) \to c_1$.

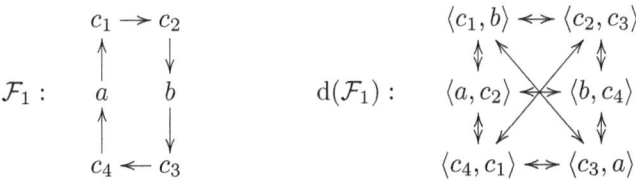

Based on Lemma 1, under complete semantics, we have the following theorem.

Theorem 1. *For all $D \in \mathrm{CO}(d(\mathcal{F}))$, $\mathrm{def}(D) \in \mathrm{co}(\mathcal{F})$.*

Theorem 1 makes clear that for each complete defense extension D of a defense graph, there exists a complete argument extension E of the corresponding argument graph such that E is equal to def(D). On the other hand, the following theorem says that for each complete argument extension E of an argument graph, there exists a complete defense extension D of the corresponding defense graph such that $D = \mathrm{d}(E)$ where $\mathrm{d}(E) = \{\langle x, y \rangle \in \mathcal{F}^{\mathrm{DEF}} \mid x \in E \cup \{\emptyset\}, y \in E\}$.

Theorem 2. *For all $E \in \mathrm{co}(\mathcal{F})$, $\mathrm{d}(E) \in \mathrm{CO}(\mathrm{d}(\mathcal{F}))$.*

The relation between argument extensions and defense extensions under other semantics is presented in the following corollaries.

Corollary 1. $\forall \Sigma \in \{\mathrm{GR}, \mathrm{PR}, \mathrm{ST}\}$, *it holds that* $\forall D \in \Sigma(\mathrm{d}(\mathcal{F}))$, $\mathrm{def}(D) \in \sigma(\mathcal{F})$.

Proofs for Lemma 1, Theorem 1, 2 and Corollary 1 are presented in the Appendix. In the following theorems and corollaries, when we say $\Sigma \in \{\mathrm{CO}, \mathrm{GR}, \mathrm{PR}, \mathrm{ST}\}$, σ is referred to co, gr, pr and st, correspondingly. Meanwhile, when we say $\sigma \in \{\mathrm{co}, \mathrm{gr}, \mathrm{pr}, \mathrm{st}\}$, Σ is referred to CO, GR, PR and ST, correspondingly.

Corollary 2. $\forall \Sigma \in \{\mathrm{GR}, \mathrm{PR}, \mathrm{ST}\}$ *it holds that* $\forall E \in \sigma(\mathcal{F})$, $\mathrm{d}(E) \in \Sigma(\mathrm{d}(\mathcal{F}))$.

Proof. Under grounded semantics, we need to verify that $\mathrm{d}(E)$ is minimal (w.r.t. set-inclusion). Assume the contrary. Then $\exists D' \subsetneq \mathrm{d}(E)$ such that D' is a grounded extension. According to theorem 1, $\mathrm{def}(D')$ is a complete extension. It follows that $\mathrm{def}(D') \subsetneq \mathrm{def}(\mathrm{d}(E)) = E$. It turns out that E is not a grounded extension. Contradiction.

Under preferred semantic, it is easy to verify that $\mathrm{d}(E)$ is maximal (w.r.t. set-inclusion).

Under stable semantics, we need to prove that for all $[x, \alpha] \in \mathcal{F}^d \setminus \mathrm{d}(E)$: $\mathrm{d}(E) \to [x, \alpha]$. Assume the contrary. Then, $\exists [x, \alpha] \in \mathcal{F}^d \setminus \mathrm{d}(E)$ such that $\mathrm{d}(E)$ does not attack $[x, \alpha]$. So, E does not attack x and α. Since E is stable, it holds that $\{x, \alpha\} \setminus \{\emptyset\} \subseteq E$. So, $[x, \alpha] \in E$. Contradiction.

Theorems 1 and 2 and Corollaries 1 and 2 describe the relation between argument extensions and defense extensions under various semantics. This relation can be further described by two equations in the following two corollaries. First, by overloading the notation, let $\mathrm{d}(\sigma(\mathcal{F})) = \{\mathrm{d}(E) \mid E \in \sigma(\mathcal{F})\}$, where $\sigma \in \{\mathrm{co}, \mathrm{gr}, \mathrm{pr}, \mathrm{st}\}$.

Corollary 3. *For all $\sigma \in \{\mathrm{co}, \mathrm{pr}, \mathrm{gr}, \mathrm{st}\}$, it holds that $\mathrm{d}(\sigma(\mathcal{F})) = \Sigma(\mathrm{d}(\mathcal{F}))$.*

Proof. For all $\mathrm{d}(E) \in \mathrm{d}(\sigma(\mathcal{F}))$, according to Theorem 2 and Corollary 2, $\mathrm{d}(E) \in \Sigma(\mathrm{d}(\mathcal{F}))$. For all $D \in \Sigma(\mathrm{d}(\mathcal{F}))$, according to Theorem 1 and Corollary 1, $\mathrm{def}(D) \in \sigma(\mathcal{F})$. Since $\mathrm{d}(\mathrm{def}(D)) = \{\langle \beta, \alpha \rangle \in \mathcal{F}^{\mathrm{DEF}} \mid \alpha, \beta \in \mathrm{def}(D)\} \cup \{\langle \emptyset, \alpha \rangle \in \mathcal{F}^{\mathrm{DEF}} \mid \alpha \in \mathrm{def}(D)\} = D$, it holds that $D \in \mathrm{d}(\sigma(\mathcal{F}))$.

Example 7. Consider $\mathcal{F}_{10}$ and $\mathrm{d}(\mathcal{F}_{10})$ below. Under complete semantics, we have:

- $\text{co}(\mathcal{F}_{10}) = \{E_1, E_2\}$, where $E_1 = \{\}$, $E_2 = \{b\}$;
- $\text{d}(\text{co}(\mathcal{F}_{10})) = \{\text{d}(E_1), \text{d}(E_2)\}$, where $\text{d}(E_1) = \{\}$, $\text{d}(E_2) = \{\langle b, b \rangle\}$;
- $\text{CO}(\text{d}(\mathcal{F}_{10})) = \{D_1, D_2\}$, where $D_1 = \{\}$, $D_2 = \{\langle b, b \rangle\}$.

So, it holds that $\text{d}(\text{co}(\mathcal{F}_{10})) = \text{CO}(\text{d}(\mathcal{F}_{10}))$.

$\mathcal{F}_{10}:$ $a \longleftrightarrow b$ $\quad$ $\text{d}(\mathcal{F}_{10}):$ $\quad \langle a, a \rangle \longleftrightarrow \langle b, b \rangle$
$\qquad\qquad\qquad\quad\; c$ $\qquad\qquad\qquad\quad (a, c) \longleftrightarrow (c, b) \longleftrightarrow (b, a)$

Second, by overloading the notation, let $\text{def}(\Sigma(\text{d}(\mathcal{F}))) = \{\text{def}(D) \mid D \in \Sigma(\text{d}(\mathcal{F}))\}$, where $\Sigma \in \{\text{CO}, \text{GR}, \text{PR}, \text{ST}\}$.

Corollary 4. *For all $\Sigma \in \{\text{CO}, \text{GR}, \text{PR}, \text{ST}\}$, it holds that $\sigma(\mathcal{F}) = \text{def}(\Sigma(\text{d}(\mathcal{F})))$.*

Proof. For all $E \in \sigma(\mathcal{F})$, according to Theorem 2 and Corollary 2, $\text{d}(E) \in \Sigma(\text{d}(\mathcal{F}))$. Since $\text{def}(\text{d}(E)) = E$, $E \in \text{def}(\Sigma(\text{d}(\mathcal{F})))$. For all $\text{def}(D) \in \text{def}(\Sigma(\text{d}(\mathcal{F})))$, since $D \in \Sigma(\text{d}(\mathcal{F}))$, according to Theorem 1 and Corollary 1, $\text{def}(D) \in \text{co}(\mathcal{F})$.

Example 8. Under compete semantics, continue Example 7, $\text{def}(\text{CO}(\text{d}(\mathcal{F}_{10}))) = \{\text{def}(D_1), \text{def}(D_2)\}$, where $\text{def}(D_1) = \{\}$, $\text{def}(D_2) = \{b\}$. It holds that $\text{co}(\mathcal{F}_{10}) = \text{def}(\text{CO}(\text{d}(\mathcal{F}_{10})))$.

The second property formulated in Theorems 3, 4 is about the equivalence of argument graphs under defense semantics, called *defense equivalence of argument graphs*.

Definition 8 (Defense equivalence of AFs). *Let $\mathcal{F}$ and $\mathcal{G}$ be two argument graphs. $\mathcal{F}$ and $\mathcal{G}$ are of defense equivalence w.r.t. a semantics Σ, denoted as $\mathcal{F} \equiv_{\text{d}}^{\Sigma} \mathcal{G}$, iff $\Sigma(\text{d}(\mathcal{F})) = \Sigma(\text{d}(\mathcal{G}))$.*

Concerning the relation between defense equivalence and standard equivalence of argument graphs, we have the following theorem.

Theorem 3. *Let $\mathcal{F}$ and $\mathcal{G}$ be two argument graphs, and $\Sigma \in \{\text{CO}, \text{GR}, \text{PR}, \text{ST}\}$ be a semantics. If $\mathcal{F} \equiv_{\text{d}}^{\Sigma} \mathcal{G}$, then $\mathcal{F} \equiv^{\sigma} \mathcal{G}$.*

Proof. If $\mathcal{F} \equiv_{\text{d}}^{\Sigma} \mathcal{G}$, then $\Sigma(\text{d}(\mathcal{F})) = \Sigma(\text{d}(\mathcal{G}))$. According to Corollary 4, it follows that $\sigma(\mathcal{F}) = \text{def}(\Sigma(\text{d}(\mathcal{F}))) = \text{def}(\Sigma(\text{d}(\mathcal{G}))) = \sigma(\mathcal{G})$. Since $\sigma(\mathcal{F}) = \sigma(\mathcal{G})$, $\mathcal{F} \equiv^{\sigma} \mathcal{G}$.

Note that $\mathcal{F} \equiv^{\sigma} \mathcal{G}$ does not imply $\mathcal{F} \equiv_{\text{d}}^{\Sigma} \mathcal{G}$ in general. Consider the following example under complete semantics.

Example 9. Since co($\mathcal{F}_3$) = co($\mathcal{F}_4$) = {{a, c}}, it holds that $\mathcal{F}_3 \equiv^{co} \mathcal{F}_4$. Since CO(d($\mathcal{F}_3$)) = {{⟨ø, a⟩}, ⟨a, c⟩} and CO(d($\mathcal{F}_4$)) = {{⟨ø, a⟩}, ⟨ø, c⟩}, CO(d($\mathcal{F}_3$)) ≠ CO(d($\mathcal{F}_4$)). So, it is not the case that $\mathcal{F}_3 \equiv_d^{CO} \mathcal{F}_4$.

$\mathcal{F}_3$: $a \twoheadrightarrow b \twoheadrightarrow c$ d($\mathcal{F}_3$):   ⟨ø, a⟩   ⟨a, c⟩

$\mathcal{F}_4$: $a \twoheadrightarrow b$ c d($\mathcal{F}_4$):   ⟨ø, a⟩   ⟨ø, c⟩

About the relation between defense equivalence and strong equivalence of argument graphs, under complete semantics, we have the following lemma and theorem.

Lemma 2 *It holds that* CO(d($\mathcal{F}$)) = CO(d($\mathcal{F}^{ck}$)).

Proof. According to Corollary 3, d(co($\mathcal{F}$)) = CO(d($\mathcal{F}$)), d(co($\mathcal{F}^{ck}$)) = CO(d($\mathcal{F}^{ck}$)). Since co($\mathcal{F}$) = co($\mathcal{F}^{ck}$), CO(d($\mathcal{F}$)) = d(co($\mathcal{F}$)) = d(co($\mathcal{F}^{ck}$)) = CO(d($\mathcal{F}^{ck}$)).

Theorem 4. *Let $\mathcal{F}$ and $\mathcal{G}$ be two argument graphs. If $\mathcal{F} \equiv_s^{co} \mathcal{G}$, then $\mathcal{F} \equiv_d^{CO} \mathcal{G}$.*

Proof. If $\mathcal{F} \equiv_s^{co} \mathcal{G}$, then $\mathcal{F}^{ck} = \mathcal{G}^{ck}$. So, CO(d($\mathcal{F}^{ck}$)) = CO(d($\mathcal{G}^{ck}$)). According to Lemma 2, CO(d($\mathcal{F}$)) = CO(d($\mathcal{F}^{ck}$)), CO(d($\mathcal{G}$)) = CO(d($\mathcal{G}^{ck}$)). So, we have CO(d($\mathcal{F}$)) = CO(d($\mathcal{G}$)), i.e., $\mathcal{F} \equiv_d^{CO} \mathcal{G}$.

Note that $\mathcal{F} \equiv_d^{CO} \mathcal{G}$ does not imply $\mathcal{F} \equiv_s^{co} \mathcal{G}$ in general. Consider the following example.

Example 10. Since CO(d($\mathcal{F}_3$)) = CO(d($\mathcal{F}_{11}$)) = {{⟨ø, a⟩}, ⟨a, c⟩}, $\mathcal{F}_3 \equiv_d^{CO} \mathcal{F}_{11}$. However, since $\mathcal{F}_3^{ck} \neq \mathcal{F}_{11}^{ck}$, $\mathcal{F}_3 \not\equiv_s^{co} \mathcal{F}_{11}$.

$\mathcal{F}_{11}$: $a \twoheadrightarrow b \twoheadrightarrow c$ d($\mathcal{F}_{11}$):   ⟨ø, a⟩   ⟨a, c⟩
            $\searrow \nearrow$
             d

5 Encoding reasons for accepting arguments

Defense semantics can be used to encode reasons for accepting arguments. Consider the following example.

Example 11. CO(d($\mathcal{F}_{12}$)) = {D_1, D_2}, where D_1 = {⟨b, b⟩, ⟨b, d⟩, ⟨g, d⟩, ⟨e, g⟩, ⟨ø, e⟩}, D_2 = {⟨a, a⟩, ⟨g, d⟩, ⟨e, g⟩, ⟨ø, e⟩}. One way to capture reasons for accepting arguments is to relate each reason to an extension of defenses. For instance, concerning the reasons for accepting d w.r.t. D_1, we differentiate the following reasons:

– Direct reason: accepting {b, g} is a direct reason for accepting d. This reason can be identified in terms of defenses ⟨b, d⟩ and ⟨g, d⟩ in D_1.

- Root reason: accepting $\{e, b\}$ is a root reason for accepting d, in the sense that each element of a root reason is either an initial argument, or an argument without further defenders except itself. This reason can be identified by means of viewing each defense as a binary relation, and allowing this relation to be transitive. Given $\langle e, g \rangle$ and $\langle g, d \rangle$ in D_1, we have $\langle e, d \rangle$. Since e is an initial argument, it is an element of the root reason. Given $\langle b, d \rangle$ in D_1, since b's defender is b itself, b is an element of the root reason.

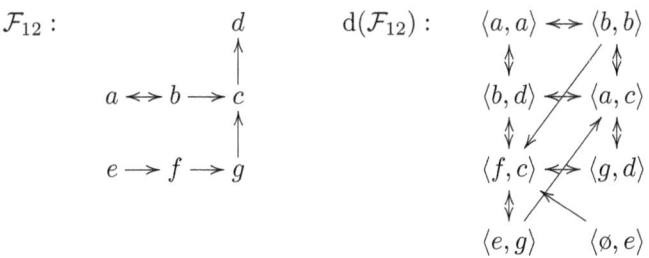

The informal notions in Example 11 are formulated as follows.

Definition 9 (Direct reasons for accepting arguments). *Let $\mathcal{F} = (\mathcal{A}, \rightarrow)$ be an argument graph. Direct reasons for accepting arguments in $\mathcal{F}$ under a semantics Σ is a function, denoted $\mathrm{dr}^{\mathcal{F}}_{\Sigma}$, mapping from $\mathcal{F}$ to sets of arguments, such that for all $\alpha \in \mathcal{A}$,*

$$\mathrm{dr}^{\mathcal{F}}_{\Sigma}(\alpha) = \{\mathcal{DR}(\alpha, D) \mid D \in \Sigma(\mathrm{d}(\mathcal{F}))\} \tag{2}$$

where $\mathcal{DR}(\alpha, D) = \{\beta \mid \langle \beta, \alpha \rangle \in D\}$, if α is not an initial argument; otherwise, $\mathcal{DR}(\alpha, D) = \{\emptyset\}$.

Example 12. Continue Example 11. According to Definition 9, $\mathrm{dr}^{\mathcal{F}_{12}}_{\mathrm{CO}}(d) = \{R_1, R_2\}$, where $R_1 = \{b, g\}$, $R_2 = \{g\}$. $\mathrm{dr}^{\mathcal{F}_{12}}_{\mathrm{CO}}(f) = \{R_3, R_4\}$, where $R_3 = R_4 = \{\}$.

For all $D \in \Sigma(\mathrm{d}(\mathcal{F}))$, we view D as a transitive relation, and let D^+ be the transitive closure of D.

Definition 10 (Root reasons for accepting arguments). *Let $\mathcal{F} = (\mathcal{A}, \rightarrow)$ be an argument graph. Root reasons for accepting arguments in $\mathcal{F}$ under a semantics Σ is a function, denoted $\mathrm{rr}^{\mathcal{F}}_{\Sigma}$, mapping from $\mathcal{F}$ to sets of arguments, such that for all $\alpha \in \mathcal{A}$,*

$$\mathrm{rr}^{\mathcal{F}}_{\Sigma}(\alpha) = \{\mathcal{RR}(\alpha, D) \mid D \in \Sigma(\mathrm{d}(\mathcal{F}))\} \tag{3}$$

where $\mathcal{RR}(\alpha, D) = \{\beta \in \mathcal{A} \mid \langle \beta, \beta \rangle \in D^+, \beta = \alpha\} \cup \{\beta \in \mathcal{A} \mid (\langle \beta, \alpha \rangle \in D^+), (\langle \beta, \beta \rangle \in D^+ \vee \beta^{\leftarrow} = \emptyset)\}$, if α is not initial; otherwise, $\mathcal{RR}(\alpha, D) = \{\emptyset\}$.

According Definition 10, we say that a set of arguments $\mathcal{RR}(\alpha, D)$ is a root reason of an argument α iff for all $\beta \in \mathcal{RR}(\alpha, D)$, β is either equal to α when α (partially) defends itself directly or indirectly through a transitive relation of defenses in D, or an initial argument, or an argument that can (partially) defend itself directly or indirectly.

Example 13. Continue Example 11. $D_1^+ = D_1 \cup \{\langle e, d \rangle, \langle \emptyset, g \rangle, \langle \emptyset, d \rangle\}$; $D_2^+ = D_2 \cup \{\langle e, d \rangle, \langle \emptyset, g \rangle, \langle \emptyset, d \rangle\}$. According to Definition 10, $\text{rr}_{CO}^{\mathcal{F}_{12}}(d) = \{R_1, R_2\}$, where $R_1 = \{b, e\}$, $R_2 = \{e\}$. $\text{rr}_{CO}^{\mathcal{F}_{12}}(f) = \{R_3, R_4\}$, where $R_3 = R_4 = \{\}$.

Motivated by the first example in Section 1 (regarding $\mathcal{F}_1$), based on the notion of root reasons, we propose as follows a notion of root equivalence of AFs.

Definition 11 (Root equivalence of AFs). *Let $\mathcal{F} = (\mathcal{A}_1, \rightarrow_1)$ and $\mathcal{H} = (\mathcal{A}_2, \rightarrow_2)$ be two argument graphs. For all $B \subseteq \mathcal{A}_1 \cap \mathcal{A}_2$, if $B \neq \emptyset$, we say that $\mathcal{F}$ and $\mathcal{H}$ are equivalent w.r.t. the root reasons for accepting B under semantics Σ, denoted $\mathcal{F}|B \equiv_{rr}^{\Sigma} \mathcal{H}|B$, iff for all $\alpha \in B$, $\text{rr}_{\Sigma}^{\mathcal{F}}(\alpha) = \text{rr}_{\Sigma}^{\mathcal{H}}(\alpha)$.*

When $B = \mathcal{A}_1 = \mathcal{A}_2$, we write $\mathcal{F} \equiv_{rr}^{\Sigma} \mathcal{H}$ for $\mathcal{F}|B \equiv_{rr}^{\Sigma} \mathcal{H}|B$.

Example 14. Consider $\mathcal{F}_1$ and $\mathcal{F}_2$ in Section 1 again. Under complete semantics, $\text{CO}(d(\mathcal{F}_1)) = \{D_1, D_2, D_3\}$ where $D_1 = \{\}$, $D_2 = \{\langle a, c_2 \rangle, \langle c_2, c_3 \rangle, \langle c_3, a \rangle\}$, $D_3 = \{\langle b, c_4 \rangle, \langle c_1, b \rangle, \langle c_4, c_1 \rangle\}$. $\text{CO}(d(\mathcal{F}_2)) = \{D_4, D_5, D_6\}$ where $D_4 = \{\}$, $D_5 = \{\langle a, a \rangle\}$, $D_6 = \{\langle b, b \rangle\}$. Let $B = \{a, b\}$. $\text{rr}_{CO}^{\mathcal{F}_1}(a) = \{\{\}, \{a\}, \{\}\}$, $\text{rr}_{co}^{\mathcal{F}_2}(a) = \{\{\}, \{a\}, \{\}\}$, $\text{rr}_{CO}^{\mathcal{F}_1}(b) = \{\{\}, \{\}, \{b\}\}$, $\text{rr}_{co}^{\mathcal{F}_2}(b) = \{\{\}, \{\}, \{b\}\}$. So, it holds that $\mathcal{F}_1|B \equiv_{rr}^{CO} \mathcal{F}_2|B$.

Theorem 5. *Let $\mathcal{F} = (\mathcal{A}_1, \rightarrow_1)$ and $\mathcal{H} = (\mathcal{A}_2, \rightarrow_2)$ be two argument graphs. If $\mathcal{F} \equiv_{rr}^{CO} \mathcal{H}$, then $\mathcal{F} \equiv^{co} \mathcal{H}$.*

Proof. According to Definition 10, the number of extensions of $co(\mathcal{F})$ is equal to the number of $\text{rr}_{CO}^{\mathcal{F}}(\alpha)$, where $\alpha \in \mathcal{A}_1$. Since $\text{rr}_{CO}^{\mathcal{F}}(\alpha) = \text{rr}_{CO}^{\mathcal{H}}(\alpha)$, $\mathcal{A}_1 = \mathcal{A}_2$. Let $\text{rr}_{CO}^{\mathcal{F}}(\alpha) = \text{rr}_{CO}^{\mathcal{H}}(\alpha) = \{R_1, \ldots, R_n\}$. Let $co(\mathcal{F}) = \{E_1, \ldots, E_n\}$ be the set of extensions of $\mathcal{F}$, where $n \geq 1$. For all $\alpha \in \mathcal{A}_1$, for all R_i, $i = 1, \ldots, n$, we have $\alpha \in E_i$ iff $R_i \neq \{\}$, in that in terms of Definition 10, when $R_i \neq \{\}$, there is a reason to accept α. On the other hand, let $co(\mathcal{H}) = \{S_1, \ldots, S_n\}$ be the set of extensions of $\mathcal{H}$. For all $\alpha \in \mathcal{A}_2 = \mathcal{A}_1$, for all R_i, $i = 1, \ldots, n$, for the same reason, we have $\alpha \in S_i$ iff $R_i \neq \{\}$. So, it holds that $E_i = S_i$ for $i = 1, \ldots, n$, and hence $co(\mathcal{F}) = co(\mathcal{H})$, i.e., $\mathcal{F} \equiv^{co} \mathcal{H}$.

Note that $\mathcal{F} \equiv^{co} \mathcal{H}$ does not imply $\mathcal{F} \equiv_{rr}^{CO} \mathcal{H}$ in general. This can be easily verified by considering $\mathcal{F}_3$ and $\mathcal{F}_4$ in Example 9.

The notion of root equivalence of argument graphs can be used to capture a kind of summarization in the graphs. Consider the following example borrowed from [8].

Example 15. Let $\mathcal{F}_{13} = (\mathcal{A}, \rightarrow)$ and $\mathcal{F}_{13}' = (\mathcal{A}', \rightarrow')$, illustrated below. Under complete semantics, $\mathcal{F}_{13}'$ is a summarization of $\mathcal{F}_{13}$ in the sense that $\mathcal{A}' \subseteq \mathcal{A}$, and the root reason of each argument in $\mathcal{F}_{13}'$ is the same as that of each corresponding argument in $\mathcal{F}_{13}$. More specifically, it holds that $\mathrm{rr}_{CO}^{\mathcal{F}_{13}}(e_3) = \mathrm{rr}_{CO}^{\mathcal{F}_{13}'}(e_3) = \{\{e_1, e_2\}\}$, $\mathrm{rr}_{CO}^{\mathcal{F}_{13}}(e_2) = \mathrm{rr}_{CO}^{\mathcal{F}_{13}'}(e_2) = \{\{\emptyset\}\}$, and $\mathrm{rr}_{CO}^{\mathcal{F}_{13}}(e_1) = \mathrm{rr}_{CO}^{\mathcal{F}_{13}'}(e_1) = \{\{\emptyset\}\}$.

$$\mathcal{F}_{13}: \quad e_1 \twoheadrightarrow a_1 \twoheadrightarrow a_2 \twoheadrightarrow o \twoheadrightarrow e_3 \qquad \mathcal{F}_{13}': \quad e_1 \twoheadrightarrow o \twoheadrightarrow e_3$$
$$\qquad\qquad e_2 \twoheadrightarrow b_1 \twoheadrightarrow b_2 \qquad\qquad\qquad e_2$$

Definition 12 (Summarization of AFs). *Let $\mathcal{F} = (\mathcal{A}_1, \rightarrow_1)$ and $\mathcal{H} = (\mathcal{A}_2, \rightarrow_2)$ be two argument graphs. $\mathcal{F}$ is a summarization of $\mathcal{H}$ under a semantics σ iff $\mathcal{A}_1 \subset \mathcal{A}_2$, and $\mathcal{F}|\mathcal{A}_1 \equiv_{rr}^{\sigma} \mathcal{H}|\mathcal{A}_1$.*

Now, a property of summarization of argument graphs under complete semantics is as follows.

Theorem 6. *Let $\mathcal{F} = (\mathcal{A}_1, \rightarrow_1)$ and $\mathcal{H} = (\mathcal{A}_2, \rightarrow_2)$ be two argument graphs. If $\mathcal{F}$ is a summarization of $\mathcal{H}$ under complete semantics CO, then $\mathrm{CO}(\mathcal{F}) = \{E \cap \mathcal{A}_2 \mid E \in \mathrm{CO}(\mathcal{H})\}$.*

Proof. Let $\mathrm{co}(\mathcal{F}) = \{E_1, \ldots, E_n\}$, $\mathrm{co}(\mathcal{H}) = \{S_1, \ldots, S_n\}$. According to the proof of Theorem 5, $E_1 = S_1 \cap \mathcal{A}_2$. Therefore, we have $\mathrm{co}(\mathcal{F}) = \{E \cap \mathcal{A}_2 \mid E \in \mathrm{co}(\mathcal{H})\}$.

The property looks similar to that of directionality of argumentation [12]. However, they are conceptually different. Specifically, it is said that if a semantics σ satisfies the property of directionality iff $\forall \mathcal{F} = (\mathcal{A}, \rightarrow)$, $\forall U \subseteq \mathcal{A}$, if U is an unattacked set, then $\sigma(\mathcal{F} \downarrow U) = \{E \cap U \mid E \in \sigma(\mathcal{F})\}$ where $\mathcal{F} \downarrow U = (U, \rightarrow \cap (U \times U))$. So, the property of directionality is about the relation between an argument graph and its subgraph induced by an unattacked set . By contrast, the property of summarization of argument graphs is about the relation between two root equivalent argument graphs.

6 Conclusions

In this paper, we have proposed a defense semantics of argumentation based on a novel notion of defense graphs, and used it to encode reasons for accepting arguments. By introducing two new kinds of equivalence relation between argument graphs, i.e., defense equivalence and root equivalence, we have shown that defense semantics can be used to capture the equivalence of argument graphs from the perspective of reasons for accepting arguments. In addition, we have defined a notion of summarization of argument graphs by exploiting root equivalence.

Under complete semantics, defense equivalence is located inbetween strong and standard equivalence. It is interesting to further investigate its position in the so-called equivalence zoo where further equivalence notions inbetween the two extremal versions are compared too [13], and to study how defense equivalence, root equivalence and strong equivalence are related. We will present this part of work in an extended version of the present paper.

Since defense semantics explicitly represents a defense relation in extensions and can be used to encoded reasons for accepting arguments, it provides a new way to investigate topics such as summarization in argumentation [8], dynamics of argumentation [9, 14, 10], dialogical argumentation [15, 16], etc. Further work on these topics is promising. Meanwhile, it might be interesting to study defense semantics beyond Dung's argumentation, including ADFs [17], bipolar frameworks [18], structured argumentation [19], etc. In particular, it would be interesting to apply defense semantics to modeling the explanation of why a conclusion can be reached. In [21], in order to increase the trust of the users for the Semantic Web applications, a system was proposed to automatically generate an explanation for every answer about why the answer has been produced. The notion of proof trace in [21] for explanation is closer to the notion of support relation between arguments. So, combining the defense relation (which is based on attack relation) and support relation would be useful to model the explanation of conclusions of a structured argumentation system.

Acknowledgements

The research reported in this paper was partially supported by the National Research Fund Luxembourg (FNR) under grant INTER/MOBILITY/14/8813732 for the project FMUAT: Formal Models for Uncertain Argumentation from Text, and the European Union's Horizon 2020 research and innovation programme under the Marie Sklodowska-Curie grant agreement No 690974 for the project MIREL: MIning and REasoning with Legal texts.

References

1. Dunne,P.E.,Dvorák,W.,Linsbichler,T.,Woltran,S.: Characteristics of multiple viewpoints in abstract argumentation. Artif. Intell. 228 (2015) 153–178
2. Baroni, P., Caminada, M., Giacomin, M.: An introduction to argumentation semantics. The Knowledge Engineering Review 26(4) (2011) 365–410
3. Charwat, G., Dvorák, W., Gaggl, S.A., Wallner, J.P., Woltran, S.: Methods for solving reasoning problems in abstract argumentation - A survey. Artif. Intell. 220 (2015) 2863
4. Dung, P.M.: On the acceptability of arguments and its fundamental role in nonmonotonic reasoning, logic programming and n-person games. Artificial Intelligence 77(2) (1995) 321 358
5. Oikarinen,E.,Woltran,S.: Characterizing strong equivalence for argumentation frameworks. In: Principles of Knowledge Representation and Reasoning: Proceedings of the Twelfth International Conference, KR 2010, Toronto, Ontario, Canada, May 9-13, 2010. (2010)

6. Baumann, R.: Characterizing equivalence notions for labelling-based semantics. In: Principles of Knowledge Representation and Reasoning: Proceedings of the Fifteenth International Conference, KR 2016, Cape Town, South Africa, April 25-29, 2016. (2016) 22–32
7. Baumann, R., Strass, H.: An abstract logical approach to characterizing strong equivalence in logic-based knowledge representation formalisms. In: Principles of Knowledge Representation and Reasoning: Proceedings of the Fifteenth International Conference, KR 2016, Cape Town, South Africa, April 25-29, 2016. (2016) 525–528
8. Baroni, P., Boella, G., Cerutti, F., Giacomin, M., van der Torre, L., Villata, S.: On the input/output behavior of argumentation frameworks. Artificial Intelligence 217 (2014) 144–197
9. Cayrol, C., de Saint-Cyr, F.D., Lagasquie-Schiex, M.: Change in abstract argumentation frameworks: Adding an argument. J. Artif. Intell. Res. (JAIR) 38 (2010) 49–84
10. Ferretti, E., Tamargo, L.H., García, A.J., Errecalde, M.L., Simari, G.R.: An approach to decision making based on dynamic argumentation systems. Artif. Intell. 242 (2017) 107–131
11. Boella,G.,Gabbay,D.M.,vanderTorre,L.W.N.,Villata,S.: Meta-argumentation modelling I: methodology and techniques. Studia Logica 93(2-3) (2009) 297–355
12. Baroni, P., Giacomin, M.: On principle-based evaluation of extension-based argumentation semantics. Artif. Intell. 171(10-15) (2007) 675–700
13. Baumann, R., Brewka, G.: The equivalence zoo for Dung-style semantics. J. Log. Comput. (2015) doi.org/10.1093/logcom/exv001.
14. Liao, B.S., Jin, L., Koons, R.C.: Dynamics of argumentation systems: A division-based method. Artif. Intell. 175(11) (2011) 1790–1814
15. Hunter, A., Thimm, M.: Optimization of dialectical outcomes in dialogical argumentation. Int. J. Approx. Reasoning 78 (2016) 73–102
16. Fan,X.,Toni,F.: On the interplay between games, argumentation and dialogues. In:Proceedings of the 2016 International Conference on Autonomous Agents & Multiagent Systems, Singapore, May 9-13, 2016. (2016) 260268
17. Brewka, G., Woltran, S.: Abstract dialectical frameworks. In: Principles of Knowledge Representation and Reasoning: Proceedings of the Twelfth International Conference, KR 2010, Toronto, Ontario, Canada, May 9-13, 2010. (2010)
18. Cayrol, C., Lagasquie-Schiex, M.: Bipolarity in argumentation graphs: Towards a better understanding. Int. J. Approx. Reasoning 54(7) (2013) 876–899
19. Besnard, P., Garcáa, A.J., Hunter, A., Modgil, S., Prakken, H., Simari, G.R., Toni, F.: Introduction to structured argumentation. Argument & Computation 5(1) (2014) 1–4
20. Bench-Capon, T. J. M., Prakken, H., Sartor, G.: Argumentation in Legal Reasoning. Argumentation in Artificial Intelligence, 2009, 363–382.
21. Antoniou, G., Bikakis, A., Dimaresis, N., Genetzakis, M., Georgalis, G., Governatori, G., Karouzaki, E., Kazepis, N., Kosmadakis, D., Kritsotakis, M., Lilis, G., Papadogiannakis, A., Pediaditis, P.,Terzakis, C., Theodosaki, R., Zeginis, D.: Proof explanation for a nonmonotonic Semantic Web rules language. Data & Knowledge Engineering, 2008, 64(3): 662–687.

Appendix

1. Proof of Lemma 1

Proof. For all $\Sigma \in \{\text{CO}, \text{GR}, \text{PR}, \text{ST}\}$, it holds that $D \in \Sigma(\text{d}(\mathcal{F}))$ is a complete extension. With respect to γ there are the following four possible cases. Let us analyze them one by one. First, γ is initial. In this case, $\langle x, y \rangle$ is attacked by $\langle \emptyset, \gamma \rangle$ that is unattacked. So, D cannot defend $\langle x, y \rangle$, contradicting D being a complete extension. Second, γ is self-attacked. In this case, $\langle \emptyset, \gamma \rangle \in \mathcal{F}^{\text{DOD}}$, and $\langle \emptyset, \gamma \rangle \rightarrow^{\text{d}} \langle x, y \rangle$. Since $\langle x, y \rangle$ is defended by D, $\exists \langle \eta, \eta' \rangle \in D$ such that $\eta \rightarrow \gamma$ or $\eta' \rightarrow \gamma$. In other word, it holds that $\text{def}(D) \rightarrow \gamma$. Third, γ is attacked by $\eta \in \mathcal{A} \setminus \{\gamma\}$, there are the following situations:

- η is initial or all attackers of η are attacked by $\text{def}(D)$: In this case, η does not attack x or y. Otherwise, $\langle x, y \rangle \notin D$. Contradiction. Meanwhile, since $\{\langle \eta, x \rangle\} \cup D$ (reps., $\{\langle \eta, y \rangle\} \cup D$) is conflict-free, and D defends $\langle \eta, x \rangle$ (reps., $\{\langle \eta, y \rangle\}$). Since D is complete, $\langle \eta, x \rangle \in D$ (reps., $\langle \eta, y \rangle \in D$). So, it holds that $\text{def}(D) \rightarrow \gamma$.
- η is self-attacked. In this case, $\langle \emptyset, \gamma \rangle \in \mathcal{F}^{\text{DOD}}$. According to the second point above, it holds that $\text{def}(D) \rightarrow \gamma$.
- η is attacked by $\eta' \in \mathcal{A} \setminus \{\eta\}$ such that η' is not attacked by $\text{def}(D)$: In this case, $(\eta', \gamma) \in \mathcal{F}^{\text{d}}$, and $(\eta', \gamma) \rightarrow^{\text{d}} \langle x, y \rangle$. Since $\langle x, y \rangle$ is defended by D, $\exists \langle \theta, \theta' \rangle \in D$ such that $\langle \theta, \theta' \rangle \rightarrow^{\text{d}} (\eta', \gamma)$. Since η' is not attacked by θ or θ', γ is attacked by θ or θ'. In other words, it holds that $\text{def}(D) \rightarrow \gamma$.

2. Proof of Theorem 1

Proof. Let $E = \text{def}(D)$. Under complete semantics, we need to prove: 1) E is conflict-free, 2) E defends each member of E, and 3) each argument in $\mathcal{A}$ that is defended by E is in E. Details:

- For all $\alpha, \beta \in E$, α and β are defenders or defendees of defenses in D. Since D is conflict-free, according to Definition 5, it is obvious that E is conflict-free.
- For all $\alpha \in E$, $\exists \langle x, \alpha \rangle \in D$ or $\langle \alpha, x \rangle \in D$ where $x \in \mathcal{A} \cup \{\emptyset\}$. For all $\gamma \in \mathcal{A}$, if γ attacks α, according to Lemma 2, $E \rightarrow \gamma$. So, α is defended by E.
- For all $\alpha \in \mathcal{A}$, if α is defended by E, we have the following possible cases:
  - α is unattacked: In this case, $\langle \emptyset, \alpha \rangle$ is in D. That is, $\alpha \in E$.
  - α is attacked by some arguments in $\mathcal{F}$: For all $\gamma \in \alpha^{\leftarrow}$, since α is defended by E, there exists $\delta \in E$ such that $\delta \rightarrow \gamma$. It follows that $\langle \delta, \alpha \rangle \in \mathcal{F}^{\text{DEF}}$, and $\exists \langle x, \delta \rangle \in D$ or $\langle \delta, x \rangle \in D$ where $x \in E \cup \{\emptyset\}$. . Then, we have the following:
    * if $\langle \delta, \alpha \rangle$ is unattacked, then since D is complete, $\langle \delta, \alpha \rangle \in D$, i.e., $\alpha \in E$; otherwise,
    * for all $[u, \eta] \in \mathcal{F}^{\text{d}}$: $[u, \eta] \rightarrow^{\text{d}} \langle \delta, \alpha \rangle$, if u or η attacks α, then since α is defended by E, there exists $\langle \psi, \psi' \rangle \in D$ such that ψ or ψ' attacks u or η. In other words, $\langle \psi, \psi' \rangle$ attacks $[u, \eta]$; if u or η attacks δ, then $[u, \eta]$ attacks $\langle x, \delta \rangle$ or $\langle \delta, x \rangle$. Since D is complete, there exists $\langle \theta, \theta' \rangle \in D$ such that $\langle \theta, \theta' \rangle$ attacks $[u, \eta]$. So, $\langle \delta, \alpha \rangle$ is defended by D. Since D is complete, it holds that $\langle \delta, \alpha \rangle \in D$, and therefore $\alpha \in E$.

3. Proof of Theorem 2

Proof. For all $E \in \text{co}(\mathcal{F})$, since it is obvious that $d(E)$ is conflict-free, we need to verify: 1) $d(E)$ defends each member of $d(E)$, and 2) each defense in $\mathcal{F}^{\text{DEF}}$ that is defended by $d(E)$ is in $d(E)$. Details:

- For all $\langle \beta, \alpha \rangle \in d(E)$, for all $[x, y] \in \mathcal{F}^d$, if $[x, y]$ attacks $\langle \beta, \alpha \rangle$ such that $x \to \beta$ or $y \to \beta$, or $x \to \alpha$ or $y \to \alpha$, since E is a complete extension, $\exists \eta \in E$ such that $\eta \to x$ or $\eta \to y$. So, $\langle \eta, \beta \rangle$ or $\langle \eta, \alpha \rangle$ is in $d(E)$, and $\langle \eta, \beta \rangle$ or $\langle \eta, \alpha \rangle$ attacks $[x, y]$. In other words, $d(E)$ defends each member of $d(E)$.
- For all $\langle \alpha, \beta \rangle \in \mathcal{F}^{\text{DEF}}$, if $\langle \alpha, \beta \rangle$ is defended by $d(E)$, then both α and β are defended by $\text{def}(d(E)) = E$. Since E is a complete extension, $\alpha, \beta \in E$. So, $\langle \alpha, \beta \rangle \in d(E)$.

4. Proof of Corollary 1

Proof. For $E = \text{def}(D)$, under grounded semantics, we need to prove that E is minimal (w.r.t. set-inclusion). Assume the contrary. Then $\exists E' \subsetneq E$ such that E' is a grounded extension. According to Theorem 2, it holds that $d(E) \in \text{CO}(d(\mathcal{F}))$ and $d(E') \in \text{CO}(d(\mathcal{F}))$. Since $E' \subsetneq E$, it holds that $d(E') \subsetneq d(E)$. Since $d(E) = d(\text{def}(D)) = D$, $d(E') \subsetneq D$. It turns out that D is not a minimal complete extension, contradicting $D \in \text{GR}(d(\mathcal{F}))$.

Under preferred semantic, similarly, it is easy to verify that E is maximal (w.r.t. set-inclusion). So, for all $D \in \text{PR}(d(\mathcal{F}))$, $\text{def}(D) \in \text{pr}(\mathcal{F})$.

Under stable semantics, we need to prove that for all $\alpha \in \mathcal{A} \setminus E$: $E \to \alpha$. Assume the contrary. Then, $\exists \alpha \in \mathcal{A} \setminus E$ such that E does not attack α. There are the following possible cases:

- α self-attacks. In this case, $(\emptyset, \alpha) \in \mathcal{F}^d \setminus D$. Since D is a stable extension, D attacks $(\emptyset, \alpha)$. So, E attacks α. Contradiction.
- α does not self-attack. Since α can not be initial, α is attacked by some argument $\beta \in \mathcal{A}$. It follows that $\beta \notin E$ and β does not self-attack. So, β is attacked by some argument $\gamma \in \mathcal{A}$. So, $[\gamma, \alpha] \in \mathcal{F}^d$. Since $\alpha \notin E$, $[\gamma, \alpha] \notin D$. Since D is a stable extension, D attacks $[\gamma, \alpha]$. Since $E = \text{def}(D)$ does not attack α, $\exists \eta \in E$ such that η attacks γ, and η does not attack α. Since α, β and η do not self-attack, we have the following possible cases:
  - $\{\eta, \beta\}$ is conflict-free: In this case, $\langle \eta, \beta \rangle \in \mathcal{F}^{\text{DEF}}$. Since E cannot attack η, if $\langle \eta, \beta \rangle \notin D$, $\exists \psi \in E$ such that ψ attacks β. So, $[\psi, \alpha] \in \mathcal{F}^d$. Since $[\psi, \alpha] \notin D$, $[\psi, \alpha]$ is attacked by D. Since E does not attack ψ, E attacks α. Contradiction.
  - $\{\eta, \beta\}$ is not conflict-free: If η attacks β, $[\eta, \alpha] \in \mathcal{F}^d$. It follows that E attacks α. Contradiction. If η does not attack β, but β attacks η, $\langle \eta, \beta \rangle \in \mathcal{F}^{\text{DEF}}$. This case also leads to a contradiction.

Two Short Notes on Argument: (1) Corrected Specificity (2) Dialectical Refinement

Ronald P. Loui

Springfield, IL and Bay Village, OH
r.p.loui@gmail.com

Abstract. In honor and celebration of Guillermo Simari's milestone two ideas are presented: one old, and one new. The first is the central, revised definition of specificity, from the unpublished *Corrigendum to Poole's Rules and a Lemma of Simari-Loui*. This fulfills a promise made in the footnote of Simari-Loui, in [19]. The second is a new start on formalizing the logic of dialectical dialogue, assuming argument on defeasible reasons as settled work. The phenomena of concern are (1) refinement of predicate sense and (2) refinement of reference, in the face of counter-argument. This is the first attempt to commit these thoughts to paper after the problem was raised during a visit to Universitad Nacional del Sur in Bahía Blanca, Argentina in the early 1990s.
[1]

1 Specificity

Poole's rule holds one argument, which he calls a theory, to be more specific than another if there is some way of activating the first without activating the second; and not vice versa.

Simari-Loui [19] made central use of this rule, but as it went to press, an important lemma of the article is wrong. It states that checking specificity is equivalent to checking that the antecedents of the less specific theory can be derived from the antecedents of the more specific theory.

The lemma states that a theory T_1 is more specific than T_2 just in case for every x, an antecedent of a rule used in T_2, x can be defeasibly derived from K_N, the necessary evidence, T_2's rules, and the antecedents of T_1's rules:

---

[1] The authors on the original report from which the first note is derived were: R. Loui, Supported by NSF R-9008012; J. Norman, Seyfarth, Shaw, Fairweather, and Geraldson, Chicago, IL; also fellow of The Center for Intelligent Computer Systems, Washington University; K. Stiefvater, Supported by NSF CDA-9102090; A. Merrill, Supported by NSF CDA-9123643; A. Costello, Supported by NSF R-9008012 and CDA-9102090; J. Olson, Supported by NSF R-57135A; Department of Computer Science, Washington University, St. Louis; Guillermo Simari had already returned to Argentina when this work began, but was engaged in the discussion.

$$(\forall x \in An(T_2))(K_N \cup An(T_1) \cup T_2 \mathrel{\mid\!\sim} x).$$

In the case of

$$< \{Penguin(O) \succ\!\!\!\!- \neg Flies(O)\}, \neg Flies(O) >$$

versus

$$< \{Bird(O) \succ\!\!\!\!- Flies(O)\}, Flies(O) >,$$

for example, $Bird(O)$ can be defeasibly derived from $Penguin(O)$. And in

$$< \{Cat(G) \succ\!\!\!\!- Aloof(G),$$
$$Aloof(G) \succ\!\!\!\!- \neg LikesPeople(G)\}, \neg LikesPeople(G) >$$

versus

$$< \{Cat(G) \succ\!\!\!\!- LikesPeople(G)\}, LikesPeople(G) >,$$

$Cat(G)$ defeasibly derives both $Cat(G)$ and $Aloof(G)$.

This last example is a simple counterexample to the lemma. If the lemma were right, then not only is the theory for $LikesPeople(G)$ more specific than the theory for $\neg LikesPeople(G)$ which we have just said is desirable, but also vice versa, which is not desirable. Not only is this intuitively undesirable, but it also violates the antisymmetry of specificity. The required relationship between antecedents is necessary for specificity, but not sufficient.[2]

Correcting this error requires [3] first that the definition of specificity be repaired. As it stands, it is vulnerable to some counter-intuitive behavior. This counter-intuitive behavior plagues Poole's rule generally, not just our use of it. The argument, for example,

$$A \prec B \wedge C$$
$$B \prec D$$
$$C \prec E$$

should be more specific than

$$\neg A \prec B$$
$$B \prec D$$

---

[2] Where the alleged proof goes wrong is quite easy to see. When there is specificity, every antecedent of the weaker theory can be derived from every antecedent of the stronger theory, but not necessarily from an activator of the stronger theory (an activator is a sentence of the proscribed contingent kinds, which allows a theory's conclusion to be derived, using the necessary evidence, K_N, and the theory's rules). That is, it may be possible to activate the theory without allowing *all* of its antecedents to be defeasibly derived, which is just plain to see.

Henry Prakken [15] has noticed that Theorem 4.16 of the paper is also in error. An example can be given in which two arguments defeat each other. Specificity is antisymmetric, but not defeat. The proof goes sour at "the same is true for defeat."

[3] Prakken (in [15]) corrects the error in an equally intuitive way. He suggests restricting the search for activators to those that activate theories in exactly the same way that the evidence does (this must be done carefully if there are multiple derivations). This fix for Poole also appears to allow a pruning lemma; however, it would involve cutsets of derivation trees, which are combinatorial in number).

But it is not more specific according to the unadulterated Poole definition, because of two separate flaws in the definition.

The first reason is that $E \wedge (B \vee \neg C)$ allows the former theory to be activated without activating the latter. This prevents the desired conclusion that the former theory is more specific. What is basically wrong here is that the weaker theory should be allowed to use the defeasible rule $E \succ\!\!\!- C$, in order to derive (defeasibly) B.

To see the second flaw, consider $E \wedge (A \vee \neg C)$, which is also disjunctive, but this time uses the disjunction to derive the theory's ultimate conclusion. This is essentially a side-stepping of the non-triviality condition for activators. The non-triviality condition should be strengthened.

Fixing the flaws in Poole's rule is important because this kind of comparison is the comparison used in the Yale Shooting Problem arguments (Hanks-McDermott [6]), as exhibited among the examples in the paper by Simari-Loui [19].

The rule also suffers in an example reminiscent of Royal Elephants (Simonet [20]), consider

$$D \prec\!\!\!- B \wedge C$$
$$B \wedge C \prec\!\!\!- A$$

compared with what should be an inferior theory:

$$\neg D \prec\!\!\!- B$$
$$B \prec\!\!\!- E$$

Neither is more specific by Poole's rule. The example appears to require right-weakening of rules: allowing rules to be derived from rules by weakening the consequent.[4]

But right-weakening is notoriously problematic. [5]

---

[4] For example,
$$\text{If } A \succ\!\!\!- B \wedge C, \text{ then } A \succ\!\!\!- C;$$
then the theory
$$D \prec\!\!\!- B \wedge C$$
$$B \prec\!\!\!- E$$
$$C \prec\!\!\!- A$$
would be the defeater of the weaker theory. The first theory does not defeat the third, but with right-weakening, it allows the construction of a third theory which directly accounts for the considerations used in the weaker theory, and which ought to be more specific.

[5] When there are competing arguments, such as
$$B \prec\!\!\!- A$$
versus
$$\neg B \prec\!\!\!- A,$$
a rule can be right-weakened with an arbitrary dilution, such as
$$A \succ\!\!\!- B, \text{ therefore, } A \succ\!\!\!- B \vee C,$$
allowing an argument for C. This cannot be done without right-weakening, since arguments must have consistent intermediate claims. This particular case is not so

The proposed development for specificity, which fixes both kinds of counterintuitive behavior due to disjunction, which allows proper treatment of the last example, and which allows a proper pruning lemma, is as follows.

Let $\Delta^{\downarrow} \subseteq L^2$, $K_N \subseteq L$, and $K_C \subseteq L$ be, respectively, instantiated defeasible rules, necessary (background) knowledge, and contingent knowledge (evidence), referring to a first-order language L.

An argument, $<T, h>$ has $T \subseteq \Delta^{\downarrow}$ and h derivable from $T \cup K_N \cup K_C$, where derivation may use rules of FOL, and a modus ponens for defeasible rules, i.e.

$$\frac{\vdash p}{\mathrel{\mid\!\sim} p}$$

$$\frac{\mathrel{\mid\!\sim} p, \quad p \mathrel{\succ\!\!\!-} q}{\mathrel{\mid\!\sim} q}$$

Also, T is minimal; no proper subset of T allows derivation of h.

A rule, R, is a *top rule* of argument $<T, h>$ just in case its consequent, $Con(R)$, is not needed for the derivation of anything but the argument's conclusion. Because the rules used in arguments are a minimal set, that is equivalent to saying that the antecedent of any rule can be derived (from evidence) using rules other than this top rule.[6]

Definition 1. $\text{Top}(R, <T, h>)$ *iff for every* r *in* T, $An(r)$ *can be defeasibly derived from* $K_N \cup K_C$ *using* $T - \{R\}$.

Example 1. $B \mathrel{\succ\!\!\!-} C$ is a top rule in the argument from A to C, using: $A \mathrel{\succ\!\!\!-} B$, and $B \mathrel{\succ\!\!\!-} C$.

Let Δ be a set of defeasible rules, let h be a sentence in the language, L, and let A be a set of sentences in L.

A finite sequence of sentences, $<B_1, \ldots, B_n>$ is a *consistent defeasible derivation (CD-derivation)* of h from A using rules Δ just in case h is derived by A's activating ground instances of rules, and the set of all intermediate sentences is consistent in L.

---

bad, since the argument

$$C \prec (B \vee C) \wedge \neg B$$
$$B \vee C \prec A$$
$$\neg B \prec A$$

has counterargument

$$B \prec A,$$

but if $A \wedge D \mathrel{\succ\!\!\!-} \neg B$, the arbitrary dilution can actually be supported. Instead, fix the rule for specificity so that it treats the problem properly without requiring right-weakening.

[6] It is not sufficient to say that a rule is top just in case it participates in eliminating a literal from the goal clause: consider $<\{Q \mathrel{\succ\!\!\!-} R, S \mathrel{\succ\!\!\!-} T\}, R \wedge T>$ which is an argument for $R \wedge T$; only the latter is a top rule; otherwise, the argument would not defeat $<\{Q \mathrel{\succ\!\!\!-} \neg R\}, \neg R>$, which it should.

Definition 2. $< B_1, \ldots, B_n >$ *is a* **CD-derivation** *of* h *from* A *iff*

1. $B_n = h$;
2. For each B_i, either
    a. $\{B_j \mid j < i\} \cup A \vdash B_i$; or
    b. *for some ground instance of a rule* R *in* Δ, $An(R) = B_j$ *for some* $j < i$ *and* $B_i = Con(R)$;
3. $\{B_i \mid i \leq n\} \not\vdash \bot$.

Example 2 (continued). $< A, B, C >$ is a CD-derivation of C from A using: $A \succ\!\!\!- B$ and $B \succ\!\!\!- C$.

Definition 3. *There is a* **CD-derivation** *of* h *from* A *using* Δ **with all top rules** *of* $< T, h >$ *just in case*

1. *there is a CD-derivation of* h *from* A *using* Δ;

2. *for each* x *such that* $Top(x, < T, h >)$, *there is no CD-derivation of* h *from* A *using* $\Delta - \{x\}$.

$< T_1, h_1 >$ is more specific than $< T_2, h_2 >$ just in case some legitimate sentence activates T_2 for h_2 without activating T_1 for h_1, using CD-derivations from *the two theories' combined set of rules*, using *every top rule* of T_2; and there is no such *asymmetric activator* of T_1 for h_1 that does not also activate T_2 for h_2.

The requirement to use top rules is just a strengthening of Poole's nontriviality condition that asymmetric activators do not activate the theory simply by FOL rules, side-stepping the defeasible rules. The combination of theories is more profound. It signals the importance of pairwise comparison as opposed to an n-wise, holistic evaluation of merit (which is what Geffner-Pearl tends toward [5]) on one extreme, or a conception of specificity as intrinsic, perhaps even measurable (which is what the algebra of Simari-Loui suggests), on the other extreme. Transitivity no longer holds of specificity.

That is,

Definition 4. e *is an* **asymmetric activator** *of* $< T_1, h_1 >$ *but not* $< T_2, h_2 >$ *just in case*

1. *there is some CD-derivation of* h_1 *from* $K_N \cup \{e\}$ *using* $T_1 \cup T_2$ *with all top rules of* $< T_1, h_1 >$;

2. *there is no CD-derivation of* h_2 *from* $K_N \cup \{e\}$ *using* $T_1 \cup T_2$ *with all top rules of* $< T_2, h_2 >$.

Let eAA_{inotj} symbolize that e is an asymmetric activator of $< T_i, h_i >$ but not $< T_j, h_j >$.

Definition 5. $<T_1, h_1>$ *is* **more specific** *than* $<T_2, h_2>$ *just in case*

1. *there is some asymmetric activator in* S_C *of* $<T_2, h_2>$ *but not of* $<T_1, h_1>$;
   i.e., there is some $e \in S_C$ *s.t.* eAA_{2not1}; *and*

2. *there is no asymmetric activator in* S_C *of* $<T_1, h_1>$ *but not* $<T_2, h_2>$;
   i.e., there is no $e \in S_C$ *s.t.* eAA_{1not2}.

Given $<T_1, h_1>$ and $<T_2, h_2>$, use the symbolization $e \mapsto_{T_i} f$ to assert the existence of a CD-derivation of f from $K_N \cup e$ using $T_1 \cup T_2$ with all top rules of $<T_i, h_i>$. $e \mapsto f$ if there is a CD-derivation at all from $K_N \cup e$ using $T_1 \cup T_2$, not requiring use of top rules. Note that $\mapsto$ and $\mapsto_{T_i}$ are defined only for a pair of theories being compared.

Note also that

$$eAA_{1not2} \text{ just in case}$$

1. $e \mapsto_{T_1} h_1$;
2. $e \not\mapsto_{T_2} h_2$;

The new pruning lemma makes use of both the top rule restriction and the union of theories when checking activation.

To find whether there is an asymmetric activator of $<T_1, h_1>$ but not $<T_2, h_2>$ it is usually sufficient to check whether the conjoined antecedents of the top rules of $<T_1, h_1>$ is an asymmetric activator of $<T_1, h_1>$ but not $<T_2, h_2>$. For simplicity, first assume that h_1 can be derived from the conjoined consequents of top rules in $<T_1, h_1>$, the last step in deriving the theory's conclusion uses just K_N and the consequents of top rules; that is, intermediate conclusions from consequents of non-top rules are used only to derive antecedents of later rules.

Lemma 1 (restricted pruning). *For any arguments* $<T_1, h_1>$ *and* $<T_2, h_2>$, *there exists an asymmetric activator of* $<T_1, h_1>$ *but not* $<T_2, h_2>$ *just in case the following has the property* AA_{1not2}:
$Conjoin_\Gamma(An(R_i))$,
where $\Gamma = \{R_i : Top(R_i, <T_1, h_1>)\}$;
under the assumption that h_1 *can be derived from the conjoined consequents of top rules in* $<T_1, h_1>$, *i.e.*,
$Conjoin_\Gamma(Con(R_i)) \mapsto h_1$.

Proof.   1. First consider the case where $<T_1, h_1>$ has a single top rule.
Suppose there is an e such that eAA_{1not2}. That is, $e \mapsto_{T_1} h_1$ and $e \not\mapsto_{T_2} h_2$. Let $e' = An(R)$, where by assumption, R is the only rule in T_1 such that $Top(R, <T_1, h_1>)$.
Clearly, $e' \mapsto_{T_1} h_1$, since we assume $Conjoin_\Gamma(Con(R_i)) \mapsto h_1$. So e' is also an asymmetric activator if $e' \not\mapsto_{T_2} h_2$.

Assume to the contrary that $e' \mapsto_{T_2} h_2$. Recall that $e \mapsto_{T_1} h_1$, so $e \mapsto e'$, since e' is just the antecedent of the top rule which must be used. Chain this with the assumption that $e' \mapsto_{T_2} h_2$, and get $e \mapsto_{T_2} h_2$. But e is supposed to be an asymmetric activator of $<T_1, h_1>$ but not $<T_2, h_2>$. This is a contradiction. So e' must be an asymmetric activator.

2. Next, consider the case where $<T_1, h_1>$ has multiple top rules. Let e' be $Conjoin_\Gamma(An(R_i))$. The same argument applies, but it is no longer obvious that $e \mapsto e'$.

Assume $e \mapsto_{T_1} h_1$. Consider any rule, $R \in \Gamma$, i.e., any R such that $Top(R, <T_1, h_1>)$. Show that $e \mapsto An(R)$. This suffices to show that $e \mapsto e'$.

Assume to the contrary that $e \not\mapsto An(R)$. This is a contradiction, because $e \mapsto_{T_1} h_1$ requires that any CD-derivation of h_1 from $K_N \cup e$ using $T_1 \cup T_2$ use all of the top rules of $<T_1, h_1>$, including R, and it is impossible to use R without deriving $An(R)$.

Now relax the assumption regarding the derivability of h_1 from consequents of top rules. The full lemma is that whenever there is an asymmetric activator of $<T_1, h_1>$ but not $<T_2, h_2>$, an asymmetric activator is formed by conjoining (1) the antecedents of top rules with (2) the conditional whose antecedent (2a) conjoins the consequents of non-top rules, and whose consequent (2b) is the claim supported by the argument.

Lemma 2. *For any arguments $<T_1, h_1>$ and $<T_2, h_2>$, there exists an asymmetric activator of $<T_1, h_1>$ but not $<T_2, h_2>$ just in case the following has the property AA_{1not2}:*

$$Conjoin_\Gamma(An(R_i)) \land (Conjoin_\Gamma(Con(R_i)) \supset h_1)$$

where Γ is as before.

Proof. The proof again begins: suppose some eAA_{1not2}; i.e., $e \mapsto_{T_1} h_1$, $e \not\mapsto_{T_2} h_2$. e' is the conjunction of top rule antecedents, and e'' is e' conjoined with the material conditional as above. We want $e''AA_{1not2}$. Clearly $e'' \mapsto_{T_1} h_1$. Show $e'' \not\mapsto_{T_2} h_2$. Suppose it did, i.e., $e'' \mapsto_{T_2} h_2$: $e \mapsto_{T_1} h_1$, so $e \mapsto e''$ since all top rules must be used and e'' is the weakest sentence that activates all top rules and also derives h_1. This property of being weakest is the key observation. Chaining, $e \mapsto_{T_2} h_2$, but this is a contradiction. So $e''AA_{1not2}$ if for any e, eAA_{1not2}.

The revised rules have been implemented twice. A version with an underlying first-order logic, on which this section focuses, was implemented in C and is quick; it is primarily limited by the underlying resolution theorem-prover.[7]

---

[7] The main programmers were Adam Costello, Andrew Merrill, and Ronald Loui; the 92k, 3300 lines of source code contain a dedication to the late computer scientist, Eugene Nathan Johnson, c. 1944 – 1984; the program is called "nathan".

The second version is in LISP with a restricted propositional language (just negation of atomic formulae and conjunction), with provision for analogical (case-based) reasoning, and with additional features peculiar to certain forms of legal reasoning.[8]

2 Refinement

One of the clear purposes of dialogical dialectic is the clarification of reference and restriction of predication. A mathematical model of such dialogue is sketched here.

Twenty-five years after Simari-Loui and related works, the AI logic community has capably formalized the dynamics of arguments built upon a given set of claims and rules. This was the inherited framework, mainly from Doyle [4] and Reiter [18]. Specificity, priority, undercutting and reinstatement occupied much of the attention of research for decades. First-order predicate quantification was also a hindrance. It took some time to gain widespread acceptance of the dialectical pro-con process, its ampliativity and potential non-determinism, its dependence on search, and the non-monotonicity of process that was distinct from non-monotonicity of syntax. ([11], [9], see e.g., Baroni, Cerutti, Giacomin, and Simari, [2])

The phenomena that are well researched concern the rules 'if p then q', 'if p and r then defeasibly not q', 'if s then not(if p then defeasibly q)', represented in various ways.

Even as these investigations were gaining speed, this author was puzzling over what seemed to be a more common phenomenon in dialogical argument. That phenomenon was the clarification of claims, the refinement of parts of claims, through argument.

There are two distinct kinds of clarification considered next. A third kind is notationally a bit farther afield.

Of the two we can easily address, first is the clarification of reference.

If a person makes a claim '$P(a)$', or 'All $x \in A$ $P(x)$', or 'Asians do well at Harvard', another person may respond in dialogue that 'A' is not referentially specific (we can quibble over whether this is reference or antecedent predication, but anyone reading this note should have already read Quine on this subject, [17]). Or a person might respond that 'P' is not sufficiently qualified or restricted in extent. The idea is that if 'Asians' were more carefully tailored, or 'does well at Harvard' were more carefully tailored, then the claim could be provisionally accepted and dialogue could continue. If not refined, however, the dialogue would enter an argument subdialogue.

The purpose of adversarial engagement is to improve the claim. It is a language game on the semantics, essentially the translation of the shorthand

---

[8] The main programmer was Jon Olson; its 52k, 2000 lines of source code are called "lmnop" after the initials of the last names of its designers. See Loui-Norman et al.[12]

natural language into formal logical symbols. David Lewis is associated with this kind of dialogical referential refinement, e.g., in [8], and it is related to the non-dialectical desire in natural language dialogue to agree on anaphora (consider Webber [21] and the literature that followed on the subject).

Note that $A(x) \succ P(x)$ can be refined simply by qualification in existing rule-based frameworks, $A(x) \wedge Q(x) \succ P(x)$, with "Asians" now eliding the qualified "Asians who are American-born" though $P(x)$ is not as easily refined with an antecedent qualifier, e.g. "Asians do well at Harvard academically". Apparently, one simply has to rewrite the predicate, $A(x) \succ Q(x)$, and $P(x) \succ Q(x)$, perhaps even $Q(x) \succ P(x)$.

Although this appears at first to be a problem of language rather than logic, it may provide a model of one of the more elusive aspects of logic: namely, its interface with language. First, the initial expression of claims may be limited by the finiteness of locution or representation. HP Grice comes to mind here, as there may be a limit to what one can say during one's turn. But it may simply be an idealization that all formal expression of claims be maximally precise at initial claiming. Second, there may be a legitimate logical dynamics in the interpersonal agreement over semantics when two persons, or more, enter into the meeting of minds. Predicate refinement, or referential refinement, may be an important part of that meeting, whether as revision of representation or revision of the content of claims. Third, there is the concept of open-texture, what AI might call underspecification and ex-post learning, where the assumption is that revision of initial representation will take place, even if the initial representation leaves no gap between natural language and formal symbolization. This may be because of finite expression, finite envisionment, or the limit of multi-party agreement. For the latter, it may be that underspecification is necessary in order to reach initial agreement, i.e., the basis of the claim from which it derives its truth-like authority or assertability.

In tandem with the development of formal defeasible dialectical argument in AI, the AI and Law community has investigated the nature of open texture of predicates. There the semantics of a linguistic element ("vehicle in the park") undergoes revision as cases are decided in time that affect what is considered a 'vehicle' and what is considered to be 'in the park'. (Hart [7]) A good example of this is the original desire to prohibit vehicles from the park, with the explicit exception of parade vehicles, hence, 'no vehicles in the park' with some exceptions. But over time, vehicles may include drones, and being in the park may implicate the airspace over the park at some lower altitudes. Further specification is part of the semantics of the edict, and the semantics forsees a process of revision over time as the world evolves, not just revision through argument, as hard cases are considered with the originally existing concepts.

Regardless of how one wants to divide the responsibility between logic and linguistics, there is clearly a challenge to find a framework for thinking about dialectical dialogue that results in reference and predication refinement.

The basic addition is a sequential index to a predicate and a term, so P becomes P_1, and A becomes A_1, a becomes a_1. The initial claim $P_1(A_1)$. So P', P'', P''' can simply be P_1, P_2, P_3.

But refinement is forced by the adversarial dialogue. In the case of "Asians do well at Harvard",[9] one counter move would be "Foreign-born Asians do not always do well at Harvard." $not(A(x) \wedge FB(x) \succ P(x))$. This could be an observation that may not be disputed, or can be subject to further dispute. If it is sufficiently justified or jointly presumed, it forces revision of the original claim. Note that a single counterexample might not force refinement, since the claim is presumably an expression of a defeasible rule.

In any case, $A_2 = A_1 - FB$. We can express A_2 alternately as $A_{(-FB)}$.

A slightly different refinement would be restriction to a subclass, i.e., "US-born Asians do well at Harvard" in which case we have $A(x) \wedge US(x) \succ P(x)$, where we could write the refinement as $A_{(\wedge US)}$.

If the counter move is that "Asians do well at Harvard only in terms of academics", the refinement is on the predicate in the consequent. The move from $A(x) \succ P(x)$ to $A(x) \succ Q(x)$ is not interesting except insofar as $P(x)$ and $Q(x)$ were so closely related that they could have been mistaken in the initial claim.

A predicate P is refined to a **polysemic correlate** Q, when $P(x) \succ Q(x)$ and $Q(x) \succ P(x)$.

This could be statistical, but as we know in defeasible reasoning, not all defeasible rule connections are adopted on statistical grounds, nor are all statistically strong correlations accepted as defeasible rules in an argument game. We can express P_2 in this case as $P_{(>Q)}$.

A related counter move is negative rather than positive, e.g., "Asians do not do well at Harvard socially". Here, the relation between P and Q contains a negative assertion, $P(x) \succ Q(x)$ where $Q(x) = R(x) \wedge S(x)$, but $A(x) \succ R(x)$, $not(A(x) \succ S(x))$. We could express this as $P_{(-S)}$, but there is a more general phenomenon here.

An **implicature of the consequent** is a property (predicate) immediately and solely (defeasibly) derivable from the consequent of a defeasible rule.

There are senses of "doing well at Harvard" that are normally implied, and are not correlates, but are essential parts of each sense of the concept or phrase that may be in play. One can do well musically. One can do well relative to predicted performance. One can do well in the sense of doing good, i.e., doing charitable works, which may be the way that one does well in an ethics-based assessment of performance. The logician of course simply requires that the correct sense be attached to a predicate Q^*, where Q^* is ideally the result of the dialectic that forces refinements Q_1, Q_2, Q_3 ... Q^*. But such a requirement does not permit discussion of how one gets from Q_1 to Q^*.

For our $Q(x)$, there may be several implicatures of the consequent:

$$Q(x) \succ T(x),$$

---

[9] This example is chosen with careful consideration of the author's own experience.

$Q(x) \succ U(x),$
$Q(x) \succ V(x),$
etc.

The adversarial argument may sever each individually: "Asians at Harvard do not generally ski well", "Asians at Harvard do not normally write senior theses well", etc. The response really does depend on dialectic, or at least a process of statistical argument, because it may in fact be defensible that Asians at Harvard write senior theses well (numerically, stereotypically, or even axiomatically[10]).

In any case, the refinement from these kinds of move is $Q_{(-T)(-U)(-V)}$, where we choose not to distinguish between a refinement that loses a logical entailment and a refinement that loses implicature. The refinement of the consequent in this way also could have an effect on the set of rules upon which arguments are constructed.

One way to do a revision would be to retain the symbol Q, but remove the severed implicatures from the rule-base. But a more literal and manipulable notation would add undercutters to the refined predicate:

$P(x) \succ Q_{(-T)(-U)(-V)}(x)$ is accepted, so long as
$not(Q_{(-T)(-U)(-V)}(x) \succ T(x))$
$not(Q_{(-T)(-U)(-V)}(x) \succ U(x))$
$not(Q_{(-T)(-U)(-V)}(x) \succ V(x)).$

are added to the rule base.

A third kind of refinement is refinement of a referential term, rather than predicate. So a particular Asian at Harvard, e.g. $JeremyLin$, may be asserted to have done well at Harvard. If the objection is that there are a few people named 'Jeremy Lin', that may not be an interesting argument. If the objection is that we know Mr. Lin excelled on the basketball court at Harvard, but we have no idea what his academics were like, then we can write $P(x) \succ Q_{(-T)}(x)$, with $not(Q_{(-T)}(x) \succ T(x))$, a removal of the sense of doing well normally associated. But if the objection is that $JeremyLin$, the freshman, was different from $JeremyLin$, the senior, we have a different sort of problem. What is needed here is the indexicals that are further associated with context refinement. To provide a notation that would support such indexicals, we might need to represent individuals as fluents (e.g., McCarthy [13], McCarthy and Hayes [14], Baker [1]), then represent the restriction to slices of fluents.

For decades, it has bothered this author that most of what dialectic seems to focus on, in real life conversation, bears no relation to the logic of argument celebrated in AI knowledge representation and reasoning, and in AI and Law research.

---

[10] Axiomatically, for example, if one wanted to take the definition of doing well at Harvard to be the specific identifiable way that Asians do: a jarring but not uncommon maneuver in the web of temporarily constructed dialogical meaning.

There has always been work on argument and dialectic in AI and natural language, as well as in the wider world of informal logic. The question of semantics is ably approached through the competing analogies and disanalogies that formalize reasoning from precedent on open-textured predicates (see Loui-Norman et al.[12], Loui-Norman [10], Praken-Sartor [16], Loui [3]). But there, the dominant dynamics is from $P(x) \succ\!\!- Q(x)$ to $not(P(x) \succ\!\!- Q(x))$ with no representation, as far as this author has seen, of the refinement that results from dialectic. The impact of adversarial dialogue on claims may be constructive: it often makes the claims sharper, the concepts finer, and the references more specific. It is not just rule qualification, though some of it could be represented that way. It is not just unrelated symbol substitution. Because the symbol substituted bears, at the very least, correlate and implicature relations to the original symbol. One can look at the failure to sustain $P(x) \succ\!\!- Q(x)$ and try a new argument based on $P(x) \succ\!\!- Q'(x)$, i.e., $Q_{(-T)}(x)$, but understanding Q's relation to T may be helpful in seeing how the whole dialogue fits together.[11]

References

1. Andrew B Baker. Nonmonotonic reasoning in the framework of situation calculus. *Artificial Intelligence*, 49(1-3):5–23, 1991.
2. Pietro Baroni, Federico Cerutti, Massimiliano Giacomin, and Guillermo Ricardo Simari. *Computational Models of Argument: Proceedings of COMMA 2010*, volume 216. Ios Press, 2010.
3. Peter Boltuc, Ronald P. Loui, Felmon Davis, and D. E. Wittkower. Scientific and legal theory formation in an era of machine learning: remembering background rules, coherence, and cogency in induction. 2015.
4. Jon Doyle. A truth maintenance system. *Artificial intelligence*, 12(3):231–272, 1979.
5. Hector Geffner and Judea Pearl. Conditional entailment: Bridging two approaches to default reasoning. *Artificial Intelligence*, 53(2-3):209–244, 1992.
6. Steve Hanks and Drew McDermott. Nonmonotonic logic and temporal projection. *Artificial intelligence*, 33(3):379–412, 1987.
7. Herbert Lionel Adolphus Hart, Herbert Lionel Adolphus Hart, and Leslie Green. *The concept of law*. Oxford University Press, 2012.
8. David Lewis. Convention cambridge. *Mass.: Harvard UP*, 1969.
9. Ronald P. Loui. Process and policy: Resource-bounded nondemonstrative reasoning. *Computational Intelligence*, 14(1):1–38.
10. Ronald P Loui and Jeff Norman. Eliding the arguments of cases. 1997.
11. Ronald Prescott Loui. Defeat among arguments: a system of defeasible inference. *Computational Intelligence*, 3:100–106, 1987.
12. Ronald Prescott Loui, Jeff Norman, Jon Olson, and Andrew Merrill. A design for reasoning with policies, precedents, and rationales. In *ICAIL*, 1993.
13. J Mac Carthy. Situations and actions and causal laws. standford artificial intelligence project memo 2, 1963.

---

[11] As David Makinson once told me in a cafe in Paris named after St. Louis's own Josephine Baker, the important step is to invent the notation.

14. John McCarthy and Patrick J Hayes. Some philosophical problems from the standpoint of artificial intelligence. In *Readings in nonmonotonic reasoning*, pages 26–45. Morgan Kaufmann Publishers Inc., 1987.
15. Henry Prakken. An argumentation framework in default logic. *Annals of Mathematics and Artificial Intelligence*, 9:93–132, 1993.
16. Henry Prakken and Giovanni Sartor. Modelling reasoning with precedents in a formal dialogue game. In *Judicial applications of artificial intelligence*, pages 127–183. Springer, 1998.
17. Willard Van Orman Quine et al. Logic and the reification of universals. *From a logical point of view*, 6, 1953.
18. Raymond Reiter. A logic for default reasoning. *Artif. Intell.*, 13:81–132, 1980.
19. Guillermo Ricardo Simari and Ronald Prescott Loui. A mathematical treatment of defeasible reasoning and its implementation. *Artif. Intell.*, 53:125–157, 1992.
20. Geneviève Simonet. Nonmonotonic inference rules for multiple inheritance with exceptions. *Proceedings of the IEEE*, 74:1345–1353, 1986.
21. Bonnie Lynn Webber. *A formal approach to discourse anaphora*. Routledge, 2016.

On the Relationship between DeLP and ASPIC+

Simon Parsons[1] and Andrea Cohen[2]

[1] Department of Informatics, King's College London
simon.parsons@kcl.ac.uk
[2] Institute for Computer Science and Engineering, CONICET-UNS
Department of Computer Science and Engineering, Universidad Nacional del Sur
ac@cs.uns.edu.ar

Abstract. In this chapter we consider the relationship between DeLP and ASPIC+. The fact that these systems are different is well known, but what is less well known is exactly how these systems differ, and, perhaps more interestingly, the ways in which they are the similar. We do not get to the bottom of the relationship between the systems in this chapter, but we do at least set the foundations for a detailed exploration.

Keywords: structured argumentation, defeasible logic programming, DeLP, ASPIC+

1 Introduction

As discussed in [8], work on argumentation can be traced back at least as far as the mid-1980s, where it grew out of attempts to create logics that were capable of defeasible reasoning. Early work on argumentation produced many different systems with different ways of representing knowledge, with different ways of constructing arguments, with different ways of identifying conflicts between arguments, and with different methods for identifying what conclusions were *acceptable*, that is which conclusions could reasonably be drawn from a given knowledge base. In the early 1990s, Dung's introduction of abstract argumentation [12, 13] changed the study of argumentation in two ways. First, Dung's work led researchers to separate the process of determining acceptable conclusions from the process of determining what arguments could be constructed from a knowledge base, and what conflicts existed between them. The first of these processes was the domain of abstract argumentation, the second of these processes was the domain of *structured argumentation*. Second, Dung's work let to a form of standardization of work on argumentation. Because his approach to determining acceptable arguments — and hence the *justified* conclusions of the acceptable arguments — became widely adopted, it led to most work on argumentation having a common theme in its use of the Dung semantics.

One indication of the extent to which the Dung semantics came to dominate work on argumentation is their adoption by the ASPIC+ framework. ASPIC+

[17,20] was conceived as a general abstract model of structured argumentation, and reading through the detail of the way it represents knowledge, constructs arguments, and identifies conflicts (see Section 2), it clearly draws in elements of many different argumentation systems. However, it only considers one method for establishing which arguments are acceptable, the Dung semantics.

Now, widely used as they are, the Dung semantics are not the only way to establish the conclusions of an argumentation system. Indeed, there are approaches to doing just this which have been around longer. One of these was proposed by Guillermo Simari in his PhD thesis in 1989 [21,22], and this method for establishing the *warranted* conclusions of a knowledge base was later integrated into the Defeasible Logic Programming (DELP) approach proposed in [14]. Because DELP does not make use of the Dung semantics, it cannot be thought of as a specialization of the ASPIC$^+$ framework. However, that does not, on its own, mean that the frameworks are particularly different. Indeed, as we show in this chapter, the two approaches are similar in many regards.

In the body of this chapter we seek to revisit several aspects that differentiate DELP from ASPIC$^+$, analyze the common grounds between the two approaches, and study the possibility of establishing conditions, either on ASPIC$^+$ or on DELP, that would help bridge the gap between them. We start with a brief introduction of the two, ASPIC$^+$ in Section 2 and DELP in Section 3. Then, in Section 4 we move on to comparing the two approaches. We discuss the similarities and differences between them under four headings: their knowledge representation capabilities, the mechanism they adopt for argument construction, the different kinds of attack and defeat they consider, and the way in which they select accepted arguments and justified conclusions.

2 ASPIC$^+$ Background

ASPIC$^+$ is deliberately defined in a rather abstract way, as a system with a minimal set of features that can capture the notion of argumentation. This is done with the intention that it can be instantiated by a number of concrete systems that then inherit all of the properties of the more abstract system. ASPIC$^+$ starts from a logical language $\mathcal{L}$ with a notion of negation. A given instantiation will then be equipped with inference rules, and ASPIC$^+$ distinguishes two kinds of inference rules: strict rules and defeasible rules. Strict rules, denoted using $\rightarrow$, are rules whose conclusions hold without exception. Defeasible rules, denoted $\Rightarrow$, are rules whose conclusions hold unless there is an exception.

The language and the set of rules define an *argumentation system*:

Definition 1 (Argumentation System). *An* argumentation system *is a tuple* $AS = \langle \mathcal{L}, ^-, \texttt{Rules}, n \rangle$ *where:*

- $\mathcal{L}$ *is a logical language.*
- $^-$ *is a function from $\mathcal{L}$ to $2^{\mathcal{L}}$, such that:*

- φ is a contrary of ψ if $\varphi \in \overline{\psi}$, $\psi \notin \overline{\varphi}$;
- φ is a contradictory of ψ if $\varphi \in \overline{\psi}$, $\psi \in \overline{\varphi}$;
- each $\varphi \in \mathcal{L}$ has at least one contradictory.

– Rules = $\text{Rules}_s \cup \text{Rules}_d$ is a set of strict (Rules_s) and defeasible (Rules_d) inference rules of the form $\phi_1, \ldots, \phi_n \to \phi$ and $\phi_1, \ldots, \phi_n \Rightarrow \phi$ respectively (where ϕ_i, ϕ are meta-variables ranging over wff in $\mathcal{L}$), and $\text{Rules}_s \cap \text{Rules}_d = \emptyset$.

– $n : \text{Rules}_d \mapsto \mathcal{L}$ is a naming convention for defeasible rules.

The function $^-$ generalizes the usual symmetric notion of negation to allow non-symmetric conflict between elements of $\mathcal{L}$. The contradictory of some $\varphi \in \mathcal{L}$ is close to the usual notion of negation, and we denote that φ is a *contradictory* of ψ by "$\varphi = \neg \psi$". Note that, given the characterization of $^-$, elements in $\mathcal{L}$ may have multiple contraries and contradictories. As we will see below, the naming convention for defeasible rules is necessary because there are cases in which we want to write rules that deny the applicability of certain defeasible rules. Naming the rules, and having those names be in $\mathcal{L}$ makes it possible to do this, and the denying applicability makes use of the contraries of the rule names.

An argumentation system, as defined above, is just a language and some rules which can be applied to formulae in that language. To provide a framework in which reasoning can happen, we need to add information that is known, or believed, to be true. In ASPIC$^+$, this information makes up a *knowledge base*:

Definition 2 (Knowledge Base). *A* knowledge base *in an argumentation system* $\langle \mathcal{L}, ^-, \text{Rules}, n \rangle$ *is a set* $\mathcal{K} \subseteq \mathcal{L}$ *consisting of two disjoint subsets* $\mathcal{K}_n$ *and* $\mathcal{K}_p$.

We call $\mathcal{K}_n$ the axioms and $\mathcal{K}_p$ the ordinary premises. We make this distinction between the elements of the knowledge base for the same reason that we make the distinction between strict and defeasible rules. We are distinguishing between those elements — axioms and strict rules — which are definitely true and allow truth-preserving inferences to be made, and those elements — ordinary premises and defeasible rules — which can be disputed.

Combining the notions of argumentation system and knowledge base gives us the notion of an *argumentation theory*:

Definition 3 (Argumentation Theory). *An* argumentation theory AT *is a pair* $\langle AS, \mathcal{K} \rangle$ *of an argumentation system* AS *and a knowledge base* $\mathcal{K}$.

We are now nearly ready to define an argument. But first we need to introduce some notions that can be defined just by understanding that an argument is made up of some subset of the knowledge base $\mathcal{K}$, along with a sequence of rules, that lead to a conclusion. Given this, $\texttt{Prem}(\cdot)$ returns all the premises, $\texttt{Conc}(\cdot)$ returns the conclusion and $\texttt{TopRule}(\cdot)$ returns the last rule in the argument. $\texttt{Sub}(\cdot)$ returns all the sub-arguments of a given argument, that is all the arguments that are contained in the given argument. In addition, given

$A' \in \text{Sub}(A)$ such that $A' \neq A$, we will say that A' is a *proper sub-argument* of A.

Definition 4 (Argument). *An argument A from an argumentation theory $AT = \langle \langle \mathcal{L}, \bar{\ }, \text{Rules}, n \rangle, \mathcal{K} \rangle$ is:*

1. *ϕ if $\phi \in \mathcal{K}$ with:* $\text{Prem}(A) = \{\phi\}$; $\text{Conc}(A) = \phi$; $\text{Sub}(A) = \{A\}$; *and* $\text{TopRule}(A) = $ *undefined.*
2. *$A_1, \ldots, A_n \to \phi$ if A_i, $1 \leq i \leq n$, are arguments and there exists a strict rule of the form* $\text{Conc}(A_1), \ldots, \text{Conc}(A_n) \to \phi$ *in* Rules_s. $\text{Prem}(A) = \text{Prem}(A_1) \cup \ldots \cup \text{Prem}(A_n)$; $\text{Conc}(A) = \phi$; $\text{Sub}(A) = \text{Sub}(A_1) \cup \ldots \cup \text{Sub}(A_n) \cup \{A\}$; *and* $\text{TopRule}(A) = \text{Conc}(A_1), \ldots, \text{Conc}(A_n) \to \phi$.
3. *$A_1, \ldots, A_n \Rightarrow \phi$ if A_i, $1 \leq i \leq n$, are arguments and there exists a defeasible rule of the form* $\text{Conc}(A_1), \ldots, \text{Conc}(A_n) \Rightarrow \phi$ *in* Rules_d. $\text{Prem}(A) = \text{Prem}(A_1) \cup \ldots \cup \text{Prem}(A_n)$; $\text{Conc}(A) = \phi$; $\text{Sub}(A) = \text{Sub}(A_1) \cup \ldots \cup \text{Sub}(A_n) \cup \{A\}$; *and* $\text{TopRule}(A) = \text{Conc}(A_1), \ldots, \text{Conc}(A_n) \Rightarrow \phi$.

We write $\mathcal{A}(AT)$ to denote the set of arguments from the theory AT.

In other words, an argument is either an element of $\mathcal{K}$, or it is a rule and its conclusion such that each premise of the rule is the conclusion of an argument. Note that, as stated by the authors in [17]: "Note that all premises in ASPIC+ arguments are used in deriving its conclusion, so enforcing a notion of relevance analogous to the subset minimality condition requirement on premises in classical logic approaches to argumentation".

A key concept in argumentation is the idea that even if there is an argument for some conclusion, indicating that there is a *prima facie* case for the conclusion, the conclusion may not be reasonable because there is a stronger argument that it does not hold. This notion is particularly natural in a multiagent setting, where different agents have different viewpoints, leading to conflicting arguments. However, it is perfectly possible for a single argumentation theory, representing the information held by a single individual, to be the basis of conflicting arguments. We capture this kind of interaction through the idea that one argument can attack and defeat another.

An argument can be attacked in three ways: on its ordinary premises, on its conclusion (either final or intermediate), or on its defeasible inference rules. These three kinds of attack are called *undermining*, *rebutting* and *undercutting* attacks, respectively.

Definition 5 (Attack). *An argument A attacks an argument B iff A undermines, rebuts or undercuts B, where:*

- *A undermines B (on B') iff $\text{Conc}(A) \in \overline{\phi}$ for some $B' = \phi \in \text{Prem}(B)$ and $\phi \in \mathcal{K}_p$.*
- *A rebuts B (on B') iff $\text{Conc}(A) \in \overline{\phi}$ for some $B' \in \text{Sub}(B)$ of the form $B_1'', \ldots, B_2'' \Rightarrow \phi$.*
- *A undercuts B (on B') iff $\text{Conc}(A) \in \overline{n(r)}$ for some $B' \in \text{Sub}(B)$ such that $\text{TopRule}(B')$ is a defeasible rule r of the form $\phi_1, \ldots, \phi_n \Rightarrow \phi$.*

We denote "A attacks B" by (A, B).

In all these cases, the idea is that an attack can be made on an element of an argument that is not known for sure to hold. An attack can thus be made on an ordinary premise — which might be an assumption or a belief — rather than an axiom, and both the other forms of attack involve defeasible rules. The difference between strict rules, using $\rightarrow$, and defeasible rules, using $\Rightarrow$, is nicely summarized by [14]. A defeasible rule captures "tentative information that may be used if nothing (can) be posed against it". The fact that "nothing can be posed against" the use of a defeasible rule is established by a proof mechanism that looks for arguments against conclusions established using defeasible rules [14]:

> (a) defeasible rule represents a weak connection between the head and the body of the rule. The effect of a defeasible rule comes from a dialectical analysis ... which involves the consideration of arguments and counter-arguments where that rule is included.

ASPIC$^+$ allows defeasible rules to be undercut, in which case the application of the rule is attacked by an argument that states the rule does not hold[3]. Similarly, since defeasible rules are tentative, ASPIC$^+$ allows the conclusions of such rules to be rebutted. The particular notion of rebutting used in ASPIC$^+$ is said to be *restricted*, meaning that an argument with a strict TopRule($\cdot$) can rebut an argument with a defeasible TopRule($\cdot$), but not vice versa. Rebutting is thus asymmetric[4].

Typically we want to model information that is believed to different degrees, and within ASPIC$^+$ we do this using a preference order over the elements of $\mathcal{R}_d$ and $\mathcal{K}_p$. The question then is how these preferences combine into an ordering $\preceq$ over arguments:

Definition 6 (Preference Ordering). *A preference ordering $\preceq$ is a binary relation over arguments, i.e., $\preceq \, \subseteq \mathcal{A} \times \mathcal{A}$, where $\mathcal{A}$ is the set of all arguments from an argumentation theory. Given $A, B \in \mathcal{A}$, we say A's preference level is less than or equal to that of B iff $A \preceq B$.*

ASPIC$^+$ does not make any assumption about the properties of the preference ordering, but as an example of a property one might use to establish $\preceq$, consider the *weakest link* principle from [17]. This assumes two pre-orderings $\leq, \leq'$ over $\mathcal{R}_d$ and $\mathcal{K}_p$ respectively, and combines them into $A \prec B$ as follows:

- the defeasible rules in A include a rule which is weaker than (strictly less than according to $\leq$) all the defeasible rules in B, and

---

[3] The canonical example here comes from [19] via [17], and is the rule that normally objects that appear red, are red. However, in the situation that everything is illuminated with red light, this rule no longer holds since under red light everything, including things that are not red, will appear to be red.

[4] This asymmetry is not uncontroversial, see [4, 16] for arguments against it.

– the ordinary premises in A include an ordinary premise which is weaker (strictly less than according to $\leq'$) all the ordinary premises in B.

$A \prec B$ is then defined as usual as $A \preceq B$ and $B \not\preceq A$.

Given $A \prec B$, we can then use this to factor the preference over arguments into the notion of attack. Attacks can be distinguished as to whether they are preference-dependent (rebutting and undermining) or preference-independent (undercutting). The former succeed only when the attacker is preferred. The latter succeed whether or not the attacker is preferred.

By combining the definition of arguments, attack relation and preference ordering, we have the following definitions:

Definition 7 (Structured Argumentation Framework). *A Structured Argumentation Framework (SAF) is a triple $\langle \mathcal{A}, Att, \preceq \rangle$, where $\mathcal{A}$ is the set of all arguments from an argumentation theory, Att is the attack relation, and $\preceq$ is a preference ordering on $\mathcal{A}$.*

Definition 8 (Defeat). *A defeats B iff A undercuts B, or if A rebuts/undermines B on B' and A's preference level is not less than that of B' ($A \not\prec B'$).*

Then the idea of an argumentation framework follows from Definitions 7 and 8.

Definition 9 (Argumentation Framework). *An Argumentation Framework (AF) corresponding to a structured argumentation framework $SAF = \langle \mathcal{A}, Att, \preceq \rangle$ is a pair $\langle \mathcal{A}, Defeats \rangle$ such that $Defeats$ is the defeat relation on $\mathcal{A}$ determined by SAF.*

In the general case, argumentation frameworks will include a defeat relation between arguments, and a natural question is what arguments are considered reasonable given those defeats. Now, argumentation frameworks as defined in Definition 9 correspond to the abstract argumentation frameworks of [13]. As a result, all the mechanisms that are defined in [13], and in later work such as [2, 5, 6, 23, 24], for establishing the *acceptability* of a set of arguments — that is identifying various mutually coherent subsets of arguments — can be employed.

Consider this example of an ASPIC$^+$ argumentation framework, adapted from [17]:

Example 1. Consider the argumentation system $AS_1 = \langle \mathcal{L}_1, \bar{\ }, \texttt{Rules}_1, n \rangle$, where:
$$\mathcal{L}_1 = \{a, b, c, d, e, f, nd, \neg a, \neg b, \neg c, \neg d, \neg e, \neg f, \neg nd\}$$
$\texttt{Rules}_1 = \mathcal{R}_{s_1} \cup \mathcal{R}_{d_1}$, with $\mathcal{R}_{s_1} = \{d, f \rightarrow \neg b\}$ and $\mathcal{R}_{d_1} = \{a \Rightarrow b; \neg c \Rightarrow d; e \Rightarrow f; a \Rightarrow \neg nd\}$, and the function $n(\cdot)$ gives $n(\neg c \Rightarrow d) = nd$. We then add the knowledge base $\mathcal{K}_1$ such that $\mathcal{K}_{n_1} = \emptyset$ and $\mathcal{K}_{p_1} = \{a; \neg c; e; \neg e\}$

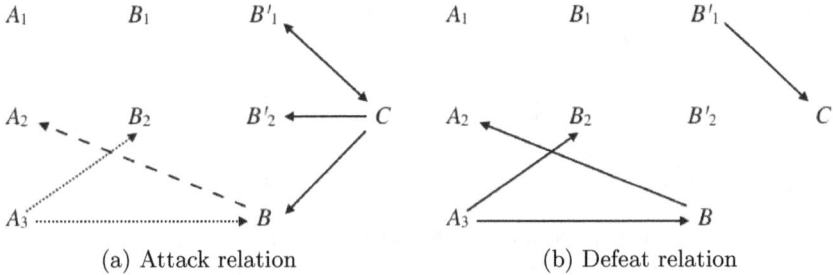

Fig. 1. Attack relations and defeat relations from Example 1. In (a), the solid arrows denote undermining attacks, the dashed arrow denotes a rebutting attack, and dotted arrows denote undercutting attacks.

to get the argumentation theory $AT_1 = \langle AS_1, \mathcal{K}_1 \rangle$. From this we can construct the arguments:

$A_1 = [a]; A_2 = [A_1 \Rightarrow b]; A_3 = [A_1 \Rightarrow \neg nd];$
$B_1 = [\neg c]; B_2 = [B_1 \Rightarrow d]; B'_1 = [e]; B'_2 = [B'_1 \Rightarrow f]; B = [B_2, B'_2 \to \neg b];$
$C = [\neg e];$

Let us call this set of arguments $\mathcal{A}_1$, so that: $\mathcal{A}_1 = \{A_1, A_2, A_3, B_1, B_2, B'_1, B'_2, B, C\}$. Note that $\text{Prem}(B) = \{\neg c; e\}$, $\text{Sub}(B) = \{B_1; B_2; B'_1; B'_2; B\}$, $\text{Conc}(B) = \neg b$, and $\text{TopRule}(B) = d, f \to \neg b$. The attacks between these arguments are shown in Figure 1 (a). These make up the set $Att_1 = \{(C, B'_1), (B'_1, C), (C, B'_2), (C, B), (B, A_2), (A_3, B_2), (A_3, B)\}$. With a preference order $\preceq$ defined by: $A_2 \prec B; C \prec B; C \prec B'_1; C \prec B'_2$, we have the structured argumentation framework $\langle \mathcal{A}_1, Att_1, \preceq \rangle$. This structured argumentation framework establishes a defeat relation $Defeats_1 = \{(B'_1, C), (B, A_2), (A_3, B), (A_3, B_2)\}$ which is shown in Figure 1 (b). With this, we can finally write down the argumentation framework $\langle \mathcal{A}_1, Defeats_1 \rangle$.

This completes a standard description of ASPIC$^+$. In addition, in [15] we introduced some ways of thinking about ASPIC$^+$ which will be useful here.

Definition 10. *Let $AT = \langle AS, \mathcal{K} \rangle$ be an argumentation theory, where AS is the argumentation system $AS = \langle \mathcal{L}, \bar{\ }, \text{Rules}, n \rangle$. We define the closure of a set of propositions $P \subseteq \mathcal{K}$ under a set of rules $R \subseteq \text{Rules}$ as $Cl(P)_R$, where:*

1. $P \subseteq Cl(P)_R$;
2. *if $p_1, \ldots, p_n \in Cl(P)_R$ and $p_1, \ldots, p_n \to p \in R_S$, then $p \in Cl(P)_R$;*
3. *if $p_1, \ldots, p_n \in Cl(P)_R$ and $p_1, \ldots, p_n \Rightarrow p \in R_D$, then $p \in Cl(P)_R$; and*
4. *$\nexists S \subset Cl(P)_R$ such that S satisfies the previous conditions.*

We use this notion of closure to establish a notion of inference in systems like ASPIC$^+$ and DeLP:

Definition 11. *Let $AT = \langle AS, \mathcal{K} \rangle$ be an argumentation theory, where AS is the argumentation system $AS = \langle \mathcal{L}, \bar{\ }, \text{Rules}, n \rangle$. Given a set of propositions $P \subseteq \mathcal{K}$, a set of rules $R \subseteq \text{Rules}$ and a proposition $p \in \mathcal{K}$, we say that p is inferred from P and R, noted as $P \vdash_R p$, if $p \in Cl(P)_R$.*

Finally, we made use of the idea of the set of rules in an argument:

Definition 12 (Argument Rules). *Let $AT = \langle AS, \mathcal{K} \rangle$ be an argumentation theory and $A \in \mathcal{A}(AT)$. We define the set of rules of A as follows:*

$$\text{Rules}(A) = \begin{cases} \emptyset & A \in \mathcal{K} \\ \{\text{TopRule}(A)\} \cup \bigcup_{i=1}^{n} \text{Rules}(A_i) & A = A_1, \ldots, A_n \rightsquigarrow \text{Conc}(A) \end{cases}$$

This allows us to describe an argument A as a triple:

$$(G, R, p)$$

where $G = \text{Prem}(A)$ are the *grounds* on which A is based, $R = \text{Rules}(A)$ is the set of rules that are used to construct A from G, and $p = \text{Conc}(A)$ is the conclusion of A. Moreover, for any an argument (G, R, p), it holds that $G \vdash_R p$.

We can also identify the sets of strict and defeasible rules of an argument A as $\text{Rules}_s(A) = \text{Rules}(A) \cap \text{Rules}_s$ (respectively, $\text{Rules}_d(A) = \text{Rules}(A) \cap \text{Rules}_d$), where $\text{Rules} = \text{Rules}_s \cup \text{Rules}_d$ is the set of rules of the argumentation system AS in the argumentation theory AT. In addition, if $\text{Rules}_d(A) = \emptyset$, argument A is said to be *strict*; otherwise, if $\text{Rules}_d(A) \neq \emptyset$, A is a *defeasible* argument. Also, given $S \subseteq \mathcal{L}$, $S \models^5 \phi$ iff there exists a strict argument A such that $\text{Conc}(A) = \phi$ and $\text{Prem}(A) \subseteq S$ (i.e. if there exists a strict argument for ϕ with all its premises taken from S).

3 DeLP Background

Defeasible Logic Programming (DeLP, for short) [14] is a formalism that combines results of Logic Programming and Defeasible Argumentation. As expressed by the authors in [14], DeLP extends logic programming with the possibility of representing information in the form of weak rules (referred to as *defeasible rules*) in a declarative manner. Then, it makes use of a defeasible argumentation inference mechanism to determine the warranted conclusions and, as a result, provide answers to queries.

The following description of DeLP is drawn from [14]. The basic unit in DeLP is a defeasible logic program:

---

[5] The authors in [17] use the symbol $\vdash$. We replaced it with $\models$ in order to avoid confusions with the notion of inference introduced in Definition 11.

Definition 13 (Defeasible Logic Program). *A Defeasible Logic Program $\mathcal{P}$, abbreviated* de.l.p.*, is a possibly infinite set of facts, strict rules and defeasible rules. In a program $\mathcal{P}$, we will distinguish the subset Π of facts and strict rules, and the subset Δ of defeasible rules. When required, we will denote $\mathcal{P}$ as (Π, Δ).*

The elements of a DeLP program are written in logic-programming style, where facts, strict rules and defeasible rules are defined in [14] as follows:

Definition 14 (Fact). *Let $\mathcal{L}$ be a set of ground atoms. A fact is a literal, i.e. a ground atom "A" or a negated ground atom "$\neg A$", where $A \in \mathcal{L}$ and "$\neg$" represents strong negation*[6].

In particular, any pair of literals "A" and "$\neg A$" are said to be *complementary*.

Definition 15 (Strict Rule). *A Strict Rule is an ordered pair, denoted "Head $\leftarrow$ Body", whose first member, Head, is a literal, and whose second member, Body, is a finite non-empty set of literals. A strict rule with head L_0 and body $\{L_1, \ldots, L_n\}$ can also be written as: $L_0 \leftarrow L_1, \ldots, L_n$ $(n > 0)$.*

It should be noted that, although the initial characterization of DeLP given in [14] requires defeasible rules to have a non-empty body, at the end of the paper the authors discuss some extensions for DeLP, among which they consider the inclusion of *presumptions* [18], which can be considered as "defeasible facts". Specifically, in [14] it is mentioned that:

> In our approach, a rule like "$a \prec$ " would express that *"there are (defeasible) reasons to believe in a."*

Next, we present a generalized definition of defeasible rule, which accounts for presumptions:

Definition 16 (Defeasible Rule). *A Defeasible Rule is an ordered pair, denoted "Head $\prec$ Body", whose first member, Head, is a literal, and whose second member, Body, is a finite set of literals. A defeasible rule with head L_0 and body $\{L_1, \ldots, L_n\}$ can also be written as: $L_0 \prec L_1, \ldots, L_n$ $(n > 0)$. A defeasible rule with head L and empty body (i.e. a presumption) can also be written as: $L \prec$.*

Given a de.l.p., we are interested in what can be derived from it:

Definition 17 (Defeasible Derivation). *Let $\mathcal{P} = (\Pi, \Delta)$ be a de.l.p. and L a ground literal. A defeasible derivation of L from $\mathcal{P}$, denoted $\mathcal{P} \mathrel{|\!\sim} L$, consists of a finite sequence $L_1, L_2, \ldots, L_n = L$ of ground literals, and each literal L_i is in the sequence because:*

---

[6] In [14] the authors use "$\sim$" to denote strong negation. However, in order to harmonize notation, in this chapter we will adopt the notation "$\neg$" introduced for ASPIC$^+$.

a) L_i is a fact in Π or a presumption in Δ; or
b) there exists a rule R_i in $\mathcal{P}$ (strict or defeasible) with head L_i and body $B_1, B_2, \ldots, B_k$ and every literal of the body is an element L_j of the sequence appearing before L_i, $(j < i.)$

As [14] say:

> Given a de.l.p. $\mathcal{P}$, a derivation for a literal L from $\mathcal{P}$ is called "defeasible", because as we will show next, there may exist information in contradiction with L that will prevent the acceptance of L as a valid conclusion.

In other words, a defeasible derivation may contain strict rules, but a derivation that only contains strict rules is not considered to be a defeasible derivation.

The idea of a defeasible derivation is then used to define an *argument* in DELP. Intuitively, an argument is a minimal set of rules used to derive a conclusion:

Definition 18 (Argument Structure). *Let L be a literal, and $\mathcal{P} = (\Pi, \Delta)$ a de.l.p.. We say that $\langle A, L \rangle$ is an argument structure for L, if A is a set of defeasible rules of Δ, such that:*

1. *there exists a defeasible derivation for L from $\Pi \cup A$;*
2. *the set $\Pi \cup A$ is non-contradictory; and*
3. *A is minimal: there is no proper subset A' of A such that A' satisfies conditions (1) and (2).*

From here on, we will sometimes refer to an argument structure simply as an argument. To complete the definition, we have to define the term "non-contradictory". [14] gives the definition:

Definition 19 (Contradictory Set of Rules). *A set of rules is contradictory if and only if, there exists a defeasible derivation for a pair of complementary literals from this set.*

and the paper takes the idea as applying to de.l.p.s as well (that is both sets of rules and facts can be non-contradictory). Moreover, the authors in [14] impose the requirement that the set Π of a de.l.p. $\mathcal{P}$ has to be non-contradictory. Specifically, this choice has to do with the meaning associated with facts and strict rules, as they represent domain information that is indisputable.

Note that the existence of a DELP argument (with a non-contradictory and minimal set of rules) does not guarantee either its acceptance, or that its conclusion is justified. This is because the argument may be in contradiction with other arguments, which may in turn be accepted. The requirement that an argument is non-contradictory, does, however, rule out the fact that the argument is in conflict with itself. Furthermore, it rules out the possibility of building an argument that contradicts the strict knowledge of a de.l.p..

Conflicts between arguments are characterized in [14] via the notion of *counter-argument*, which relies on the notion of *disagreement* between literals.

Definition 20 (Disagreement). Let $\mathcal{P} = (\Pi, \Delta)$ be a de.l.p.. We say that two literals h_1 and h_2 disagree, if and only if the set $\Pi \cup \{h_1, h_2\}$ is contradictory.

Definition 21 (Attack). We say that $\langle A_1, h_1 \rangle$ counter-argues, rebuts, or attacks $\langle A_2, h_2 \rangle$ at literal h, if and only if there exists a sub-argument $\langle A, h \rangle$ of $\langle A_2, h_2 \rangle$ such that h_1 and h disagree.

Note that attacks in DELP can be aimed not only at the final conclusion of an argument, but also at its intermediate conclusions. Such intermediate conclusions correspond to the conclusions of its proper *sub-arguments* where, as defined in [14], an argument $\langle B, q \rangle$ is a sub-argument of $\langle A, h \rangle$ if $B \subseteq A$.

Given an attack from argument $\langle A_1, h_1 \rangle$ to $\langle A_2, h_2 \rangle$, these two arguments can be compared in order to determine which one prevails. Briefly, if argument $\langle A_2, h_2 \rangle$ is not better than $\langle A_1, h_1 \rangle$ with respect to a comparison criterion, noted $\langle A_1, h_1 \rangle \not\prec \langle A_2, h_2 \rangle$, $\langle A_1, h_1 \rangle$ will be called a defeater of $\langle A_2, h_2 \rangle$.

Definition 22 (Defeat). Let $\langle A_1, h_1 \rangle$ and $\langle A_2, h_2 \rangle$ be two argument structures such that $\langle A_1, h_1 \rangle$ counter-argues $\langle A_2, h_2 \rangle$ at literal h. We say that $\langle A_1, h_1 \rangle$ is a defeater for $\langle A_2, h_2 \rangle$ if and only if either:

a) the attacked sub-argument $\langle A, h \rangle$ of $\langle A_2, h_2 \rangle$ is such that $\langle A, h \rangle \prec \langle A_1, h_1 \rangle$, in which case $\langle A_1, h_1 \rangle$ is a proper defeater of $\langle A_2, h_2 \rangle$; or
b) the attacked sub-argument $\langle A, h \rangle$ of $\langle A_2, h_2 \rangle$ is such that $\langle A, h \rangle \not\prec \langle A_1, h_1 \rangle$ and $\langle A_1, h_1 \rangle \not\prec \langle A, h \rangle$, in which case $\langle A_1, h_1 \rangle$ is a blocking defeater of $\langle A_2, h_2 \rangle$.

Note that the second case in the previous definition does not only account for the case where the compared arguments are considered to be equivalent by the adopted comparison criterion, but also for the case where the attacking argument and the attacked sub-argument are incomparable (i.e. they are not related by the comparison criterion).

Example 2. Consider the de.l.p. $\mathcal{P}_2 = (\Pi_2, \Delta_2)$, with the same facts and rules as the argumentation system and knowledge base of the argumentation theory AT_1 from Example 1:

$$\Pi_2 = \{\neg b \leftarrow d, f\} \qquad \Delta_2 = \begin{cases} a \prec & b \prec a \\ \neg c \prec & d \prec \neg c \\ e \prec & f \prec e \\ \neg e \prec & \neg nd \prec a \end{cases}$$

Note that the ordinary premises of the knowledge base $\mathcal{K}_1$ are represented as presumptions in Δ_2. Furthermore, in Example 1, the defeasible rule "$a \Rightarrow \neg nd$" refers to the name associated to the defeasible rule "$\neg c \Rightarrow d$" in the argumentation system AS_1. In contrast, in DELP rules do not have associated names and thus, the literal "$\neg nd$" in the head of the defeasible rule "$\neg nd \prec a$" will not appear in any other rule or fact of the de.l.p. $\mathcal{P}_2$.[7]

Finally, from the DELP program $\mathcal{P}_2$ we can build the following arguments, similarly to those obtained in Example 1:

$\langle A_1, a \rangle$, with $A_1 = \{a \prec \ \}$
$\langle A_2, b \rangle$, with $A_2 = \{(a \prec \ ), (b \prec a)\}$
$\langle A_3, \neg nd \rangle$, with $A_3 = \{(a \prec \ ), (\neg nd \prec a)\}$
$\langle B_1, \neg c \rangle$, with $B_1 = \{\neg c \prec \ \}$
$\langle B_2, d \rangle$, with $B_2 = \{(\neg c \prec \ ), (d \prec \neg c)\}$
$\langle B'_1, e \rangle$, with $B'_1 = \{e \prec \ \}$
$\langle B'_2, f \rangle$, with $B'_2 = \{(e \prec \ ), (f \prec e)\}$
$\langle B, \neg b \rangle$, with $B = \{(\neg c \prec \ ), (d \prec \neg c), (e \prec \ ), (f \prec e)\}$
$\langle C, \neg e \rangle$, with $C = \{\neg e \prec \ \}$

Note that the only difference between the set of ASPIC+ arguments from Example 1 and the one listed above is that the DELP argument $\langle B, \neg b \rangle$ does not include the strict rule "$\neg b \leftarrow d, f$", whereas the ASPIC+ argument B includes the strict rule "$d, f \rightarrow \neg b$". Let us call the set of DELP arguments $\mathcal{A}_2$, so that: $\mathcal{A}_2 = \{\langle A_1, a \rangle, \langle A_2, b \rangle, \langle A_3, \neg nd \rangle, \langle B_1, \neg c \rangle, \langle B_2, d \rangle, \langle B'_1, e \rangle, \langle B'_2, f \rangle, \langle B, \neg b \rangle, \langle C, \neg e \rangle\}$. The attacks between these arguments make up the relation $Att_2 = \{(\langle C, \neg e \rangle, \langle B'_1, e \rangle), (\langle B'_1, e \rangle, \langle C, \neg e \rangle), (\langle C, \neg e \rangle, \langle B'_2, f \rangle), (\langle C, \neg e \rangle, \langle B, \neg b \rangle), (\langle B, \neg b \rangle, \langle A_2, b \rangle), (\langle A_2, b \rangle, \langle B, \neg b \rangle)\}$. This set of attacks is depicted in Figure 2 (a) where, for illustration purposes, the conclusions of the arguments are omitted. In particular, differently from Example 1, the attack involving arguments $\langle B, \neg b \rangle$ and $\langle A_2, b \rangle$ is symmetric; furthermore, since DELP does not account for undercutting attacks, the attacks from $\langle A_3, \neg nd \rangle$ to $\langle B_2, d \rangle$ and $\langle B, \neg b \rangle$ in Example 1 no longer exist.[8] Finally, if we resolve the attacks using a comparison criterion such that: $\langle A_2, b \rangle \prec \langle B, \neg b \rangle$; $\langle C, \neg e \rangle \prec \langle B, \neg b \rangle$; $\langle C, \neg e \rangle \prec \langle B'_1, e \rangle$; $\langle C, \neg e \rangle \prec \langle B'_2, f \rangle$, we obtain the following defeat relation $Defeats_2 = \{(\langle B'_1, e \rangle, \langle C, \neg e \rangle), (\langle B, \neg b \rangle, \langle A_2, b \rangle)\}$, shown in Figure 2 (b).

In order to determine the justified conclusions of accepted arguments built from a DELP program (referred to in [14] as warranted literals), we need to account for the defeat relation between arguments. In particular, given an argument $\langle A_0, h_0 \rangle$, all defeaters for $\langle A_0, h_0 \rangle$ have to be considered. Let $\langle A_1, h_1 \rangle$ be one of such defeaters. Then, since $\langle A_1, h_1 \rangle$ is an argument structure, defeaters for it may also exist, and so on. As a result, in order to determine the

---

[7] In [11] an extension of DELP was proposed, where defeasible rules have labels associated with them, and those labels (as well as their complement with respect to the strong negation "$\neg$") can appear in the head of defeasible rules. We will come back to this point later in Section 4.3.

[8] In particular, the extension of DELP introduced in [9] and [11] accounts for the existence of undercutting attacks. As mentioned before, we will come back to this point later in Section 4.3.

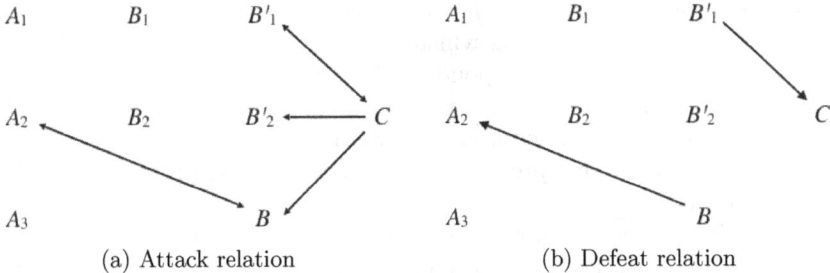

Fig. 2. (a) Attacks and (b) defeats between the arguments built from the DeLP program from Example 2.

acceptance status of argument $\langle A_0, h_0 \rangle$ (thus, the warrant status of its conclusion h_0) [14] introduces the notion of *dialectical tree*, a tree structure that gathers all sequences of defeaters starting from a given argument (the root of the tree). Then, once built, the dialectical tree is *marked* according to the following criterion: (1) leaf nodes are marked as undefeated and (2) a non-leaf node (i.e. an inner node or the root) is marked as undefeated if all its children are marked as defeated; otherwise it is marked as defeated. Finally, a literal h is said to be warranted from a de.l.p. $\mathcal{P}$ if there exists an argument $\langle A, h \rangle$ obtained from $\mathcal{P}$ such that $\langle A, h \rangle$ is the root of a marked dialectical tree and is marked as undefeated; moreover, in that case, argument $\langle A, h \rangle$ will be considered as accepted. For full details on the construction of dialectical trees and their marking criterion, we refer the reader to [14].

4 Comparison and Discussion

In this section we will study and compare different features of ASPIC$^+$ and DeLP, identifying the commonalities and differences between them. In addition, for those elements where the two systems differ, we will try to provide alternative characterizations with the aim of either bridging the gap between them, or pointing towards a way in which the gap might be bridged.

4.1 Knowledge Representation

There are clearly some commonalities between ASPIC$^+$ and DeLP. Both start with the same raw materials, a set of facts, $\mathcal{F}$, and a set of rules, $\mathcal{R}$. We can imagine both of these being partitioned into strict and defeasible parts:

$$\mathcal{F} = \mathcal{F}_S \cup \mathcal{F}_D$$
$$\mathcal{R} = \mathcal{R}_S \cup \mathcal{R}_D$$

where the subscript S denotes the strict part and D denotes the defeasible part (and there are no elements which are both strict and defeasible). Then, given sets $\mathcal{F}$ and $\mathcal{R}$, the corresponding ASPIC$^+$ argumentation system AS is $\langle \mathcal{L}, \bar{}, R_S \cup R_D, n \rangle$ and the corresponding knowledge base is $\mathcal{K} = F_S \cup F_D$, to then make up the argumentation theory AT $\langle AS, \mathcal{K} \rangle$. Similarly, in DeLP, this knowledge would be represented as a defeasible logic program $\mathcal{P} = (F_S \cup R_S, F_D \cup R_D)$.

Recall that, as noted above, even though the same elements are used in ASPIC$^+$ and DeLP, these systems represent defeasible facts differently. On the one hand, ASPIC$^+$ explicitly accounts for defeasible facts within the knowledge base $\mathcal{K}$: the set F_D corresponds to the set of ordinary premises $\mathcal{K}_p$. On the other hand, as mentioned before, DeLP accounts for defeasible facts under the notion of presumption; hence, defeasible facts are represented as a special case of defeasible rules. This is also how defeasible facts are represented in ASPIC-[4].

4.2 Argument Construction

There are several differences between the notion of argument in ASPIC$^+$ and in DeLP, and these lead to three main points of comparison:

1. Argument structure: arguments in DeLP and ASPIC$^+$ differ in structure even when built from the same knowledge, though they can be related at a rather abstract level;
2. Minimality: ASPIC$^+$ arguments are not explicitly minimal sets of facts and rules (though they contain no irrelevant information), whereas DeLP arguments are, and even when the obvious notion of minimality is applied to ASPIC$^+$ arguments it differs from that imposed in DeLP. Finally;
3. Consistency: DeLP arguments are required to conform to a kind of consistency, whereas ASPIC$^+$ arguments are not. However, ASPIC$^+$ does make use of a notion of c-consistency that can be considered to be complementary to the use of consistency in DeLP.

Argument Structure There are several ways in which the structure of DeLP and ASPIC$^+$ arguments are different.

Firstly, the two systems rely on different mechanisms for the construction of arguments. As expressed in Definition 4, an ASPIC$^+$ argument corresponds to a tree structure, where the root node is the argument itself and every other node corresponds to one of its proper sub-arguments. Furthermore, any node in the tree is connected to its children through the application of a strict or defeasible rule — specifically, the conclusions of the children are the premises of the rule and the conclusion of the parent is the conclusion of the rule. In contrast, a DeLP argument is characterized by a set of rules and a conclusion, so there is no explicit structure to a DeLP argument in the same way that there is to an ASPIC$^+$ argument. Moreover, unlike an ASPIC$^+$ argument, a

DeLP argument does not contain every piece of information (facts and rules) used in its construction process. Specifically, an argument only includes the defeasible knowledge (defeasible rules and presumptions) used for building it. This does not mean that DeLP ignores strict information when constructing an argument. Rather strict knowledge is taken into account when considering the notion of defeasible derivation, a key element in the argument construction process in DeLP.

Secondly, ASPIC+ does not impose restrictions on arguments, other than the implicit requirement that their construction as tree structures results from the application of strict and defeasible rules on their sub-arguments. In other words, ASPIC+ arguments will not include irrelevant elements. In contrast, DeLP explicitly constrains the definition of an argument by imposing two requirements on its set of defeasible rules: minimality and consistency.

This leads us to the third difference between the ways that arguments are constructed in ASPIC+ and DeLP. This is that the check that an argument is non-contradictory, performed when constructing it, effectively considers *all* the strict knowledge in the de.l.p.. When considering whether $\langle A, h \rangle$ is an argument, it is necessary to check that there is no combination of strict information that could, with the rules in A, be used to derive two complementary literals. This contrasts with ASPIC+ which only takes into account the elements explicitly recorded in the argument structure.

Given these structural differences between arguments in ASPIC+ and DeLP, we need to find a unifying mechanism for bringing them closer. For that purpose, we can consider a notion of derivation, similar to the one proposed in Definition 17 for DeLP. In particular, given an ASPIC+ argument A, there exists a defeasible derivation of $\text{Conc}(A)$ from S, where $S = \text{Prem}(A) \cup \text{Rules}(A)$. Furthermore, the conclusion of every sub-argument of A will appear in the corresponding derivation.

However, since defeasible derivations are sequences of literals, the same argument (either in ASPIC+ or in DeLP) could be associated with multiple derivations which result from the permutation of elements in the sequence (while maintaining the condition *b)* from Definition 17). Moreover, the notion of defeasible derivation makes it possible to include literals in the sequence that are not needed to derive the conclusion of an argument. Thus, by including irrelevant literals, a potentially infinite number of defeasible derivations could be associated with the same argument.

The above mentioned issue suggests the need to find an alternative unifying mechanism for ASPIC+ and DeLP arguments. If we consider the sets of rules and facts used for deriving the conclusions of arguments, referred to as *deriving sets*, then we have a common basis. On the one hand, the deriving set of an ASPIC+ argument A will be $\text{Prem}(A) \cup \text{Rules}(A)$. On the other hand, the deriving set of a DeLP argument $\langle A', h \rangle$ will contain every defeasible rule and presumption in A', and should also include the facts and strict rules used in the derivation of h. However, it could be the case that some literal l in the derivation of h is associated with multiple strict derivations. In such a case, the same DELP argument would have multiple deriving sets associated with it. As a result, we can conclude that given the existence of a DELP argument $\langle A', h \rangle$, there will exist an ASPIC$^+$ argument A with $\texttt{Conc}(A) = h$ and whose deriving set coincides with one of the deriving sets of $\langle A', h \rangle$. Conversely, given an ASPIC$^+$ argument A such that $\texttt{Conc}(A) = h$, if there exists a DELP argument $\langle A', h \rangle$ such that A' contains the DELP counterpart of every defeasible rule in the ASPIC$^+$ argument A, then the deriving set of A will coincide with one of the deriving sets of $\langle A', h \rangle$.

Minimality As we pointed out above, one difference between ASPIC$^+$ and DELP is that Definition 4 does not impose any minimality requirement on the grounds or the set of rules of an ASPIC$^+$ argument while Definition 18 does. Nevertheless, as already mentioned and as discussed in [17] and [15], every premise and rule of an ASPIC$^+$ argument is used for deriving its conclusion, meaning that the argument does not contain any extraneous propositions or rules. In particular, in [15] it was formally shown that, given the characterization of an ASPIC$^+$ argument A as a triple (G, R, c), every element in G is the conclusion of a sub-argument A' of A and is the premise of a rule in R, and that every rule in R is the $\texttt{TopRule}$ of a sub-argument A'' of A. This feature of ASPIC$^+$ arguments can be considered as a form of minimality, as irrelevant elements are not introduced within an argument.

For DELP arguments, the subset-minimality requirement imposed in the third clause of Definition 18 ensures that irrelevant elements will not be included in an argument; otherwise, there would be a smaller set of defeasible rules satisfying the first two clauses of Definition 18 and the argument in question would not be an argument. In addition, the subset-minimality requirement in DELP avoids introducing redundant elements to an argument. To illustrate the notion of *redundancy*, let us consider the following example.

Example 3. Given the `de.l.p.` $\mathcal{P}_3 = (\Pi_3, \Delta_3)$:

$$\Pi_3 = \begin{Bmatrix} p \\ q \end{Bmatrix} \qquad \Delta_3 = \begin{Bmatrix} r \prec p \\ r \prec q \\ s \prec r \\ t \prec r, s \end{Bmatrix}$$

we can build an argument $\langle A, t \rangle$, with $A = \{(r \prec p), (s \prec r), (t \prec r, s)\}$. Alternatively, we can build an argument $\langle B, t \rangle$ with $B = \{(r \prec q), (s \prec r), (t \prec r, s)\}$. However, there is no argument $\langle C, t \rangle$ with $C = \{(r \prec p), (r \prec q), (s \prec r), (t \prec r, s)\}$, because C is a superset of A and B.

Let us now consider the knowledge represented in Example 3 in the context of an ASPIC$^+$ argumentation system:

Example 4. Consider the argumentation system $AS_4 = \langle \mathcal{L}_4, ^-, \mathcal{R}_4, n \rangle$, where $\mathcal{L}_4 = \{p, q, r, s, t, \neg p, \neg q, \neg r, \neg s, \neg t\}$ and $\mathcal{R}_4 = \{p \Rightarrow r; q \Rightarrow r; r \Rightarrow s; r, s \Rightarrow t\}$.

By adding the knowledge base $\mathcal{K}_4 = \{p, q\}$ we obtain the argumentation theory $AT_4 = \langle AS_4, \mathcal{K}_4 \rangle$, from which we can construct (among others) the following arguments:

$$A_1 = [p]; A_2 = [A_1 \Rightarrow r]; A_3 = [A_2 \Rightarrow s]; A = [A_2, A_3 \Rightarrow t];$$
$$B_1 = [q]; B_2 = [B_1 \Rightarrow r]; B_3 = [B_2 \Rightarrow s]; B = [B_2, B_3 \Rightarrow t];$$
$$C = [B_2, A_3 \Rightarrow t]$$

Argument C from Example 4 is redundant because its set of rules provides two ways to derive r, one that relies on p and another that relies on q, and r appears twice in the derivation of t: once to produce s, and once when the rule $r, s \Rightarrow t$ is applied. Then C, the redundant argument, uses both rules for deriving r while arguments A and B use just one of them, providing a more compact derivation.

Another situation that can occur in ASPIC$^+$, which in DeLP is prevented by the third clause of Definition 18 is the existence of *circularity* within an argument. This is illustrated by the following example:

Example 5. Consider the argumentation system $AS_1 = \langle \mathcal{L}_5, ^-, \mathcal{R}_5, n \rangle$, where $\mathcal{L}_5 = \{a, b, c, d, e, \neg a, \neg b, \neg c, \neg d, \neg e\}$ and $\mathcal{R}_5 = \{a, b \Rightarrow d; d \Rightarrow b; b, c \Rightarrow e\}$. By adding the knowledge base $\mathcal{K}_5 = \{a, b, c\}$ we obtain the argumentation theory $AT_5 = \langle AS_5, \mathcal{K}_5 \rangle$, from which we can construct the following arguments:

$$D_1 = [a]; D_2 = [b]; D_3 = [D_1, D_2 \Rightarrow d]; D_4 = [D_3 \Rightarrow b]; D_5 = [c];$$
$$D = [D_4, D_5 \Rightarrow e]; E = [D_2, D_5 \Rightarrow e]$$

such that $D = (\{a, b, c\}, \{a, b \Rightarrow d; d \Rightarrow b; b, c \Rightarrow e\}, e)$ and $E = (\{b, c\}, \{b, c \Rightarrow e\}, e)$.

In the context of Example 5, argument E is circular because it starts with a and b as premises, from which it derives c. Then, it uses c to (again) derive b and, finally, use c and the second derivation of b to obtain the conclusion e. This loop is removed in E to give a more compact argument for the conclusion e. Moreover, this circularity becomes evident when observing the characterization of D and E as a triple, since the sets of grounds and rules of argument E are proper subsets of those of argument D.

In contrast to these examples of circularity and redundancy in ASPIC$^+$, because of the subset-minimality requirement established by the third clause in Definition 18, an argument like D would not exist in DeLP.

One way to bridge this gap between ASPIC$^+$ and DeLP is to impose some form of minimality on an ASPIC$^+$ argument. In [15] we showed that defining ASPIC$^+$ arguments to eliminate circularity and redundancy was equivalent to enforcing minimality on the set of grounds or rules used to construct the argument. (To be precise, if we describe an ASPIC$^+$ argument A as a triple (G, R, c), then if there is no argument (G', R, c) such that $G' \subset G$ and no argument (G, R', c) such that $R' \subset R$, then A is not redundant or circular.) However, note that this is a less restrictive form of minimality than the one enforced in DELP, since DELP necessarily requires the set of defeasible rules, $R_D \subseteq R$ in the notation we introduced in Definition 10, to be minimal. Furthermore, as grounds (facts) are not included within a DELP argument, the minimality check on that set is not necessary. As a result, it could be the case that an argument is minimal in ASPIC$^+$ but not in DELP. In contrast, for every argument in DELP (which, by definition, is minimal) there exists a minimal argument in ASPIC$^+$ having the same set of defeasible rules. This difference between the notion of minimality in ASPIC$^+$ and in DELP is illustrated in the following example.

Example 6. Let $AT_6 = \langle AS_6, \mathcal{K}_6 \rangle$ be an argumentation theory, where $AS_6 = \langle \mathcal{L}_6, ^-, \mathcal{R}_6, n \rangle$, $\mathcal{R}_6 = \{d \Rightarrow b; b \Rightarrow c; b, c \Rightarrow a\}$ and $\mathcal{K}_6 = \mathcal{K}_{n_6} = \{b, d\}$. From AT we can construct the following arguments:

$$A_1 = [d]; A_2 = [A_1 \Rightarrow b]; A_3 = [A_2 \Rightarrow c]; A_4 = [b]; A = [A_4, A_3 \Rightarrow a];$$
$$B = [A_2, A_3 \Rightarrow a]; A_5 = [A_4 \Rightarrow c]; C = [A_4, A_5 \Rightarrow a]$$

Here, $A = (G, R, a)$, with $G = \{b, d\}$ and $R = \mathcal{R}_6$. In this case, A is not minimal since there exists $B = (G', R, a)$ with $G' = \{d\} \subset G$. On the other hand, argument C is represented by the triple (G'', R', a), with $G'' = \{b\}$ and $R' = \{b \Rightarrow c; b, c \Rightarrow a\}$. In particular, argument C is minimal. Furthermore, B is also minimal since, even though $R' \subset R$, (G', R', a) is not an argument for a.

Let us now consider the DELP program $\mathcal{P}_6 = (\Pi_6, \Delta_6)$:

$$\Pi_6 = \{(b \prec d), (c \prec b), (a \prec b, c)\}$$

Here, there is only one argument whose conclusion is the literal a; this is the argument $\langle C', a \rangle$, with $C' = \{(c \prec b), (a \prec b, c)\}$. In particular, argument $\langle C', a \rangle$ in DELP would correspond to argument C in ASPIC$^+$. Note that, even though there exists a derivation for the literal a from the set $B' = \{d, (b \prec d), (c \prec b), (a \prec b, c)\}$, there is no other argument for a. In particular, B' would correspond to the ASPIC$^+$ argument B, whose set of grounds is $G' = \{d\}$ and its set of rules (following DELP's notation) is $R = \{(b \prec d), (c \prec b), (a \prec b, c)\}$.

This relates back to the point we made in the previous section about relating DELP and ASPIC$^+$ through deriving sets. Even if the sets of defeasible rules in a DELP argument A and an ASPIC$^+$ argument A' coincide, the arguments can have different conclusions, or the same conclusion and different deriving sets, because the strict part of the DELP argument is not constrained. These situations are illustrated in the following example.

Example 7. Consider the argumentation system $AS_7 = \langle \mathcal{L}_7, ^-, \mathcal{R}_7, n \rangle$, where $\mathcal{L}_7 = \{a, b, c, d, \neg a, \neg b, \neg c, \neg d\}$ and $\mathcal{R}_7 = \{a \to c; a \to b; b \to c; d \Rightarrow a\}$. Then, if we add the knowledge base $\mathcal{K}_7 = \mathcal{K}_{n_7} = \{d\}$, we get the argumentation theory $AT_7 = \langle AS_7, \mathcal{K}_7 \rangle$. From this theory, we can build the following arguments:

$$D = [d]; \quad A = [D \Rightarrow a]; \quad C = [A \to c]; \quad B = [A \to b]; \quad C' = [B \to c]$$

The DELP counterpart of the argumentation theory AT_7 would be the de.l.p. $\mathcal{P}_7 = (\Pi_7, \Delta_7)$:

$$\Pi_7 = \begin{cases} d \\ c \leftarrow a \\ c \leftarrow b \\ b \leftarrow a \end{cases} \quad \Delta_7 = \{a \prec d\}$$

from which we can build the following arguments: $\langle D, d \rangle$, with $D = \emptyset$; and $\langle A, a \rangle$, $\langle A, b \rangle$, $\langle A, c \rangle$, with $A = \{a \prec d\}$. Note that, unlike in ASPIC$^+$, since DELP arguments do not include the strict knowledge used in the derivation of their conclusions, there is only one argument for c. Furthermore, whereas in ASPIC$^+$ we have two arguments for c, one of which has a unique deriving set, argument $\langle A, c \rangle$, will have two deriving sets: one of them including the strict rule $c \leftarrow a$, and the other including the strict rules $c \leftarrow b$ and $b \leftarrow a$. As a result, the DELP argument $\langle A, c \rangle$ has the same conclusion and the same set of defeasible rules as the ASPIC$^+$ arguments C and C', but their deriving sets differ. On the other hand, even though the DELP argument $\langle A, a \rangle$ has the same set of defeasible rules as the ASPIC$^+$ arguments C and C', their conclusions differ: $\text{Conc}(C) = \text{Conc}(C') = c$ while the conclusion of $\langle A, a \rangle$ is a; also, the deriving set of $\langle A, a \rangle$ differs from that of the ASPIC$^+$ arguments C and C'.

Consistency The last aspect to consider in terms of argument construction is the requirement in DELP that the set $\Pi \cup A$ which gives rise to the defeasible derivation behind an argument $\langle A, h \rangle$ is non-contradictory. The requirement for the basis of an argument to be consistent is not uncommon in argumentation systems — see, for example, [1,3] — but this is not exactly what is required in DELP. In DELP, the consistency is between the argument structure in the form of the defeasible rules A, and the entire set of strict information in the knowledge base Π. The effect of the consistency requirement in DELP, which is encoded in clause 2 of Definition 18, is to prevent an argument from coming into existence if it derives the complement of something that is in the strict part of the knowledge base Π, or can be derived from the corresponding program $\mathcal{P}$ using the facts and rules in both Π and A. That is a rather stronger check than is imposed in systems such as [1,3], as shown in the following example:

Example 8. Let $AS_8 = \langle \mathcal{L}_8, ^-, \mathcal{R}_8, n \rangle$ be an argumentation system, where $\mathcal{L}_8 = \{a, b, \neg a, \neg b\}$ and $\mathcal{R}_8 = \{b \Rightarrow \neg a\}$. Then, if we add the knowledge base $\mathcal{K}_8 = \mathcal{K}_{n_8} = \{a, b\}$, we get the argumentation theory $AT_8 = \langle AS_8, \mathcal{K}_8 \rangle$. In ASPIC$^+$, and many other argumentation systems, from this argumentation theory we can construct the following arguments:

$$A_1 = [a]; \quad A_2 = [b]; \quad A_3 = [A_2 \Rightarrow \neg a]$$

However from the corresponding de.l.p. $\mathcal{P}_8 = (\Pi_8, \Delta_8)$:

$$\Pi_8 = \{a, b\} \quad \Delta_8 = \{\neg a \prec b\}$$

we can build the arguments $\langle \emptyset, a \rangle$ and $\langle \emptyset, b \rangle$, but are prevented from building the argument $\langle \{\neg a \prec b\}, \neg a \rangle$ because its conclusion $\neg a$ contradicts the fact a.

This difference represents a fundamental difference in viewpoint between DELP and ASPIC$^+$ (and other systems like those cited above). In ASPIC$^+$, there is a clear separation between argument construction and argument evaluation. ASPIC$^+$ specifies how to construct arguments, and how to recognize conflicts (attacks and defeats) between them. This results in an argumentation framework of arguments and defeats which is then processed in exactly the same way as a Dung abstract argumentation framework [13]. Thus ASPIC$^+$ handles the above example by constructing A_1, A_2 and A_3, recognizing the conflict between A_1 and A_3, and, with its restricted rebut, only recognizing the attack of A_1 on A_3. This will result in A_1 and its conclusion being justified, whereas A_3 and its conclusion are not (under any semantics).

In contrast, the consistency checking aspect of DELP can be seen as a combination of argument construction and evaluation. Since, like ASPIC$^+$, DELP privileges strict information in the sense that it cannot be overturned by arguments with a defeasible component, DELP refuses to allow arguments that (would) challenge this information to be brought into existence. In terms of the fate of A_3, this leaves ASPIC$^+$ and DELP drawing the same conclusion — A_3 is not justified (in particular, in DELP, because it would not even exist).

However, even though not included in Definition 4, this does not mean that ASPIC$^+$ ignores any notion of consistency *within* arguments. In [17] the authors introduce a notion of *c-consistency* that accounts only for contradictories (and not contraries); hence, it refers to "contradictory-consistency".

Definition 23 (c-consistency). *A set $S \subseteq \mathcal{L}$ is c-consistent if $\nexists \phi \in \mathcal{L}$ such that $S \models \phi, \neg \phi$. Otherwise S is c-inconsistent.*

Given the above definition, if $S \models \phi, \psi$, where $\psi \in \overline{\phi}$ but $\phi \notin \overline{\psi}$, then S can still be c-consistent. Then, the authors of [17] characterize a special class of ASPIC$^+$ arguments, whose premises are *c-consistent*.

Definition 24. *An argument A built from an argumentation theory AT on the basis of an argumentation system $\langle \mathcal{L}, ^-, \mathcal{R}, n \rangle$ and a knowledge base $\mathcal{K}$ is c-consistent iff $\mathtt{Prem}(A)$ is c-consistent.*

Note that a c-consistent argument is one with c-consistent premises. Thus c-consistency prevents the construction of arguments where the foundations, the premises, disagree amongst themselves by including a proposition and its contrary. In addition, because the premises are c-consistent, it is not possible to construct strict arguments from those premises such that the conclusions of those arguments contradict one another. Note that, an argument A with premise a and conclusion $\neg a$ may still be c-consistent, as long as there are no strict rules leading to derive contradictory conclusions, starting from the premise a. Indeed, c-consistency does not exclude the construction of an argument that contradicts something in the set of axioms (the strict part of the knowledge base). Thus, it is perfectly reasonable in ASPIC$^+$ to have a in the set of axioms and also have a strict argument for $\neg a$. Such an argument would not be permitted in DeLP, because the conclusion conflicts with the strict premise a. As an example of this, take Example 8 and make the defeasible rule strict. In such a case, argument $A'_3 = [A_2 \to \neg a]$ would still be c-consistent since it is not possible to obtain strict arguments with contradictory conclusions such that their premises are taken from $\{b\}$; specifically, even though the set $\{b\}$ makes it possible to obtain the strict argument A'_3, it is not possible to build A_1 (the strict argument whose conclusion contradicts $\text{Conc}(A'_3)$) such that its premises are taken from the set $\{b\}$. Then, because of the definition of attack in ASPIC$^+$, since the conclusion of A_1 is an axiom and $\text{TopRule}(A'_3)$ is strict, neither of the two arguments will attack the other, and so both a and $\neg a$ would be in the set of justified conclusions. This is exactly why such a theory is not *well-defined* [17], which is a requirement for the theory to obey the rationality postulates [7] (which require justified conclusions to be consistent).

Finally, we can conclude that the notion of c-consistency in ASPIC$^+$ is somehow complementary to the consistency check made in DeLP's argument construction process. On the one hand, DeLP avoids building an argument whose set of (defeasible) rules, together with the strict knowledge of a de.l.p., leads to the derivation of complementary literals. On the other hand, in a setting where only c-consistent arguments are allowed, ASPIC$^+$ prevents the construction of an argument whose set of premises, together with other strict knowledge in the theory, leads to the construction of strict arguments with contradictory conclusions.

To summarize, both DeLP and ASPIC$^+$ perform some kind of consistency check on the arguments against the strict knowledge of the program/argumentation theory: the former by considering the defeasible rules of an argument, and the latter by considering the premises of an argument. In both cases, the check is not of the consistency of the argument itself, but of the consistency of things that can be derived from it.

4.3 Attacks and Defeats

In this section we will start by contrasting the ways in which DeLP and ASPIC$^+$ account for the existence of conflicts between arguments. Then, we will turn to analyze the mechanism in which they resolve those conflicts, leading to the existence of defeats.

As we briefly mentioned in Section 3, there are some differences in the characterization of attacks in DeLP and in ASPIC$^+$. The main difference relies on the fact that ASPIC$^+$ distinguishes between undermining, rebutting and undercutting attacks. In contrast, DeLP defines a general notion of attack, which accounts for all the possible situations in which two arguments are considered to be conflicting. This difference becomes evident when looking at the attack relation in Examples 1 and 2; moreover, this can be easily observed when contrasting Figures 1(a) and 2(a). In the following, we will discuss the three kinds of attack from ASPIC$^+$, and study ways in which they can be realized in DeLP.

Recall that undermining attacks in ASPIC$^+$ are aimed at attacking the ordinary premises of an argument. On the other hand, even though undermining attacks are not explicitly accounted as such, they are contemplated within the DeLP's general notion of attack . As discussed in Section 4.1, ordinary premises in DeLP are represented in the form of presumptions. Then, when considering Definition 21, if the attacked sub-argument $\langle A, h \rangle$ is such that its conclusion corresponds to a presumption "$h \prec\,$", then the attack would be an undermining attack. To illustrate this, let us consider again Examples 1 and 2. In particular, the undermining attacks from C to B_1', B_2' and B, and the undermining attack from B_1' to C in ASPIC$^+$ still occur in DeLP.

Let us now consider rebutting attacks. To start with, even though DeLP does not explicitly distinguish between different kinds of attack, the general notion proposed in Definition 21 accounts for the situation in which rebutting attacks occur; that is, where the conclusion of the attacking argument is in conflict with the conclusion of a sub-argument of the attacked argument. However, there are several differences between the consideration of rebutting attacks in Definitions 5 and 21. On the one hand, rebutting attacks in ASPIC$^+$ are *restricted* in the sense that an argument cannot be attacked at the conclusion of a sub-argument whose TopRule is strict. In contrast, DeLP allows arguments to be attacked on the conclusions of strict rules, as long as the attacked sub-arguments make use of defeasible knowledge as well. In other words, once defeasibility is introduced within an argument, DeLP allows an attack at any literal whose derivation goes beyond the consideration of strict knowledge. This difference is evidenced in Examples 1 and 2, where ASPIC$^+$ only accounts for an attack from argument B to argument A_2, whereas in DeLP the arguments $\langle A_2, b \rangle$ and $\langle B, \neg b \rangle$ attack each other.

Note that restricted rebut was introduced [7] to ensure that the rationality postulate of closure (under strict rules) holds and one way to view what it does is to prioritize the conclusions obtained through the use of strict rules. (In the context of closure, if you prioritize the conclusions of strict rules, then inferences drawn from justified conclusions using strict rules will also be justified conclusions, and closure holds.) The idea of prioritization of strict rules is somehow accounted for in DeLP by imposing restrictions on the construction of arguments. Specifically, no argument in DeLP can be such that, when considered together with the strict knowledge of a de.l.p. allows to derive complementary literals. Then, we could say that DeLP also prioritizes the consideration of strict over defeasible knowledge. Nevertheless, we should remark that the notion of attack in DeLP can be easily modified in order to restrict rebut, similarly to ASPIC$^+$.

Another difference between the characterization of rebutting attack in ASPIC$^+$ and the general notion of attack in DeLP regards the nature of the conflict between the conclusions of the attacking argument and the attacked sub-argument. In DeLP, the existence of an attack depends on the condition that the conclusions of these two arguments disagree (i.e. when considering the two literals together with the strict knowledge of the corresponding DeLP program, complementary literals can be derived). On the other hand, ASPIC+ identifies the existence of a conflict if the conclusion of the attacking argument is a contrary of the conclusion of the attacked sub-argument. Then, we can clearly identify two differences. First, ASPIC+'s notion of conflict is more general, in the sense that it allows for non-symmetric attacks. That is, it can be the case that there exists an argument A that rebuts another argument B on B', but neither B nor B' rebut A, even in the case where TopRule(A) is defeasible. In contrast, since attacks in DeLP rely on the notion of disagreement, which is inherently symmetric (i.e. if literal L_1 disagrees with literal L_2, then L_2 also disagrees with L_1), if argument $\langle A_1, L_1 \rangle$ attacks argument $\langle A_2, L_2 \rangle$ on the sub-argument $\langle A, h \rangle$, then it holds that $\langle A, h \rangle$ also attacks $\langle A_1, L_1 \rangle$. Second, in ASPIC+, the conflict between the conclusion of the attacking argument and the attacked sub-argument is always direct, as the former is a contrary of the latter. However, in DeLP, in cases where the conclusions of the attacking and the attacked sub-argument are not complementary, the conflict between them would not be direct. As a result, there might be arguments considered to be conflicting in ASPIC+ (leading to the existence of a rebutting attack) which are not accounted as such in DeLP and, conversely, there might be arguments considered to be in conflict in DeLP (leading to the existence of an attack) but not in ASPIC+. To illustrate this, let us consider the following example:

Example 9. Consider the argumentation system $AS_9 = \langle \mathcal{L}_9, \bar{}, \mathcal{R}_9, n \rangle$, where $\mathcal{L}_9 = \{a, b, c, d, e, f, \neg a, \neg b, \neg c, \neg d, \neg e, \neg f\}$, the contrariness function is such that $a \in \overline{f}$, and $\mathcal{R}_9 = \{a \to c; b \to \neg c; d \Rightarrow a; e \Rightarrow b; e \Rightarrow f\}$. Then, if we add the knowledge base $\mathcal{K}_9 = \mathcal{K}_{n_9} = \{d, e\}$, we get the argumentation theory $AT_9 = \langle AS_9, \mathcal{K}_9 \rangle$. From this theory, we can build the following arguments:

$$D = [d]; \quad E = [e]; \quad A = [D \Rightarrow a]; \quad B = [E \Rightarrow b];$$
$$F = [E \Rightarrow f] \quad C = [A \to c]; \quad C' = [B \to \neg c]$$

Note that the only attack that arises from the consideration of these arguments is the rebutting attack from A to F (because $a \in \overline{f}$). In contrast, even though C and C' have contradictory conclusions, both arguments have a strict TopRule, so neither of them rebuts the other.

If we want to represent the knowledge within the argumentation theory AT_9 in DeLP, because the strong negation "¬" is symmetric, we will not be able to model that a is a contrary of f. So, we have two alternatives: ignore the conflict between a and f, or represent the conflict through the use of a rule like $\neg f \prec a$, which leads the literals a and f to disagree (thus, the conflict between them to become symmetric). As the second alternative loses the intuition behind the notion of *contrary* in ASPIC+, we will adopt the first one and define the de.l.p. $\mathcal{P}_9 = (\Pi_9, \Delta_9)$:

$$\Pi_9 = \{d, e, (c \leftarrow a), (\neg c \leftarrow b)\} \quad \Delta_9 = \{(a \prec d), (b \prec e), (f \prec e)\}$$

From this DELP program, we can build the following arguments: $\langle D, d \rangle$ and $\langle E, e \rangle$, with $D = E = \emptyset$; $\langle A, a \rangle$ and $\langle A, c \rangle$, with $A = \{a \prec d\}$; $\langle B, b \rangle$ and $\langle B, \neg c \rangle$, with $B = \{b \prec e\}$; and $\langle F, f \rangle$, with $F = \{f \prec e\}$. Note that arguments $\langle A, c \rangle$ and $\langle B, \neg c \rangle$ would correspond to arguments C and C' in ASPIC$^+$; in particular, the sets of defeasible rules of the two DELP arguments coincide with the sets of defeasible rules of the two corresponding ASPIC$^+$ arguments. However, unlike the two ASPIC$^+$ arguments, the two DELP arguments will attack each other. On the other hand, since the conflict between a and f in ASPIC$^+$ is not captured within $\mathcal{P}_9$, argument $\langle A, a \rangle$ will not attack argument $\langle F, f \rangle$.

Regarding the nature of conflicts in ASPIC$^+$ and in DELP, more specifically, the existence of contraries, it is worth to note the following. As mentioned in [17], the notion of contrary is somehow associated with the notion of *negation as failure*. Then, that a is a contrary of f (and not a contradictory) can be interpreted as $f = not(a)$, where "not" represents negation as failure. On the other hand, in [14] the authors discuss an extension of DELP that accounts for this kind of negation (i.e. negation as failure). As a result, ASPIC$^+$'s contraries could be represented in the extended version of DELP by making use of negation as failure. In particular, it would be possible to represent that $a \in \overline{f}$ in $\mathcal{P}_9$: the literal "f" would have to be replaced with "$not(a)$" and the defeasible rule "$f \prec e$" would be replaced with "$not(a) \prec e$". Moreover, in such a case, argument $\langle F', not(a) \rangle$, with $F' = \{not(a) \prec e\}$ would be attacked by argument $\langle A, a \rangle$.

Let us now focus on undercutting attacks. As introduced in Section 2, ASPIC$^+$ includes a naming function for defeasible rules within the characterization of an argumentation system. Then, by having the names of defeasible rules in the logical language $\mathcal{L}$, it is possible to have contraries and contradictories for them, leading to the existence of undercutting attacks. In contrast, the formalization of DELP given in [14] does not account for the existence of undercutting attacks. Notwithstanding this, there exists an extension of DELP that incorporates undercut as a type of attack between arguments [10, 11]. In [11] the set of atoms in a program includes a set of *labels*, and each defeasible rule has an associated label with the restriction that no pair of defeasible rules within a program shares the same label. Then, by allowing labels (and their negations with respect to "$\neg$") to appear in the head of other defeasible rules, it is possible to express reasons for and against the use of the corresponding defeasible rules. As a result, an argument whose conclusion is "$\neg l$", with "l" being the label of a defeasible rule R, would undercut every other argument including the rule R. A different approach is taken in [9]. This incorporates backing and undercutting rules as meta-rules, allowing to express reasons for and against the use of defeasible rules, respectively. Then, undercutting rules are used for building undercutting arguments, which lead to the existence of undercutting attacks.

Finally, having compared the types of attack accounted for (either explicitly or implicitly) in DeLP and ASPIC$^+$, let us turn our attention to the way in which the they determine the success of attacks, leading to the existence of defeats. Since it was shown that DeLP does not consider undercutting attacks, we will leave those out of the discussion. When contrasting Definitions 8 and 22, we see that both ASPIC$^+$ and DeLP make use of a preference ordering or a comparison criterion between arguments. Furthermore, even though DeLP distinguishes between two kinds of defeat (namely, proper and blocking defeat) whereas ASPIC$^+$ does not, they both consider the existence of a defeat if and only if the attacked sub-argument is not better than the attacking argument (under the adopted preference ordering or comparison criterion). As a result, we can conclude that DeLP and ASPIC$^+$ handle the resolution of undermining and rebutting attacks into defeats equivalently.

4.4 Acceptance of Arguments and Justification of Conclusions

This section will focus on contrasting the ways in which DeLP and ASPIC$^+$ select the accepted arguments and the justified conclusions, thus determining the inferences of the system.

As briefly introduced in sections 2 and 3, ASPIC$^+$ and DeLP adopt different approaches for this purpose. ASPIC$^+$ first constructs a Dung-like argumentation framework [13] consisting of the arguments and defeats obtained from a given argumentation theory. Then, by applying any of the existing semantics for Dung's abstract argumentation frameworks (see [2]) ASPIC$^+$ identifies the *extensions* of the framework, which correspond to collectively acceptable sets of arguments. The justified conclusions of the original theory can then be identified. In contrast, DeLP defines its own semantics based on a dialectical process that involves the construction and marking of dialectical trees. As a result, the accepted arguments built from a de.l.p. will be those marked as "undefeated" in their dialectical trees. Again, the conclusions can then be established from the arguments that are computed to be accepted.

One major difference that we identify between the two approaches is that ASPIC$^+$'s approach is oriented at determining the acceptance/justification status of *every* argument, and hence conclusion, in the argumentation theory. In contrast, as expressed in [14], DeLP is conceived as a query-answering tool, and its dialectical process is aimed at determining the warrant status of a queried literal "l"; hence, it only requires to consider (and analyze the acceptance status of) the arguments belonging to the dialectical tree rooted in arguments of the form $\langle A, l \rangle$ or $\langle A, \neg l \rangle$. Of course, in DeLP it is also possible to determine the acceptance status of every argument built from a de.l.p. and, consequently, the warrant status of every literal in that program. Specifically, the accepted arguments will be those that are marked as "undefeated" in their dialectical trees, and the warranted literals will be the accepted arguments' conclusions.

Another difference is that, given a query for a literal "l" in a program, DeLP will unequivocally provide an answer (i.e. the answer will always be the same): *YES* if "l" is warranted from the corresponding DELP program, *NO* if the literal "$\neg l$" is warranted instead, *UNDECIDED* if neither "l" nor "$\neg l$" are warranted, or *UNKNOWN* if "l" is not in the language of the program (i.e. neither "l" nor "$\neg l$" appear in the facts or rules of the de.l.p.). In particular, the answer *UNDECIDED* will correspond to situations in which there are no accepted arguments for the literals "l" and "$\neg l$"; an example of this situation would be the case where there exist two arguments $\langle A, l \rangle$ and $\langle B, \neg l \rangle$ that are blocking defeaters of each other, and none of them has any other defeater. Thus, we can consider DELP's dialectical process to be *cautious*. On the other hand, since ASPIC$^+$ allows the use of any semantics defined for Dung's AFs, it can be the case that the adopted semantics (e.g. preferred semantics) makes it possible to obtain multiple extensions and thus allows for multiple answers to a query about a given literal. For instance, in a situation like the one described above, ASPIC$^+$ would obtain the preferred extensions $\{A'\}$ and $\{B'\}$, where A' and B' would be the ASPIC$^+$ counterpart of the DELP arguments $\langle A, l \rangle$ and $\langle B, \neg l \rangle$. Thus, one can think of the preferred semantics as offering a choice between the alternative of accepting one argument or the other and, consequently, justifying one conclusion or the other. Note that, given the existence of multiple extensions under the adopted semantics, it is possible to make use of the *credulous* or the *skeptical* (or *cautious*) acceptance of arguments: an argument will be skeptically accepted under a semantics iff it belongs to every extension obtained under that semantics, whereas it will be credulously accepted under that semantics iff it belongs to some (but not every) extension. Nevertheless, some semantics that can be used with ASPIC$^+$, like the *grounded* semantics, do not allow for multiple extensions, and thus, they can be considered to be *cautious* as well[9].

To illustrate the way in which ASPIC$^+$ and DELP determine the accepted arguments and the justified conclusions of the system, let us consider the arguments and defeats from Examples 1 and 2.

Example 10. The arguments and defeats identified in Example 1, depicted in Figure 1(b), define the abstract argumentation framework $\langle \{A_1, A_2, A_3, B_1, B_2, B'_1, B'_2, B, C\}, \{(A_3, B), (A_3, B_2), (B, A_2), (B'_1, C)\} \rangle$. For instance, if we consider the grounded semantics, the only grounded extension will be $\{A_1, A_2, A_3, B_1, B'_1, B'_2\}$; in particular, the grounded semantics accepts all arguments having no defeaters (in this case, A_1, A_3, B_1, B'_1 and B'_2), leaves out arguments that are defeated by the undefeated arguments (here, B, B_2 and C), and includes the arguments that are defended by those already included in the extension (A_2). Therefore, the set of ASPIC$^+$ justified conclusions will be $\{a, b, \neg nd, \neg c, e, f\}$.

Let us now consider the arguments and defeats identified in Example 2 for DELP, depicted in Figure 2(b). There, the only arguments having defeaters are $\langle A_2, b \rangle$ (whose defeater is $\langle B, \neg b \rangle$) and $\langle C, \neg e \rangle$ (whose defeater

---

[9] In the labelling approach [2], the grounded labelling is the one that maximizes the UNDEC labels, and so, in a sense, is the most cautious of the possible semantics.

is $\langle B', e \rangle$). Therefore, every argument except those two will be marked as "undefeated" in their dialectical trees, whereas $\langle A_2, b \rangle$ and $\langle B, \neg b \rangle$ will be marked as "defeated". As a result, the set of accepted arguments will be $\{\langle A_1, a \rangle, \langle A_3, \neg nd \rangle, \langle B_1, \neg c \rangle, \langle B_2, d \rangle, \langle B_1', e \rangle, \langle B_2', f \rangle, \langle B, \neg b \rangle\}$, and the set of warranted literals from the de.l.p. program specified in Example 2 will be $\{a, \neg d, c, d, e, f, \neg b\}$.

It can be noted that the accepted arguments and the justified/warranted conclusions in DeLP and ASPIC$^+$ differ. In particular, the difference in this case relies on the consideration of undercutting attacks. Since A_3 undercuts B_2 and B in ASPIC$^+$, argument A_2 is defended by A_3 and, as a result, can be accepted. Furthermore, arguments B and B_2 are not accepted. In contrast, since DeLP does not account for undercutting attacks, arguments $\langle A_3, \neg nd \rangle$, $\langle B_2, d \rangle$ and $\langle B, \neg b \rangle$ will be accepted together. This difference is also observed when considering the sets of justified/warranted conclusions: whereas ASPIC$^+$ justifies "b", DeLP warrants "$\neg b$".

It should be noted that, even though the difference in the results obtained by DeLP and ASPIC$^+$ in the previous example is related to the existence of undercutting attacks/defeats in ASPIC$^+$ (which do not occur in DeLP), it can also be the case that an ASPIC$^+$ theory and a de.l.p. have the same set of arguments and defeats but different justified/warranted conclusions. An example of such a situation would be the one described in the paragraph before Example 10, where DeLP would have no accepted arguments and no warranted conclusions (literals), whereas ASPIC$^+$ (under the preferred semantics) will consider the arguments A' and B' to be credulously accepted, and their conclusions to be credulously justified.

Finally, notwithstanding the above discussed differences in the acceptance or justification process adopted by DeLP and ASPIC$^+$, as well as the difference in their consideration of attacks and defeats, there are cases in which the two systems behave alike and their outcomes coincide. This is illustrated by the following example:

Example 11. Consider the argumentation system $AS_{11} = \langle \mathcal{L}_{11}, \bar{\ }, \mathcal{R}_{11}, n \rangle$, where $\mathcal{L}_{11} = \{a, b, c, d, f, g, h, k, \neg a, \neg b, \neg c, \neg d, \neg f, \neg g, \neg h, \neg k\}$ and $\mathcal{R}_{11} = \{b, d \rightarrow a; h \rightarrow f; h, c \rightarrow \neg k; g \rightarrow d; \neg f \Rightarrow b; c, k \Rightarrow \neg b; g \Rightarrow k; d \Rightarrow h\}$. Then, if we add the knowledge base $\mathcal{K}_{11} = \mathcal{K}_{n_{11}} \cup \mathcal{K}_{p_{11}}$, with $\mathcal{K}_{n_{11}} = \{g, c\}$ and $\mathcal{K}_{p_{11}} = \{\neg f\}$, we get the argumentation theory $AT_{11} = \langle AS_{11}, \mathcal{K}_{11} \rangle$. From this theory, we can build the following arguments:

$F' = [\neg f]$; $\quad B = [F' \Rightarrow b]$; $\quad G = [g]$; $\quad D = [G \rightarrow d]$;
$A = [B, D \rightarrow a]$; $\quad C = [c]$; $\quad K = [G \Rightarrow k]$; $\quad B' = [C, K \Rightarrow \neg b]$;
$H = [D \Rightarrow h]$; $\quad K' = [H, C \rightarrow \neg k]$; $\quad F = [H \rightarrow f]$

The DeLP counterpart of the argumentation theory AT_{11} would be the de.l.p. $\mathcal{P}_{11} = (\Pi_{11}, \Delta_{11})$:

$$\Pi_{11} = \begin{cases} g \\ c \\ \end{cases} \begin{matrix} a \leftarrow b,d \\ f \leftarrow h \\ \neg k \leftarrow h,c \\ d \leftarrow g \end{matrix} \Bigg\} \quad \Delta_{11} = \begin{cases} b \prec \neg f \\ \neg b \prec c, k \\ k \prec g \\ h \prec d \\ \neg f \prec \end{cases}$$

from which we can build the following arguments:

$\langle F', \neg f \rangle$, with $F' = \{\neg f \prec \}$;
$\langle B, b \rangle$, with $B = \{(b \prec \neg f), (\neg f \prec)\}$;
$\langle G, g \rangle$, with $G = \emptyset$;
$\langle D, d \rangle$, with $D = \emptyset$;
$\langle A, a \rangle$, with $A = \{(b \prec \neg f), (\neg f \prec)\}$;
$\langle C, c \rangle$, with $C = \emptyset$;
$\langle K, k \rangle$, with $K = \{k \prec g\}$;
$\langle B', \neg b \rangle$, with $B' = \{(\neg b \prec c, k), (k \prec g)\}$;
$\langle H, h \rangle$, with $H = \{h \prec d\}$;
$\langle K', \neg k \rangle$, with $K' = \{h \prec d\}$;
$\langle F, f \rangle$, with $F = \{h \prec d\}$

Recall that, unlike arguments in ASPIC$^+$, arguments in DeLP only include the defeasible component of the argument (i.e. defeasible rules, including presumptions). As a result, there exist different arguments, with different conclusions, that have the same associated set of rules. Nevertheless, without loss of generality, we can identify each of the argument structures listed above through the name of its associated set of rules and presumptions; hence, we can refer to them as F', B, G, ..., H, K', F, respectively.

The attacks between the arguments obtained from the argumentation theory AT_{11} are depicted in Figure 3(a), whereas the attacks between the arguments built from the de.l.p. $\mathcal{P}_{11}$ are illustrated in Figure 3(b). Note that the difference between the two relies on the fact that ASPIC$^+$ considers restricted rebut, and so some rebutting attacks that are symmetrical in DeLP are not symmetrical in ASPIC$^+$.

Suppose now that we consider a preference criterion or ordering on arguments such that: $B \prec B'$, $K \prec K'$ and $F' \prec F$. Then, as shown in Figure 4, the defeats between arguments built from the argumentation theory AT_{11} and the de.l.p. $\mathcal{P}_{11}$ coincide. Furthermore, if we consider the grounded, preferred or stable semantics (in the case of ASPIC$^+$) and DeLP's dialectical process, we obtain the same outcome: the set of accepted arguments is $\{D, G, C, H, F, K', B\}$ and the set of warranted (DeLP) and justified (ASPIC$^+$) conclusions is $\{d, g, c, h, f, \neg k, b\}$.

5 Conclusion

This chapter has examined the relationship between DeLP and ASPIC$^+$. While the analysis has been largely informal, we hope that it is still clear

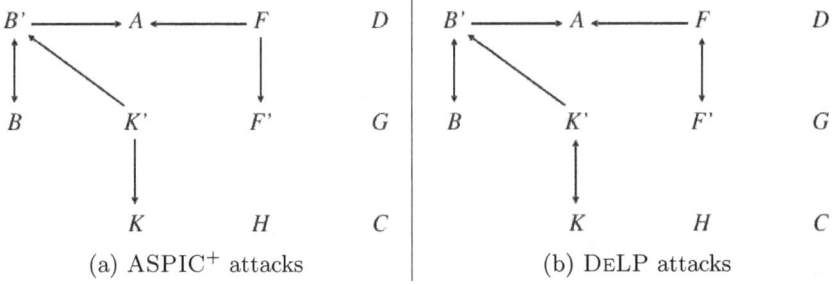

Fig. 3. Attacks between arguments from Example 11 in (a) ASPIC$^+$ and (b) DeLP.

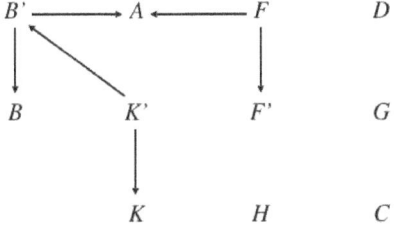

Fig. 4. Defeats between arguments built from AT_{11} and $\mathcal{P}_{11}$.

that the two approaches are very similar in many regards. Indeed, they are perhaps more similar than they are dissimilar. In our view there is certainly enough similarity to justify a more formal analysis that looks to find out, precisely where the approaches overlap, and where they differ, especially in terms of what conclusions they draw from a given knowledge base. This is work we hope to carry out in the near future.

Acknowledgment This research was partially supported by the UK Engineering & Physical Sciences Research Council (EPSRC) under grant #EP/P010105/1. In addition, both the authors acknowledge a considerable debt to Guillermo Simari.

SP: As someone who hasn't worked directly with Guillermo, I am very grateful for the fact that he introduced me to two people who have become valued collaborators: Andrea Cohen, my co-author here, and Gerardo Simari, with whom I have worked on a number of topics over the years. I particularly value the fact that in both cases Guillermo presented the opportunity for me to work with Andrea and Gerardo as an instance of me doing him a favour, when in fact it was him who did me the favour. However, I think that my greatest debt to Guillermo is in the example that he sets. I have always found him to be among the most thoughtful and gracious people that I have met in my research career. In that respect I think he embodies qualities that I strive (but regularly fail) to achieve myself, and I thank him for the ongoing example.

AC: I will always be grateful to Guillermo for being a wonderful teacher, supervisor and mentor. Every since I started my research career he was there for me, giving me advice and helping me (and everyone in our group) in every step of the way. Also, I am grateful to him for always pushing me to achieve great things, giving me the opportunity to live wonderful experiences such as my research stay in Brooklyn College-CUNY, where I had the pleasure to meet and start working with Simon Parsons. Last but not least, I will always cherish the thoughtful and kind words he had towards me, especially in moments when I felt I was not cut out for this. For all this, I feel lucky to have been able to grow and work with him. Thank you, Guillermo, for being our role-model.

References

1. L. Amgoud and C. Cayrol. A reasoning model based on the production of acceptable arguments. *Annals of Mathematics and Artifical Intelligence*, 34(3):197–215, 2002.
2. Pietro Baroni, Martin Caminada, and Massimiliano Giacomin. An introduction to argumentation semantics. *Knowledge Eng. Review*, 26(4):365–410, 2011.
3. P. Besnard and A. Hunter. A logic-based theory of deductive arguments. *Artificial Intelligence*, 128:203–235, 2001.
4. M. Caminada, S. Modgil, and N. Oren. Preferences and unrestricted rebut. *Computational Models of Argument: Proceedings of COMMA 2014*, pages 209–220, 2014.
5. M. W. A. Caminada. On the issue of reinstatement in argumentation. In *Proceedings of the 10th European Conference on Logic in Artificial Intelligence*, pages 111–123, Liverpool, UK, 2006.
6. M. W. A. Caminada. An algorithm for computing semi-stable semantics. In *Proceedings of the 9th European Conference on Symbolic and Quantitative Approaches to Reasoning with Uncertainty*, pages 222–234, Verona, Italy, 2007.
7. M. W. A. Caminada and L. Amgoud. On the evaluation of argumentation formalisms. *Artificial Intelligence*, 171(5–6):286–310, 2007.
8. C. I. Chesñevar, A. G. Maguitman, and R. P. Loui. Logical models of argument. *ACM Computing Surveys*, 32(4):337–383, 2000.
9. Andrea Cohen, Alejandro J. García, and Guillermo R. Simari. Backing and undercutting in defeasible logic programming. In *11th European Conference on Symbolic and Quantitative Approaches to Reasoning with Uncertainty*, pages 50–61, 2011.
10. Andrea Cohen, Alejandro J. García, and Guillermo R. Simari. Backing and undercutting in abstract argumentation frameworks. In *7th International Symposium on Foundations of Information and Knowledge Systems*, pages 107–123, 2012.
11. Andrea Cohen, Alejandro Javier García, and Guillermo Ricardo Simari. Extending delp with attack and support for defeasible rules. In *IBERAMIA*, volume 6433 of *Lecture Notes in Computer Science*, pages 90–99. Springer, 2010.
12. P. M. Dung. On the acceptability of arguments and its fundamental role in nonmonotonic reasoning and logic programming. In *Proceedings of the 13th International Joint Conference on Artificial Intelligence*, Chambéry, France, 1993.

13. P. M. Dung. On the acceptability of arguments and its fundamental role in nonmonotonic reasoning, logic programming and n-person games. *Artificial Intelligence*, 77:321–357, 1995.
14. A. J. García and G. Simari. Defeasible logic programming: an argumentative approach. *Theory and Practice of Logic Programming*, 4(1):95–138, 2004.
15. Z. Li, A. Cohen, and S. Parsons. Two forms of minimality in ASPIC$^+$. In *15th European Conference on Multi-Agent System*, Évry, France, 2017.
16. Zimi Li and Simon Parsons. On argumentation with purely defeasible rules. In *Scalable Uncertainty Management - 9th International Conference, SUM 2015, Québec City, QC, Canada, September 16-18, 2015. Proceedings*, pages 330–343, 2015.
17. S. Modgil and H. Prakken. A general account of argumentation with preferences. *Artificial Intelligence*, 195:361–397, 2013.
18. Donald Nute. Defeasible reasoning: a philosophical analysis in PROLOG. In J. H. Fetzer, editor, *Aspects of Artificial Intelligence*, pages 251–288. Kluwer Academic Pub., 1988.
19. J. L. Pollock. Defeasible reasoning. *Cognitive Science*, 11:481–518, 1987.
20. H. Prakken. An abstract framework for argumentation with structured arguments. *Argument and Computation*, 1:93–124, 2010.
21. G. R. Simari. *A Mathematical Treatment of Defeasible Reasoning and its Implementation*. PhD thesis, Department of Computer Science, Washington University in St Louis, 1989.
22. G. R. Simari and R. P. Loui. A mathematical treatment of defeasible reasoning and its implementation. *Artificial Intelligence*, 53:125–157, 1992.
23. B. Verheij. A labeling approach to the computation of credulous acceptance in argumentation. In *Proceedings of the 20th International Joint Conference on Aritificial Intelligence*, pages 623–628, Hyderabad, India, 2007.
24. G. Vreeswijk. An algorithm to compute minimally grounded and admissible defence sets in argument systems. In *Proceedings of the First International Conference on Computational Models of Argument*, pages 109–120, Liverpool, UK, 2006.

Characterizing Abduction with Implicit Background Theory*

María Victoria León and Ramón Pino Pérez

School of Mathematical Sciences and Information Techonology
Yachay Tech University, Urcuquí, Ecuador

{mleon,rpino}@yachaytech.edu.ec

Abstract. We study explanatory relations called also abductive relations. We characterize two families in terms of their axiomatic behavior: the weakly reflexive explanatory relations and the ordered explanatory relations. Both families present tight relationships with the framework of Credibility limited revision. These relationships allow to establish semantical representations for each family. A very interesting consequence of our representations results is that our axiomatizations allow us to overcome the background theory present in most axiomatizations of abduction.

Introduction

We study binary relations $\triangleright$ over propositional formulas built over a finite set of variables which try to capture the abductive reasoning. The expression $\alpha \triangleright \gamma$ will mean that the observation α is well explained by the explanation γ.

In the pioneer logical work of Levesque [14] there is a a background theory Σ such that when we have $\alpha \triangleright \gamma$ then the explanation γ has to entail together with Σ the observation α. Actually, the good explanations, in that framework, are in some sense some preferred formulas of the set

$$\{\gamma : \gamma \text{ consistent with } \Sigma \text{ and } \Sigma \cup \{\gamma\} \vdash \alpha\}.$$

This is the view adopted also in [16].

In this work we don't make use of an explicit background theory Σ. We take an axiomatic presentation of our explanatory relations. Following the works [16, 18] we make a systematic study of two families of explanatory relations. In particular, we show that our families are very deeply related with the credibility limited revision operators introduced by Hansson et al. [11] These operators are a generalization of the AGM revision operators introduced by Alchourron et al. in [2].

---

* This work is an extended version of a previous work by the authors presented in IBERAMIA 2016 [13].

Different families of explanatory relations and their properties have been studied in [9, 10, 16, 17, 3]. Our main goal is to continue to have a better understanding of these abstract explanatory relations (behavior, axioms, constructions) and their links with belief revision. There are works relating the explanatory reasoning with belief revision [5, 15, 16, 7, 18, 8], but to our knowledge this is the first time that some explicit relationships between abduction and credibility limited revision operators are established. However, we have to point out that in the work of Fermé and Rodríguez [8], where they establish deep relationships between belief revision and the defeasible model (DFT) of Alchourrón [1], some typical postulates of credibility limited revision appear.

In the logic of abduction, some postulates -for instance (Right strengthening) (see Section 3) or the (Reflexivity)- are not well viewed. Using (Right strengthening) we could have that if Good_coffee $\triangleright$ colombian_coffee then Good_coffee $\triangleright$ colombian_coffee_with_pepper which is not very convincing. In this paper we will see a family of explanatory relations which doesn't satisfy (Right strengthening).

Concerning the reflexivity of the explanatory relations, one can consider the relations satisfying this property like proto-explanatory relations. And then using them for constructing explanatory relations where (Reflexivity) doesn't hold. A way to do that is considering the proto-explanatory relation and then to withdraw the diagonal (*i.e.* the pairs $\alpha \triangleright \alpha$).

A natural feature to consider is the fact of having *impossible observations*, that is, observations which don't have any explanation. Imagine an observation like A_pink_elephant_driving_a_Fiat_500. One is not leaning to belief that such observation has an explanation. Considering impossible observations has interesting consequences like we will see in Section 3.

This work is organized as follows. Section 1 is devoted to fix some notation and some results about credibility limited revision operators. Section 2 is devoted to give a very abstract view of explanatory relations. Section 3 is devoted to give our two families of explanatory relations the links with credibility limited revision operators and the main representation theorems. We finish with Section 4 in which we make some final remarks and point out some future work.

1 Preliminaries

Let $\mathcal{L}$ be the set of propositional formulas built over a finite set $\mathcal{P}$ of atomic propositions. A consistent formula is a formula which does not entail the formula $\alpha \wedge \neg \alpha$ (a contradiction). $\mathcal{L}^*$ will denote the set of consistent formulas. An interpretation ω is a total function from $\mathcal{P}$ to $\{0, 1\}$. The set of all interpretations is denoted $\mathcal{W}$. An interpretation ω is a model of a formula $\varphi \in \mathcal{L}$ if and only if it makes it true in the usual truth functional way. If φ is a formula, we denote by $[\![\varphi]\!]$ the set of models of φ, *i.e.* $[\![\varphi]\!] = \{\omega \in \mathcal{W} : \omega \models \varphi\}$. As usual, $\top$ and $\bot$ denote two fixed propositional formulas such that $[\![\top]\!] = \mathcal{W}$ and $[\![\bot]\!] = \emptyset$. If I is a nonempty set of interpretations, we denote by α_I a formula such that $[\![\alpha_I]\!] = I$. We write α_ω instead of $\alpha_{\{\omega\}}$. When A is a finite set of formulas we denote by $\bigvee$ the disjunction of all the formulas of A. We denote by $\vdash$ the classical entailment relation, that is $\alpha \vdash \beta$ when α entails β (this happens exactly when $[\![\alpha]\!] \subseteq [\![\beta]\!]$). The logical equivalence will be denoted by $\equiv$, that is $\alpha \equiv \beta$ when $\alpha \vdash \beta$ and $\beta \vdash \alpha$.

Let A be a set. A binary relation $\preceq$ over A is a total preorder if it is total (therefore reflexive) and transitive. Let $\preceq$ be a total preorder over A. We define the strict relation $\prec$ and the indifference $\simeq$ associated to $\preceq$ as follows: $a \prec b$ if and only if $a \preceq b$ and $b \not\preceq a$; $a \simeq b$ if and only if $a \preceq b$ and $b \preceq a$.

Let $\preceq$ be a total preorder over A. Let C be a subset of A. We say that c is a minimal element of C with respect to $\preceq$ if $c \in C$ and for all $x \in C$, $x \not\prec c$. The set of minimal elements of C will be denoted $min(C, \preceq)$. The minimal elements of the whole set A in which the total preorder is defined will be denoted $min(\preceq)$.

Let $\triangleright$ be a binary relation over $\mathcal{L}$. This relation is supposed to represent a well behaved explanatory relation. We use infix notation $\alpha \triangleright \gamma$ to express that the pair (α, γ) is in the relation $\triangleright$. The set $\{\gamma : \alpha \triangleright \gamma\}$ is called the set of explanations of α, and it is denoted $Expl(\alpha)$. The set Expl is the set of all explanations, that is $\mathsf{Expl} = \{\gamma : \exists \alpha \text{ such that } \alpha \triangleright \gamma\}$. An explanation having one unique model is called a complete explanation.

1.1 Credibility-limited revision

We give a recall of propositional credibility-limited revision. The first version of credibility-limited revision was proposed by Hansson et al. in [11]. This is a generalization of revision operators proposed by Alchourrón et al. [2]. The version we proposed here, is a more compact version which has been presented by Booth et al. in [4] in the Katsuno-Mendelzon [12] style.

Definition 1 *An operator* $\circ : \mathcal{L}^* \times \mathcal{L} \longrightarrow \mathcal{L}$ *is called a credibility-limited revision operator (CL revision operator for short) if the following postulates*[1] *hold for any consistent formula φ and every formulas α and β:*

$\varphi \circ \alpha \equiv \varphi$ *or* $\varphi \circ \alpha \vdash \alpha$ (**Relative success**)

If $\alpha \wedge \varphi \not\vdash \bot$ *then* $\varphi \circ \alpha \equiv \varphi \wedge \alpha$ (**Vacuity**)

$\varphi \circ \alpha \not\vdash \bot$ (**Strong coherence**)

If $\varphi \equiv \psi$ *and* $\alpha \equiv \beta$ *then* $\varphi \circ \alpha \equiv \psi \circ \beta$ (**Syntax independence**)

If $\varphi \circ \alpha \vdash \alpha$ *and* $\alpha \vdash \beta$ *then* $\varphi \circ \beta \vdash \beta$ (**Success monotony**)

$\varphi \circ (\alpha \vee \beta) \equiv \begin{cases} \varphi \circ \alpha \text{ or} \\ \varphi \circ \beta \text{ or} \\ (\varphi \circ \alpha) \vee (\varphi \circ \beta) \end{cases}$ (**Trichotomy**)

Definition 2 (Assignment) *A CL-faithful assignment is a function mapping each consistent formula φ into a pair $(C_\varphi, \leq_\varphi)$ where $[\![\varphi]\!] \subseteq C_\varphi \subseteq \mathcal{W}$, $\leq_\varphi$ is a total total preorder over C_φ, and the following conditions hold for all $\omega, \omega' \in C_\varphi$:*

---

[1] As usual we use the infix notation $\alpha \circ \beta$ instead of $\circ(\alpha, \beta)$.

1. If $\omega \models \varphi$, then $\omega \leq_\varphi \omega'$
2. If $\omega \models \varphi$ and $\omega' \not\models \varphi$, then $\omega <_\varphi \omega'$
3. If $\varphi \equiv \varphi'$, then $(C_\varphi, \leq_\varphi) = (C_{\varphi'}, \leq_{\varphi'})$

The interpretations in C_φ will be called the *credible worlds* relative to φ. Next we state the Representation Theorem in [4].

Theorem 1 (Booth et al.). *A mapping $\circ : \mathcal{L}^* \times \mathcal{L} \longrightarrow \mathcal{L}$ is a CL revision operator iff there exists a CL-faithful assignment $\varphi \mapsto (C_\varphi, \leq_\varphi)$ such that*

$$[\![\varphi \circ \alpha]\!] = \begin{cases} \min([\![\alpha]\!], \leq_\varphi) & \text{if } [\![\alpha]\!] \cap C_\varphi \neq \emptyset \\ [\![\varphi]\!] & \text{otherwise} \end{cases}$$

2 An abstract approach for abduction

The main idea is to think that the meaning of $\alpha \triangleright \gamma$ is actually $\pi_1(\gamma) \vdash \pi_2(\alpha)$, where the functions $\pi_i : \mathcal{L} \longrightarrow \mathcal{L}$ are some sort of "core" functions. That is, they are functions which give the more relevant part of their input. In particular, they have to satisfy $\pi_i(\beta) \vdash \beta$. Note that, when π_1 and π_2 are the identity function the relation $\triangleright$ is simply reverse deduction which is not, in general, a very interesting explanatory relation [6].

Actually, we will give two families of explanatory relations such that $\triangleright$ satisfies a set of postulates iff π_1 and π_2 are precisely determined and moreover we have the following equivalence:

$$\alpha \triangleright \gamma \Leftrightarrow \pi_1(\gamma) \vdash \pi_2(\alpha)$$

In order to see how this relations are working, let us give the following illustrative example.

Example 1 *Consider a propositional language with four propositional variables: c, s, p, g meaning Colombian coffee, coffee with sugar, coffee with pepper and good coffee respectively. We are reasoning about good coffee. Thus, the worlds in which the coffee is not good are impossible worlds. In particular, the worlds in which there are pepper in the coffee are incompatible with good coffee. Thus, the only credible worlds are $0101, 0001, 1101, 1001$ (the bits follow the order c, s, p, g). Suppose now that the preferences of these worlds are*

$$0101$$
$$1101 \qquad 0001$$
$$1001$$

that is, the worlds in lowest levels are preferred ($1001 \prec 1101 \sim 0001 \prec 0101$). Suppose that π_1 is the identity and π_2 is "taking the minimal models". Then we would have

- $g \triangleright c \wedge \neg s \wedge \neg p$
- $g \wedge s \triangleright c \wedge s \wedge \neg p$
- $g \wedge s \wedge p$ *has no explanations!*

3 Credibility limited explanatory relations

In this section we define two classes of explanatory relations which will characterized in terms of credibility limited revision operators. Then as a corollary of our characterization we will obtain representation theorems for these two classes of explanatory relations.

3.1 Weakly reflexive explanatory relations

We begin with the definition of our first class of explanatory relations.

Definition 3 *Let $\triangleright$ be a binary relation over $\mathcal{L}$. The relation $\triangleright$ is called a weakly reflexive explanatory relation if the following postulates hold:*

$$Expl(\top) \neq \emptyset \qquad \text{(Strong non triviality)}$$
$$\alpha \triangleright \gamma \Rightarrow \gamma \nvdash \bot \qquad \text{(Coherence)}$$
$$\alpha \triangleright \gamma, \delta \vdash \gamma, \delta \nvdash \bot \Rightarrow \alpha \triangleright \delta \qquad \text{(Right strengthening)}$$
$$\alpha \wedge \beta \triangleright \delta, \exists \gamma(\alpha \triangleright \gamma \text{ and } \gamma \vdash \beta) \Rightarrow \alpha \triangleright \delta \qquad \text{(Weak cut)}$$
$$\alpha \triangleright \gamma \Rightarrow \gamma \vdash \alpha \qquad \text{(Infra-classicality)}$$
$$\alpha \triangleright \gamma, \alpha \triangleright \delta \Rightarrow \alpha \triangleright \gamma \vee \delta \qquad \text{(Right or)}$$
$$\alpha \triangleright \gamma, \gamma \vdash \beta \Rightarrow \alpha \wedge \beta \triangleright \gamma \qquad \text{(Cautious monotony)}$$
$$\alpha \equiv \alpha', \gamma \equiv \gamma' \Rightarrow (\alpha \triangleright \gamma \Leftrightarrow \alpha' \triangleright \gamma') \qquad \text{(Congruence)}$$
$$Expl(\alpha) \neq \emptyset, \alpha \vdash \beta \Rightarrow Expl(\beta) \neq \emptyset \qquad \text{(Explanatory monotony)}$$

The postulates above can be seen as a minimal set of inference rules whose meaning we explain as follows. *Strong non triviality* says (together with *Congruence*) that every tautology has at least one explanation. *Coherence* says that the explanations are consistent. *Right strengthening* says that consistent formulas stronger than one explanation are also explanations. *Weak cut* is a cut rule for the observations under some conditions. *Infra-classicality* is the way to impose that explanations have to be formulas classically entailing the observation. *Right or* says that the disjunction of explanations of one observation is still one explanation of this observation. *Cautious monotony* establish the conditions under which we can strength one original observation and maintain the same explanation of this original observation. *Congruence* says that the explanatory relation is independent of the syntax representation. *Explanatory monotony* says that weaker formulas than formulas having explanations, have also explanations.

The following Theorem can be interpreted as a justification for the choice of the set of postulates appearing in Definition 3.

Theorem 2. *$\triangleright$ is a weakly reflexive explanatory relation iff there exists a consistent formula φ and a credibility-limited revision operator $\circ$ such that*

$$\alpha \triangleright \gamma \Leftrightarrow (\gamma \vdash \varphi \circ \alpha), (\varphi \circ \alpha \vdash \alpha) \text{ and } \gamma \nvdash \bot \tag{1}$$

Definition 4 *We define φ, associated to the relation $\rhd$ as follows:*

$$\varphi \equiv \bigvee \alpha_\omega : \top \rhd \alpha_\omega$$

Observation 1 *Note that if $\rhd$ satisfies (Strong non triviality), (Coherence) and (Right strengthening) the formula φ of the Definition 4 is consistent.*

Proposition 1 *Let $\rhd$ be a weakly reflexive explanatory relation. If $\alpha \rhd \gamma$, then $\gamma \rhd \gamma$.*

Proof. If $\alpha \rhd \gamma$ then, by (Coherence) and (Infra-classicality), we have $\gamma \nvdash \bot$ and $\gamma \vdash \alpha$. By (Cautious mootony) we have $\alpha \wedge \gamma \rhd \gamma$. Since $\gamma \vdash \alpha$ we have $\alpha \wedge \gamma \equiv \gamma$. Then, by (Congruence), $\gamma \rhd \gamma$. ∎

Proof of Theorem 2: First we prove the *only if* part. Assume that $\rhd$ is a weakly reflexive explanatory relation. Define an operator $\circ$ in the following way:

$$\varphi \circ \alpha \equiv \begin{cases} \varphi & \text{si } Expl(\alpha) = \emptyset \\ \bigvee \alpha_\omega : \alpha \rhd \alpha_\omega & \text{si } Expl(\alpha) \neq \emptyset \end{cases}$$

(i) First, we prove the following equivalence:

$$\alpha \rhd \gamma \Leftrightarrow (\gamma \vdash \varphi \circ \alpha), (\varphi \circ \alpha \vdash \alpha) \text{ y } \gamma \nvdash \bot$$

($\Leftarrow$) Suppose that $\gamma \vdash \varphi \circ \alpha$ and $\varphi \circ \alpha \vdash \alpha$. If $Expl(\alpha)$ is nonempty, then, it is easy to see that there are elements of the form α_{ω_i} in $Expl(\alpha)$ and for all such elements, $\gamma \vdash \bigvee \alpha_{\omega_i}$. Then by (Right or), $\alpha \rhd \bigvee \alpha_{\omega_i}$. From this, by (Right strengthening), $\alpha \rhd \gamma$. If α has not explanations, then there are elements α_{ω_i} of $Expl(\top)$ such that $\gamma \vdash \bigvee \alpha_{\omega_i}$ (the disjunction of all such α_{ω_i}). Then, by (Right or) and (Right strengthening), we have $\top \rhd \gamma$. Since $\gamma \vdash \alpha$ and $\top \wedge \alpha \nvdash \bot$, by (Cautious monotony), we have $\top \wedge \alpha \rhd \gamma$ and therefore, by (Congruence), $\alpha \rhd \gamma$.

($\Rightarrow$) If $\alpha \rhd \gamma$, then, by (Coherence), γ is consistent. Thus,

$$[\![\gamma]\!] = \{\omega_1, \ldots, \omega_n\} \quad \text{where} \quad \gamma \equiv \bigvee_{i=1}^n \alpha_{\omega_i}$$

Then $\alpha_{\omega_i} \vdash \gamma$ for all i. By (Right strengthening) we have $\alpha \rhd \alpha_{\omega_i}$. Then $\alpha_{\omega_i} \vdash \varphi \circ \alpha$ for all i. Thus, $\bigvee_{i=1}^n \alpha_{\omega_i} \vdash \varphi \circ \alpha$, and therefore $\gamma \vdash \varphi \circ \alpha$. We have also that $\alpha \rhd \alpha_\omega$ for any α_ω satisfying $\alpha_\omega \vdash \varphi \circ \alpha$. Then, by (Infra-classicality) $\alpha_\omega \vdash \alpha$, and then $\bigvee \alpha_\omega \vdash \alpha$, that is $\varphi \circ \alpha \vdash \alpha$.

(ii) Now we prove that the operator $\circ$ satisfies all the postulates defining a credibility limited revision operator relative to φ:
(Relative success) $\varphi \circ \alpha \vdash \alpha$ or $\varphi \circ \alpha \equiv \varphi$.
If $Expl(\alpha) = \emptyset$, then $\varphi \circ \alpha \equiv \varphi$. If $Expl(\alpha) \neq \emptyset$, there exists γ such that $\alpha \rhd \gamma$ and +for (i), we have $\varphi \circ \alpha \vdash \alpha$.
(Vacuity) $\varphi \wedge \alpha \nvdash \bot \Rightarrow \varphi \circ \alpha \equiv \varphi \wedge \alpha$.

If $\alpha_w \vdash \varphi \wedge \alpha$, then $\top \triangleright \alpha_w$ and $\alpha_w \vdash \alpha$. Since $\alpha \not\vdash \bot$ then $\top \wedge \alpha \not\vdash \bot$, and, by (Cautious monotony) and (Congruence), we have $\alpha \triangleright \alpha_w$. Thus, $\alpha_w \vdash \varphi \circ \alpha$.
Now suppose that $\alpha_w \vdash \varphi \circ \alpha$. Since $Expl(\alpha) \neq \emptyset$, by definition of $\circ$, we have $\alpha \triangleright \alpha_w$, so $\top \wedge \alpha \triangleright \alpha_w$. Since $\varphi \wedge \alpha \not\vdash \bot$, there exists $\alpha_{w'}$ such that $\top \triangleright \alpha_{w'}$ and $\alpha_{w'} \vdash \alpha$. From this, by (Weak cut), we have $\top \triangleright \alpha_w$, then $\alpha_w \vdash \varphi$. From the fact that $\alpha \triangleright \alpha_w$, by (Infraclassicality), $\alpha_w \vdash \alpha$. Therefore $\alpha_w \vdash \varphi \wedge \alpha$.
(Strong coherence) $\varphi \circ \alpha \not\vdash \bot$.
If $Expl(\alpha) = \emptyset$ then $\varphi \circ \alpha \equiv \varphi$. Since $\varphi \not\vdash \bot$ (see *Observation 1*), we have $\varphi \circ \alpha \not\vdash \bot$. If $Expl(\alpha) \neq \emptyset$, there exists γ tal que $\alpha \triangleright \gamma$. By (Coherence) $\gamma \not\vdash \bot$, thus, there exists α_w such that $\alpha_w \vdash \gamma$. Then, by (Right strengthening) $\alpha \triangleright \alpha_w$; in particular, $\alpha_w \vdash \varphi \circ \alpha$. Therefore, $\varphi \circ \alpha$ is consistent.
(Syntax independence) $\alpha \equiv \alpha' \Rightarrow \varphi \circ \alpha \equiv \varphi \circ \alpha'$.
If α has not explanations, then α' has not explanations. Then $\varphi \circ \alpha \equiv \varphi$ and $\varphi \circ \alpha' \equiv \varphi$, therefore $\varphi \circ \alpha \equiv \varphi \circ \alpha'$.
If α has explanations, there exists α_w such that $\alpha_w \vdash \varphi \circ \alpha$ and $\alpha \triangleright \alpha_w$. Then, by (Congruence), $\alpha' \triangleright \alpha_w$. Therefore, $\alpha_w \vdash \varphi \circ \alpha'$. Thus, we have proven $\varphi \circ \alpha \vdash \varphi \circ \alpha'$. The converse entailment can be proven in an analogous way.
(Success monotony) $\varphi \circ \alpha \vdash \alpha$ and $\alpha \vdash \beta \Rightarrow \varphi \circ \beta \vdash \beta$.
Assume $\varphi \circ \alpha \vdash \alpha$ and $\alpha \vdash \beta$. Suppose that α has no explanations. Then $\varphi \circ \alpha \equiv \varphi$. From this and the assumptions $\varphi \vdash \alpha$. Then, if $\alpha_w \vdash \varphi$ we have $\top \triangleright \alpha_w$ and $\alpha_w \vdash \alpha$, thus $\top \wedge \alpha \not\vdash \bot$. Then, by (Cautious monotony) $\top \wedge \alpha \triangleright \alpha_w$, thus, by (Congruence) $\alpha \triangleright \alpha_w$, contradiction. Therefore, α has explanations. Then, by (Explanatory monotony), β has explanations. If $\alpha_w \vdash \varphi \circ \beta$, by definition of $\circ$, $\beta \triangleright \alpha_w$. Then, by (Infraclassicality), $\alpha_w \vdash \beta$. Therefore, $\varphi \circ \beta \vdash \beta$.
(Trichotomy)

$$\varphi \circ (\alpha \vee \beta) \equiv \begin{cases} \varphi \circ \alpha & \text{or} \\ \varphi \circ \beta & \text{or} \\ (\varphi \circ \alpha) \vee (\varphi \circ \beta) \end{cases}$$

If $Expl(\alpha \vee \beta) = \emptyset$, by (Explanatory monotony,) we have $Expl(\alpha) = \emptyset$ and $Expl(\beta) = \emptyset$. Thus, $\varphi \circ (\alpha \vee \beta) \equiv \varphi$, $\varphi \circ \alpha \equiv \varphi$ and $\varphi \circ \beta \equiv \varphi$. In particular, $\varphi \circ (\alpha \vee \beta) \equiv \varphi \circ \alpha$.
Now suppose that $Expl(\alpha \vee \beta) \neq \emptyset$. Thus, there exists α_w such that $\alpha \vee \beta \triangleright \alpha_w$ and, by (Infraclassicity), $\alpha_w \vdash \alpha \vee \beta$. Note that, by (Strong coherence), $\varphi \circ (\alpha \vee \beta) \not\vdash \bot$.

Case 1: $Expl(\alpha) \neq \emptyset$ and $Expl(\beta) = \emptyset$. In this case we prove that $\varphi \circ (\alpha \vee \beta) \equiv \varphi \circ \alpha$.
Let $\alpha_{w'}$ be such that $\alpha_{w'} \vdash \varphi \circ (\alpha \vee \beta)$, then, by definition of $\circ$, $\alpha \vee \beta \triangleright \alpha_{w'}$. Then, by (Infraclassicity), $\alpha_{w'} \vdash \alpha \vee \beta$. If $\alpha_{w'} \vdash \beta$, since $(\alpha \vee \beta) \wedge \beta \not\vdash \bot$, we have, by (Cautious monotony), $(\alpha \vee \beta) \wedge \beta \triangleright \alpha_{w'}$. From this and (Congruence), $\beta \triangleright \alpha_{w'}$, a contradiction. Therefore, $\alpha_{w'} \not\vdash \beta$, thus, $\alpha_{w'} \vdash \alpha$.

From this, (Cautious monotony) and (Congruence), we have $\alpha \triangleright \alpha_{w'}$. Therefore $\alpha_{w'} \vdash \varphi \circ \alpha$.
Now suppose that $\alpha_{w'} \vdash \varphi \circ \alpha$, then, by definition of $\circ$, $\alpha \triangleright \alpha_{w'}$. If $\alpha_w \vdash \beta$, then, since $\alpha \vee \beta \triangleright \alpha_w$, by (Cautious monotony) and (Congruence) we would have $\beta \triangleright \alpha_w$, contradiction. Therefore, $\alpha_w \nvdash \beta$ and, since $\alpha_w \vdash \alpha \vee \beta$, necessarily $\alpha_w \vdash \alpha$. Since $\alpha \triangleright \alpha_{w'}$, by (Congruence), $(\alpha \vee \beta) \wedge \alpha \triangleright \alpha_{w'}$. Moreover, since $\alpha \vee \beta \triangleright \alpha_w$ and $\alpha_w \vdash \alpha$, by (Weak cut), we have $\alpha \vee \beta \triangleright \alpha_{w'}$, that is $\alpha_{w'} \vdash \varphi \circ (\alpha \vee \beta)$.

Case 2: $Expl(\alpha) = \emptyset$ and $Expl(\beta) \neq \emptyset$. In this case we prove that $\varphi \circ (\alpha \vee \beta) \equiv \varphi \circ \beta$.
The proof is analogous to the previous case.

Case 3: $Expl(\alpha) \neq \emptyset$ and $Expl(\beta) \neq \emptyset$. Let $\alpha_{w'}$ and $\alpha_{w''}$ be such that $\alpha \triangleright \alpha_{w'}$ and $\beta \triangleright \alpha_{w''}$.
(a) Suppose $\alpha \vee \beta \triangleright \alpha_{w'}$ and $\alpha \vee \beta \not\triangleright \alpha_{w''}$. In this case we prove that $\varphi \circ (\alpha \vee \beta) \equiv \varphi \circ \alpha$.
Suppose $\alpha_w \vdash \varphi \circ (\alpha \vee \beta)$, then $\alpha \vee \beta \triangleright \alpha_w$. From this, by (Infraclassicity) $\alpha_w \vdash \alpha \vee \beta$.
Suppose $\alpha_w \vdash \beta$. Since $(\alpha \vee \beta) \wedge \beta \equiv \beta$ and $(\alpha \vee \beta) \wedge \beta \triangleright \alpha_{w''}$, by (Weak cut), we have $\alpha \vee \beta \triangleright \alpha_{w''}$, contradiction. Thus, $\alpha_w \vdash \alpha$. Since $(\alpha \vee \beta) \wedge \alpha \nvdash \bot$, by (Cautious monotony), we have $(\alpha \vee \beta) \wedge \alpha \triangleright \alpha_w$. From this and (Congruence), we obtain $\alpha \triangleright \alpha_w$. Therefore, $\alpha_w \vdash \varphi \circ \alpha$. Thus, we have proven $\varphi \circ (\alpha \vee \beta) \vdash \varphi \circ \alpha$.
Now suppose that $\alpha_w \vdash \varphi \circ \alpha$. Then $\alpha \triangleright \alpha_w$. From the assumptions and (Infraclassicity), we have $\alpha_{w'} \vdash \alpha$. Since $(\alpha \vee \beta) \wedge \alpha \equiv \alpha$, by (Congruence), $(\alpha \vee \beta) \wedge \alpha \triangleright \alpha_w$. Since $\alpha \vee \beta \triangleright \alpha_{w'}$ and $\alpha_{w'} \vdash \alpha$, by (Weak cut) $\alpha \vee \beta \triangleright \alpha_w$. Therefore, $\alpha_w \vdash \varphi \circ (\alpha \vee \beta)$. Thus, we have proven $\varphi \circ \alpha \vdash \varphi \circ (\alpha \vee \beta)$. This finishes the proof that $\varphi \circ (\alpha \vee \beta) \equiv \varphi \circ \alpha$.
(b) $\alpha \vee \beta \not\triangleright \alpha_{w'}$ and $\alpha \vee \beta \triangleright \alpha_{w''}$. In this case we prove that $\varphi \circ (\alpha \vee \beta) \equiv \varphi \circ \beta$.
The proof is similar to the one of case (a).
(c) $\alpha \vee \beta \triangleright \alpha_{w'}$ and $\alpha \vee \beta \triangleright \alpha_{w''}$. In this case we prove that $\varphi \circ (\alpha \vee \beta) \equiv (\varphi \circ \alpha) \vee (\varphi \circ \beta)$.
Suppose $\alpha_w \vdash \varphi \circ (\alpha \vee \beta)$. Then $\alpha \vee \beta \triangleright \alpha_w$ and, by (Infraclassicity) $\alpha_w \vdash \alpha \vee \beta$. Suppose that $\alpha_w \vdash \alpha$. Since $(\alpha \vee \beta) \wedge \alpha \nvdash \bot$, by (Cautious monotony) $(\alpha \vee \beta) \wedge \alpha \triangleright \alpha_w$. From this, by (Congruence), we have $\alpha \triangleright \alpha_w$. Thus, $\alpha_w \vdash (\varphi \circ \alpha)$ and therefore $\alpha_w \vdash (\varphi \circ \alpha) \vee (\varphi \circ \beta)$. Suppose $\alpha_w \nvdash \alpha$. Then $\alpha_w \vdash \beta$, with a analogous reasoning we get $\alpha_w \vdash \varphi \circ \beta$. Therefore $\alpha_w \vdash (\varphi \circ \alpha) \vee (\varphi \circ \beta)$. This proves that $\varphi \circ (\alpha \vee \beta) \vdash (\varphi \circ \alpha) \vee (\varphi \circ \beta)$. Suppose $\alpha_w \vdash (\varphi \circ \alpha) \vee (\varphi \circ \beta)$. If $\alpha_w \vdash \varphi \circ \alpha$ then $\alpha \triangleright \alpha_w$. From the assumption and (Infraclassicity), we have $\alpha_{w'} \vdash \alpha$. From this and (Weak cut), $\alpha \vee \beta \triangleright \alpha_w$. Therefore, $\alpha_w \vdash \varphi \circ (\alpha \vee \beta)$. If $\alpha_w \vdash \varphi \circ \beta$, the proof is made in a similar way. This proves that $(\varphi \circ \alpha) \vee (\varphi \circ \beta) \vdash \varphi \circ (\alpha \vee \beta)$.

The proof of the *if* part of the Theorem is easy using Theorem 1 and we leave it as an exercise for the reader. However, we recommended the reader to see the proof of the *if* part of Theorem 3 which is similar. ∎

From the previous Theorem and Theorem 1 we obtain the following result:

Corollary 1 $\triangleright$ *is a weak reflexive explanatory relation iff there exists a nonempty set $C \subseteq W$ and a total preorder $\preceq$ over C such that $\alpha \triangleright \gamma$ iff $[\![\alpha]\!] \cap C \neq \emptyset$ and $[\![\gamma]\!] \subseteq \min(\alpha \cap C, \preceq)$.*

Note that the Example 1 is an illustration of this kind of explanatory relation at work. Actually, in that example the set C is $\{1001, 1101, 0001, 0001\}$ and the total preorder over C is given by $1001 \prec 1101 \sim 0001 \prec 0001$. The only observations α having explanations are the formulas α satisfying $[\![\alpha]\!] \cap C \neq \emptyset$. Here π_1 is the identity and π_2 is a function such that $[\![\pi_2(\alpha)]\!] = \min([\![\alpha]\!] \cap C, \preceq)$ for any α having explanations.

3.2 Ordered explanatory relations

Now we define our second class of explanatory relations.

Definition 5 *Let $\triangleright$ be a binary relation over $\mathcal{L}$. The relation $\triangleright$ is called an ordered explanatory relation if the following postulates hold:*

$\triangleright \neq \emptyset$	(Non triviality)
$Expl(\alpha) \neq \emptyset \Rightarrow \alpha \triangleright \alpha$	(Limited reflexivity)
$\alpha \triangleright \gamma \Rightarrow \alpha \wedge \gamma \nvdash \bot$	(Weak infra-classicality)
$\alpha \triangleright \gamma, \delta \vdash \gamma, \delta \nvdash \bot \Rightarrow \alpha \triangleright \delta$ or $\alpha \wedge \neg \delta \triangleright \gamma$	(Weak right strengthening)
$\alpha \triangleright \gamma, \gamma \triangleright \delta \Rightarrow \alpha \triangleright \delta$	(Transitivity)
$\alpha \triangleright \gamma, \beta \triangleright \gamma \Rightarrow \alpha \wedge \beta \triangleright \gamma$	(Left and)
$\alpha \triangleright \gamma, \alpha \triangleright \delta \Rightarrow \alpha \triangleright \gamma \vee \delta$	(Right or)
$\alpha \triangleright \gamma, \gamma \vdash \beta \Rightarrow \alpha \wedge \beta \triangleright \gamma$	(Cautious monotony)
$\alpha \equiv \alpha', \gamma \equiv \gamma' \Rightarrow (\alpha \triangleright \gamma \Leftrightarrow \alpha' \triangleright \gamma')$	(Congruence)
$Expl(\alpha) \neq \emptyset, \alpha \vdash \beta \Rightarrow Expl(\beta) \neq \emptyset$	(Explanatory monotony)

As in Definition 3, the postulates above can be seen as a minimal set of inference rules whose meaning we can explain as follows. Actually, the meaning of *Right or, Cautious monotony, Congruence* and *Explanatory monotony* have been explained after Definition 3. *Non triviality* says that there exists something that can be explained. *Limited reflexivity* says that one explanation is explained by itself. *Weak infra-classicality* says that the explanations are at least consistent with the observation. *Weak right strengthening* says that if a formula stronger than one explanation of a given observation is not an explanation of the given observation, the reason is that the original explanation is also an explanation of the conjunction of the given observation and the negation of the stronger formula. The meaning of *Transitivity* is explained by itself. *Left and* says that one explanation of two observations is also an explanation of the conjunction of the two observations.

The following Theorem explain the rationale behind the choice of postulates in Definition 5.

Theorem 3. $\triangleright$ *is an ordered explanatory relation iff there exists a consistent formula φ and a credibility-limited revision operator $\circ$ such that*

$$\alpha \triangleright \gamma \Leftrightarrow (\varphi \circ \gamma \vdash \varphi \circ \alpha), (\varphi \circ \alpha \vdash \alpha) \text{ and } (\varphi \circ \gamma \vdash \gamma) \qquad (2)$$

The proof of this theorem uses the following two results.

Lemma 1 *If $\triangleright$ is an ordered explanatory relation then $Expl(\alpha) \neq \emptyset \Leftrightarrow \alpha \in$ Expl.*

Proof.
($\Rightarrow$) If $Expl(\alpha) \neq \emptyset$ then, by (Limited reflexivity), $\alpha \triangleright \alpha$. Therefore $\alpha \in$ Expl.
($\Leftarrow$) If $\alpha \in$ Expl, there exists β such that $\beta \triangleright \alpha$, since $Expl(\beta) \neq \emptyset$ and $\beta \vdash \alpha \vee \beta$, by (Explanatory monotony) $Expl(\alpha \vee \beta) \neq \emptyset$ and by (Limited reflexivity) $\alpha \vee \beta \triangleright \alpha \vee \beta$. As $\alpha \vdash \alpha \vee \beta$, by (Weak right strengthening), we have:

$$\alpha \vee \beta \triangleright \alpha \quad \text{or} \quad (\alpha \vee \beta) \wedge \neg \alpha \triangleright \alpha \vee \beta$$

If $(\alpha \vee \beta) \wedge \neg \alpha \triangleright \alpha \vee \beta$, since $(\alpha \vee \beta) \wedge \neg \alpha \equiv \beta \wedge \neg \alpha$, by (Congruence) $\beta \wedge \neg \alpha \triangleright \alpha \vee \beta$. But $\beta \vdash \alpha \vee \beta$, thus, by (Weak right strengthening)again, we have:

$$\beta \wedge \neg \alpha \triangleright \beta \quad \text{or} \quad (\beta \wedge \neg \alpha) \wedge \neg \beta \triangleright \alpha \vee \beta$$

If $\beta \wedge \neg \alpha \triangleright \beta$ and $\beta \triangleright \alpha$ we have by (Transitivity) $\beta \wedge \neg \alpha \triangleright \alpha$ and, by (Weak infra-classicality) $\beta \wedge \neg \alpha \wedge \alpha \not\vdash \bot$, contradiction.
If $(\beta \wedge \neg \alpha) \wedge \neg \beta \triangleright \alpha \vee \beta$, by (Weak infra-classicality) $(\beta \wedge \neg \alpha) \wedge \neg \beta \wedge (\alpha \vee \beta) \not\vdash \bot$, contradiction.
Thus, $(\alpha \vee \beta) \wedge \neg \alpha \not\triangleright \alpha \vee \beta$. Then, necessarily $\alpha \vee \beta \triangleright \alpha$. Since $\alpha \vdash \alpha$, we have by (Left and) $(\alpha \vee \beta) \wedge \alpha \triangleright \alpha$. But $(\alpha \vee \beta) \wedge \alpha \equiv \alpha$, then, by (Congruence), $\alpha \triangleright \alpha$. Therefore $Expl(\alpha) \neq \emptyset$.

∎

Proposition 2 *If $\triangleright$ is an ordered explanatory relation then the following postulates hold:*

(P1) $\alpha \in$ Expl $\Rightarrow \alpha \triangleright \alpha$.
(P2) $\alpha \triangleright \alpha_\omega \Rightarrow \alpha_\omega \vdash \alpha$.
(P3) $\alpha \triangleright \alpha$ and $\beta \vdash \alpha \Rightarrow \alpha \triangleright \beta$ or $\alpha \triangleright \alpha \wedge \neg \beta$.
(P4) $Expl(\alpha) \neq \emptyset \Rightarrow \alpha$ has complete explanations.
(P5) $\alpha \triangleright \gamma, \ \beta \vdash \alpha$ and $(\alpha \not\vdash \beta) \Rightarrow \alpha \wedge \neg \beta \triangleright \gamma$.

(P6) ⊤ has explanations.
(P7) $Expl(\alpha) \neq \emptyset \Rightarrow \alpha \vee \beta \triangleright \alpha \vee \beta$ for any formula β.
(P8) $\alpha \triangleright \gamma$ and $\beta \triangleright \delta \Rightarrow \alpha \vee \beta \triangleright \gamma$ or $\alpha \vee \beta \triangleright \delta$.
(P9) $\alpha \triangleright \gamma$, $\gamma \vdash \beta$ and $\alpha \wedge \beta \triangleright \delta \Rightarrow \alpha \triangleright \delta$.
(P10) $Expl(\alpha \vee \beta) \cap Expl(\alpha) = \emptyset \Rightarrow Expl(\alpha \vee \beta) = Expl(\beta)$.

Proof.

(P1) If $\alpha \in$ Expl, by Lemma 1 and (Limited reflexivity) we have $\alpha \triangleright \alpha$.

(P2) If $\alpha \triangleright \alpha_\omega$, by (Weak infra classicality), we have $\alpha \wedge \alpha_\omega \nvdash \bot$. But α_ω is a complet formula, thus, we have $\alpha_\omega \vdash \alpha$.

(P3) By hypothesis $\alpha \triangleright \alpha$ and $\beta \vdash \alpha$, by (Weak right strengthening) we have $\alpha \triangleright \beta$ or $\alpha \wedge \neg \beta \triangleright \alpha$. If $\alpha \triangleright \beta$ the proof is complete. Suppose that $\alpha \ntriangleright \beta$. We have $\alpha \triangleright \alpha$ and $\alpha \wedge \neg \beta \vdash \alpha$. Thus, by (Weak right strengthening), we have $\alpha \triangleright \alpha \wedge \neg \beta$ or $\alpha \wedge \neg(\alpha \wedge \neg \beta) \triangleright \alpha$. If $\alpha \wedge \neg(\alpha \wedge \neg \beta) \triangleright \alpha$, since $\alpha \wedge \neg(\alpha \wedge \neg \beta) \equiv \alpha \wedge \beta$, then we have, by (Congruence), $\alpha \wedge \beta \triangleright \alpha$, since $\alpha \wedge \neg \beta \triangleright \alpha$, by (Left and) we have $\alpha \wedge \beta \wedge \alpha \wedge \neg \beta \triangleright \alpha$, and, by (Weak infra-classicality), we have $\alpha \wedge \beta \wedge \alpha \wedge \neg \beta \wedge \alpha \nvdash \bot$ a contradiction. Therefore $\alpha \triangleright \alpha \wedge \neg \beta$.

(P4) If $Expl(\alpha) \neq \emptyset$ by (Limited reflexivity) we have $\alpha \triangleright \alpha$ and, by (Weak infra-classicality) $\alpha \wedge \alpha \nvdash \bot$. Thus, α is consistent. We can assume that α has n models (at least one model), that is:

$$[\![\alpha]\!] = \{\omega_1, \ldots, \omega_n\} \text{ and } \alpha \equiv \bigvee_{i=1}^{n} \alpha_{\omega_i}$$

In order to finish the proof of **(P4)**, we prove the following claim:

Claim. $\alpha \triangleright \alpha_{\omega_i}$ for some $i \in \{1, \ldots, n\}$.

We prove the claim by induction on the number of models of the formula α.

Basis: suppose that $[\![\alpha]\!] = \{\omega_1\}$, then $\alpha \equiv \alpha_{\omega_1}$. Since $\alpha \triangleright \alpha$, by (Congruence) we have $\alpha \triangleright \alpha_{\omega_1}$.

Inductive hypothesis: Suppose that, for all α whin n models and $Expl(\alpha) \neq \emptyset$, there exists $\alpha_\omega \in$ Expl such that $\alpha \triangleright \alpha_\omega$.

Now consider a formula α, such that $\alpha \triangleright \alpha$ and $[\![\alpha]\!] = \{\omega_1, \ldots, \omega_{n+1}\}$. Since $\alpha_{\omega_1} \vdash \alpha$, by P3, we have:

$$\alpha \triangleright \alpha_{\omega_1} \quad \text{or} \quad \alpha \triangleright \alpha \wedge \neg \alpha_{\omega_1}$$

If $\alpha \triangleright \alpha_{\omega_1}$ we have finished the proof. Suppose that $\alpha \not\triangleright \alpha_{\omega_1}$, then $\alpha \triangleright \alpha \wedge \neg \alpha_{\omega_1}$, by P1 we have $\alpha \wedge \neg \alpha_{\omega_1} \triangleright \alpha \wedge \neg \alpha_{\omega_1}$ and also $[\![\alpha \wedge \neg \alpha_{\omega_1}]\!] = \{\omega_2, \ldots, \omega_{n+1}\}$, by the induction hypothesis exists $\omega_j \in [\![\alpha \wedge \neg \alpha_{\omega_1}]\!]$ such that $\alpha \wedge \neg \alpha_{\omega_1} \triangleright \alpha_{\omega_j}$. From this and the fact $\alpha \triangleright \alpha \wedge \neg \alpha_{\omega_1}$, we get, by (Transitivity,) $\alpha \triangleright \alpha_{\omega_j}$.

(P5) Suppose that $\alpha \triangleright \gamma$, then by (P1) $\alpha \triangleright \alpha$. Since $\beta \vdash \alpha$, by (Weak right strengthening),

$$\alpha \triangleright \beta \quad \text{or} \quad \alpha \wedge \neg \beta \triangleright \alpha$$

But by hypothesis $(\alpha \not\triangleright \beta)$, thus, necessarily, $\alpha \wedge \neg \beta \triangleright \alpha$. Then, by (Transitivity), $\alpha \wedge \neg \beta \triangleright \gamma$.

(P6) By (Non triviality) there exist α and γ such that $\alpha \triangleright \gamma$, then $Expl(\alpha) \neq \emptyset$. Note that $\alpha \vdash \top$, then, by (Explanatory monotony), $Expl(\top) \neq \emptyset$.

(P7) Since $Expl(\alpha) \neq \emptyset$ and $\alpha \vdash \alpha \vee \beta$, by (Explanatory monotony), $Expl(\alpha \vee \beta) \neq \emptyset$. Then, by (Limited reflexivity), $\alpha \vee \beta \triangleright \alpha \vee \beta$.

(P8) Assume that $\alpha \triangleright \gamma$ and $\beta \triangleright \delta$. Since $Expl(\alpha) \neq \emptyset$ by (P7) $\alpha \vee \beta \triangleright \alpha \vee \beta$. Note that $\alpha \vdash \alpha \vee \beta$ and $\beta \vdash \alpha \vee \beta$, then by (Weak right strengthening) we have:

$(\alpha \vee \beta \triangleright \alpha \quad \text{or} \quad (\alpha \vee \beta) \wedge \neg \alpha \triangleright \alpha \vee \beta) \quad \text{and} \quad (\alpha \vee \beta \triangleright \beta \quad \text{or} \quad (\alpha \vee \beta) \wedge \neg \beta \triangleright \alpha \vee \beta)$

Suppose that neither $\alpha \vee \beta \triangleright \alpha$ nor $\alpha \vee \beta \triangleright \beta$ occurs. Then $\beta \wedge \neg \alpha \triangleright \alpha \vee \beta$ and $\alpha \wedge \neg \beta \triangleright \alpha \vee \beta$. By (Left and) $\beta \wedge \neg \alpha \wedge \alpha \wedge \neg \beta \triangleright \alpha \vee \beta$, that is, $\bot \triangleright \alpha \vee \beta$. Then, by (Weak infra-classicality), we have $\bot \wedge (\alpha \vee \beta) \not\vdash \bot$, a contradiction. Therefore, we have $\alpha \vee \beta \triangleright \alpha$ or $\alpha \vee \beta \triangleright \beta$. From this and the assumption $\alpha \triangleright \gamma$ and $\beta \triangleright \delta$, we conclude, using (Transitivity), $\alpha \vee \beta \triangleright \gamma$ or $\alpha \vee \beta \triangleright \delta$.

(P9) Suppose that $\alpha \triangleright \gamma$, $\gamma \vdash \beta$ and $\alpha \wedge \beta \triangleright \delta$. We want to prove that $\alpha \triangleright \delta$.
First, we show that $\alpha \triangleright \alpha \wedge \beta$. Towards a contradiction, suppose $\alpha \not\triangleright \alpha \wedge \beta$. Since $\alpha \triangleright \gamma$, $\alpha \wedge \beta \vdash \alpha$, by (P5), we have $\alpha \wedge \neg(\alpha \wedge \beta) \triangleright \gamma$, that is, by (Congruence), $\alpha \wedge \neg \beta \triangleright \gamma$, Then, by (Weak infra-classicality), we have $\alpha \wedge \neg \beta \wedge \gamma \not\vdash \bot$, in particular, $\neg \beta \wedge \gamma \not\vdash \bot$, contradicting the hypothesis $\gamma \vdash \beta$. Therefore $\alpha \triangleright \alpha \wedge \beta$. From this and the fact $\alpha \wedge \beta \triangleright \delta$, by (Transitivity), we have $\alpha \triangleright \delta$.

(P10) Suppose that $Expl(\alpha \vee \beta) \cap Expl(\alpha) = \emptyset$. We want to show that $Expl(\alpha \vee \beta) \cap Expl(\beta) = Expl(\beta)$.

First, suppose that $Expl(\alpha \vee \beta) = \emptyset$. Since $\beta \vdash \alpha \vee \beta$, we have, by (Explanatory monotony), $Expl(\beta) = \emptyset$, therefore $Expl(\alpha \vee \beta) = Expl(\beta)$.

Now, suppose that $Expl(\alpha \vee \beta) \neq \emptyset$. We prove $Expl(\alpha \vee \beta) \cap Expl(\beta) = Expl(\beta)$ by establishing the double inclusion.

$[Expl(\alpha \vee \beta) \subseteq Expl(\beta)]$: By Lemma 1 and P1 we have $\alpha \vee \beta \triangleright \alpha \vee \beta$; but $\alpha \vdash \alpha \vee \beta$, thus, by (Weak right strengthening) and (Congruence)

$$\alpha \vee \beta \triangleright \alpha \quad \text{or} \quad \beta \wedge \neg \alpha \triangleright \alpha \vee \beta$$

Note that if $\alpha \vee \beta \triangleright \alpha$, we have $\alpha \in \mathsf{Expl}$ and then, by P1, $\alpha \triangleright \alpha$. Therefore, $\alpha \in Expl(\alpha \vee \beta) \cap Expl(\alpha)$, a contradiction with our assumption. Thus, necessarily $\beta \wedge \neg \alpha \triangleright \alpha \vee \beta$. From this and the fact $\beta \vdash \alpha \vee \beta$ we have, by (Weak right strengthening) and (Congruence)

$$\beta \wedge \neg \alpha \triangleright \beta \quad \text{or} \quad \beta \wedge \neg \alpha \wedge \neg \beta \triangleright \alpha \vee \beta$$

If $\beta \wedge \neg \alpha \wedge \neg \beta \triangleright \alpha \vee \beta$, by (Weak infra-classicality) and (Congruence), we have $\bot \wedge (\alpha \vee \beta) \not\vdash \bot$, a contradiction. Therefore $\beta \wedge \neg \alpha \triangleright \beta$. Then $\beta \in \mathsf{Expl}$. Thus, by P1, $\beta \triangleright \beta$. Note that $\beta \wedge \neg \alpha \vdash \beta$, then, by (Weak right strengthening) and (Congruence),

$$\beta \triangleright \beta \wedge \neg \alpha \quad \text{or} \quad \beta \wedge \alpha \triangleright \beta$$

If $\beta \wedge \alpha \triangleright \beta$ holds. Then, from the fact $\beta \wedge \neg \alpha \triangleright \beta$, we get, by (Left and), $\beta \wedge \alpha \wedge \beta \wedge \neg \alpha \triangleright \beta$. From this, by (Weak infra-classicality) and (Congruence), we obtain $\bot \wedge \beta \not\vdash \bot$, a contradiction. Then, necessarily $\beta \triangleright \beta \wedge \neg \alpha$. Since $\beta \wedge \neg \alpha \triangleright \alpha \vee \beta$, we have by (Transitivity) $\beta \triangleright \alpha \vee \beta$. From this, by (Transitivity), we have $Expl(\alpha \vee \beta) \subseteq Expl(\beta)$.

$[Expl(\beta) \subseteq Expl(\alpha \vee \beta)]$: By the previous inclusion, necessarily $Expl(\beta) \neq \emptyset$. Then, by similar arguments to those in the proof of (P8), we have $\alpha \vee \beta \triangleright \alpha$ or $\alpha \vee \beta \triangleright \beta$. We have seen above that $\alpha \vee \beta \triangleright \alpha$ is impossible because it lead to a contradiction with the assumption $Expl(\alpha \vee \beta) \cap Expl(\alpha) = \emptyset$. Thus, we have $\alpha \vee \beta \triangleright \beta$ and, by (Transitivity), $Expl(\beta) \subseteq Expl(\alpha \vee \beta)$. ∎

Proof of Theorem 3: First we prove the *only if part*. Assume that $\triangleright$ is a ordered explanatory relation. Define the formula φ as in Definition 4. By P4 and P6, φ is a consistent formula. Define an operator $\circ$ in the following way:

$$\varphi \circ \alpha \equiv \begin{cases} \varphi & \text{if } Expl(\alpha) = \emptyset \\ \bigvee \{\alpha_\omega : \alpha \triangleright \alpha_\omega\} & \text{if } Expl(\alpha) \neq \emptyset \end{cases}$$

By P4, the operator $\circ$ is well defined.

(i) First, we prove the following equivalence:

$$\alpha \triangleright \gamma \Leftrightarrow (\varphi \circ \gamma \vdash \varphi \circ \alpha), (\varphi \circ \alpha \vdash \alpha) \text{ and } (\varphi \circ \gamma \vdash \gamma)$$

($\Leftarrow$) Suppose that $\varphi \circ \gamma \vdash \varphi \circ \alpha$, $\varphi \circ \alpha \vdash \alpha$ and $\varphi \circ \gamma \vdash \gamma$. If $Expl(\alpha) = \emptyset$ then $\varphi \circ \alpha \equiv \varphi$. If $\alpha_\omega \vdash \varphi$, then $\top \triangleright \alpha_\omega$ and $\alpha_\omega \vdash \alpha$, by (Left and) and (Congruence) $\alpha \triangleright \alpha_\omega$, then $\alpha_\omega \in Expl(\alpha)$, contradiction. Therefore $Expl(\alpha) \neq \emptyset$.
In a similar way, we prove that $Expl(\gamma) \neq \emptyset$.

We claim that $\alpha \triangleright \alpha \wedge \gamma$ and $\alpha \wedge \gamma \triangleright \gamma$.
By (Limited reflexivity) $\alpha \triangleright \alpha$. Since $\alpha \wedge \gamma \vdash \alpha$, by (Weak right strengthening), we have $\alpha \triangleright \alpha \wedge \gamma$ or $\alpha \wedge \neg(\alpha \wedge \gamma) \triangleright \alpha$. If $\alpha \wedge \neg(\alpha \wedge \gamma) \triangleright \alpha$, by (Congruence), we have $\alpha \wedge \neg \gamma \triangleright \alpha$. By (P4) exists α_ω such that $\gamma \triangleright \alpha_\omega$; by definition of $\circ$, $\alpha_\omega \vdash \varphi \circ \gamma$. From this and the hypotheses, we obtain $\alpha_\omega \vdash \varphi \circ \alpha$. Then, by definition of $\circ$, $\alpha \triangleright \alpha_\omega$. By (Transitivity), we obtain $\alpha \wedge \neg \gamma \triangleright \alpha_\omega$ and by (Weak infra-classicality) $\alpha \wedge \neg \gamma \wedge \alpha_\omega \not\vdash \bot$, contradicting the fact that $\alpha_\omega \vdash \gamma$. Therefore, $\alpha \triangleright \alpha \wedge \gamma$.
Now we prove the other fact of the claim: $\alpha \wedge \gamma \triangleright \gamma$. By (Limited reflexivity) $\gamma \triangleright \gamma$. Since $\neg \alpha \wedge \gamma \vdash \gamma$, we have, by (Weak right strengthening), $\gamma \triangleright \neg \alpha \wedge \gamma$ or $\gamma \wedge \neg(\neg \alpha \wedge \gamma) \triangleright \gamma$.
If $\gamma \triangleright \neg \alpha \wedge \gamma$, by Lemma 1, we have $Expl(\neg \alpha \wedge \gamma) \neq \emptyset$. By (P4), there exists α_ω such that $\neg \alpha \wedge \gamma \triangleright \alpha_\omega$. By (Weak infra-classicality), $\neg \alpha \wedge \gamma \wedge \alpha_\omega \not\vdash \bot$ and by (Transitivity) $\gamma \triangleright \alpha_\omega$. Then, by definition of $\circ$, $\alpha_\omega \vdash \varphi \circ \gamma$ and by the hypotheses, $\alpha_\omega \vdash \alpha$, a contradiction. Therefore, $\gamma \wedge \neg(\neg \alpha \wedge \gamma) \triangleright \gamma$ and, by (Congruence), $\alpha \wedge \gamma \triangleright \gamma$.
From the claim, we obtain, by (Transitivity), $\alpha \triangleright \gamma$.

($\Rightarrow$) Suppose that $\alpha \triangleright \gamma$. We have to verify that $\varphi \circ \gamma \vdash \gamma$, $\varphi \circ \gamma \vdash \varphi \circ \alpha$ and $\varphi \circ \alpha \vdash \alpha$.

By (Limited reflexivity) and (P1) we have $Expl(\alpha) \neq \emptyset$ and $Expl(\gamma) \neq \emptyset$. Let us verify $\varphi \circ \gamma \vdash \gamma$: if $\alpha_\omega \vdash \varphi \circ \gamma$, then $\gamma \triangleright \alpha_\omega$ and by (P2) $\alpha_\omega \vdash \gamma$.
Now, we verify $\varphi \circ \gamma \vdash \varphi \circ \alpha$: if $\alpha_\omega \vdash \varphi \circ \gamma$, then $\gamma \triangleright \alpha_\omega$. By the assumption $\alpha \triangleright \gamma$, then, by (Transitivity), $\alpha \triangleright \alpha_\omega$, then $\alpha_\omega \vdash \varphi \circ \alpha$.
Finally, we check $\varphi \circ \alpha \vdash \alpha$: if $\alpha_\omega \vdash \varphi \circ \alpha$, then $\alpha \triangleright \alpha_\omega$ and by (P2) $\alpha_\omega \vdash \alpha$.

(ii) Now we prove that the operator $\circ$ satisfies all the postulate defining a credibility limited revision operator relative to φ:

(Relative success) $\varphi \circ \alpha \vdash \alpha$ or $\varphi \circ \alpha \equiv \varphi$.
If $Expl(\alpha) = \emptyset$, then by definition $\varphi \circ \alpha \equiv \varphi$
If $Expl(\alpha) \neq \emptyset$, then exists γ such that $\alpha \triangleright \gamma$. Then, by the part(i), $\varphi \circ \alpha \vdash \alpha$.

(Vacuity) If $\varphi \wedge \alpha \not\vdash \bot$ then $\varphi \circ \alpha \equiv \varphi \wedge \alpha$.
First we prove $\varphi \circ \alpha \vdash \varphi \wedge \alpha$. Considerer α_ω such that $\alpha_\omega \vdash \varphi$ and $\alpha_\omega \vdash \alpha$ (α_ω exists by hypothesis). By definition of φ, we have $\top \triangleright \alpha_\omega$, then by de (Cautious monotony) and (Congruence) $\alpha \triangleright \alpha_\omega$. Then, by (i), we have $\varphi \circ \alpha \vdash \alpha$ (*).

Claim: $\top \triangleright \alpha \wedge \varphi$ and $\top \triangleright \alpha$
If $\alpha'_\omega \vdash \alpha \wedge \varphi$ then $\alpha'_\omega \vdash \varphi$ and $\top \triangleright \alpha'_\omega$. Note that $\alpha \wedge \varphi \equiv \bigvee \alpha'_\omega$ then, by (Right or), $\top \triangleright \alpha \wedge \varphi$. Moreover, if $\top \not\triangleright \alpha$, since $\alpha \vdash \top$ and $\top \triangleright \alpha \wedge \varphi$; by (P5), $\top \wedge \neg \alpha \triangleright \alpha \wedge \varphi$ and, by (Congruence), $\neg \alpha \triangleright \alpha \wedge \varphi$. From this,

by (Weak infra-classicality), we get $\neg \alpha \wedge \alpha \wedge \varphi \not\vdash \bot$, a contradiction. Therefore, $\top \triangleright \alpha$. This finish the proof of the claim.

Now, suppose that $\alpha_w \vdash \varphi \circ \alpha$. Since $\top \triangleright \alpha$, $\alpha \in \mathsf{Expl}$. Thus, by Lemma 1, $Expl(\alpha) \neq \emptyset$. By definition $\varphi \circ \alpha$, we have $\alpha \triangleright \alpha_w$. Now, we have $\top \triangleright \alpha$ and $\alpha \triangleright \alpha_w$, then, by (Transitivity), $\top \triangleright \alpha_w$ and, by definition of φ, $\alpha_w \vdash \varphi$. Thus, we have proved $\varphi \circ \alpha \vdash \varphi$ (**).
From (*) and (**) we obtain $\varphi \circ \alpha \vdash \varphi \wedge \alpha$

Now we prove $\varphi \wedge \alpha \vdash \varphi \circ \alpha$. Suppose $\alpha_w \vdash \varphi \wedge \alpha$, then $\alpha_w \vdash \alpha$ and $\alpha_w \vdash \varphi$. By definition of φ, $\top \triangleright \alpha_w$, and by (Cautious monotony) and (Congruence) we have $\alpha \triangleright \alpha_w$. Then, by definition of $\circ$, $\alpha_w \vdash \varphi \circ \alpha$.

(Strong coherence) $\varphi \circ \alpha \not\vdash \bot$.
If $Expl(\alpha) = \emptyset$, then $\varphi \circ \alpha \equiv \varphi$. Since $\varphi \not\vdash \bot$, we have $\varphi \circ \alpha \not\vdash \bot$.
If $Expl(\alpha) \neq \emptyset$, by (P4), there exists α_w such that $\alpha \triangleright \alpha_w$. Thus, $\alpha_w \vdash \varphi \circ \alpha$ and therefore, $\varphi \circ \alpha \not\vdash \bot$.

(Syntax independence) If $\varphi \equiv \varphi'$ and $\alpha \equiv \alpha'$ then $\varphi \circ \alpha \equiv \varphi' \circ \alpha'$.
If $Expl(\alpha) = \emptyset$ by (Congruence) $Expl(\alpha') = \emptyset$, by definition of $\circ$ we have $\varphi \circ \alpha \equiv \varphi$ and $\varphi' \circ \alpha' \equiv \varphi'$. Then, by the hypotheses, we have $\varphi \circ \alpha \equiv \varphi' \circ \alpha'$.

Suppose that $Expl(\alpha) \neq \emptyset$. Since $\alpha \vdash \alpha'$ by (Explanatory monotony) $Expl(\alpha') \neq \emptyset$.
First we prove $\varphi \circ \alpha \vdash \varphi' \circ \alpha'$: if $\alpha_w \vdash \varphi \circ \alpha$, then $\alpha \triangleright \alpha_w$. By hypothesis $\alpha \equiv \alpha'$, then, by (Congruence), $\alpha' \triangleright \alpha_w$, thus $\alpha_w \vdash \varphi \circ \alpha'$. Finally by definition of $\circ$ and the hypothesis $\varphi \equiv \varphi'$, we have $\alpha_w \vdash \varphi' \circ \alpha'$
In a similar way we prove that $\varphi' \circ \alpha' \vdash \varphi \circ \alpha$).

(Success monotony) If $\varphi \circ \alpha \vdash \alpha$ and $\alpha \vdash \beta$, then $\varphi \circ \beta \vdash \beta$
Case 1: $Expl(\beta) = \emptyset$
By hypothesis $\alpha \vdash \beta$. Then, by (Explanatory monotony), $Expl(\alpha) = \emptyset$. Thus $\varphi \circ \alpha \equiv \varphi$. From this and the hypotheses, $\varphi \vdash \beta$. We know also $\varphi \circ \beta \equiv \varphi$. Thus, we have $\varphi \circ \beta \vdash \beta$.

Case 2: $Expl(\beta) \neq \emptyset$
By (P4), there exists α_w such that $\beta \triangleright \alpha_w$, by (i), $\varphi \circ \beta \vdash \beta$.

(Trichotomy)
$$\varphi \circ (\alpha \vee \beta) \equiv \begin{cases} \varphi \circ \alpha & \text{or} \\ \varphi \circ \beta & \text{or} \\ (\varphi \circ \alpha) \vee (\varphi \circ \beta) \end{cases}$$

If $Expl(\alpha \vee \beta) = \emptyset$, then by (Explanatory monotony) $Expl(\alpha) = \emptyset$ and $Expl(\beta) = \emptyset$, thus $\varphi \circ (\alpha \vee \beta) \equiv \varphi$, $\varphi \circ \alpha \equiv \varphi$ and $\varphi \circ \beta \equiv \varphi$. Then $\varphi \circ (\alpha \vee \beta) \equiv \varphi \circ \alpha$.

Suppose that $Expl(\alpha \vee \beta) \neq \emptyset$.

Case 1: $Expl(\alpha) = \emptyset$, then $Expl(\alpha \vee \beta) \cap Expl(\alpha) = \emptyset$, by (P10) $Expl(\alpha \vee \beta) = Expl(\beta)$, thus $\varphi \circ (\alpha \vee \beta) = \varphi \circ \beta$.

Case 2: $Expl(\beta) = \emptyset$. The prove of $\varphi \circ (\alpha \vee \beta) \equiv \varphi \circ \alpha$ is analogous to the previous case.

Case 3: $Expl(\alpha) \neq \emptyset$ and $Expl(\beta) \neq \emptyset$.

Consider three subcases:

(i) $(\varphi \circ (\alpha \vee \beta)) \wedge \beta \vdash \bot$. In this case we will prove $\varphi \circ (\alpha \vee \beta) \equiv \varphi \circ \alpha$.
Suppose that $\alpha_w \vdash \varphi \circ (\alpha \vee \beta)$. Since $Expl(\alpha) \neq \emptyset$, we have, by (Explanatory monotony), $Expl(\alpha \vee \beta) \neq \emptyset$. Then, by definition of $\circ$, we have $\alpha \vee \beta \triangleright \alpha_w$. By (P2), $\alpha_w \vdash \alpha \vee \beta$. By hypothesis $\alpha_w \nvdash \beta$ thus $\alpha_w \vdash \alpha$. Since $(\alpha \vee \beta) \triangleright \alpha_w$ and $\alpha_w \vdash \alpha$ by (Cautious monotony) and (Congruence) $\alpha \triangleright \alpha_w$. Thus, by definition of $\circ$, we have $\alpha_w \vdash \varphi \circ \alpha$. Therefore $\varphi \circ (\alpha \vee \beta) \vdash \varphi \circ \alpha$.
Now, suppose that $\alpha_w \vdash \varphi \circ \alpha$. we have to prove $\alpha_w \vdash \varphi \circ (\alpha \vee \beta)$. Since $\alpha_w \vdash \varphi \circ \alpha$, we have $\alpha \triangleright \alpha_w$. By hypothesis, $Expl(\beta) \neq \emptyset$. Then, by (Limited reflexivity), $\beta \triangleright \beta$. We have $\alpha \triangleright \alpha_w$ and $\beta \triangleright \beta$, then, by (P8),
$$\alpha \vee \beta \triangleright \alpha_w \quad \text{or} \quad \alpha \vee \beta \triangleright \beta$$
Suppose that $\alpha \vee \beta \triangleright \beta$, by (Transitivity) and definition of $\circ$, we have $\varphi \circ \beta \vdash \varphi \circ (\alpha \vee \beta)$, thus $\beta \wedge (\varphi \circ (\alpha \vee \beta))$ is consistent, because the models of $\varphi \circ \beta$ are models of β. But this is in contradiction with the hypothesis $\varphi \circ (\alpha \vee \beta) \wedge \beta \vdash \bot$. Then, necessarily $\alpha \vee \beta \triangleright \alpha_w$. Then, by definition of $\circ$ $\alpha_w \vdash \varphi \circ (\alpha \vee \beta)$. Therefore, $\varphi \circ \alpha \vdash \varphi \circ (\alpha \vee \beta)$.

(ii) $\varphi \circ (\alpha \vee \beta) \wedge \alpha \vdash \bot$. In this case we will prove $\varphi \circ (\alpha \vee \beta) \equiv \varphi \circ \beta$. The proof is similar to the previous one by exchanging the role of α and β.

(iii) $(\varphi \circ (\alpha \vee \beta)) \wedge \alpha \nvdash \bot$ and $(\varphi \circ (\alpha \vee \beta)) \wedge \beta \nvdash \bot$. In this case, we will prove $\varphi \circ (\alpha \vee \beta) \equiv (\varphi \circ \alpha) \vee (\varphi \circ \beta)$.
First we check that $\varphi \circ (\alpha \vee \beta) \vdash (\varphi \circ \alpha) \vee (\varphi \circ \beta)$. If $\alpha_w \vdash \varphi \circ (\alpha \vee \beta)$, like we have seen before, $\alpha_w \vdash \alpha \vee \beta$. Thus $\alpha_w \vdash \alpha$ or $\alpha_w \vdash \beta$.
If $\alpha_w \vdash \alpha$, as before, we get $\alpha \triangleright \alpha_w$ and by definition $\alpha_w \vdash \varphi \circ \alpha$.
If $\alpha_w \vdash \beta$ we have $\beta \triangleright \alpha_w$ and, by definition, $\alpha_w \vdash \varphi \circ \beta$.
Thus $\alpha_w \vdash (\varphi \circ \alpha) \vee (\varphi \circ \beta)$. This proves, $\varphi \circ (\alpha \vee \beta) \vdash (\varphi \circ \alpha) \vee (\varphi \circ \beta)$.
Now we chech that $(\varphi \circ \alpha) \vee (\varphi \circ \beta) \vdash \varphi \circ (\alpha \vee \beta)$. We have to prove $\varphi \circ \alpha \vdash \varphi \circ (\alpha \vee \beta)$ and $\varphi \circ \beta \vdash \varphi \circ (\alpha \vee \beta)$ We prove the first fact; the proof of the second fact is analogous.
By hypothesis, there exists α_w such that $\alpha \vee \beta \triangleright \alpha_w$ and $\alpha_w \vdash \alpha$.

By hypothesis $Expl(\alpha) \neq \emptyset$, thus, by (Limited reflexivity), $\alpha \triangleright \alpha$. By (Congruence), $(\alpha \vee \beta) \wedge \alpha \triangleright \alpha$. Then, by (P9) $\alpha \vee \beta \triangleright \alpha$. Thus, por (Right or), (Transitivity) and definition of $\circ$, we have $\varphi \circ \alpha \vdash \varphi \circ (\alpha \vee \beta)$.

Now we prove the *if part* of the Theorem.

Note that if $\gamma \vdash \varphi \circ \alpha$ and $\varphi \circ \alpha \vdash \alpha$, then $\gamma \vdash \alpha$, by (Success monotony) $\varphi \circ \gamma \vdash \gamma$.

(**Non triviality**) By *(Vacuity)* $\varphi \circ \varphi \equiv \varphi$, thus $\varphi \vdash \varphi \circ \varphi$ and $\varphi \circ \varphi \vdash \varphi$. Then $\varphi \triangleright \varphi$, therefore $\triangleright \neq \emptyset$. By Theorem 1, $\triangleright$ can be defined equivalently as

$$\alpha \triangleright \gamma \Leftrightarrow [\![\alpha]\!] \cap C_\varphi \neq \emptyset \text{ and } [\![\gamma]\!] \cap C_\varphi \neq \emptyset \text{ and } min([\![\gamma]\!] \cap C_\varphi) \subseteq min([\![\alpha]\!] \cap C_\varphi)$$

With this new definition, it must be demonstrated that $\triangleright$ satisfies the postulates of the ordered explanatory relationship.

(**Limited reflexivity**) If $Expl(\alpha) \neq \emptyset$, exists γ such that $\alpha \triangleright \gamma$, by definition $[\![\alpha]\!] \cap C_\varphi \neq \emptyset$ and $min([\![\alpha]\!] \cap C_\varphi) \subseteq min([\![\alpha]\!] \cap C_\varphi)$, thus $\alpha \triangleright \alpha$.

(**Weak infra-classicality**) If $\alpha \triangleright \gamma$, then $\emptyset \neq min([\![\gamma]\!] \cap C_\varphi) \subseteq min([\![\alpha]\!] \cap C_\varphi) \subseteq [\![\alpha]\!]$, since $min([\![\gamma]\!] \cap C_\varphi) \subseteq [\![\gamma]\!]$, then $[\![\alpha]\!] \cap [\![\gamma]\!] \neq \emptyset$, thus $\alpha \wedge \gamma \nvdash \bot$. See next picture.

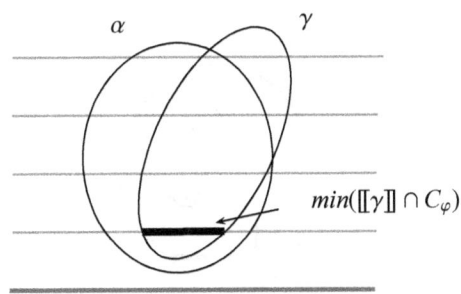

If γ explain α, like in previous picture, the minimal models of γ are minimal models of α, in particular, α and γ share models, thus $\alpha \wedge \gamma \nvdash \bot$.

(**Weak right strengthening**) By hypothesis $min([\![\gamma]\!] \cap C_\varphi) \subseteq min([\![\alpha]\!] \cap C_\varphi)$, $[\![\alpha]\!] \cap C_\varphi \neq \emptyset$, $[\![\gamma]\!] \cap C_\varphi \neq \emptyset$, $[\![\delta]\!] \subseteq [\![\gamma]\!]$ and $[\![\delta]\!] \neq \emptyset$.

Case 1: $[\![\delta]\!] \cap min([\![\gamma]\!] \cap C_\varphi) \neq \emptyset$. Since $[\![\delta]\!] \neq \emptyset$ and $[\![\delta]\!] \subseteq [\![\gamma]\!]$, it's easy to see $\emptyset \neq min([\![\delta]\!] \cap C_\varphi) \subseteq min([\![\gamma]\!] \cap C_\varphi)$, by transitivity $min([\![\delta]\!] \cap C_\varphi) \subseteq min([\![\alpha]\!] \cap C_\varphi)$, thus, $\alpha \triangleright \delta$. The following picture illustrates the behavior in this case:

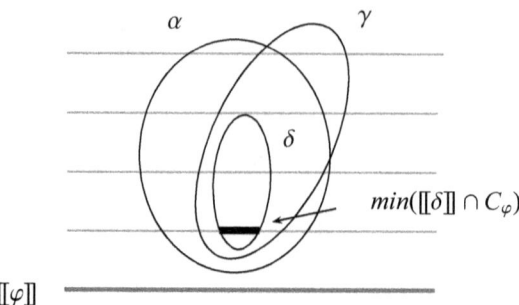

If every model of δ is a model of γ and at least one of its models is a minimal model of γ, then the minimal models of δ are minimal models of γ, in particular, they are minimal models of α.

Case 2: If $[\![\delta]\!] \cap min([\![\gamma]\!] \cap C_\varphi) = \emptyset$, then $min([\![\alpha]\!] \cap C_\varphi) \subseteq [\![\neg \delta]\!]$. Since $min([\![\alpha]\!] \cap C_\varphi) \subseteq [\![\neg \delta]\!]$ it is clear that $min([\![\alpha]\!] \cap [\![\neg \delta]\!] \cap C_\varphi) = min([\![\alpha]\!] \cap C_\varphi)$. Then, by transitivity $\emptyset \neq min([\![\gamma]\!] \cap C_\varphi) \subseteq min([\![\alpha]\!] \cap [\![\neg \delta]\!] \cap C_\varphi)$, that is to say, $\emptyset \neq min([\![\gamma]\!] \cap C_\varphi) \subseteq min([\![\alpha \wedge \neg \delta]\!] \cap C_\varphi)$. Thus, $\alpha \wedge \neg \delta \rhd \gamma$. The following picture illustrates this case:

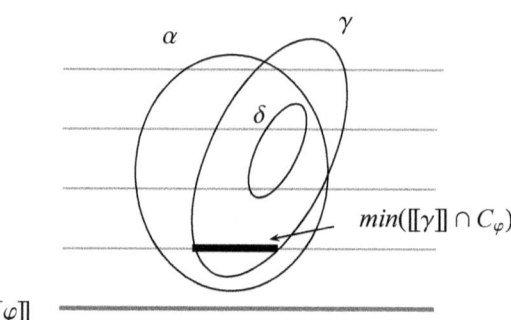

If every model of δ is a model of γ but it is not minimal model of γ and the minimal models of γ are also minimal models of α, then $\neg \delta$ contains the minimum models of α.

(Right or) By hypothesis and the property $min(A \cup B) \subseteq min(A) \cup min(B)$ we have
$min([\![\gamma \vee \delta]\!] \cap C_\varphi) \subseteq min([\![\gamma]\!] \cap C_\varphi) \cup min([\![\delta]\!] \cap C_\varphi) \subseteq min([\![\alpha]\!] \cap C_\varphi)$, thus, $\alpha \rhd \gamma \vee \delta$. The following picture illustrates this behavior:

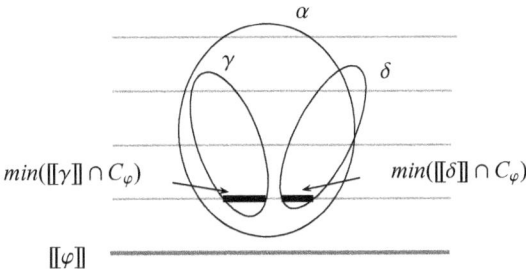

If the minimal models of γ and the minimal models of δ are minimal models of α, then the minimum models of the disjunction $\gamma \vee \delta$ are minimal models of α.

(**Cautious monotony**) By hypothesis, $min([\![\gamma]\!] \cap C_\varphi) \subseteq min([\![\alpha]\!] \cap C_\varphi)$, and $min([\![\gamma]\!] \cap C_\varphi) \subseteq [\![\gamma]\!] \subseteq [\![\beta]\!]$. Thus, $min([\![\gamma]\!] \cap C_\varphi) \subseteq min([\![\alpha]\!] \cap C_\varphi) \cap [\![\beta]\!]$. Note that we have the following general property $min(A) \cap B \subseteq min(A \cap B)$. From this and the previous facts we have $min([\![\alpha]\!] \cap C_\varphi) \cap [\![\beta]\!] \subseteq min([\![\alpha \wedge \beta]\!] \cap C_\varphi)$. Then $min([\![\gamma]\!] \cap C_\varphi) \subseteq min([\![\alpha \wedge \beta]\!] \cap C_\varphi)$. Thus, $\alpha \wedge \beta \triangleright \gamma$. The following picture illustrates this situation:

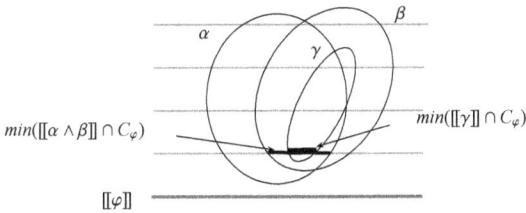

If the minimal models of γ are minimal models of α and every model of γ is a model of β, then the minimal models of γ are minimal models of conjunction $\alpha \wedge \beta$.

(**Transitivity**) If $min([\![\gamma]\!] \cap C_\varphi) \subseteq min([\![\alpha]\!] \cap C_\varphi)$, and $min([\![\delta]\!] \cap C_\varphi) \subseteq min([\![\gamma]\!] \cap C_\varphi)$, by transitivity $\emptyset \neq min([\![\delta]\!] \cap C_\varphi) \subseteq min([\![\alpha]\!] \cap C_\varphi)$. Thus, $\alpha \triangleright \delta$. The following picture illustrates this situation:

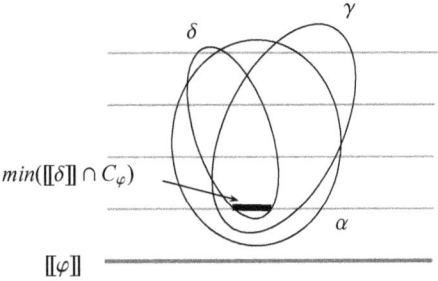

If the minimal models of δ are minimal models of γ and these in turn are also minimal models of α, then the minimal models of δ are minimal models of α.

(**Left and**) By hypothesis, $min(\llbracket\gamma\rrbracket \cap C_\varphi) \subseteq min(\llbracket\alpha\rrbracket \cap C_\varphi)$ and $min(\llbracket\gamma\rrbracket \cap C_\varphi) \subseteq min(\llbracket\beta\rrbracket \cap C_\varphi)$. Thus, $min(\llbracket\gamma\rrbracket \cap C_\varphi) \subseteq min(\llbracket\alpha\rrbracket \cap C_\varphi) \cap min(\llbracket\beta\rrbracket \cap C_\varphi)$. Note that we have the following general property: $min(A) \cap min(B) \neq \emptyset$, then $min(A \cap B) = min(A) \cap min(B)$. From this and the previous facts, we have $min(\llbracket\alpha\rrbracket \cap C_\varphi) \cap min(\llbracket\beta\rrbracket \cap C_\varphi) = min(\llbracket\alpha\rrbracket \cap \llbracket\beta\rrbracket \cap C_\varphi)$. Then $min(\llbracket\gamma\rrbracket \cap C_\varphi) \subseteq min(\llbracket\alpha \wedge \beta\rrbracket \cap C_\varphi)$. Thus, $\alpha \wedge \beta \triangleright \gamma$. The following picture illustrates this situation:

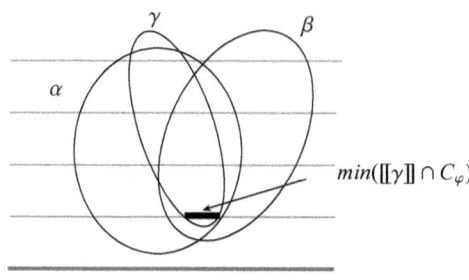

If minimal models of γ are minimal models of α and β, then they are minimal models of $\alpha \wedge \beta$.

(**Congruence**) If $\alpha \triangleright \gamma$ then, by hypothesis $min(\llbracket\gamma'\rrbracket \cap C_\varphi) \subseteq min(\llbracket\alpha'\rrbracket \cap C_\varphi)$ thus $\alpha' \triangleright \gamma'$. The reciprocal is analogous.

(**Explanatory monotony**) If $\text{Expl}(\alpha) \neq \emptyset$ exists γ such that $\alpha \triangleright \gamma$. Then $\emptyset \neq min(\llbracket\gamma\rrbracket \cap C_\varphi) \subseteq min(\llbracket\alpha\rrbracket \cap C_\varphi)$, since $\llbracket\alpha\rrbracket \subseteq \llbracket\beta\rrbracket$ we have $\llbracket\beta\rrbracket \cap C_\varphi \neq \emptyset$ and $min(\llbracket\beta\rrbracket \cap C_\varphi) \neq \emptyset$, then $\beta \triangleright \beta$, thus $\text{Expl}(\beta) \neq \emptyset$.

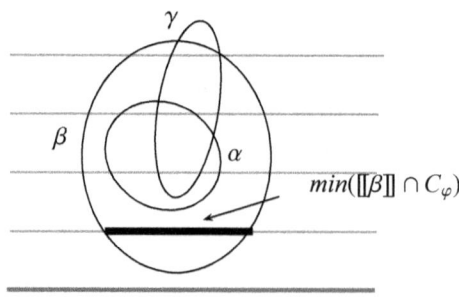

If γ explain α, its minimal models are minimal models of α, in particular, they are models of β. Thus, β has models in C_φ, therefore he has minimal models and for that reason he explains himself. ∎

From the previous Theorem and Theorem 1 we obtain the following result:

Corollary 2 $\triangleright$ *is an ordered explanatory relation iff there exists a nonempty set $C \subseteq \mathcal{W}$ and a total preorder $\preceq$ over C such that $\alpha \triangleright \gamma$ iff $[\![\alpha]\!] \cap C \neq \emptyset$, $[\![\gamma]\!] \cap C \neq \emptyset$ and $\min(\gamma \cap C, \preceq) \subseteq \min(\alpha \cap C, \preceq)$.*

Example 2 *We take the same data that in Example 1, that is, four propositional variables: c, s, p, g meaning Colombian coffee, coffee with sugar, coffee with pepper and good coffee respectively. The set C is $\{1001, 1101, 0001, 0001\}$ and the total preorder over C is given by $1001 \prec 1101 \sim 0001 \prec 0001$. The only observations α having explanations are the formulas α satisfying $[\![\alpha]\!] \cap C \neq \emptyset$. The only possible explanations γ are also formulas γ satisfying $[\![\gamma]\!] \cap C \neq \emptyset$. Here π_1 and π_2 have the same behavior, that is, $[\![\pi_i(\beta)]\!] = \min([\![\beta]\!] \cap C, \preceq)$ for any β having explanations. Thus, taking $\triangleright$ the ordered explanatory relation given by $(C, \preceq)$, using Corollary 2, we have good_coffee $\triangleright$ (Colombian_coffee _without_pepper); that is, the coffee could have or not sugar but if it is a Colombian one and it has not pepper this is still an explanation of the coffee goodness. However, this explanation of good_coffee is not an explanation in the framework of the weakly reflexive explanatory relation associated to same pair $(C, \preceq)$.*

It is interesting to note that the set C and the total preorder $\preceq$ given by corollaries 1 and 2 can be completely different in case we have two explanatory relations $\triangleright_1$ and $\triangleright_2$, the first one a weakly reflexive explanatory relation and the second one an ordered explanatory relation. But if we start from a subset C and a total preorder $\preceq$ corollaries 1 and 2 define explanatory relations $\triangleright_1$ and $\triangleright_2$, the first one weakly reflexive and the second one ordered, such that $\triangleright_1 \subseteq \triangleright_2$, that is for all observation α if $\alpha \triangleright_1 \gamma$ then $\alpha \triangleright_2 \gamma$. Thus, for a given pair $(C, \preceq)$ the weakly reflexive explanatory relation associated is more precise than the ordered explanatory relation associated. Moreover, the weakly reflexive explanatory relation associated is an explanatory relation in a classical sense, that is the explanation entails the observation. That is not, in general, the case with the ordered explanatory relations. The ordered explanatory relations are more flexible than the weakly reflexive relations. They accept more explanations. Some explanations can be conceived as imprecise explanations. A complete knowledge of the one ordered explanatory relation can lead to a more precise explanatory relation: a weakly reflexive explanatory relation.

4 Final remarks

We have obtained two families of explanatory relations, namely the weakly reflexive explanatory relations and the ordered explanatory relations. We have established tight links with credibility limited revision operators, given by Theorems 2 and 3. As corollaries we obtain semantical representations.

In both families of explanatory relations studied, the background theory Σ is, actually, implicit. It is, in fact, the theory of C, where the set C is the set given by Corollaries 1 and 2. The weakly reflexive explanatory relations are in fact an alternative view of the E-rational relations defined in [16].

The fact that the theory Σ is implicit allows us to give a simpler and purer logical presentation of the explanatory relations than the logical presentations where the postulates have to mention the background theory.

Note that, there are, in general, formulas without explanations. In particular the formulas whose set of models has no intersection with C.

By the semantical representation, it is easy to see that the family of ordered explanatory relations doesn't satisfy (Right strengthening).

Due to the fact that CL revision operators are generalizations of AGM revision operators, our Theorems 2 and 3 are, actually, generalizations of Walliser et al. results in [18].

There are many future research lines in perspective. In particular, to study more schemas (π_1, π_2) in order to define more families of explanatory relations. To introduce dynamics in the explanatory relations. To study the minimality of the set of postulates defining the families of explanatory relations considered in this work.

Acknowledgements

The second author was partially supported by the research project CDCHT-ULA N° C-1451-07-05- A.

References

1. C. E. Alchourrón. Detachment and defeasibility in deontic logic. *Studia Logica*, 51:5–18, 1996.
2. C. E. Alchourrón, P. Gärdenfors, and D. Makinson. On the logic of theory change: Partial meet contraction and revision functions. *Journal of Symbolic Logic*, 50:510–530, 1985.
3. I. Bloch, R. Pino Pérez, and C. Uzcátegui. Explanatory relations based on mathematical morphology. *In ECSQARU2001, Toulouse, France*, pages 736–747, 2001.
4. R. Booth, E. Fermé, S. Konieczny, and R. Pino Pérez. Credibility limited revision operators in propositional logic. *In Proceedings of the Thirteenth International Conference on Principles of Knowledge Representation And Reasoning*, pages 116–125, 2012.
5. C. Boutilier and V. Becher. Abduction as belief revision. *Artificial Intelligence*, 77:43–94, 1995.
6. M. Cialdea Mayer and F. Pirri. Abduction is not deduction-in-reverse. *Journal of the IGPL*, 4(1):1–14, 1996.
7. M. A. Falappa, G. Kern-Isberner, and G. R. Simari. Explanations, belief revision and defeasible reasoning. *Artificial Intelligence*, 143:1–28, 2002.
8. E. Fermé and R. Rodríguez. DFT and belief revision. *Análisis Filosófico*, XXVII(2):373–393, 2006.
9. P. A. Flach. Rationality postulates for induction. *Y. Shoam (Ed.), Proc. Sixth Conference of Theoretical Aspects of Rationality and Knowledge (TARK-96)*, pages 267–281, 1996.
10. P. A. Flach. Logical characteristics of inductive learning. *In D. M. Gabbay and R. Kruse, editors, Abductive Reasoning and Learning*, pages 155–196, 2000.

11. S. O. Hansson, E. L. Fermé, J. Cantwell, and M. A. Falappa. Credibility limited revision. *Journal of Symbolic Logic*, 66:1581–1596, 2001.
12. H. Katsuno and A. Mendelzon. Propositional knowledge base revision and minimal change. *Artificial Intelligence*, 52:263–294, 1991.
13. María Victoria León and Ramón Pino Pérez. Explanatory relations revisited: Links with credibility-limited revision. In Manuel Montes-y-Gómez, Hugo Jair Escalante, Alberto Segura, and Juan de Dios Murillo, editors, *Advances in Artificial Intelligence - IBERAMIA 2016 - 15th Ibero-American Conference on AI, San José, Costa Rica, November 23-25, 2016, Proceedings*, volume 10022 of Lecture Notes in Computer Science, pages 25–36, 2016.
14. H. J. Levesque. A knowledge level account of abduction. *In procedings of the eleventh International Joint Conference on Artificial Intelligence, Detroit*, pages 1061–1067, 1989.
15. M. Pagnucco. *The Role of Abductive Reasoning Within the Process of Belief Revision*. PhD thesis, Department of Computer Science, University of Sydney, February 1996.
16. R. Pino Pérez and C. Uzcátegui. Jumping to explanations vs jumping to conclusions. *Artificial Intelligence*, 111(2):131–169, 1999.
17. R. Pino Pérez and C. Uzcátegui. Preferences and explanations. *Artificial Intelligence*, 149(1):1–30, 2003.
18. B. Walliser, D. Zwirn, and H. Zwirn. Abductive logic in a belief revision framework. *J. Logic Language Inform*, 14:87–117, 2005.

Modelling Support Relations between Arguments in Debates

Henry Prakken[1,2]

[1] Department of Information and Computing Sciences,
Utrecht University, Utrecht, The Netherlands
[2] Faculty of Law, University of Groningen, Groningen, The Netherlands

Abstract. Many formal modellings of structured argumentation presuppose a knowledge base from which arguments are constructed. However, in debate contexts there usually is no global knowledge base from which the debate participants construct their arguments. The question then arises how these formalisms can be used for evaluating debates. On issue here is how support relations between arguments put forward in a debate should be modelled. This paper develops a formal approach within the $ASPIC^+$ framework and compares it to approaches using bipolar abstract argumentation frameworks. It is argued that for a proper model of debate evaluation it is crucial to look at the structure of arguments, which casts doubt on the benefits of purely abstract models of debate evaluation.

1 Introduction

Imagine John Doe watching or reading a debate on a topic like 'Does global global warming exist?', 'Should the west bomb the IS?' or 'Should the Schengen area be terminated?'. After the debate is finished, John wants to determine which arguments and claims put forward in the debate are acceptable. He does not care about how the debate evolved over time or who said what, he just wants to look at the contents of the arguments. He wants to reconstruct how the various arguments support or attack each other, he wants to express whether he accepts their premises or inferences, he wants to express his preferences between conflicting arguments, and he may want to add some arguments of his own. This is what in this paper will be called 'evaluating a debate'. Evaluating a debate in this sense is largely a subjective matter, since different people can make different choices on all the points just mentioned. However, there are still rational constraints, namely, those formulated by informal and formal argumentation theory. In this paper the focus is on how the current formal models of argumentation constrain debate evaluation in the sense just explained. To this end, it will be assumed that a formalised version of the arguments put forward in a debate already exists. In practice the step from a natural-language debate to a formalised version is far from trivial but this step is not what this paper is about.

At first sight, there would seem to be no problem: if a formalised set of arguments in some argumentation logic already exists, then it would seem to suffice to simply apply the argumentation logic to the set of arguments. However, there is still a problem here, since many formal modellings of structured argumentation presuppose a knowledge base and a set of rules from which arguments have to be constructed (e.g. [22, 17, 9, 7, 12, 19, 13]). The problem is that in debate contexts there usually is no global knowledge base or agreed set of rules from which the debaters construct their arguments. One problem in particular is how to deal with arguments put forward by different participants of the debate and where the conclusion of one argument provides a premise of the other argument. Should the two arguments be regarded as a single complex argument on the basis of some global background assumed in the debate, where the inferential support relations inside the single combined argument captures the relation of support, or should the arguments be regarded as separate entities related through another notion of support? Cayrol & Lagasquie-Schiex argue in [3] for the latter solution on the ground that it would be less natural to regard the situation where one agent supports an argument stated by another agent as a case of revising the supported argument by the supporter. Instead it would be more natural to model both arguments as separate entities. This view on debate evaluation was one motivation for the development of so-called bipolar argumentation frameworks (BAFS), which add abstract support relations to the theory of abstract argumentation frameworks (AFs) originating from [5].

One aim of this paper is to argue that this criticism is not justified and that it is still better to model the support relation between such arguments by combining them into a single complex argument and capturing their support relation in the inferential support relations within the argument. To account for the fact that the two original arguments were possibly stated by different debate participants, these two arguments will not be ignored but will also be considered in the evaluation process. A second aim of this paper is to show how this evaluation process allows for a more subtle evaluation of arguments than in the theory of BAFs.

Since BAFs are defined on top of AFs, the present investigations will also assume the theory of AFs. A third aim of this paper then is to show that analysing support relations between arguments in debates cannot be done without an account of the structure of arguments and the nature of attacks. As such an account, the $ASPIC^+$ framework [19, 13–15] will be used, which allows a modelling of inferential support relations with its notion of a subargument. However, the main ideas of this paper also apply to formalisms for structured argumentation that do not precisely instantiate the $ASPIC^+$ framework.

This paper is organised as follows. After presenting the formal preliminaries in Section 2, a way to reconstruct the arguments put forward in a debate in $ASPIC^+$ will be proposed in Section 3. Then in Section 4 it will be shown that a combination of $ASPIC^+$ and a BAF approach with premise support for modelling debate evaluation has some disadvantages and that it is better to combine supporting and supported arguments in a single compound argument by using $ASPIC^+$'s subargument relation. This method will be formalised in Section 5, after which it will be shown in Section 6 how the method can be applied to evaluating debates. The paper will conclude in Section 7.

2 Formal preliminaries

In this section the formal frameworks used or discussed in this paper are reviewed.

An *abstract argumentation framework* (AF) is a pair $\langle \mathcal{A}, \mathcal{D} \rangle$, where $\mathcal{A}$ is a set of arguments and $\mathcal{D} \subseteq \mathcal{A} \times \mathcal{A}$ is a relation of defeat. The theory of AFs [5] identifies sets of arguments (called *extensions*) which are internally coherent and defend themselves against defeaters. An argument $A \in \mathcal{A}$ is *defended* by a set by $S \subseteq \mathcal{A}$ if for all $B \in \mathcal{A}$: if B defeats A, then some $C \in S$ defeats B. Then relative to a given AF, $E \subseteq \mathcal{A}$ is *admissible* if E is conflict-free and defends all its members; E is a *complete extension* if E is admissible and $A \in E$ iff A is defended by E; E is a *preferred extension* if E is a $\subseteq$-maximal admissible set; E is a *stable extension* if E is admissible and attacks all arguments outside it; and $E \subseteq \mathcal{A}$ is the *grounded extension* if E is the least fixpoint of operator F, where $F(S)$ returns all arguments defended by S. It holds that any preferred, stable or grounded extension is a complete extension. Finally, for $T \in \{$complete, preferred, grounded, stable$\}$, X is *sceptically* or *credulously* justified under the T semantics if X belongs to all, respectively at least one, T extension.

Several proposals exist for adding support relations to abstract argumentation frameworks, the best-known being [3]'s bipolar argumentation frameworks (BAFs). The literature on BAFs contains several proposals for definitions of conflict-freeness and admissibility. For now it is not necessary to commit to any specific proposal; therefore for now simply 'abstract argumentation frameworks with support (SuppAFs)' (a term borrowed from [21]) will be considered, which add a binary support relation to AFs. Thus SuppAFs are a triple $(\mathcal{A}, \mathcal{D}, \mathcal{S})$ where $\mathcal{D}$ and $\mathcal{S}$ are binary relations over a set $\mathcal{A}$. Depending on the definitions of conflict-freeness and admissibility and on which further constraints are added, SuppAFs may or may not be BAFs.

The **$ASPIC^+$ framework** [19, 13, 14] gives structure to Dung's arguments and defeat relation. It defines arguments as directed acyclic graphs formed by applying strict or defeasible inference rules to premises formulated in some logical language. Arguments can be attacked on their (non-axiom) premises and on their applications of defeasible inference rules. Some attacks succeed as *defeats*, as partly determined by preferences. The acceptability status of arguments is then defined by applying any of [5]'s semantics for abstract argumentation frameworks to the resulting set of arguments with its defeat relation. Since the initial paper [19] on $ASPIC^+$, several variants of the framework have been proposed. For present purposes, their differences do not matter. In this paper we will use the variant with symmetric negation and so-called 'defeat conflict-freeness' as presented in [15].

$ASPIC^+$ is not a system but a framework for specifying systems. It defines the notion of an abstract *argumentation system* as a structure consisting of a logical language $\mathcal{L}$ with a binary negation symbol $\neg$, a set $\mathcal{R}$ consisting of two subsets $\mathcal{R}_s$ and $\mathcal{R}_d$ of strict and defeasible inference rules, and a naming convention n in $\mathcal{L}$ for defeasible rules in order to talk about the applicability of defeasible rules in $\mathcal{L}$. Informally, $n(r)$ is a wff in $\mathcal{L}$ which says that rule $r \in \mathcal{R}$ is applicable.

Definition 1. *[Argumentation systems] An* argumentation system *is a triple* $AS = (\mathcal{L}, \mathcal{R}, n)$ *where:*

- $\mathcal{L}$ *is a logical language with a binary negation symbol* $\neg$.
- $\mathcal{R} = \mathcal{R}_s \cup \mathcal{R}_d$ *is a finite set of strict* ($\mathcal{R}_s$) *and defeasible* ($\mathcal{R}_d$) *inference rules of the form* $\{\varphi_1, \ldots, \varphi_n\} \to \varphi$ *and* $\{\varphi_1, \ldots, \varphi_n\} \Rightarrow \varphi$ *respectively (where* φ_i, φ *are meta-variables ranging over wff in* $\mathcal{L}$*), such that* $\mathcal{R}_s \cap \mathcal{R}_d = \emptyset$. $\varphi_1, \ldots, \varphi_n$ *are called the* antecedents *and* φ *the* consequent *of the rule.*[3]
- n *is a partial function from* $\mathcal{R}_d$ *to* $\mathcal{L}$*, which to rules in* $\mathcal{R}_d$*, a naming convention for defeasible rules.*

We write $\psi = -\varphi$ *just in case* $\psi = \neg \varphi$ *or* $\varphi = \neg \psi$.

Definition 2. *[Knowledge bases] A* knowledge base *in an* $AS = (\mathcal{L}, \mathcal{R}, n)$ *is a set* $\mathcal{K} \subseteq \mathcal{L}$ *consisting of two disjoint subsets* $\mathcal{K}_n$ *(the* axioms*) and* $\mathcal{K}_p$ *(the* ordinary premises*).*

Arguments can be constructed step-by-step from knowledge bases by chaining inference rules into directed acyclic graphs (which are trees if no premise is used more than once). In what follows, for a given argument the function Prem returns all its premises, Conc returns its conclusion, Prop and Rules return, respectively, all wff and all rules occurring in it, Sub returns all its sub-arguments and TopRule returns the last inference rule applied in the argument.

Definition 3. *[Arguments] An* argument A *on the basis of a knowledge base* $\mathcal{K}$ *over an argumentation system* AS *is any structure obtainable by applying one or more of the following steps finitely many times:*

1. φ *if* $\varphi \in \mathcal{K}$ *with:* $\text{Prem}(A) = \{\varphi\}$; $\text{Conc}(A) = \varphi$; $\text{Prop}(A) = \{\varphi\}$; $\text{Sub}(A) = \{\varphi\}$; $\text{TopRule}(A) = $ *undefined;* $\text{Rules}(A) = \emptyset$.
2. $A_1, \ldots, A_n \to/\Rightarrow \psi$ *if* $A_1, \ldots, A_n$ *are arguments such that there exists a strict/defeasible rule* $\text{Conc}(A_1), \ldots, \text{Conc}(A_n) \to/\Rightarrow \psi$ *in* $\mathcal{R}_s/\mathcal{R}_d$.
   $\text{Prem}(A) = \text{Prem}(A_1) \cup \ldots \cup \text{Prem}(A_n)$;
   $\text{Conc}(A) = \psi$;
   $\text{Prop}(A) = \text{Prop}(A_1) \cup \ldots \cup \text{Prop}(A_n) \cup \{\psi\}$;
   $\text{Sub}(A) = \text{Sub}(A_1) \cup \ldots \cup \text{Sub}(A_n) \cup \{A\}$;
   $\text{TopRule}(A) = \text{Conc}(A_1), \ldots, \text{Conc}(A_n) \to/\Rightarrow \psi$;
   $\text{Rules}(A) = \text{Rules}(A_1) \cup \ldots \cup \text{Rules}(A_n) \cup \{\text{Conc}(A_1), \ldots, \text{Conc}(A_n) \to/\Rightarrow \psi\}$.

---

[3] Below the brackets around the antecedents will be omitted.

For any argument A we define $\text{Prem}_n(A) = \text{Prem}(A) \cap \mathcal{K}_n$ and $\text{Prem}_p(A) = \text{Prem}(A) \cap \mathcal{K}_p$. Moreover, for any set $\mathcal{A}$ of arguments, $\text{Prem}(\mathcal{A}) = \{\varphi \mid \varphi \in \text{Prem}(A) \text{ for some } A \in \mathcal{A}\}$. The notations $\text{Conc}(\mathcal{A})$, $\text{Prop}(\mathcal{A})$ and $\text{Rules}(\mathcal{A})$ are defined likewise while $\mathcal{R}_s(\mathcal{A}) = \text{Rules}(\mathcal{A}) \cap \mathcal{R}_s$ and $\mathcal{R}_d(\mathcal{A}) = \text{Rules}(\mathcal{A}) \cap \mathcal{R}_d$.

Arguments can be attacked in three ways: on their premises (undermining attack), on their conclusion (rebutting attack) or on an inference step (undercutting attack). The latter two are only possible on applications of defeasible inference rules.

Definition 4. *[Attack]* A attacks B iff A undercuts, rebuts or undermines B, where:
- A undercuts argument B (on B') iff $\text{Conc}(A) = -n(r)$ and $B' \in \text{Sub}(B)$ such that B''s top rule r is defeasible.
- A rebuts argument B (on B') iff $\text{Conc}(A) = -\varphi$ for some $B' \in \text{Sub}(B)$ of the form $B''_1, \ldots, B''_n \Rightarrow \varphi$.
- Argument A undermines B (on B') iff $\text{Conc}(A) = -\varphi$ for some $B' = \varphi$, $\varphi \notin \mathcal{K}_n$.

Argumentation systems plus knowledge bases form argumentation theories, which induce structured argumentation frameworks.

Definition 5. *[Structured Argumentation Frameworks]* Let AT be an argumentation theory $(AS, \mathcal{K})$. A structured argumentation framework (SAF) defined by AT, is a triple $\langle \mathcal{A}, \mathcal{C}, \preceq \rangle$ where $\mathcal{A}$ is the set of all arguments on the basis of $\mathcal{K}$ in AS, $\preceq$ is an ordering on $\mathcal{A}$, and $(X, Y) \in \mathcal{C}$ iff X attacks Y.

The $ASPIC^+$ notion of *defeat* can then be defined as follows. Undercutting attacks succeed as *defeats* independently of preferences over arguments, since they express exceptions to defeasible inference rules. Rebutting and undermining attacks succeed only if the attacked argument is not stronger than the attacking argument ($A \prec B$ is defined as usual as $A \preceq B$ and $B \npreceq A$).

Definition 6. *[Defeat]* A defeats B iff:A undercuts B, or; A rebuts/undermines B on B' and $A \nprec B'$.

Abstract argumentation frameworks are then generated from SAFs as follows:

Definition 7 (Argumentation frameworks). *An abstract argumentation framework (AF) corresponding to a SAF $= \langle \mathcal{A}, \mathcal{C}, \preceq \rangle$ is a pair $(\mathcal{A}, \mathcal{D})$ such that $\mathcal{D}$ is the defeat relation on $\mathcal{A}$ determined by SAF.*

Then several ways are possible of using the theory of AFs for evaluating conclusions of arguments. One is to say that a formula φ is a skeptical (credulous) consequence of a SAF iff an argument with conclusion φ is in all (some) extensions of the AF corresponding to SAF. Other definitions are possible; for present purposes their differences do not matter.

3 Formalising debates

Assume that in a debate the participants construct the arguments in their own internal $ASPIC^+$ argumentation theory. It cannot in general be assumed that these theories have the same language, rules and knowledge base. All an outsider can observe is the arguments, that is, their premises and inferences. The task then is to construct a global argumentation theory that generates these arguments and that does not contain any information that is not contained in these arguments. To make this well-defined, such an AT will be defined indirectly in the notion of a debate-generated SAF, otherwise the set of arguments that generates the AT cannot be easily stated in a well-defined way.

Let a partial SAF be defined as a SAF except that $\mathcal{A}$ is any set of arguments constructible on the basis of $\mathcal{K}$ in AS (so not all constructible arguments need to be in $\mathcal{A}$). Then:

Definition 8. *Let a debate-generated SAF be any partial $SAF = \langle \mathcal{A}, \mathcal{C}, \preceq \rangle$ defined by an $AT = ((\mathcal{L}, \mathcal{R}, n), \mathcal{K}_n \cup \mathcal{K}_p)$ such that*

1. $\mathcal{L} = \text{Prop}(\mathcal{A})$;
2. $\mathcal{K}_n = \emptyset$;
3. $\mathcal{K}_p = \mathcal{A} \cap \mathcal{L}$;
4. $\mathcal{R}_s(\mathcal{A}) = \{\text{Conc}(A_1), \ldots, \text{Conc}(A_n) \to \psi \mid \text{there exists an } A = [A_1], \ldots, [A_n] \to \psi \in \mathcal{A}\}$;
5. $\mathcal{R}_d(\mathcal{A}) = \{\text{Conc}(A_1), \ldots, \text{Conc}(A_n) \Rightarrow \psi \mid \text{there exists an } A = [A_1], \ldots, [A_n] \Rightarrow \psi \in \mathcal{A}\}$;
6. *n is any partial function from $\mathcal{R}_d$ to $\mathcal{L}$;*
7. *$\preceq$ is any ordering on $\mathcal{A}$.*

This definition is in fact a fixpoint construction and it may have multiple fixpoints, as will be shown below. The idea is that it formally reconstructs the set $\mathcal{A}$ of arguments stated in the debate and adds no further arguments but that the argument preferences are provided by an evaluator of the debate. As for the n function it is simply assumed that it can be sensibly identified from the debate, without going into details. The following proposition states that the construction of Definition 8 is well-defined.

Proposition 1. *For any debate-generated $SAF = \langle \mathcal{A}, \mathcal{C}, \preceq \rangle$ defined by an $AT = ((\mathcal{L}, \mathcal{R}, n), \mathcal{K}_n \cup \mathcal{K}_p)$ it holds that all elements of $\mathcal{A}$ are constructible on the basis of $\mathcal{K}_n \cup \mathcal{K}_p$ in $(\mathcal{L}, \mathcal{R}, n)$.*

Proof. The proof is by induction on the construction of arguments. If $A \in \mathcal{L}$ then $A \in \mathcal{K}_p$ by clause (3) of Definition 8, so A is constructible. Otherwise A is of the form $A = [A_1], \ldots, [A_n] \to/\Rightarrow \psi \in \mathcal{A}\}$. By the induction hypothesis, $A_1, \ldots, A_n$ are constructible. Moreover, $\text{Toprule}(A)$ is in $\mathcal{R}_s$ or $\mathcal{R}_d$ by clause (4) or (5) of Definition 8. So A is constructible.

Corollary 1. *For any debate-generated SAF the set $\mathcal{A}$ is closed under the subargument relation.*

The converse of Proposition 1 does not hold in general. Consider the following debate:

$A_1: p$ $B_1: s$
$A_2: A_1 \Rightarrow q$ $B_2: B_1 \Rightarrow q$
$A_3: A_2 \Rightarrow r$ $B_3: B_2 \Rightarrow t$

Then the following argument C is constructible on the basis of the debate-generated SAF:

$A_1: p$
$A_2: A_1 \Rightarrow q$
$C: \;\; A_2 \Rightarrow t$

However, C is not in $\mathcal{A}$. This also shows that the construction of Definition 1 may have multiple fixpoints, since there exist a debate-generated SAF with $\mathcal{A} = \{A_1, A_2, A_3, B_1, B_2\}$ and one with $\mathcal{A} = \{A_1, A_2, A_3, B_1, B_2, C\}$.

Now the problem motivating this paper arises if for some arguments $A_i \in \mathcal{A}_i$ and $A_j \in \mathcal{A}_j (i \neq j)$ we have that A_i's conclusion is an ordinary premise of A_j, where neither argument is atomic (so not an element of $\mathcal{L}$). For example, consider a debate variant of a well-known example from the literature on non-monotonic logic, with a propositional language. John says "Nixon was a pacifist (p) since he was a Quaker (q) and Quakers are usually pacifists ($q \Rightarrow p$)", while Bob says "Nixon was not a pacifist ($\neg p$) since he was a republican (r) and republicans are usually not pacifists ($r \Rightarrow \neg q$)" (thus generalisations are modelled as defeasible inference rules). Formally:

$A_1: q$ $B_1: r$
$A_2: A_1 \Rightarrow p$ $B_2: B_2 \Rightarrow \neg p$

Suppose now that Mary supports John's argument by saying "Nixon regularly attended service in a Quaker church (c), people who are regularly seen in a Quaker church usually are a Quaker ($c \Rightarrow q$), so Nixon was a Quaker". Formally:

$A_3: c$
$A_4: A_3 \Rightarrow q$

It is this kind of situation that is of special interest in this paper. On the one hand, we want to capture that in some sense argument A_4 supports argument A_2. For example, the evaluator might not be prepared to accept q if it is not supported by some argument. On the other hand, we do not want to simply replace these two arguments with the following combined argument, while deleting q from $\mathcal{K}$ of the debate-generated SAF:

$A_3: c$
$A_4: A_2 \Rightarrow q$
$A_5: A_4 \Rightarrow p$

The reason why we do not want to do this is that there can also be situations in which the evaluator does not accept argument A_5 for q but still accepts q as a premise: so argument A_5 should be part of the debate on its own, at least initially.

This paper's solution will be that both the two individual arguments A_2 and A_4 and their combination A_5 will initially be part of the debate and that the evaluator should for each premise of A_2 decide whether to accept it without further argument. If so, then A_5 is irrelevant for the issue p whether Nixon was a pacifist, otherwise, A_2 will be removed from the debate and A_5 becomes relevant. It is this approach that will be formalised in Sections 5 and 6. But first the alternative solution presented by [3] will be discussed.

4 The BAF approach with premise support

In $ASPIC^+$ the only support relation between arguments is the subargument relation as defined in Definition 3, where each argument contains all its subarguments as part of itself. Let us now examine Cayrol & Lagasquie-Schiex's claim in [3] that to model debates, instead support relations between separate arguments are needed. To this end, consider a version of $ASPIC^+$ that generates SuppAFs, leaving the definitions of conflict-freeness and admissibility as they are. Two ways of defining the support relation suggest themselves: the original subargument relation from $ASPIC^+$ and the definition that A supports B if the conclusion of A is a premise of B (henceforth called *premise support*). $ASPIC^+$-SuppAFs with subargument support were in [21], building on [20], shown to be equivalent to $ASPIC^+$ when generating AFs as defined above in Definition 7 (this result was reproved by [4]). However, the same does not hold for $ASPIC^+$-SuppAFs with premise support. The crucial difference between premise support and ASPIC's subargument relation is that while each argument contains all subarguments as part of itself, premise-support relations can hold between different arguments of which neither contains the other.

To illustrate the difference, consider again the above modelling the Nixon example. Argument A_5 for the conclusion that Nixon was a pacifist contains the arguments A_3 that Nixon was regularly seen in a Quaker church and A_4 that therefore he was a pacifist as part of itself. This modelling is by Cayrol & Lagasquie-Schiex considered less natural. They want that John's, Bob's and Mary's arguments are three individual arguments, which all stand on their own but can be related by support relations. For instance, John's argument has a premise q and conclusion p and is supported by Mary's argument which has premise c and conclusion q: John's argument does not contain Mary's argument as part of itself. Likewise, Bob's argument stands on its own, with premise r and conclusion $\neg p$.

Let us see how Definition 8 can be applied to respect this view. There exists a debate-generated SAF which only contains the arguments $A_1, A_2, A_3, A_4, B_1, B_2$ and not A_5. This (partial) SAF is defined as follows.

- $\mathcal{L} = \{c, p, q, r, \neg p\}$;
- $\mathcal{K}_p = \{c, q, r\}$;
- $\mathcal{R}_s(\mathcal{A}) = \emptyset$;
- $\mathcal{R}_d(\mathcal{A}) = \{c \Rightarrow q; q \Rightarrow p; r \Rightarrow \neg p\}$.

- $n = \emptyset$;
- $\preceq$ is any;
- $\mathcal{A} = \{A_1, A_2, A_3, A_4, B_1, B_2\}$.

With premise support we have that the support relations are that A_1 supports A_2, B_1 supports B_2, A_3 supports A_4 and A_4 supports A_2.

Note that Definition 8 also allows a debate-generated SAF with q deleted from $\mathcal{K}_p$ and A_5 added to $\mathcal{A}$ (recall that Definition 8 is a fixpoint construction that may have multiple fixpoints). However, this is not a problem, since the point is that an evaluator can, following [3], decide that the first debate-generated SAF is the one corresponding to the debate s/he is analysing.

Is this then the way $ASPIC^+$ can be used for evaluating debates, by letting it generate SuppAFs as just sketched with the notion of premise support? The first issue here is whether defeating a supporter of an argument should have an impact on the supported argument. So far this is with premise support not guaranteed. Suppose in the Nixon example that argument A_3 is defeated by an argument C. Then C does not indirectly defeat A_1 or A_2, since A_3 is not a subargument of these arguments (recall that A_5, of which A_3 is a subargument, is not in the SuppAF with premise support). So in SuppAFs as defined thus far, defeating a supporter of an argument does not have any logical effect on the status of the supported argument.

For contexts in which arguments are generated from a given knowledge base this is clearly undesirable. However, at first sight, it might be argued that for debate contexts this is otherwise. Suppose that Bob in our example debate succeeds in defeating Mary's argument by arguing that Nixon only attended service in a Quaker church to please his wife, who was a Quaker. It could reasonably be argued that this does not knock down John's argument, since why should John be blamed for Mary's flawed attempt to support his argument? On the other hand, it is still unsatisfactory that there is no logical relation at all between defeating a supporter and the status of a supported argument. If support means anything at all, then surely defeating a supporter should have some logical effect on the status of an argument supported by it. The problem then is how this view can be reconciled with the view that in debate contexts this logical effect cannot simply be that the supported argument is always defeated.

At first sight, a possible solution would be to close the defeat relation in a SuppAF under the constraint (adopted from [8, 16]) that if A supports B and C defeats A, then C also defeats B. This constraint (in the literature on BAFs called 'secondary attack') is both necessary and sufficient to prove that $ASPIC^+$-SuppAFs with subargument support are equivalent to $ASPIC^+$-$SAFs$ [21]. Can $ASPIC^+$-SuppAFs with premise support and with secondary attacks be used for evaluating debates? This is still not true, for two reasons. The first reason is that then argument B is (indirectly) defeated by C regardless whether the evaluator wants to accept its premise q without further support. The second reason was given in [21], namely, that such SuppAFs cannot distinguish between the following two situations:

Situation 1: A has premises p and q, B has conclusion p, C has conclusion q, D undercuts C.
Situation 2: A has premise p and both B and C have conclusion p, D undercuts C.

Both situations induce the same SuppAF with premise support and secondary attack, in which both B and C support A and D defeats both A and C. However, this is counter-intuitive, since in the second situation A should not be defeated, since its premise p is still provided by an undefeated argument, namely, B. With support as subargument support instead the intuitive outcome is obtained, since then there are in situation 2 two 'versions' of A, one with B as subargument for q and the other with C as subargument for q; and only the second version of A is defeated by D. It can be concluded that the notion of secondary attack is not suitable for premise support.

More generally, this analysis shows that support relations between arguments in debates after all have to be modelled by combining the two arguments into a single complex one, that is, as $ASPIC^+$-style subargument relations. But how can this be done in a way that still respects that the combined arguments may have been stated by different debate participants? This paper's solution to this problem will be based on the idea that debates are always evaluated relative to a given or assumed 'basis for discussion' (an idea also underlying the Carneades framework of [10]). This basis can be objective or authoritative, such as when a judge or jury in a legal proceeding has to evaluate a dispute between opposing parties, but it can also be purely subjective, such as when an ordinary citizen evaluates a political debate in light of his or her subjective opinions. In the Nixon example, if the evaluator accepts the premise of John's argument that Nixon was a Quaker, then defeating Mary's supporting argument has no effect on the status of John's argument. But if the evaluator does not accept this premise but only accepts Mary's premise that Nixon regularly attended service in a Quaker church, then defeating Mary's argument also defeats John's argument. To make this work, both the original arguments and their combined version should be considered in the evaluation process.

5 Combining arguments

Recall that in debate evaluation we are not interested in the sources of the arguments but only in their logical and dialectical relations. We therefore face the problem of combining arguments that may come from different sources. Since, as just shown, in the present formal context the notion of premise support cannot be used for this purpose but the $ASPIC^+$ subargument relation has to be used. The problem now is how to do this, that is, how to extend debate-generated SAFs as defined in Definition 8 with further arguments and attack relations.

The idea is twofold. First, if the conclusion of an argument A provides a premise of an argument B, then the arguments are combined into a new argument by replacing the premise in B by A as a subargument of B. This is in fact the notion of 'backwards extending' an argument from [18] and the notion of 'weakening' an argument from [6]. But this is not all. It may happen that an argument A has the same conclusion as a non-premise proper subargument B' of B. For example, suppose argument A_5 in our Nixon example was stated by a debate participant and now another participant states a new argument with conclusion q, for instance, that an inhabitant of Nixon's home town says that Nixon was a Quaker (h).

A_6: h
A_7: $A_6 \Rightarrow q$

Then it seems reasonable to create a new version of A_5 with its subargument A_4 replaced by A_7. It turns out that both cases (see Figure 1) can be combined into a single inductive definition.

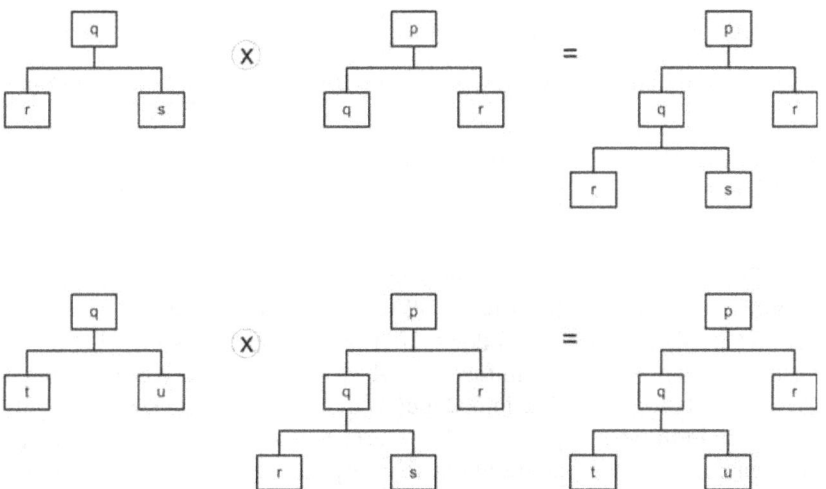

Fig. 1. Two ways to modify an argument

Definition 9. *For any argument B and wff φ, $B^\varphi = \varphi$ iff $\mathrm{Conc}(B) = \varphi$, otherwise $B^\varphi = B$.*

For any pair of arguments $A \notin \mathcal{L}$ and B such that $A \notin \mathrm{Sub}(B)$ the modification $A \otimes B$ of B by A is inductively defined as follows:

1. *if $\mathrm{Conc}(A) \neq \mathrm{Conc}(B')$ for all $B' \in \mathrm{Sub}(B)$ then $A \otimes B = B$, else:*
2. *if $B \in \mathcal{L}$ then $A \otimes B = A$;*
3. *if $B \notin \mathcal{L}$ then we have that B is of the form $B_1, \ldots, B_n \to / \Rightarrow \psi$: then $A \otimes B = A \otimes B_1^\varphi, \ldots, A \otimes B_n^\varphi \to / \Rightarrow \psi$, where $\varphi = \mathrm{Conc}(A)$.*

In case (1) A does not add anything to B so B remains unchanged. In case (2) B is an item from the knowledge base which is equal to A's has conclusion; then A replaces B. Case (3) is the inductive clause. For any $B_i (1 \leq i \leq n)$ such that A's conclusion equals the conclusion of B_i, B_i is replaced with A in B by first replacing B_i with A's conclusion and then applying case (2) to the resulting premise argument. For all remaining $B_j (j \neq i)$ case (3) is inductively applied. In sum, case (2) replaces premises of B with an argument for the premise (the first case in Figure 1) while case (3) replaces subarguments of B with alternative subarguments for the same conclusion (the second case in Figure 1) and takes care of the recursion.

Definition 10. *The* closure *of a set $\mathcal{A}$ of arguments is the smallest set $\mathcal{A}^\otimes$ such that:*

1. *If $A \in \mathcal{A}$ then $A \in \mathcal{A}^\otimes$;*
2. *If $A \in \mathcal{A}^\otimes$ and $B \in \mathcal{A}^\otimes$ then $A \otimes B \in \mathcal{A}^\otimes$.*

A set $\mathcal{A}$ of arguments is closed *if $\mathcal{A}^\otimes = \mathcal{A}$.*

A debate-generated SAF with a closed set of arguments is always a non-partial SAF.

Proposition 2. *For any debate-generated $SAF = \langle \mathcal{A}, \mathcal{C}, \preceq \rangle$ defined by an AT where $\mathcal{A}$ is closed, it holds that $\mathcal{A}$ is the set of all arguments on the basis of AT.*

Proof. The only-if part (all arguments in $\mathcal{A}$ are constructible on the basis AT) is Proposition 1, while the if-part (all arguments constructible on the basis of AT are in $\mathcal{A}$) is proven as follows. If A is constructible since $A \in \mathcal{K}_p$, then $A \in \mathcal{A} \cap \mathcal{L}$ by clause (3) of Definition 8, so $A \in \mathcal{A}$. Then if A is constructible since $A_1, \ldots, A_n$ are constructible and $\mathsf{Conc}(A_1), \ldots, \mathsf{Conc}(A_n) \to/\Rightarrow \psi \in \mathcal{R}_s/\mathcal{R}_d$ then by the induction hypothesis $A_1, \ldots, A_n$ are in $\mathcal{A}$. By clause (4) or (5) of Definition 8 there exists an argument $A' = A'_1, \ldots, A'_n \to/\Rightarrow \psi$ in $\mathcal{A}$. Note that $A \otimes A' = A_1, \ldots, A_n \to/\Rightarrow \psi$. Then $A \in \mathcal{A}$ by clause (3) of Definition 9 and clause (2) of Definition 10.

Looking back at the counterexample to the if-part of Proposition 1, it can be seen that it is now excluded since $C = A_2 \otimes B_3$.

Some but in general not all attack and defeat relations are preserved under $\otimes$-closure.

Proposition 3. *For any triple of arguments A, B and C it holds that*

1. *if C attacks A then C attacks $A \otimes B$;*
2. *if $A \otimes B$ only replaces a premise of B with A, then if C rebuts or undercuts B then C attacks $A \otimes B$;*
3. *if C undermines B and A has a defeasible top rule, then C rebuts $A \otimes B$;*
4. *if C rebuts or undercuts B on its subargument B' and B' is also a subargument of $A \otimes B$, then C also rebuts or undercuts $A \otimes B$ on B';*

5. if C defeats A then C defeats $A \otimes B$.

Proof. (1) If C undermines A on φ then C undermines $A \otimes B$ on φ since by construction all premises of A are also premises of $A \otimes B$. If C rebuts or undercuts A on its subargument A' then C also rebuts or undercuts $A \otimes B$ on A' since by construction A' is also a subargument of $A \otimes B$.

(2) If C rebuts or undercuts B on its subargument B' then C also rebuts or undercuts $A \otimes B$ on B' since by construction A' is also a subargument of $A \otimes B$.

(3) Suppose C undermines B on φ. Then A has a defeasible top rule with consequent φ so C rebuts $A \otimes B$ on A.

(4) obvious.

(5) If C undermines A on φ then C undermines $A \otimes B$ on φ since by construction all premises of A are also premises of $A \otimes B$. Then C defeats B since $C \not\prec \varphi$. If C undercuts A then C also undercuts $A \otimes B$ by (1), so C defeats $A \otimes B$. Finally, if C rebuts A on A' then C also rebuts $A \otimes B$ on A' by (1). Then C defeats $A \otimes B$ since $C \not\prec A'$.

Properties (2-4) do not in general hold for defeat, since in general an argument may become 'weaker'[4] according to $\preceq$ if one of its premises or non-premise subarguments is replaced by another argument. For example, with [13]'s weakest-link ordering the following may happen (with the rule names attached to $\Rightarrow$ for clarity).

$A_1: r$ $B_1: p$ $D_1: s$
$A_2: A_1 \Rightarrow_{r1} p$ $B_2: B_1 \Rightarrow_{r2} q$ $D_2: D_1 \Rightarrow_{r3} \neg q$
$C: A_2 \Rightarrow_{r2} q$

If we have that $r_1 < r_3 < r_2$ then B defeats D but $C = A \otimes B$ does not defeat D. Even with [13]'s last-link ordering properties (2-4) do not in general. For example, if $A \otimes B$ replaces a non-premise subargument B' of B with A and C rebuts B on B', then it may be that $C \not\prec B'$ but $C \prec B$. Likewise if B' is a premise of B.

6 Applying the definitions to debate evaluation

The approach proposed in the previous section solves the problems with the approach discussed in Section 4 to generate SuppAFs. When an argument in a debate has multiple supports on the same premise, it will be multiplied as desired. The following example debate illustrates this. Let $\mathcal{A} = \{A_1, A_2, B_1, B_2, C_1, C_2\}$ where:

$A_1: p$ $B_1: r$ $C_1: s$
$A_2: A_1 \Rightarrow q$ $B_2: B_1 \Rightarrow p$ $C_2: C_1 \Rightarrow p$

Then $\mathcal{A}^\otimes$ adds the following arguments to $\mathcal{A}$:

---

[4] In this context an argument B is said to be 'weaker' than an argument A (and A 'stronger' than B) if there exists an argument C such that $B \prec C$ but not $A \prec C$.

$D_1\colon B_1 \Rightarrow p \quad E_1\colon C_1 \Rightarrow p$
$D_2\colon D_1 \Rightarrow q \quad E_2\colon E_1 \Rightarrow q$

Here:
$D_1 = B_2 \otimes A_1$
$D_2 = D_1 \otimes A_2 = (B_2 \otimes A_1) \otimes A_2$
$E_1 = C_2 \otimes A_1$
$E_2 = E_1 \otimes A_2 = (C_2 \otimes A_1) \otimes A_2$

So $\mathcal{A}^\otimes$ contains three different arguments for q: the original argument A_2 from $\mathcal{A}$ and the new arguments D_2 and E_2.

Now an evaluator who accepts statement p without further argument, can modify the AT of the SAF by moving p from $\mathcal{K}_p$ to $\mathcal{K}_n$. If, furthermore, the argument ordering is such that an argument can never be 'strengthened' by replacing a necessary premise with a non-premise subargument, then the two arguments for p become irrelevant to the issue q and only argument A_2 counts for this issue. By contrast, an evaluator who does not accept p without further argument can delete p from $\mathcal{K}$ after which A_2 is not constructible any more and the arguments D_2 and E_2 become relevant to the issue q. (An alternative approach is to add 'issue premises' to $ASPIC^+$ as in [19] and to move p to the issue premises). This in fact embodies a dynamic view on debate evaluation, where evaluators can not only provide preferences for given arguments, but can also modify or discard arguments and perhaps add arguments of their own. This is arguably a realistic view on debate evaluation.

Let us briefly discuss some alternative ways to define the closure of a set of arguments resulting from a debate, which further illustrate that debate evaluation need not constrain itself to given arguments. Consider first the just-given example. The original set $\mathcal{A}$ contains two arguments for p: one with r and the other with s as a defeasible reason for p. An evaluator of the debate might then wish to aggregate these reasons in a new argument for p based on a rule $q, r \Rightarrow p$.

Second, a debate might be evaluated against the background of some assumed set of inference rules which were not necessarily used in the debate to construct arguments; that is, they need not be part of the AT of a debate-generated SAF. For example, a set of strict rules generated by a monotonic logic for $\mathcal{L}$ might be assumed (such as classical logic) or a set of defeasible argument schemes might be assumed. In these cases the closure of a set $\mathcal{A}$ of arguments might be defined to contain 'implied' arguments. For example, if $\mathcal{A}$ contains $A = p \Rightarrow q$ and $B = r \Rightarrow s$ and the evaluator assumes a rule $p, r \Rightarrow \neg q$, then $\mathcal{A}^\otimes$ might be defined to also contain $C = p, r \Rightarrow \neg q$, which attacks A even though A was not actually attacked in the debate.

7 Conclusion

In this paper a formal model was proposed of support relations between arguments put forward in debates in the context of the theory of abstract argumentation frameworks [5]. This context was chosen in order to evaluate claims from the literature on bipolar argumentation frameworks that just having attack relations between arguments would be insufficient for modelling debates. We learned that there are problems with the idea that arguments that support each other can be evaluated as independent entities and that it is better to combine supporting and supported arguments in a single compound argument by using $ASPIC^+$'s subargument relation. A method was defined to formally reconstruct the set of arguments put forward in a debate in a way that respects these observations. Then several ways were sketched in which debates can be evaluated in terms of the formal reconstruction. An important insight was that to correctly model support relations between arguments put forward in debates, an explicit formal account was needed of the structure of arguments and the nature of attack and defeat relations between arguments. This casts doubt on the applicability of purely abstract models of debate evaluation.

This paper thus also provides an answer to Betz's criticism in [1] that the theory of abstract argumentation frameworks would not be suitable for modelling debates. This criticism might be justified if these frameworks are applied on their own, but in this paper we have seen that if they are combined with accounts of the structure of argumentation and of how debates with structured arguments can be evaluated, then the theory of abstract argumentation frameworks is still a useful component of an adequate model of debate evaluation.

As for related research, this paper's approach is most closely related to the Carneades system of [10, 11], which was originally proposed as a formalism for evaluating debates. The present idea to let the evaluation of arguments partly depend on whether the evaluator accepts its premises without further support was taken from the 2007 version of Carneades. A main difference in approach between $ASPIC^+$ (and other instantiations of the theory of abstract argumentation frameworks) on the one hand and Carneades (and also e.g. [2]'s Abstract Dialectical Frameworks, ADFs) on the other is that while in $ASPIC^+$ the main focus is on evaluating *arguments*, in Carneades it is on evaluating *statements*. Because of this difference, the issues discussed in this paper do not arise in the same way in Carneades or ADFs. Future research should shed further light on the relative merits of both approaches as regards these issues.

Acnowledgement

I thank Tom Gordon for his comments on an earlier version of this paper.

References

1. G. Betz. Evaluating dialectical structures. *Journal of Philosophical Logic*, 38:283–312, 2009.
2. G. Brewka and S. Woltran. Abstract dialectical frameworks. In *Principles of Knowledge Representation and Reasoning: Proceedings of the Twelfth International Conference*, pages 102–111. AAAI Press, 2010.

3. C. Cayrol and M.-C. Lagasquie-Schiex. Bipolar abstract argumentation systems. In I. Rahwan and G.R. Simari, editors, *Argumentation in Artificial Intelligence*, pages 65–84. Springer, Berlin, 2009.
4. A. Cohen, S. Parsons, E. Sklar, and P. McBurney. A characterization of types of support between structured arguments and their relationship with support in abstract argumentation. *International Journal of Approximate Reasoning*, 94:76–104, 2018.
5. P.M. Dung. On the acceptability of arguments and its fundamental role in non-monotonic reasoning, logic programming, and n–person games. *Artificial Intelligence*, 77:321–357, 1995.
6. P.M. Dung. An axiomatic analysis of structured argumentation with priorities. *Artificial Intelligence*, 231:107–150, 2016.
7. P.M. Dung, P. Mancarella, and F. Toni. Computing ideal sceptical argumentation. *Artificial Intelligence*, 171:642–674, 2007.
8. P.M. Dung and P.M. Thang. Closure and consistency in logic-associated argumentation. *Journal of Artificial Intelligence Research*, 49:79–109, 2014.
9. A.J. Garcia and G.R. Simari. Defeasible logic programming: An argumentative approach. *Theory and Practice of Logic Programming*, 4:95–138, 2004.
10. T.F. Gordon, H. Prakken, and D.N. Walton. The Carneades model of argument and burden of proof. *Artificial Intelligence*, 171:875–896, 2007.
11. T.F. Gordon and D.N. Walton. Formalizing balancing arguments. In P. Baroni, T.F. Gordon, T. Scheffler, and M. Stede, editors, *Computational Models of Argument. Proceedings of COMMA 2016*, pages 327–338. IOS Press, Amsterdam etc, 2016.
12. N. Gorogiannis and A. Hunter. Instantiating abstract argumentation with classical-logic arguments: postulates and properties. *Artificial Intelligence*, 175:1479–1497, 2011.
13. S. Modgil and H. Prakken. A general account of argumentation with preferences. *Artificial Intelligence*, 195:361–397, 2013.
14. S. Modgil and H. Prakken. The ASPIC+ framework for structured argumentation: a tutorial. *Argument and Computation*, 5:31–62, 2014.
15. S. Modgil and H. Prakken. Abstract rule-based argumentation. In P. Baroni, D. Gabbay, M. Giacomin, and L. van der Torre, editors, *Handbook of Formal Argumentation*, volume 1, pages 73–141. College Publications, London, 2018.
16. F. Nouioua and V. Risch. Argumentation frameworks with necessities. In *Proceedings of the 4th International Conference on Scalable Uncertainty Management (SUM'11)*, number 6929 in Springer Lecture Notes in AI, pages 163–176, Berlin, 2011. Springer Verlag.
17. J.L. Pollock. Justification and defeat. *Artificial Intelligence*, 67:377–408, 1994.
18. H. Prakken. Coherence and flexibility in dialogue games for argumentation. *Journal of Logic and Computation*, 15:1009–1040, 2005.
19. H. Prakken. An abstract framework for argumentation with structured arguments. *Argument and Computation*, 1:93–124, 2010.
20. H. Prakken. Relating ways to instantiate abstract argumentation frameworks. In K.D. Atkinson, H. Prakken, and A.Z. Wyner, editors, *From Knowledge Representation to Argumentation in AI, Law and Policy Making. A Festschrift in Honour of Trevor Bench-Capon on the Occasion of his 60th Birthday*, pages 167–189. College Publications, London, 2013.

21. H. Prakken. On support relations in abstract argumentation as abstractions of inferential relations. In *Proceedings of the 21st European Conference on Artificial Intelligence*, pages 735–740, 2014.
22. G.R. Simari and R.P. Loui. A mathematical treatment of defeasible argumentation and its implementation. *Artificial Intelligence*, 53:125–157, 1992.

Towards Conditional Logic Semantics for Abstract Dialectical Frameworks

Gabriele Kern-Isberner[1] and Matthias Thimm[2]

[1] Department of Computer Science, Technical University Dortmund, Germany
[2] Institute for Web Science and Technologies (WeST), University of Koblenz-Landau, Germany

Abstract. We take first steps towards an integrative approach of combining conditional logic semantics with abstract dialectical frameworks. More precisely, we interpret an abstract dialectical frameworks as a conditional logic knowledge base and apply the Z-inference relation in order to obtain a new semantics for abstract dialectical frameworks. We discuss some example translations and obtain a first result pertaining to a characterisation of the different notions of consistency.

1 Introduction

It is well-known that argumentation and nonmonotonic resp. default logics are closely connected: In [7] it is shown that Reiter's default logic can be implemented by abstract argumentation frameworks, a most basic form of computational model of argumentation to which many existing approaches to formal argumentation refer. On the other hand, it is clear that argumentation allows for nonmonotonic, defeasible reasoning, and in [22] computational models of argumentation are assessed by formal properties that have been adapted from nonmonotonic logics. Furthermore, answer set programming [11] as one of the most successful nonmonotonic logics has often been used to implement argumentation [8, 5]. Nevertheless, argumentation and nonmonotonic reasoning are perceived as two different fields which do not subsume each other, and indeed, often attempts to transform reasoning systems from one side into systems of the other side have been revealing gaps that could not be closed (cf., e.g., [26, 15]. While one might argue that this is due to the seemingly richer, dialectical structure of argumentation, in the end the evaluation of arguments often boils down to comparing arguments with their attackers, and comparing degrees of belief is a basic operation in qualitative nonmonotonic reasoning. Therefore, in spite of the abundance of existing work studying connections between the two fields, the true nature of the relationship between argumentation and nonmonotonic reasoning has not been fully understood.

We aim at deepening the understanding of the relationships between argumentation and nonmonotonic logics and establishing a theoretical basis for integrative approaches by focusing on most fundamental approaches on either side: *Abstract Dialectical Frameworks* (ADFs) [4] for argumentation, and *Conditional Logics* (CL) [20, 12, 23] for nonmonotonic logics. ADFs are an approach to formal argumentation, which subsumes many other argumentative formalisms in a generic, logic-based way. On the side of nonmonotonic logics, conditionals have been shown (and often used) to implement nonmonotonic inferences and provide expressive formalisms to represent knowledge bases; some of the best nonmonotonic inference systems (e. g., System Z [12]) make use of conditionals. Furthermore, they also play a basic role for belief revision which is often considered to be a dynamic counterpart to nonmonotonic logics [10]. Both ADFs and CL can be considered as high-level formalisms implementing properly the basic nature of the respective field without being restricted too much by subtleties of specific approaches, and both are based on 3-valued logics.

In this work we take a first step towards an integrative approach by using conditional logic semantics for abstract dialectical frameworks. Syntactically, both frameworks focus on pairs of objects such as (ϕ, ψ). In conditional logic, these pairs are interpreted as conditionals with the informal meaning "if ϕ is true then, usually, ψ is true as well" and written as $(\psi|\phi)$. In abstract dialectical frameworks, these pairs are interpreted as acceptance conditions, with the constraint that $\psi = a$ is a single statement, and interpreted as "if ϕ is accepted then a is accepted as well". The resemblance of these informal interpretations is striking, but both approaches use fundamentally different semantics to formalise these interpretations. Here we ask the question of whether, and how we can interpret abstract dialectical frameworks in terms of conditional logic so that acceptance in the argumentative system is defined by a nonmonotonic inference relation based on conditionals. We take some first steps towards answering this question by translating ADFs into conditional knowledge bases and applying the Z-inference relation [12] to these knowledge bases. We exemplify this translation with several examples and compare the resulting acceptance relation to the usual ADF semantics. The main theoretical contribution of this paper is a characterisation result of the applicability of conditional logic semantics, stating that an abstract dialectical framework is consistent wrt. conditional logic semantics if and only if the conjunction of each acceptance function with its conclusion is satisfiable.

With this paper, we continue work relating argumentation and conditional reasoning that was started in [14] where DeLP rules [9] are interpreted as conditionals; hence, a DeLP program immediately corresponds to a conditional knowledge base, and nice relationships between DeLP acceptance and system Z inference could be shown. The results of that paper encouraged us to broaden these investigations to other argumentation systems. Due to their generality and their non-classical, 3-valued semantics, ADFs seem to be perfectly suited for taking first steps towards a more general framework.

The rest of this paper is organised as follows. In Section 2 we provide some necessary preliminaries. Section 3 contains our main contribution by formalising our approach to applying conditional logic semantics on abstract dialectical frameworks, discussing several examples on this application, and stating our main result. In Section 4 we discuss some related works and we conclude in Section 5 with a summary.

2 Preliminaries

In the following, we we briefly recall some general preliminaries on propositional logic (Section 2.1), as well as technical details on ADFs [4] (Section 2.2) and conditional logic (Section 2.3).

2.1 Propositional Logic

For a set At of atoms let $\mathcal{L}(\mathsf{At})$ be the corresponding propositional language constructed using the usual connectives $\wedge$ (*and*), $\vee$ (*or*), and $\neg$ (*negation*). A (classical) *interpretation* (also called *possible world*) ω for a propositional language $\mathcal{L}(\mathsf{At})$ is a function $\omega : \mathsf{At} \to \{\mathsf{T},\mathsf{F}\}$. Let $\Omega(\mathsf{At})$ denote the set of all interpretations for At. We simply write Ω if the set of atoms is implicitly given. An interpretation ω *satisfies* (or is a *model* of) an atom $a \in \mathsf{At}$, denoted by $\omega \models a$, if and only if $\omega(a) = \mathsf{T}$. The satisfaction relation $\models$ is extended to formulas as usual.

As an abbreviation we sometimes identify an interpretation ω with its *complete conjunction*, i.e., if $a_1,\ldots,a_n \in \mathsf{At}$ are those atoms that are assigned T by ω and $a_{n+1},\ldots,a_m \in \mathsf{At}$ are those propositions that are assigned F by ω we identify ω by $a_1\ldots a_n\overline{a_{n+1}}\ldots\overline{a_m}$ (or any permutation of this). For example, the interpretation ω_1 on $\{a,b,c\}$ with $\omega(a) = \omega(c) = \mathsf{T}$ and $\omega(b) = \mathsf{F}$ is abbreviated by $a\overline{b}c$.

For $\Phi \subseteq \mathcal{L}(\mathsf{At})$ we also define $\omega \models \Phi$ if and only if $\omega \models \phi$ for every $\phi \in \Phi$. Define the set of models $\mathsf{Mod}(X) = \{\omega \in \Omega(\mathsf{At}) \mid \omega \models X\}$ for every formula or set of formulas X. A formula or set of formulas X_1 *entails* another formula or set of formulas X_2, denoted by $X_1 \models X_2$, if $\mathsf{Mod}(X_1) \subseteq \mathsf{Mod}(X_2)$.

Finally, let $Cn(\Phi)$ denote the deductive closure of a set $\Phi \subseteq \mathcal{L}(\mathsf{At})$, i.e., $Cn(\Phi) = \{\phi \mid \Phi \models \phi\}$.

2.2 Abstract Dialectical Frameworks

Abstract Dialectical Frameworks generalise abstract argumentation frameworks [7] and provide a general framework to discuss various issues in formal argumentation such as preferences [4] and support [25]. An ADF D is a tuple $D = (S, L, C)$ where S is a set of statements, $L \subseteq S \times S$ is a set of links, and $C = \{C_s\}_{s \in S}$ is a set of total functions $C_s : 2^{par_D(s)} \to \{\mathsf{T},\mathsf{F}\}$ for each $s \in S$ with $par_D(s) = \{s' \in S \mid (s',s) \in L\}$ (acceptance functions). An acceptance function C_s defines the cases when the statement s can be accepted (truth value T), depending on the acceptance status of its parents in D. By abuse of notation, we will often identify an acceptance function C_s by its equivalent *acceptance condition* which models the acceptable cases as a propositional formula. In other words, we assume $C_s \in \mathcal{L}(S)$.

Example 1. Consider an ADF $D_1 = (S_1, L_1, C_1)$ with

$$S_1 = \{a, b, c, d\}$$
$$L_1 = \{(a,b), (b,a), (b,d), (c,d)\}$$
$$C_1 = \{C_a, C_b, C_c, C_d\}$$

with $C_a = \neg b$, $C_b = \neg a$, $C_c = \mathsf{T}$, and $C_d = c \wedge \neg b$. The framework D is depicted as a graph in Figure 1. Informally speaking, the acceptance conditions can be

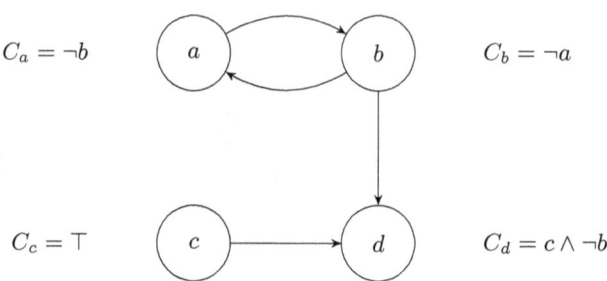

Fig. 1. A abstract dialectical framework

read as "a is accepted if b is not accepted", "b is accepted if a is not accepted", "c is always accepted", and "d is accepted if c is accepted and b is not accepted". Figure 1 shows a graphical depiction of the ADF D_1, where next to each node its acceptance condition is given.

An ADF $D = (S, L, C)$ is interpreted through 3-valued interpretations $v : S \to \{\mathsf{T}, \mathsf{F}, \mathsf{U}\}$, which assign to each statement in S either the value T (accepted), F (rejected), or U (undecided, unknown). A 3-valued interpretation v can be extended to arbitrary propositional formulas over S via

1. $v(\neg \phi) = \mathsf{F}$ iff $v(\phi) = \mathsf{T}$, $v(\neg \phi) = \mathsf{T}$ iff $v(\phi) = \mathsf{F}$, and $v(\neg \phi) = \mathsf{U}$ iff $v(\phi) = \mathsf{U}$;
2. $v(\phi \wedge \psi) = \mathsf{T}$ iff $v(\phi) = c(\psi) = \mathsf{T}$, $v(\phi \wedge \psi) = \mathsf{F}$ iff $v(\phi) = \mathsf{F}$ or $v(\psi) = \mathsf{F}$, and $v(\phi \wedge \psi) = \mathsf{U}$ otherwise;
3. $v(\phi \vee \psi) = \mathsf{T}$ iff $v(\phi) = \mathsf{T}$ or $v(\psi) = \mathsf{T}$, $v(\phi \vee \psi) = \mathsf{F}$ iff $v(\phi) = c(\psi) = \mathsf{F}$, and $v(\phi \vee \psi) = \mathsf{U}$ otherwise.

Then v is a *model* of D if for all $s \in S$, $v(s) \neq \mathsf{U}$ implies $v(s) = v(C_s)$.

Example 2. We continue Example 1 and consider the three-valued interpretations v_1, v_2, v_3 defined via

$v_1(a) = \mathsf{T}$ $v_1(b) = \mathsf{F}$ $v_1(c) = \mathsf{T}$ $v_1(d) = \mathsf{T}$
$v_2(a) = \mathsf{F}$ $v_2(b) = \mathsf{T}$ $v_2(c) = \mathsf{T}$ $v_2(d) = \mathsf{F}$
$v_3(a) = \mathsf{U}$ $v_3(b) = \mathsf{U}$ $v_3(c) = \mathsf{T}$ $v_3(d) = \mathsf{F}$

Observe that both v_1 and v_2 are models of D_1 (e.g., it holds $\mathsf{T} = v_1(d) = v_1(C_d) = v_1(c \wedge \neg b) = \mathsf{T} \wedge \mathsf{T} = \mathsf{T}$). Observe also that v_3 is not a model as, e.g., $v_3(d) = \mathsf{F}$ but $v_3(C_d) = v_3(c \wedge \neg b) = \mathsf{T} \wedge \mathsf{U} = \mathsf{U}$.

On top of the notion of a model, various semantics can be defined for ADFs such as the grounded, complete, preferred, and stable semantics [4]. These semantics constrain the set of models further by imposing additional constraints. For that, let $\leq_i$ be the *information order* on truth values defined via $\mathsf{U} \leq_i \mathsf{T}$ and $\mathsf{U} \leq_i \mathsf{F}$. Then $(\{\mathsf{T}, \mathsf{F}, \mathsf{U}\}, \leq_i)$ is a complete meet-semi-lattice [4] where the meet operator $\sqcap$ is defined via $\mathsf{T} \sqcap \mathsf{T} = \mathsf{T}$, $\mathsf{F} \sqcap \mathsf{F} = \mathsf{F}$, and $\alpha \sqcap \beta = \mathsf{U}$ otherwise. For two interpretations v_1, v_2 be write $v_1 \leq_i v_2$ iff $v_1(s) \leq_i v_2(s)$ for all $s \in S$ and

$$(v_1 \sqcap v_2)(s) = v_1(s) \sqcap v_2(s)$$

for all $s \in S$. For an interpretation v let $[v]_2$ be the set of interpretations v' with $v \leq_i v'$ and $\mathsf{U} \notin \mathrm{im}\, v'$.[3] Define the operator Γ_D on interpretations via

$$\Gamma_D(v)(s) = \sqcap\{w(C_s) \mid w \in [v]_2\}$$

for all interpretations v and statements s. In other words, $\Gamma_D(v)$ is an interpretation, which maps every statement to the consensus of all two-valued extensions of v.

Definition 1. *Let $D = (S, L, C)$ be an ADF and v an interpretation.*

1. *v is a complete model of D iff $\Gamma_D(v) = v$.*
2. *v is the grounded model of D if it is complete and $v \leq_i v'$ for all complete models v'.*
3. *v is a preferred model of D if it is complete and there is no complete model v' with $v <_i v'$.*

Brewka and Woltran also define stable models [4], which we do not consider here.

For $\sigma \in \{gr, co, pr\}$ (grounded, complete, preferred semantics, respectively) define an inference relation $\mathrel{\vert\!\sim}^{\sigma}_{\mathrm{cr}}$ via $D \mathrel{\vert\!\sim}^{\sigma}_{\mathrm{cr}} a$ iff $a \in S$ is mapped to T in some σ-model of D. Considering a skeptical reasoning perspective, we can define an inference relation $\mathrel{\vert\!\sim}^{\sigma}_{\mathrm{sk}}$ via $D \mathrel{\vert\!\sim}^{\sigma}_{\mathrm{sk}} a$ iff $a \in S$ is mapped to T in all σ-models of D. As there is a uniquely defined grounded model, both inference relations collapse for grounded semantics and we simply write $\mathrel{\vert\!\sim}^{gr}$ instead of $\mathrel{\vert\!\sim}^{gr}_{\mathrm{cr}}$ or $\mathrel{\vert\!\sim}^{gr}_{\mathrm{sk}}$.

2.3 Conditional Logic

A conditional of the form $(\psi|\phi)$ connects two formulas, the antecedence ϕ and the conclusion ψ (often in a meaningful way) and represents a rule "*If ϕ then (usually, probably) ψ*". An important property of conditionals in general is

---

[3] $\mathrm{im}\, f$ is the image of f

their *defeasibility*, i.e., the possibility of the conclusion of a conditional being overruled when more information becomes available. Formalising defeasibility is one of the core challenges in knowledge representation and reasoning, and there are plenty of approaches that aim at addressing this, see e.g. [21, 17, 16] for some examples. We will take conditionals as basic formal entities for (nonmonotonic, plausible) reasoning. There are many different conditional logics (cf., e.g., [17, 20]), we will just use basic properties of conditionals that are common to many conditional logics and are especially important for nonmonotonic reasoning: Basically, we follow the approach of de Finetti [6] who considered conditionals as *generalized indicator functions* for possible worlds ω and define:

$$(\psi|\phi)(\omega) = \begin{cases} 1 & : \ \omega \models \phi \land \psi \\ 0 & : \ \omega \models \phi \land \neg \psi \\ u & : \ \omega \models \neg \phi \end{cases} \quad (1)$$

where u stands for *unknown* or *indeterminate* (not to be confused with the truth value U from the previous section). In other words, a propositional interpretation ω *verifies* a conditional $(\psi|\phi)$ iff it satisfies both antecedence and conclusion $((\psi|\phi)(\omega) = 1)$; it *falsifies* it iff it satisfies the antecedence but not the conclusion $((\psi|\phi)(\omega) = 0)$; otherwise the conditional is *not applicable*, i.e., the interpretation does not satisfy the antecedence $((\psi|\phi)(\omega) = u)$. If $(\psi|\phi)(\omega) \neq 0$, we also say that ω satisfies $(\psi|\phi)$. Hence, conditionals are three-valued logical entities and thus extend the binary setting of classical logics substantially in a way that is compatible with the probabilistic interpretation of conditionals as conditional probabilities. Such a conditional $(\psi|\phi)$ can be accepted as plausible if its verification $\phi \land \psi$ is more plausible than its falsification $\phi \land \neg \psi$, where plausibility is often modelled by a total preorder on possible worlds. This is in full compliance with nonmonotonic inference relations $\phi \hspace{0.1em}\mid\hspace{-0.5em}\sim \psi$ [18] expressing that from ϕ, ψ may be plausibly derived. A particular convenient implementation of total preorders are *ordinal conditional functions (OCFs)*, (also called *ranking functions*) $\kappa : \Omega \to \mathbb{N} \cup \{\infty\}$ [23] on the set of possible worlds Ω. They express degrees of (im)plausibility of possible worlds and propositional formulas ϕ by setting $\kappa(\phi) := \min\{\kappa(\omega) \mid \omega \models \phi\}$. OCFs κ are a very popular formal environment for nonmonotonic and conditional reasoning, allowing for simply expressing the acceptance of conditionals and nonmonotonic inferences via stating that $(\psi|\phi)$ is accepted by κ iff $\phi \hspace{0.1em}\mid\hspace{-0.5em}\sim_\kappa \psi$ iff $\kappa(\phi \land \psi) < \kappa(\phi \land \neg \psi)$, implementing formally the intuition of conditional acceptance based on plausibility mentioned above. For an OCF κ, $Bel(\kappa)$ denotes the propositional beliefs that are implied by all most plausible worlds, i.e., $Bel(\kappa) = Cn(\kappa^{-1}(0))$. A set Δ of conditionals is *consistent* if there is an OCF accepting all conditionals in Δ.

We denote with CL the framework of reasoning from conditional knowledge bases Δ based on OCFs that are so-called (ranking) models of Δ, i.e., which accept all conditionals in Δ. Specific examples of such ranking models are system Z yielding the inference relation $\hspace{0.1em}\mid\hspace{-0.5em}\sim^Z$ [12] and c-representations providing the basis for c-inference relations [13, 1]. In this paper, we consider the relation $\hspace{0.1em}\mid\hspace{-0.5em}\sim^Z$ defined as follows. A conditional $(\psi|\phi)$ is *tolerated* by Δ if there is a possible world ω with $(\psi|\phi)(\omega) = 1$ and $(\psi'|\phi')(\omega) \neq 0$ for all $(\psi'|\phi') \in \Delta$, i. e., ω verifies $(\psi|\phi)$ and does not falsify any (other) conditional in Δ. The Z-partitioning $(\Delta_0, \ldots, \Delta_n)$ of Δ is defined as

1. $\Delta_0 = \{\delta \in \Delta \mid \Delta \text{ tolerates } \delta\}$,
2. $\Delta_1, \ldots, \Delta_n$ is the Z-partitioning of $\Delta \setminus \Delta_0$.

For $\delta \in \Delta$ define furthermore

$$Z_\Delta(\delta) = i \quad \Longleftrightarrow \quad \delta \in \Delta_i \text{ and } (\Delta_0, \ldots, \Delta_n) \text{ is the Z-partitioning of } \Delta$$

Finally, the ranking function κ_Δ^z is defined via

$$\kappa_\Delta^z(\omega) = \max\{Z(\delta) \mid (\psi|\phi)(\omega) = 0, (\psi|\phi) \in \Delta\} + 1$$

with $\max \emptyset = -1$. Define then $\Delta \vdash^Z (\psi|\phi)$ iff $\phi \vdash_{\kappa_\Delta^z} \psi$. For a propositional formula ϕ, we have $\Delta \vdash^Z \phi$ iff $\Delta \vdash^Z (\phi|\top)$ iff $\phi \in Bel(\kappa_\Delta^z)$ iff $\kappa_\Delta^z(\neg\phi) > 0$.

3 Interpreting ADFs in Conditional Logic

Given an ADF $D = (S, L, C)$, it is straightforward to derive a conditional-logic knowledge base $\Theta(D)$ defined via

$$\Theta(D) = \{(s|C_s) \mid s \in S\}$$

from D. In other words, every acceptance function of D is interpreted as a conditional. Now we can use inference relations for conditional logic on $\Theta(D)$ and see how the inferences compare to inferences made using e. g. $\vdash^{gr}$ directly on D.

As a first application, in this paper we use the Z-inference relation to define an inference relation on D via

$$D \vdash^Z a \quad \text{iff} \quad \Theta(D) \vdash^Z a. \tag{2}$$

Note that $\Theta(D) \vdash^Z a$ is equivalent to stating $a \in Bel(\kappa_\Delta^z)$. The three-valued model v_Z that can be associated with this inference relation is defined by

$$v_Z(s) = \begin{cases} \mathsf{T} & \text{if } s \in Bel(\kappa_\Delta^z), \\ \mathsf{F} & \text{if } \overline{s} \in Bel(\kappa_\Delta^z), \\ \mathsf{U} & \text{otherwise.} \end{cases} \tag{3}$$

We illustrate this new inference relation with some examples which are taken from [4].

Example 3. Let $D_2 = (S_2, L_2, C_2)$ with

$$S_2 = \{a, b, c\} \qquad L_2 = \{(a, b)\} \qquad C_2 = \{C_a = \top, C_b = \neg a \vee c, C_c = b\}$$

$$C_a = \neg a \qquad C_b = \neg c \qquad C_c = \top$$

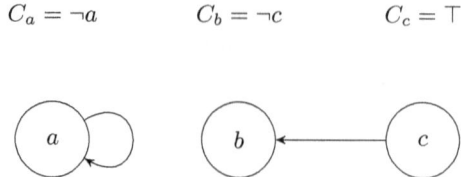

Fig. 2. The abstract dialectical framework from Example 6.

Then $\Delta = \Theta(D_2) = \{(a|\top), (b|\neg a \vee c), (c|b)\}$, and it is easily checked that the tolerance partitioning is just $\Delta_0 = \Delta$. This means $\kappa_\Delta^z(abc) = \kappa_\Delta^z(a\bar{b}\bar{c}) = 0$ and $\kappa_\Delta^z(w) = 1$ for all other worlds w because each such w falsifies at least one conditional from $\Theta(D)$. Therefore, $Bel(\kappa_\Delta^z) = Cn(a(bc \vee \bar{b}\bar{c}))$, In terms of arguments, this yields $D_2 \mathrel{\vert\!\sim}^Z a$, and $v_Z(a) = \top, v_Z(b) = v_Z(c) = \mathsf{U}$.

Example 4. Let $D_3 = (S_3, L_3, C_3)$ with

$$S_3 = \{a, b, c\} \qquad L_3 = \{(a, c), (b, c)\} \qquad C_3 = \{C_a = c, C_b = c, C_c = a \Leftrightarrow b\}$$

We obtain $\Delta = \Theta(D_3) = \{(a|c), (b|c), (c|a \Leftrightarrow b)\}$, and again, the tolerance partitioning is just $\Delta_0 = \Delta$. This means that κ_Δ^z assigns the value 0 exactly to the worlds $abc, a\bar{b}\bar{c}$, and $\bar{a}\bar{b}\bar{c}$, and 1 to all other worlds. in this case, none of a, b, c is in $Bel(\kappa_\Delta^z)$. The model v_Z assigns U to all arguments.

Finally, we apply the Z-inference relation to our Example 1.

Example 5. Consider the ADF D_1 from Example 1. The Θ-translation yields the conditional knowledge base $\Delta = \Theta(D_1) = \{(a|\bar{b}), (b|\bar{a}), (c|\top), (d|\bar{b}c)\}$ with $\Delta_0 = \Delta$. Therefore, exactly the worlds $abcd, ab\bar{c}d, \bar{a}bcd, \bar{a}b\bar{c}d$, and $a\bar{b}cd$ are assigned 0 by κ_Δ^z, and we have $Bel(\kappa_\Delta^z) = Cn(bc \vee a\bar{b}cd) = Cn(c(b \vee a\bar{b}d))$. In terms of arguments, this means $D_1 \mathrel{\vert\!\sim}^Z c$, and $v_Z(a) = v_Z(b) = v_Z(d) = \mathsf{U}$, while $v_Z(c) = \top$.

Note that in all three examples, the $\mathrel{\vert\!\sim}^Z$-inferred arguments coincide with the grounded semantics. This is, in general, not true as the following example shows.

Example 6. Let $D_4 = (S_4, L_4, C_4)$ with

$$S_4 = \{a, b, c\} \qquad L_4 = \{(a, a), (c, b)\} \qquad C_4 = \{C_a = \neg a, C_b = \neg c, C_c = \top\}$$

which is also depicted in Figure 2. The Θ-translation yields the conditional knowledge base $\Theta(D_4) = \{(a|\bar{a}), (b|\bar{c}), (c|\top)\}$ which is inconsistent as there can be no ranking function κ that accepts the first conditional. Therefore we obtain $\Theta(D) \mathrel{\vert\!\not\sim}^Z a$, $\Theta(D) \mathrel{\vert\!\not\sim}^Z b$, and $\Theta(D) \mathrel{\vert\!\not\sim}^Z c$. However, the grounded model of the initial ADF is able to infer c.

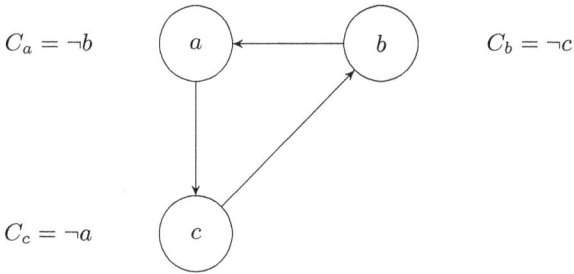

Fig. 3. The abstract dialectical framework from Example 7

We say that an ADF $D = (S, L, C)$ is *sound* iff $s \wedge C_s$ is satisfiable for every $s \in S$. Then the observation from the previous example can be generalised as follows.

Proposition 1. *If ADF $D = (S, L, C)$ is not sound then $\Theta(D)$ is inconsistent.*

Proof. If D is not sound then there is an acceptance condition C_s such that $s \wedge C_s$ is unsatisfiable. For the conditional $(s|C_s) \in \Theta(D)$ this means that there is no world verifying $(s|C_s)$ and therefore $\kappa(s \wedge C_s) = \infty$ for every OCF κ. Therefore κ cannot accept $(s|C_s)$.

Example 7. Let $D_5 = (S_5, L_5, C_5)$ with

$$S_5 = \{a, b, c\} \quad L_5 = \{(b,a), (c,b), (a,c)\} \quad C_5 = \{C_a = \neg b, C_b = \neg c, C_c = \neg a\}$$

This ADF models a classical issue in argumentation: a cycle with an odd number of arguments (see Figure 3). In this case, no argument can be defended and for $\sigma \in \{gr, co, pr\}$ no argument can be inferred credulously nor skeptically. The Θ-translation yields the conditional knowledge base $\Delta = \{(a|\bar{b}), (b|\bar{c}), (c|\bar{a})\}$ with $\Delta_0 = \Delta$. Therefore, the worlds abc, $ab\bar{c}$, $a\bar{b}c$, and $\bar{a}bc$ are assigned 0 by κ^z_Δ, yielding $\Theta(D) \not\hspace{-2pt}\mid\hspace{-2pt}\sim^Z a$, $\Theta(D) \not\hspace{-2pt}\mid\hspace{-2pt}\sim^Z b$, and $\Theta(D) \not\hspace{-2pt}\mid\hspace{-2pt}\sim^Z c$ as well.

So far, all examples gave us a trivial Z-partitioning (or none at all). However, this is also not always the case as the following example shows.

Example 8. Let $D_6 = (S_6, L_6, C_6)$ with

$$S_6 = \{a, b, c\} \quad L_6 = \{(b,a), (c,b), (a,c)\} \quad C_6 = \{C_a = \neg b \wedge \neg c, C_b = \neg c, C_c = \neg a\}$$

Then $\Theta(D_6) = \{(a|\bar{b}\bar{c}), (b|\bar{c}), (c|\bar{a})\}$ and the Z-partitioning of $\Theta(D_6)$ is (Δ_0, Δ_1) with

$$\Delta_0 = \{(b|\bar{c}), (c|\bar{a})\}$$
$$\Delta_1 = \{(a|\bar{b}\bar{c})\}$$

Accordingly, we get

$$\kappa_\Delta^z(abc) = 0$$
$$\kappa_\Delta^z(ab\bar{c}) = 0$$
$$\kappa_\Delta^z(a\bar{b}c) = 0$$
$$\kappa_\Delta^z(a\bar{b}\bar{c}) = 1$$
$$\kappa_\Delta^z(\bar{a}bc) = 0$$
$$\kappa_\Delta^z(\bar{a}b\bar{c}) = 1$$
$$\kappa_\Delta^z(\bar{a}\bar{b}c) = 0$$
$$\kappa_\Delta^z(\bar{a}\bar{b}\bar{c}) = 2$$

and therefore $\Theta(D_6) \not\hspace{-2pt}\sim^Z a$, $\Theta(D) \not\hspace{-2pt}\sim^Z b$, and $\Theta(D_6) \not\hspace{-2pt}\sim^Z c$.

Let us now consider the converse of Proposition 1; we first prove a lemma that will be helpful to show that sound ADFs induce consistent conditional knowledge bases.

Lemma 1. *Let Δ be some conditional knowledge base and ψ, ϕ, ξ formulas. If $\Delta \cup \{(\psi|\phi), (\psi|\xi)\}$ is consistent then $\Delta \cup \{(\psi|\phi \vee \xi)\}$ is consistent.*

Proof. Assume $\Delta_1 = \Delta \cup \{(\psi|\phi), (\psi|\xi)\}$ is consistent. In order to show consistency of $\Delta_2 = \Delta \cup \{(\psi|\phi \vee \xi)\}$ it suffices to show that there is at least one conditional in every $\emptyset \neq \Delta' \subseteq \Delta_2$, which is tolerated by Δ', cf. Theorem 4 in [12].

1. Case 1 $(\psi|\phi \vee \xi) \notin \Delta'$: Then $\Delta' \subseteq \Delta_1$ and the claim follows from the consistency of Δ_1.
2. Case 2 $(\psi|\phi \vee \xi) \in \Delta'$: Define $\Delta'' = \Delta' \setminus \{(\psi|\phi \vee \xi)\} \cup \{(\psi|\phi), (\psi|\xi)\} \subseteq \Delta_1$. Due to the consistency of Δ_1 there is a conditional $\delta \in \Delta''$ which is tolerated by Δ''.
   (a) Case 2.1 $\delta \neq (\psi|\phi), \delta \neq (\psi|\xi)$: Let ω be the world that verifies δ and satisfies all conditionals in Δ''. Then either $\omega \models \psi$ (in that case ω necessarily satisfies $(\psi|\phi \vee \xi)$ as well) or $\omega \not\models \phi$ and $\omega \not\models \xi$. In the latter case it follows $\omega \not\models \phi \vee \xi$ and therefore ω also satisfies $(\psi|\phi \vee \xi)$. It follows that δ is also tolerated by Δ'.
   (b) Case 2.2 $\delta = (\psi|\phi)$ (analogously for $\delta = (\psi|\xi)$). Let ω be the world that verifies δ and satisfies all conditionals in Δ''. Then $\omega \models \psi \wedge \phi$ and therefore also accepts $(\psi|\phi \vee \xi)$. It follows that $(\psi|\phi \vee \xi)$ is also tolerated by Δ'. □

Theorem 1. *If the ADF D is sound then $\Theta(D)$ is consistent.*

Proof. Let $D = (S, L, C)$ be sound. Let $\Theta(D) = \{\delta_1, \ldots, \delta_n\}$ with $\delta_i = (s_i|C_i)$, $i = 1, \ldots, n$, for $S = \{s_1, \ldots, s_n\}$. As D is sound, $C_i \not\models \neg s_i$ for $i = 1, \ldots, n$. We can also assume $C_i \not\models s_i$ for $i = 1, \ldots, n$, otherwise δ_i would be verified for trivial reasons. From both statements we can therefore assume that C_i does not mention s_i at all.

In order to show consistency of $\Theta(D)$ it suffices to show that there is at least one conditional in every $\emptyset \neq \Delta \subseteq \Theta(D)$, which is tolerated by Δ, cf. Theorem 4 in [12]. Without loss of generality assume $\Delta = \{\delta_1, \ldots, \delta_m\}$ for some $1 \leq m \leq n$ (e. g. after reordering the conditionals).

Without loss of generality, we now assume that each condition C_i, $i = 1, \ldots, m$, is a conjunction of literals. This is justified due to the following:

1. We can first safely assume that each C_i is in disjunctive normal form as the syntactic representation does not influence semantic evaluation.
2. Due to Lemma 1 we can split each conditional $(\phi|\psi_1 \vee \ldots \vee \psi_k)$ into $\{(\phi|\psi_1), \ldots, (\phi|\psi_k)\}$. As we will show that the conditional knowledge base consisting of the latter conditionals is consistent, it follows by Lemma 1 that the original knowledge base is consistent as well.

Define
$$\#neg(C_i) = \{s \in S \mid C_i \models \neg s\}$$
to be the set of statements that occur negatively in each conjunction C_i. Let $j \in \{1, \ldots, m\}$ be such that $\#neg(C_j)$ is minimal wrt. set inclusion among all $\#neg(C_i)$. Define the possible world $\hat{\omega}'$ through assigning F to all atoms in $\#neg(C_j)$ and T to all remaining ones. We claim that $\hat{\omega}'$ verifies δ_j and satisfies all other conditionals in Δ.

1. First we show that $\hat{\omega}'$ verifies δ_j: As $\hat{\omega}'$ assigns F exactly to all atoms negatively occurring in C_j we have $\hat{\omega}' \models C_j$. Furthermore, as $C_j \not\models \neg s_j$, s_j is assigned to T in $\hat{\omega}'$. Therefore $\hat{\omega}' \models s_j \wedge C_j$.
2. Now, we show that $\hat{\omega}'$ satisfies every $\delta_i \in \Delta$, $i = 1, \ldots, m$: If $\hat{\omega}' \not\models C_i$ then $\hat{\omega}'$ trivially satisfies δ_i. So assume $\hat{\omega}' \models C_i$. We first show $\#neg(C_i) = \#neg(C_j)$:
   (a) We have $\#neg(C_i) \subseteq \#neg(C_j)$ for the following reasons: Let $s \in \#neg(C_i)$, i.e., $C_i \models \neg s$. As $\hat{\omega}' \models C_i$ we have $\hat{\omega}' \models \neg s$. As an atom is set to F in $\hat{\omega}'$ only if $s \in \#neg(C_j)$ the claim follows.
   (b) $\#neg(C_i) \not\subset \#neg(C_j)$: this is clear as $\#neg(C_i) \subset \#neg(C_j)$ would violate the minimality of $\#neg(C_j)$.
   Due to $\#neg(C_j) = \#neg(C_i)$ and the fact that $C_i \not\models \neg s_i$ it follows that $\hat{\omega}' \models s_i$ (recall that all atoms are set to T except the ones in $\#neg(C_j)$). Therefore, $\hat{\omega}'$ satisfies δ_i.

As we showed that δ_j is tolerated by Δ, we have proven the general claim of the theorem. □

Taking Proposition 1 and Theorem 1 together, we obtain the following corollary.

Corollary 1. *An ADF D is sound iff $\Theta(D)$ is consistent.*

The result above characterises consistency in the CL framework by a simple assumption on ADFs and is a first step towards a deeper understanding of the relationships between these two approaches.

4 Related Works

Our aim in this paper is to lay foundations of integrative techniques for argumentative and conditional reasoning. There are previous works, which have similar aims or are otherwise related to this endeavour. We will discuss those in the following.

First, there is huge body of work on *structured argumentation*, i.e., approaches to argumentative reasoning that build on rule-based knowledge bases and construct arguments from chains of reasoning. Examples of such approaches are ASPIC$^+$ [19], *Assumption-based Argumentation* (ABA) [27], deductive argumentation [3], and *Defeasible Logic Programming* (DeLP) [9]. Roughly, these approaches work as follows. Starting from a knowledge base consisting of facts and rules, arguments are identified as minimal consistent derivations of their respective claims. One argument attacks another if the claim of the former somehow contradicts the contents of the latter. Building on this notion of conflict, an abstract argumentation framework can be constructed (in e.g. the case of ASPIC$^+$) and acceptable arguments are identified using some semantics for those [7]. The claims of the acceptable arguments are then regarded as justified. As one can see, structured argumentation approaches provide a *stacked* view on formal argumentation and rule-based reasoning: syntactically, structured argumentation approaches use rule-based knowledge representation components but, semantically, rely on argumentative notions.

In [15] conditional reasoning based on System Z [12] and DeLP are combined in a novel way. Roughly, the paper provides a novel semantics for DeLP by borrowing concepts from System Z that allows using *plausibility* as a criterion for comparing the strength of arguments and counterarguments. Besnard et al. [2] develop a structured argumentation approach where general conditional logic is used as the base knowledge representation formalism. Their framework is constructed in a similar fashion as the deductive argumentation approach [3] but they also provide with *conditional contrariety* a new conflict relation for arguments, based on conditional logical terms. In [29] a new semantics for abstract argumentation is presented, which is also rooted in conditional logical terms. Building on the ranking semantics System J for conditional logic [28] a ranking interpretation for extensions is provided when arguments can be instantiated by strict and defeasible rules. In [24] Strass presents a translation from an ASPIC-style defeasible logic theory to ADFs. While actually Strass embeds one argumentative formalism (the ASPIC-style theory) into another argumentative formalism (ADFs) and shows how the latter can simulate the former, the process of embedding is similar to our approach.

5 Summary

In this paper we took some first steps towards a deeper understanding of the relationship between conditional logic and abstract dialectical frameworks, thus broadening our understanding of the fields of argumentation and nonmonotonic reasoning in general. By means of examples we showed how the Z-inference relation can be applied to dialectical frameworks and we discovering striking similarities between this approach of reasoning and classical argumentation semantics (though no formal relationship has been shown yet). Our first result concerning the characterisation of consistency in conditional logic via soundness of abstract dialectical frameworks opens the way for more investigation.

References

1. Christoph Beierle, Christian Eichhorn, Gabriele Kern-Isberner, and Steven Kutsch. Skeptical, weakly skeptical, and credulous inference based on preferred ranking functions. In G.A. Kaminka, M. Fox, Paolo Bouquet, E. Hüllermeier, V. Dignum, F. Dignum, and F. van Harmelen, editors, *Proceedings of the 22nd European Conference on Artificial Intelligence, ECAI 2016*, volume 285 of *Frontiers in Artificial Intelligence and Applications*, pages 1149–1157, Amsterdam, NL, 2016. IOS Press.
2. Philippe Besnard, Eric Grégoire, and Badran Raddaoui. A conditional logic-based argumentation framework. In *Proceedings of the 7th International Conference on Scalable Uncertainty Management (SUM 2013)*, pages 44–56. Springer, 2013.
3. Philippe Besnard and Anthony Hunter. *Elements of Argumentation*. The MIT Press, 2008.
4. Gerhard Brewka, Stefan Ellmauthaler, Hannes Strass, Johannes Peter Wallner, and Stefan Woltran. Abstract dialectical frameworks revisited. In *Proceedings of the 23rd International Joint Conference on Artificial Intelligence (IJCAI'13)*, 2013.
5. Federico Cerutti, Sarah A. Gaggl, Matthias Thimm, and Johannes P. Wallner. Foundations of implementations for formal argumentation. In Pietro Baroni, Dov Gabbay, Massimiliano Giacomin, and Leendert van der Torre, editors, *Handbook of Formal Argumentation*, chapter 15. College Publications, February 2018.
6. B. DeFinetti. *Theory of Probability*, volume 1,2. John Wiley and Sons, New York, 1974.
7. Phan Minh Dung. On the Acceptability of Arguments and its Fundamental Role in Nonmonotonic Reasoning, Logic Programming and n-Person Games. *Artificial Intelligence*, 77(2):321–358, 1995.
8. Uwe Egly, Sarah Alice Gaggl, and Stefan Woltran. Answer-set programming encodings for argumentation frameworks. Technical Report DBAI-TR-2008-62, Technische Universität Wien, 2008.
9. A. Garcia and Guillermo R. Simari. Defeasible Logic Programming: An Argumentative Approach. *Theory and Practice of Logic Programming*, 4(1–2):95–138, 2004.
10. P. Gärdenfors. Belief revision and nonmonotonic logic: Two sides of the same coin? In *Proceedings European Conference on Artificial Intelligence, ECAI'92*, pages 768–773. Pitman Publishing, 1992.
11. M. Gelfond and N. Leone. Logic programming and knowledge representation – the A-prolog perspective. *Artificial Intelligence*, 138:3–38, 2002.
12. Moises Goldszmidt and Judea Pearl. Qualitative probabilities for default reasoning, belief revision, and causal modeling. *Artificial Intelligence*, 84:57–112, 1996.

13. G. Kern-Isberner. *Conditionals in nonmonotonic reasoning and belief revision*. Springer, Lecture Notes in Artificial Intelligence LNAI 2087, 2001.
14. G. Kern-Isberner and G. Simari. A default logical semantics for defeasible argumentation. In *Proceedings of the 24th Florida Artificial Intelligence Research Society Conference FLAIRS-24*. AAAI Press, 2011.
15. Gabriele Kern-Isberner and Guillermo R. Simari. A default logical semantics for defeasible argumentation. In *Proceedings of the Twenty-Fourth International Florida Artificial Intelligence Research Society Conference (FLAIRS'11)*, 2011.
16. Sarit Kraus, Daniel J. Lehmann, and Menachem Magidor. Nonmonotonic reasoning, preferential models and cumulative logics. *Artificial Intelligence*, 44(1-2):167–207, 1990.
17. D. Lehmann. What Does a Conditional Knowledge Base Entail? In *Proceedings of KR'89*, Toronto, Canada, 1989.
18. D. Makinson. General theory of cumulative inference. In M. Reinfrank et al., editors, *Non-monotonic Reasoning*, pages 1–18. Springer Lecture Notes on Artificial Intelligence 346, Berlin, 1989.
19. Sanjay Modgil and Henry Prakken. The aspic+ framework for structured argumentation: a tutorial. *Argument and Computation*, 5:31–62, 2014.
20. Donald Nute and Charles Cross. Conditional logic. In D. Gabbay and F. Guenther, editors, *Handbook of Philosophical Logic*, volume 4, pages 1–98. Kluwer Academic Publishers, second edition edition, 2002.
21. R. Reiter. A logic for default reasoning. *Artificial Intelligence*, 13:81–132, 1980.
22. Tjitze Rienstra, Chiaki Sakama, and Leendert van der Torre. Persistence and monotony properties of argumentation semantics. In *Proceedings of the 2015 International Workshop on Theory and Applications of Formal Argument (TAFA'15)*, july 2015.
23. W. Spohn. Ordinal conditional functions: a dynamic theory of epistemic states. In W.L. Harper and B. Skyrms, editors, *Causation in Decision, Belief Change, and Statistics, II*, pages 105–134. Kluwer Academic Publishers, 1988.
24. Hannes Strass. Instantiating rule-based defeasible theories in abstract dialectical frameworks and beyond. *Journal of Logic and Computation*, 2015. In press.
25. Hannes Strass and Johannes Peter Wallner. Analyzing the computational complexity of abstract dialectical frameworks via approximation fixpoint theory. *Artificial Intelligence*, 226:34–74, 2015.
26. Matthias Thimm and Gabriele Kern-Isberner. On the relationship of defeasible argumentation and answer set programming. In Philippe Besnard, Sylvie Doutre, and Anthony Hunter, editors, *Proceedings of the 2nd International Conference on Computational Models of Argument (COMMA'08)*, number 172 in Frontiers in Artificial Intelligence and Applications, pages 393–404. IOS Press, May 2008.
27. Francesca Toni. A tutorial on assumption-based argumentation. *Argument & Computation*, 5(1):89–117, 2014.
28. Emil Weydert. System J - revision entailment. default reasoning through ranking measure updates. In *Practical Reasoning, International Conference on Formal and Applied Practical Reasoning, FAPR '96, Bonn, Germany, June 3-7, 1996, Proceedings*, pages 637–649, 1996.
29. Emil Weydert. On the plausibility of abstract arguments. In *Proceedings of the 12th European Conference on Symbolic and Quantitative Approaches to Reasoning with Uncertainty (ECSQARU'13)*, 2013.

Argument Graphs
for Defeasible Logic Programming

Paolo Mancarella[1] and Francesca Toni[2]

[1] Dipartimento di Informatica, Università di Pisa, Italy
paolo.mancarella@unipi.it
[2] Department of Computing, Imperial College London, UK
ft@imperial.ac.uk

Abstract. Both Defeasible Logic Programming (DeLP) and standard Assumption-based Argumentation (ABA) obtain arguments from rules (albeit of different kinds), and define these arguments in terms of deductions (albeit of different kinds). A recent variant of ABA uses instead a notion of argument graphs, representing "non-bloated" sets of "rule-minimal" arguments only. In this paper we consider a similar variant of DeLP, where arguments are graphs, and discuss the potential benefits of using this variant.

1 Introduction

Argumentation in Artificial Intelligence, as overviewed in [15] and more recently in [1], relies on a plethora of computational formalisms, differing in level of abstraction and design choices but all allowing to represent and reason with conflicts and equipped with mechanisms of resolving these conflicts (called semantics). The leanest form of argumentation formalism is *abstract argumentation* [8], exclusively focusing on conflict (attack) between arguments, seen as black-boxes: an *AA framework* is simply a pair $(Args, \rightarrow)$ with a set $Args$ of arguments and a binary attack relation $\rightarrow$ on $Args$; a mechanism to resolve conflicts in an AA framework is to identify, for example, so-called *stable* sets of arguments (i.e. $S \subseteq Args$ such that $S = \{\alpha \in Args| \not\exists \beta \in S$ such that $(\beta, \alpha) \in \rightarrow\}$.

AA is non-committal as to what arguments may actually amount to, or how attack between them is defined, taking this attack relationship as a starting point. This lack of committment affords broad applicability to AA, from non-monotonic reasoning and games [8] to case-based reasoning [7] and machine learning [4], to mention a few. Other argumentation formalisms instead choose to commit to specific ways to obtain arguments from other information, notably logical sentences and rules, and derive notions of attack based on the resulting *structure* of arguments, again using other information, notably some form of negation of logical sentences. In these *structured argumentation* formalisms, arguments and attacks are automatically obtained from the other information, and thus, when this information is naturally and readily available, these formalisms require no effort from humans to define by hand arguments and attacks, as in abstract argumentation.

Several such formalisms exist, as recently overviewed in [2]. In this paper, we focus in particular on Defeasible Logic Programming (DeLP) [10], building upon [16] and recently surveyed in [11,12], and Assumption-based Argumentation (ABA) [3], building upon [13] and recently surveyed in [17,6]. Both formalisms are equipped with computational engines that supports their deployment in concrete applications. In both formalisms, arguments are defined as (forms of) deductions from rules (of different kinds), and, as in the case of AA, conflict resolution amounts to identifying (dialectically) acceptable *sets* of arguments (in different ways in the two formalisms). In a recent variant of ABA, however, arguments as deductions as well as sets of these arguments are replaced by so-called *argument graphs*, a more "economical"' representation which leads to conceptual and computational advantages (see [5] for a discussion), and, spefically, to more efficient computational engines than for standard ABA (see [5]).

In this paper we explore the use of a form of argument graphs, instead of sets of arguments defined as deductions, for DeLP, and discuss some potential benefits of the resulting formalism, variant of DeLP.

The paper is organised as follows. In Section 2, we recap DeLP, focusing on the notion of argument. In Section 3, we recap ABA and its variant based on argument graphs, again focusind on the respective notions of argument, and focusing on desirable properties of *rule-minimality* of arguments (as deductions) when these arguments are represented by argument graphs and of *non-bloatedness* of sets of arguments (again as deductions) when these sets of arguments are represented by argument graphs. In Section 4, we explore the graph-variant of DeLP. We conclude in Section 5.

2   Arguments in Defeasible Logic Programming: a Recap

Defeasible logic programs (see [10,11,12]) are defined in terms of ground literals, namely ground atoms (e.g. $born(1991, jo)$) or their negation (e.g. $\neg female(jo)$), and may also include *negation as failure* literals, of the form $not\ l$, for l a ground literal. Thus, double negation (as in $not\ \neg born(1991, jo)$) may occur within defeasible logic programs. Negation as failure is understood in DeLP as in standard logic programming: $not\ l$ holds by default, in the absence of evidence to the contrary l. Formally, a *defeasible logic program* is a pair (Π, Δ) where

- Π is a set of *facts* and *strict rules*, where
  - facts are ground literals;
  - strict rules are of the form

$$\varphi_0 \leftarrow \varphi_1, \ldots, \varphi_m$$

    with $m > 0$ and φ_i ground literals, for $i \in \{0, \ldots, m\}$; φ_0 is the *head* and $\varphi_1, \ldots, \varphi_m$ the *body* of the strict rule;

- Δ is a set of *defeasible rules* of the form

$$\varphi_0 \Leftarrow \varphi_1, \ldots, \varphi_m$$

with $m \geqslant 0$, φ_0 a ground literal and each φ_i either a ground literal or a negation as failure literal *not l*, for ground literal l, for $i \in \{1, \ldots, m\}$; φ_0 is the *head* and $\varphi_1, \ldots, \varphi_m$ the *body* of the defeasible rule.

Overall, facts may be understood as strict rules with an empty (true) body. Facts and strict rules represent "certain" information, that is considered to hold in the given application domain, so that when their body holds, their head can be inferred. Instead, defeasible rules represent "uncertain" information, in that their head may not hold even when their body does. Negation as failure literals in the body of defeasible rules introduce a further form of defeasibility (of the conditions for the applicability of the rule, rather than of the rule itself). Defeasible rules may have an empty body, in which case they may be deemed to be *presumptions*.

Arguments can be extracted from defeasible logic programs. Given a defeasible logic program (Π, Δ) and a ground literal l, a *DeLP argument for l* is a set of defeasible rules $D \subseteq \Delta$ that is minimal (with respect to $\subseteq$) such that

1. there exists a *defeasible derivation* of l from $\Pi \cup D$, i.e. a finite sequence of ground literals $l_1, \ldots, l_n$ where $n \geqslant 1$, $l_n = l$ and each l_i, for $i \in \{1, \ldots, n\}$, is either a fact in Π or there exists a rule in $\Pi \cup D$ with head l_i and body $\varphi_1, \ldots, \varphi_m$ such that for each φ_j, for $j \in \{1, \ldots, m\}$, either $\varphi_j = l_k$ for some $k \leq i$ or φ_j is a negation as failure literal;
2. there exists no pair of "contradictory" literals $a, \neg a$ with defeasible derivations of both a and $\neg a$ from $\Pi \cup D$; and
3. for every *not l'* in the body of rules in D, $l' \neq l_i$ for all $i \in \{1, \ldots, n\}$ and $l_1, \ldots, l_n$ the defeasible derivation at 1.

Example 1. For illustration, given (Π, Δ) with[3]

- $\Pi = \{p \leftarrow a, q;\ q \leftarrow p;\ q\}$ (with two strict rules and one fact, q) and
- $\Delta = \{a \Leftarrow\}$ (with a single presumption)

$\{a \Leftarrow\}$ is a DeLP argument for p and no other DeLP argument for p exists given this simple defeasible logic program. However, note that this DeLP argument can be obtained by means of infinitely many defeasible derivations, including a, q, p (using the presumption in Δ and the fact in Π, as well as the strict rule $p \leftarrow a, q$) and a, q, p, q, p (using the full $\Pi \cup \Delta$).

The semantics of defeasible logic programs is given in terms a notion of *warrant*, whereby a literal is warranted from a defeasible logic program if there is an argument for the literal and a *dialectical tree* for that argument whose branches are *acceptable argumentation lines* from this argument where supporting and

---
[3] Throughout, we use ';' as separator between elements of sets.

interfering arguments are interleaved until no more interfering arguments can be identified. We omit the formal definition of these notions, as they play no role in this paper, but the interested reader can find all formal definitions in [10,11,12]. Here, we underline that these notions require a definition of *defeat* between DeLP arguments, taking into account *preferences between arguments* possibly obtained from *preferences between rules*. Moreover, the notion of acceptable argumentation line can be seen as amounting to identifying *sets of supporting arguments* for the argument to be warranted, and including the latter. Finally, no arguments within an acceptable argumentation line may use a subset of the set of rules in an earlier argument in the line.

3 Arguments and Argument Graphs in Assumption-Based Argumentation: a Recap

An *ABA framework* (see [3,17,6]) is a tuple $(\mathcal{L}, \mathcal{R}, \mathcal{A}, ^-)$, where:

- $(\mathcal{L}, \mathcal{R})$ is a deductive system with $\mathcal{L}$ a language (i.e. a set of sentences) and $\mathcal{R}$ a set of rules of the form $\varphi_0 \leftarrow \varphi_1, \ldots, \varphi_m$ with $m \geq 0$ and $\varphi_i \in \mathcal{L}$ for $i \in \{0, \ldots, m\}$; φ_0 is the *head* and $\varphi_1, \ldots, \varphi_m$ the *body* of the rule; if $m = 0$, then $\varphi_0 \leftarrow \varphi_1, \ldots, \varphi_m$, written as $\varphi_0 \leftarrow$, has an empty body;
- $\mathcal{A} \subseteq \mathcal{L}$ is a non-empty set of *assumptions*;
- $^- : \mathcal{A} \to \mathcal{L}$ is a total map: for $\alpha \in \mathcal{A}$, the $\mathcal{L}$-sentence $\overline{\alpha}$ is referred to as the *contrary* of α.

Differently from DeLP, ABA keeps the choice of underlying language open, but, as a special choice, $\mathcal{L}$ may be a set of standard literals and negation as failure literals, as in logic programming instances of ABA (see [3,6]). Also differently from DeLP, in ABA there is only one type of rules, and these can be all interpreted as strict (as discussed in [14]), but defeasibility is supported by assumptions, which thus can be understood as presumptions in DeLP. Thus, the simple example defeasible logic program (Π, Δ) in Example 1, with $\Pi = \{p \leftarrow a, q; \; q \leftarrow p; \; q\}$ and $\Delta = \{a \Leftarrow\}$, can be represented in ABA as $(\mathcal{L}, \mathcal{R}, \mathcal{A}, ^-)$ with $\mathcal{R} = \{p \leftarrow a, q; \; q \leftarrow p; \; q \leftarrow\}$ and $\mathcal{A} = \{a\}$.

In the logic programming instance of ABA, assumptions amount to negation as failure literals, and these cannot be the head of any (strict) rules (thus the logic programming instance of ABA is so-called *flat*). This restriction is analogous to the restriction, in DeLP, that negation as failure literals only occur in the body of (defeasible) rules.

In ABA [17,6], arguments are standardly understood as deductions, which can be defined "backwards" [9], similarly to DeLP, or, equivalently, as trees:

- a *deduction for* $\varphi \in \mathcal{L}$ *supported by* $S \subseteq \mathcal{L}$ *and* $R \subseteq \mathcal{R}$, denoted $S \vdash^R \varphi$, is a finite tree with: the root labelled by φ; leaves labelled by τ or sentences, with S being the set of all such sentences; the children of non-leaves ψ labelled by the elements of the body of some ψ-headed rule in $\mathcal{R}$, with R being the set of all such rules.

– an *argument for* $\varphi \in \mathcal{L}$ *supported by* $A \subseteq \mathcal{A}$ *and* $R \subseteq \mathcal{R}$ is a deduction $A \vdash^R \varphi$.

Example 2. For illustration, given $(\mathcal{L}, \mathcal{R}, \mathcal{A},\overline{})$ with

$$\mathcal{R} = \{p \leftarrow a, q;\ q \leftarrow p;\ q \leftarrow\}, \quad \mathcal{A} = \{a\}$$

the following are two different arguments for p:

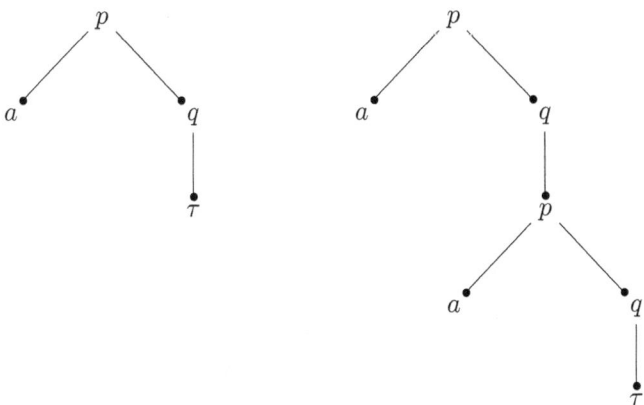

denoted $\{a\} \vdash^{\{p \leftarrow a, q; q \leftarrow\}} p$ (argument on the left) and $\{a\} \vdash^{\mathcal{R}} p$ (argument on the right). As in the case of the DeLP version of this ABA framework given earlier, there are infinitely many arguments for p in this example, all denoted $\{a\} \vdash^{\mathcal{R}} p$.

The latter argument(s) in this example can be deemed to be *non-rule-minimal*, where

– an argument $A \vdash^R \varphi$ is *rule-minimal* iff for any two nodes $n \neq n'$ in (the tree denoted by) $A \vdash^R \varphi$ such that n and n' are labelled by the same sentence, the children of n and n' are labelled by the same sentences (in $\mathcal{L}$ or τ).

Thus, $\{a\} \vdash^{\mathcal{R}} p$ above is non-rule-minimal as the two nodes labelled by q have children differently labelled (by p in one case and by τ in the other). Instead, the argument denoted $\{a\} \vdash^{\{p \leftarrow a, q; q \leftarrow\}} p$ is rule-minimal.

Recently, a "graph-based" variant of (flat) ABA has been proposed [5], where (sets of) arguments are defined as graphs. Let $v(G)$ and $e(G)$ denote, respectively, the vertices and edges of a graph G (thus $e(G) \subseteq v(G) \times v(G)$), and let $sinks(G)$ and $sources(G)$ denote, respectively, all vertices in $v(G)$ with no outgoing edges and with no incoming edges. Then

– an *argument graph* G is a directed, acyclic graph where $v(G) \subseteq \mathcal{L}$ and for all $\varphi \in v(G)$:

i) if $\varphi \in \mathcal{A}$, then $\varphi \in \mathit{sinks}(G)$;
ii) if $\varphi \notin \mathcal{A}$, then there is a rule $(\varphi \leftarrow \varphi_1, \ldots, \varphi_m) \in \mathcal{R}$ such that there is an edge (φ, φ') in $e(G)$ iff $\varphi' \in \{\varphi_1, \ldots, \varphi_m\}$;

the *support* of an argument graph G is $v(G) \cap \mathcal{A}$; if $\mathit{sources}(G) = \{\varphi\}$ for an argument graph G, namely G has a unique source, then this is called the *claim* of G.

Example 3. For illustration, again given $(\mathcal{L}, \mathcal{R}, \mathcal{A}, \overline{})$ in Example 2, the only argument graph with claim p is the following:

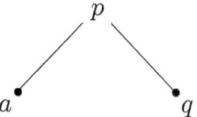

Argument graphs afford a representations of *rule-minimal* (tree) arguments only, as in Example 3, where the single argument graph represents only the rule-minimal (tree) argument (denoted $\{a\} \vdash^{\{p \leftarrow a,q;\ q \leftarrow\}} p$) in Example 2. Argument graphs can also represent compactly sets of rule-minimal (tree) arguments. For instance, in Example 3, the argument graph represents $\{\{a\} \vdash^{\{p \leftarrow a,q;\ q \leftarrow\}} p, \{a\} \vdash^{\{\}} a, \{\} \vdash^{\{q \leftarrow\}} q\}$ (namely $\{a\} \vdash^{\{p \leftarrow a,q;\ q \leftarrow\}} p$ and all its "sub-arguments").

Example 4. As a further illustration, given $(\mathcal{L}, \mathcal{R}, \mathcal{A}, \overline{})$ with

$$\mathcal{R} = \{p \leftarrow a, q;\ q \leftarrow p;\ q \leftarrow;\ r \leftarrow q\}, \quad \mathcal{A} = \{a\}$$

the argument graph

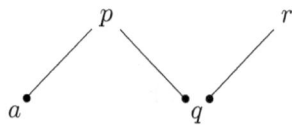

represents $\{a\} \vdash^{\{p \leftarrow a,q;q \leftarrow\}} p$, $\{a\} \vdash^{\mathcal{R}} p$ and $\{\} \vdash^{\{r \leftarrow q;q \leftarrow\}} r$, amongst others.

Argument, given as trees, that are represented by argument graphs are guaranteed to be *rule-minimal*. In addition to forcing the choice of rule-minimal (tree) arguments only, argument graphs also force rule-minimality across different arguments (represented as trees).

- a set S of (tree) arguments is *bloated* iff there exist arguments $A \vdash^R \varphi$ and $A' \vdash^{R'} \varphi'$ in S and nodes n in $A \vdash^R \varphi$ and n' in $A' \vdash^{R'} \varphi'$ such that n and n' are labelled by the same sentence but the children of n and n' are labelled by different sentences (in $\mathcal{L}$ or τ).

Trivially, if a set includes a non-rule-minimal argument then it is bloated. However, a set of rule-minimal arguments may be bloated nonetheless.

Example 5. For illustration, consider $(\mathcal{L}, \mathcal{R}, \mathcal{A}, \bar{})$ with $\mathcal{R} = \{p \leftarrow a;\ p \leftarrow b;\ q \leftarrow p\}$. Then the set of (tree) arguments $\{\{a\} \vdash^{\{p \leftarrow a\}} p;\ \{b\} \vdash^{\{q \leftarrow p;\ p \leftarrow b\}} q\}$ is bloated. These arguments cannot be represented by the same argument graph. Indeed, for this ABA frameworks, argument graphs include

$$q \longrightarrow \bullet p \longrightarrow \bullet a \quad \text{and} \quad q \longrightarrow \bullet p \longrightarrow \bullet b$$

representing the non-bloated sets of (rule-minimal tree) arguments $\{\{a\} \vdash^{\{\}} a;\ \{a\} \vdash^{\{p \leftarrow a\}} p;\ \{a\} \vdash^{\{q \leftarrow p;\ p \leftarrow a\}} p\}$ (for the argument graph on the left), and $\{\{b\} \vdash^{\{\}} b;\ \{b\} \vdash^{\{p \leftarrow b\}} p;\ \{b\} \vdash^{\{q \leftarrow p;\ p \leftarrow b\}} p\}$ (for the argument graph on the right).

The semantics of ABA, using the notion of argument as tree is defined in terms of (dialectically) acceptable sets of arguments (or equivalently sets of assumptions, see the discussion in [2]), defined in turn using a notion of *attack* between arguments (or equivalently between sets of arguments). Instead, using the notion of argument as tree, the semantics is defined in terms of (dialectically) acceptable argument graphs (where a single argument graph may represent a set of (tree) arguments, as illustrated above) and a suitably defined notion of *attack* between argument graphs. The definitions of these notions are omitted as they do not play a role in this paper, but the interested reader is referred to [5].

More efficient computational machinery for the version of ABA using argument graphs exists than for the version of ABA using sets of (tree) arguments (see [5] for definitions and experiments).

4 Deploying Argument Graphs in DeLP

The definition of DeLP argument imposes (amongst other requirements) that it is "defeasible rule"-minimal. However, it imposes no requirements on the use of strict rules. Thus, as we argued earlier in Example 1, there may be infinitely many ways to obtain the same argument, leading to a computational burden. This may happen when the use of different strict rules for the same literal gives rise to *potentially redundant DeLP arguments*, where, given a defeasible logic program (Π, Δ)

- a DeLP argument D for l is potentially redundant iff there exist multiple defeasible derivations of l from $\Pi \cup D$, in the form of finite sequences of ground literals $l_1, \ldots, l_n$.

Thus, given (Π, Δ) in Example 1, $\{a \Leftarrow\}$ is a potentially redundant DeLP argument for p, e.g. because of the defeasible derivation a, q, p, q, p. Thus, when extracting DeLP arguments automatically for the purposes of automated reasoning, it may be helpful to impose "strict rule"-minimality too. There may be other, conceptual rather than computational, advantages to doing so, e.g. that reasoning should be parsimonious. A notion of DeLP argument graph may thus be useful, where the first condition in the definition of DeLP argument D for l is replaced with the requirement:

1'. there exists a *DeLP argument graph* G with $sources(G) = \{l\}$, where G is a directed, acyclic graph with $v(G)$ a set of literals or negation as failure literals and for all $\varphi \in v(G)$:
   i) if $\varphi \in D$ (namely φ is a presumption in D) or φ is a fact in Π or φ is a negation as failure literal, then $\varphi \in sinks(G)$;
   ii) if $\varphi \notin \Delta$ and φ is neither a fact nor a negation as failure literal, then there is a rule $\varphi \leftarrow \varphi_1, \ldots, \varphi_m \in \Pi$ or $\varphi \Leftarrow \varphi_1, \ldots, \varphi_m \in D$ such that there is an edge (φ, φ') in $e(G)$ iff $\varphi' \in \{\varphi_1, \ldots, \varphi_m\}$.

Minimality of D is still imposed, as are conditions 2. and 3 in the definition of DeLP argument. Minimality of use of strict rules results from the use of a graph as the underlying argument structure. For illustration, in the case of the defeasible logic program of Example 1, the only DeLP argument graph is $D = \{a\}$ with underlying graph:

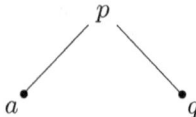

Note that, differently from the global requirement of minimality on D, in both definitions of DeLP argument and DeLP argument graph, the minimality requirement on strict rules is imposed locally, in DeLP argument graph.

The notion of warrant in DeLP may also suffer from problems of "bloatedness" similar to those suffered by sets of (tree) arguments in ABA. For instance, DeLP arguments on different acceptable argumentation lines may may use different (defeasible or strict) rules to prove the same literal. It would be interesting to modify these notions so that sets of supporting arguments are substituted for by DeLP argument graphs with non-singleton sources. This may have computational impacts, as in the case of ABA, but potentially also determine a different semantics behaviour when preferences over rules are taken into account.

5 Conclusions

We proposed a novel notion of argument in DeLP, adapted from ABA argument graphs, whereby the notion of defeasible deduction is replaced by a notion of graph. We have discussed some of its potential benefits as well as potential future directions of use, especially to guarantee sets of rule-minimal supporting arguments in determining whether arguments are warranted.

Overall, it would be interesting to study properties of the DeLP framework resulting from the use of these novel notions of arguments and sets of arguments, in relation to standard DeLP and properties of minimality and computational efficiency.

References

1. Pietro Baroni, Dov M Gabbay, Massimiliano Giacomin, and Leendert van der Torre, editors. *Handbook Of Formal Argumentation*, volume 1. College Publications, 2018.
2. Philippe Besnard, Alejandro Javier García, Anthony Hunter, Sanjay Modgil, Henry Prakken, Guillermo Ricardo Simari, and Francesca Toni. Introduction to structured argumentation. *Arg. & Comp.*, 5(1):1–4, 2014.
3. Andrei Bondarenko, Phan Minh Dung, Robert Kowalski, and Francesca Toni. An abstract, argumentation-theoretic approach to default reasoning. *Art. Intelligence*, 93(97):63–101, 1997.
4. Oana Cocarascu, Kristijonas Čyras, and Francesca Toni. Explanatory predictions with artificial neural networks and argumentation. In *IJCAI/ECAI Workshop on Explainable Artificial Intelligence (XAI-18)*, 2018.
5. Robert Craven and Francesca Toni. Argument graphs and assumption-based argumentation. *Artificial Intelligence*, 233:1–59, 2016.
6. Kristijonas Čyras, Xiuyi Fan, Claudia Schulz, and Francesca Toni. Assumption-Based Argumentation: Disputes, Explanations, Preferences. In Pietro Baroni, Dov M Gabbay, Massimiliano Giacomin, and Leendert van der Torre, editors, *Handbook Of Formal Argumentation*, volume 1. College Publications, 2018.
7. Kristijonas Čyras, Ken Satoh, and Francesca Toni. Abstract argumentation for case-based reasoning. In Chitta Baral, James P. Delgrande, and Frank Wolter, editors, *Principles of Knowledge Representation and Reasoning: Proceedings of the Fifteenth International Conference, KR 2016, Cape Town, South Africa, April 25-29, 2016.*, pages 549–552. AAAI Press, 2016.
8. Phan Minh Dung. On the acceptability of arguments and its fundamental role in nonmonotonic reasoning, logic programming and n-person games. *Art. Intelligence*, 77:321–357, 1995.
9. Phan Minh Dung, Robert A. Kowalski, and Francesca Toni. Dialectic proof procedures for assumption-based, admissible argumentation. *Artificial Intelligence*, 170(2):114–159, 2006.
10. Alejandro Javier García and Guillermo Ricardo Simari. Defeasible logic programming: An argumentative approach. *Theory and Practice of Logic Programming*, 4(1-2):95–138, 2004.
11. Alejandro Javier García and Guillermo Ricardo Simari. Defeasible Logic Programming: DeLP-servers, Contextual Queries, and Explanations for Answers. *Argument & Computation*, 5(1):63–88, 2014.
12. Alejandro Javier García and Guillermo Ricardo Simari. Argumentation based on logic programming. In Pietro Baroni, Dov M Gabbay, Massimiliano Giacomin, and Leendert van der Torre, editors, *Handbook Of Formal Argumentation*, volume 1. College Publications, 2018.

13. Antonis C. Kakas, Robert A. Kowalski, and Francesca Toni. Abductive logic programming. *Journal of Logic and Computation*, 2(6):719–770, 1992.
14. Robert A. Kowalski and Francesca Toni. Abstract argumentation. *Artificial Intelligence and Law*, 4(3–4):275–296, 1996. Also published in the book "Logical Models of Argumentation".
15. Iyad Rahwan and Guillermo R. Simari, editors. *Argumentation in Artificial Intelligence*. Springer, 2009.
16. Guillermo Ricardo Simari and Ronald Prescott Loui. A mathematical treatment of defeasible reasoning and its implementation. *Artificial Intelligence*, 53(2-3):125–157, 1992.
17. Francesca Toni. A tutorial on assumption-based argumentation. *Arg. & Comp.*, 5(1):89–117, 2014.

www.ingramcontent.com/pod-product-compliance
Lightning Source LLC
Chambersburg PA
CBHW071327190426
43193CB00041B/918